16th Russian Youth Conference on Physics and Astronomy

(PhysicA.SPb/2013)

Journal of Physics: Conference Series Volume 572

St. Petersburg, Russia
23 – 24 October 2013

Editors:

N. S. Averkiev
S. Poniaev
G. Sokolovskii

ISBN: 978-1-63439-701-8
ISSN: 1742-6588

Printed from e-media with permission by:

Curran Associates, Inc.
57 Morehouse Lane
Red Hook, NY 12571

Some format issues inherent in the e-media version may also appear in this print version.

Printed by Curran Associates, Inc. (2015)

Institute of Physics
Dirac House, Temple Back
Bristol BS1 6BE UK

Phone: 44 1 17 929 7481
Fax: 44 1 17 920 0979

techtracking@iop.org

Additional copies of this publication are available from:

Curran Associates, Inc.
57 Morehouse Lane
Red Hook, NY 12571 USA
Phone: 845-758-0400
Fax: 845-758-2634
Email: curran@proceedings.com
Web: www.proceedings.com

TABLE OF CONTENTS

OTHER TOPICS OF PHYSICS

PHYSICS AND TECHNOLOGY OF ENERGY CONVERSION

PHYSICS OF QUANTUM STRUCTURES

PLASMA PHYSICS, HYDRO-AND AERO-DYNAMICS

Journal of Physics: Conference Series **572** (2014) 011001 doi:10.1088/1742-6596/572/1/011001

XVI Russian Conference on Physics and Astronomy PhysicA.SPb

The sixteenth Russian Conference on Physics and Astronomy PhysicA.SPb was held 23-24 October 2013 in Saint-Petersburg, Russia. The Conference continues the tradition of Saint-Petersburg Seminars on Physics and Astronomy originating from mid-90s. Since then PhysicA.SPb maintains both scientific and educational quality of contributions delivered to the young audience. This is the main feature of the Conference that makes it possible to combine the whole spectrum of modern Physics and Astronomy within one event.

PhysicA.SPb/2013 has brought together about 200 students, young scientists and their colleague professors from many universities and research institutes across whole Russia as well as from Belarus, Ukraine, Switzerland, Turkey, Finland and France. Oral and poster presentations were combined into a few well-defined sections among which one should name Astronomy and Astrophysics, Plasma physics, hydro- and aero-dynamics, Physics of quantum-sized structures, Nanostructured and thin-film materials, Biophysics, THz and UHF materials and devices, Optoelectronic devices, Optics and spectroscopy, Atomic and elementary particles physics, Defects and impurities in solid state, Physics and technology of the alternative energetics.

This issue of the *Journal of Physics: Conference Series* presents the extended contributions from participants of PhysicA.SPb/2013 that were peer-reviewed by expert referees through processes administered by the Presiders of the Organising and Programme Committees to the best professional and scientific standards.

The Editors: **Nikita S. Averkiev, Sergey A. Poniaev and Grigorii S. Sokolovskii**

Peer review statement

All papers published in this volume of *Journal of Physics: Conference Series* have been peer reviewed through processes administered by the proceedings Editors. Reviews were conducted by expert referees to the professional and scientific standards expected of a proceedings journal published by IOP Publishing.

Journal of Physics: Conference Series **572** (2014) 012001 doi:10.1088/1742-6596/572/1/012001

Diffusive currents and Coulomb separation of ions in dense matter

M V Beznogov[1] **and D G Yakovlev**[2]

[1] St. Petersburg Academic University, 8/3 Khlopina Street, St. Petersburg 194021, Russia
[2] Ioffe Physical Technical Institute, 26 Politekhnicheskaya, St. Petersburg 194021, Russia

E-mail: mikavb89@gmail.com

Abstract. We derive diffusive currents in strongly coupled Coulomb ion mixtures in dense stellar matter. Coulomb coupling in the presence of gravity and electric fields, induced by gravity owing to plasma polarization, creates a specific diffusion current which separates ions with the same A/Z (mass to charge number ratios) but different Z. This Coulomb separation can be important in white dwarfs and neutron stars.

1. Introduction

We consider diffusion of ions in a multicomponent ion plasma of white dwarfs (WDs) and neutron stars (NSs). As an example we can mention gravitational settling of ^{22}Ne in ^{12}C–^{16}O cores of WDs (see, e.g., [1–5]). This settling is thought to reheat old WDs and delay their cooling in accordance with observations. Diffusion of ions affects also chemical evolution and nuclear reactions in the envelopes of NSs (e.g., [6–8]).

Diffusion formalism is well developed for almost ideal plasma of ions [9, 10]. However, the ions in WDs and NSs are usually strongly coupled by Coulomb forces. The diffusion coefficients of ions in such plasmas have been thoroughly investigated, mainly by molecular dynamics simulations (e.g., [11–14]). We derive a new expression for diffusive currents. Our first results on this subject were published in [15].

2. Diffusion currents of ions

Let a plasma be composed of electrons and a mixture of ion species $j = 1, 2, \ldots$ (whose atomic numbers are A_j and charge numbers are Z_j). Let n_j denote the number density of ions j. Then the electron number density is $n_e = \sum_j Z_j n_j$ because of charge neutrality.

The state of ions j can be described (e.g., [16]) by the Coulomb coupling parameter Γ_j,

$$\Gamma_j = Z_j^2 e^2/(a_j kT) = Z_j^{5/3} e^2/(a_e kT), \tag{1}$$

where T is the temperature, k is the Boltzmann constant, $a_e = (4\pi n_e/3)^{-1/3}$ is the electron-sphere radius, and $a_j = a_e Z_j^{1/3}$ is the ion-sphere radius (for a sphere around an ion j, containing an electron charge which compensates the ion charge). If $\Gamma_j \ll 1$ then the ion species j form nearly ideal Boltzmann gas, while for $\Gamma_j \gg 1$ they form either Coulomb liquid or solid. One component plasma of ions solidifies at $\Gamma \approx 175$. We consider gaseous or liquid ion plasma only.

Journal of Physics: Conference Series **572** (2014) 012001 doi:10.1088/1742-6596/572/1/012001

The diffusion currents in nearly ideal plasma are well studied [9, 10]. However, the case of non-ideal plasmas requires special treatment. We introduce diffusion currents for non-ideal plasma following [17]. Let the plasma slightly deviate from thermodynamic equilibrium under the action of forces $\boldsymbol{F}_\alpha$ [on all particles α, electrons ($\alpha = e$) and ions ($\alpha = j$)] and gradients of number density $\boldsymbol{\nabla} n_\alpha$. For simplicity, we assume thermal equilibrium ($\boldsymbol{\nabla} T = 0$). The forces $\boldsymbol{F}_\alpha$ and gradients $\boldsymbol{\nabla} n_\alpha$ create weak gradients of chemical potentials $\boldsymbol{\nabla}\mu_\alpha$ of particles α. Let $\widetilde{\boldsymbol{F}}_\alpha = \boldsymbol{F}_\alpha - \boldsymbol{\nabla}\mu_\alpha = e_\alpha \boldsymbol{E} + m_\alpha \boldsymbol{g} - \boldsymbol{\nabla}\mu_\alpha$, where $\boldsymbol{F}_\alpha = e_\alpha \boldsymbol{E} + m_\alpha \boldsymbol{g}$ (with e_α and m_α being electric charge and mass of particles α, respectively). In stars, the force $\boldsymbol{F}_\alpha$ is produced by gravity $\boldsymbol{g}$ (that can be treated as locally constant) and by the polarization electric field $\boldsymbol{E}$.

Notice that $\sum_\alpha n_\alpha \widetilde{\boldsymbol{F}}_\alpha = \rho \boldsymbol{g} - \boldsymbol{\nabla} P$, where $\rho = \sum_\alpha m_\alpha n_\alpha$ is the mass density of the plasma, $\boldsymbol{\nabla} P = \sum_\alpha n_\alpha \boldsymbol{\nabla}\mu_\alpha$ [18], and P is the total plasma pressure; the $\boldsymbol{E}$ field drops off due to electric neutrality.

If particle species α are in hydrostatic balance, then $\widetilde{\boldsymbol{F}}_\alpha$=0. If the plasma is in hydrostatic equilibrium as a whole, then $\sum_\alpha n_\alpha \widetilde{\boldsymbol{F}}_\alpha = 0$ and $\rho \boldsymbol{g} = \boldsymbol{\nabla} P$. Hydrostatic equilibration ($\rho \boldsymbol{g} = \boldsymbol{\nabla} P$) in NSs is established in milliseconds and in WDs in a few minutes [19] but the diffusion of ions can take Gyrs (see, e.g., [3]). This diffusion can be analyzed assuming $\rho \boldsymbol{g} = \boldsymbol{\nabla} P$.

A deviation of particles α from hydrostatic equilibrium in a mixture can be measured by

$$\boldsymbol{d}_\alpha = \frac{\rho_\alpha}{\rho} \sum_\beta n_\beta \widetilde{\boldsymbol{F}}_\beta - n_\alpha \widetilde{\boldsymbol{F}}_\alpha, \tag{2}$$

where $\rho_\alpha = m_\alpha n_\alpha$ is the mass density of these particles α. Evidently, $\sum_\alpha \boldsymbol{d}_\alpha = 0$.

Let $\boldsymbol{J}_\alpha = \rho_\alpha \boldsymbol{V}_\alpha$ be the diffusive flux of particle species α, and $\boldsymbol{V}_\alpha$ the diffusion velocity of these particles [9, 10]. Phenomenological relations for the diffusive currents can be written as

$$\boldsymbol{J}_\alpha = \Phi \sum_{\beta \neq \alpha} m_\alpha m_\beta D_{\alpha\beta} \boldsymbol{d}_\beta. \tag{3}$$

Here, $D_{\alpha\beta}$ [cm^2 s^{-1}] is a generalized diffusion coefficient of particles α relative to β; Φ is a normalization function to be specified later. The diffusion coefficients obey the relation

$$\sum_\alpha \boldsymbol{J}_\alpha = 0. \tag{4}$$

In a rarefield, nearly ideal plasma, $\widetilde{\boldsymbol{F}}_\alpha = \boldsymbol{F}_\alpha - n_\alpha^{-1} \boldsymbol{\nabla} P_\alpha$, where P_α is the partial pressure of particle species α. Then (3) reproduces the standard definition of diffusion coefficients in rarefied plasmas [9, 10]. For strongly coupled plasmas, partial pressures P_α are ambiguous, while the definition (3) is not.

Since the electrons are light and mobile they are always in hydrostatic quasiequilibrium responding almost instantly to the motion of the ions; it is sufficient to set $m_e \to 0$. Then from (2) one has $\boldsymbol{d}_e = -n_e \widetilde{\boldsymbol{F}}_e$. This electron hydrostatic quasiequilibrium implies

$$\boldsymbol{d}_e = 0, \quad \widetilde{\boldsymbol{F}}_e = -e\boldsymbol{E} - \boldsymbol{\nabla}\mu_e = 0. \tag{5}$$

It allows one to exclude electrons from the ion transport problem. In this case (3) and (4) retain their form but indices α and β run only over ion species ($j = 1, 2, \ldots$). Strictly speaking, (3) is valid for non-relativistic particles, while the electrons in stars can be relativistic. However, the results are applicable even in this case as long as the electrons can be considered as massless.

In a binary ion plasma (j=1, 2) $\boldsymbol{J}_1 = -\boldsymbol{J}_2$, $\boldsymbol{d}_1 = -\boldsymbol{d}_2$, and $D_{12} = D_{21} \equiv D$. Therefore, there is actually one diffusion coefficient D,

$$\boldsymbol{J}_2 = -\boldsymbol{J}_1 = \frac{n m_1 m_2}{\rho k T} D \boldsymbol{d}_1. \tag{6}$$

Here we set $\Phi = n/(\rho kT)$ ($n = n_1 + n_2$ is the total number density of the ions). Then D becomes the standard diffusion coefficient [9, 10] for a binary plasma of ions.

Now we simplify (6). From (2) we obtain

$$\boldsymbol{d}_j = -\frac{\rho_j}{\rho}\boldsymbol{\nabla} P - n_j e Z_j \boldsymbol{E} + n_j \boldsymbol{\nabla}\mu_j, \tag{7}$$

where $j = 1$ or 2. Because $\boldsymbol{d}_1 + \boldsymbol{d}_2 = 0$, we have $en_e\boldsymbol{E} = -\boldsymbol{\nabla}P + n_1\boldsymbol{\nabla}\mu_1 + n_2\boldsymbol{\nabla}\mu_2$. Substituting this into (7) with $m_j = A_j m_u$ (m_u being the atomic mass unit), we get

$$\boldsymbol{d}_1 = \frac{n_1 n_2}{n_e}\left[m_u(Z_1A_2 - Z_2A_1)\frac{\boldsymbol{\nabla}P}{\rho} + Z_2\boldsymbol{\nabla}\mu_1 - Z_1\boldsymbol{\nabla}\mu_2\right]. \tag{8}$$

It is well known (e.g., [16]) that $\mu_j = \mu_j^{(id)} + \mu_j^{(C)}$, where upperscripts (id) and (C) label the ideal gas and Coulomb contributions, respectively. Then $\boldsymbol{d}_1 = \boldsymbol{d}_a + \boldsymbol{d}_b + \boldsymbol{d}_c$, with

$$\boldsymbol{d}_a = m_u Z_1 Z_2 \frac{n_1 n_2}{n_e}\left(\frac{A_2}{Z_2} - \frac{A_1}{Z_1}\right)\frac{\boldsymbol{\nabla}P}{\rho}, \tag{9}$$

$$\boldsymbol{d}_b = \frac{n_1 n_2}{n_e}\left[Z_2\boldsymbol{\nabla}\mu_1^{(id)} - Z_1\boldsymbol{\nabla}\mu_2^{(id)}\right] = \frac{kT}{n_e}\left(Z_2 n_2\boldsymbol{\nabla}n_1 - Z_1 n_1\boldsymbol{\nabla}n_2\right), \tag{10}$$

$$\boldsymbol{d}_c = \frac{n_1 n_2}{n_e}\left[Z_2\boldsymbol{\nabla}\mu_1^{(C)} - Z_1\boldsymbol{\nabla}\mu_2^{(C)}\right]. \tag{11}$$

In (10) we have made use of the well known relation $\boldsymbol{\nabla}\mu_j^{(id)} = kT\, n_j^{-1}\boldsymbol{\nabla}n_j$.

Combined with (6), these equations define the diffusion current $\boldsymbol{J}_2$. It consists of three terms labeled by subscripts a, b and c. The terms a and b are known but the term c is new.

(a). Consider the matter which is in hydrostatic equilibrium as a whole ($\boldsymbol{\nabla}P = \rho\boldsymbol{g}$). Then $\boldsymbol{d}_a$ describes gravitational sedimentation of the ion species 2 (provided their effective "molecular weight" A_2/Z_2 is larger than "molecular weight" A_1/Z_1 for ion species 1).

(b). The term $\boldsymbol{d}_b$ has especially transparent form in the limit of $n_2 \ll n_1$. Then $n_e \approx Z_1 n_1$ and we obtain $\boldsymbol{d}_b = -k_B T\,\boldsymbol{\nabla}n_2$ which corresponds to ordinary diffusion of ion species 2. Therefore, $\boldsymbol{d}_b$ describes diffusive motion of the ions if their number densities deviate from equilibrium ones.

(c) The term $\boldsymbol{d}_c$ is most important in the regime of strong ion coupling. In this case it can be accurately described in the ion-sphere approximation supplemented by the linear mixing rule (e.g., [16]):

$$\mu_j^{(C)} = -0.9\,\frac{Z_j^{5/3}e^2}{a_e}, \qquad \boldsymbol{\nabla}\mu_j^{(C)} = -0.3\,\frac{Z_j^{5/3}e^2}{a_e}\,\frac{\boldsymbol{\nabla}n_e}{n_e}, \tag{12}$$

$$\boldsymbol{d}_c = 0.3\,\frac{n_1 n_2}{n_e}\,\frac{Z_1 Z_2 e^2}{a_e}\left(Z_2^{2/3} - Z_1^{2/3}\right)\frac{\boldsymbol{\nabla}n_e}{n_e}. \tag{13}$$

The structure of $\boldsymbol{d}_c$ is similar to that of $\boldsymbol{d}_a$; it describes the specific 'Coulomb' sedimentation of ion species 2 (provided $Z_2 > Z_1$) under the effect of Coulomb coupling in gravitational field. Its most interesting feature is that it separates even the ions with $A_1/Z_1 = A_2/Z_2$. Such ions possess the same "molecular weights." Then $\boldsymbol{d}_a = 0$ (as long as we neglect small mass defects of ions 1 and 2) and it is widely thought that the ions are not separated. This is seen to be not true.

For illustration, let us rewrite (6) for the degenerate cores of WDs and outer envelopes of NSs. To this aim, we assume that $\boldsymbol{\nabla}P = \rho\boldsymbol{g}$ and the pressure is provided by degenerate electrons,

$P \approx P_e(n_e)$, which allows us to express ∇n_e in (13) through ∇P. Then the diffusion flux can be expressed in the standard form

$$\boldsymbol{J}_2 = D\,\frac{m_1 m_2 n}{\rho n_e}\,(Z_2 n_2 \nabla n_1 - Z_1 n_1 \nabla n_2) + (\boldsymbol{u}_a + \boldsymbol{u}_c) m_2 n_2. \tag{14}$$

In this case

$$\boldsymbol{u}_a = \frac{\rho_1 n D}{\rho n_e k_B T}\, Z_1 Z_2 m_u \boldsymbol{g} \left(\frac{A_2}{Z_2} - \frac{A_1}{Z_1}\right), \tag{15}$$

$$\boldsymbol{u}_c = \frac{\rho_1 n D}{n_e k_B T}\, Z_1 Z_2 \boldsymbol{g} \left(Z_2^{2/3} - Z_1^{2/3}\right) \frac{0.3 e^2}{a_e P \gamma} \tag{16}$$

are the velocities of gravitational settling of ion species 2 under the effects of "molecular weight" difference and Coulomb separation, respectively; $\gamma = \partial \ln P / \partial \ln \rho$.

Let us remark that, when the matter is in hydrostatic equilibrium, the gravitational settling of ion species 2 is accompanied by "lifting" of ion species 1 (with $\boldsymbol{J}_1 = -\boldsymbol{J}_2$). This diffusive motion of ions generates the collisional production of the specific entropy ($\dot{S}_{\rm coll}$) which leads to the thermal energy release with the rate Q [erg cm^{-3} s^{-1}] (e.g., [9, 10])

$$Q = T\,\dot{S}_{\rm coll} = \boldsymbol{J}_2 \cdot \boldsymbol{d}_1\, \rho/(\rho_1 \rho_2) \tag{17}$$

that is easily computed.

3. Observational consequences and summary

Although the diffusion flux (14) retains its standard form, it contains a new settling term (16) owing to Coulomb separation. That separation was predicted by Chang et al. [8] who analyzed equilibrium distributions of ion mixtures including the Coulomb interaction term. We extend their work to non-equilibrium ion mixtures and show that the Coulomb separation affects the diffusion flux (14) and gravitational settling of ions.

The strongest effect occurs at temperatures at which the ions form strongly coupled Coulomb liquid. At lower temperatures ions solidify; in the solid state they diffuse but much slower [13]. At higher temperatures, Coulomb coupling is available but less efficient. The Coulomb sedimentation is most important for the mixtures of ions with the same A/Z (e.g., ^{4}He, ^{12}C, and ^{16}O). The traditional gravitational sedimentation (15) in these mixtures is greatly suppressed [8]; the Coulomb settling (16) in these mixtures is much stronger. The ions with larger Z must diffuse to deeper layers. The effect is stronger for larger difference of Z values in the mixture.

The effect is basically regulated by gravity. It is most pronounced in compact stars (WDs, and NSs) with the strongest gravity. First of all we mean ^{4}He–^{12}C cores of low mass WDs and ^{12}C–^{16}O cores of more massive WDs, and similar mixtures in the envelopes of NSs. The velocity of sedimentation is given by (16) with appropriate diffusion coefficients (e.g., [11–13, 20]).

Figure 1 shows $\boldsymbol{u}_c$ in the ^{12}C–^{16}O cores of medium mass and massive WDs and in the ^{4}He–^{12}C cores of low mass WDs. The adopted temperate range $T \sim (0.5 - 5) \times 10^7$ K is appropriate to old isolated WDs [21]. The settling velocities are higher for massive WDs (with larger g). The velocity profile throughout the WD core has a maximum at the core boundary $r = R_{core}$. We have $\boldsymbol{u}_c \to 0$ as $r \to 0$ because $g(r) \to 0$ at the star's center. The maximum velocity in $1.2\,M_\odot$ WD reaches $\sim$ 100 km Gyr^{-1}; the ^{12}C–^{16}O separation in the outer core occurs in a few Gyrs. The velocity of Coulomb separation of ^{12}C and ^{16}O ions is typically lower than the settling velocity of ^{22}Ne ions in the ^{12}C–^{16}O core [1–5] but the fraction of ^{22}Ne ions is much smaller than the fractions of ^{12}C and ^{16}O. Using (17) and (16) we have estimated the thermal energy generation rate $Q(r)$ which accompanies this separation and found it insufficient to noticeably reheat old WDs. The profile $Q(r)$ has a maximum in the outer part of the WD core.

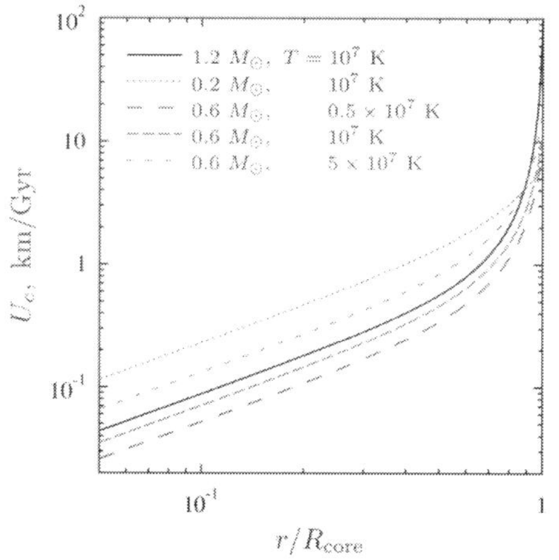

Figure 1: (Color online) Velocity (16) of Coulomb settling in degenerate cores ($r < R_{core}$) of WDs with masses $M = 0.2$, 0.6, and $1.2\,M_\odot$ ($M_\odot$ being the Sun's mass) and internal temperatures 0.5, 1, and 5×10^7 K. We show settling of ^{12}C in mixture with ^{4}He in the $0.2\,M_\odot$ WD. In other cases it is settling of ^{16}O mixed with ^{12}C. In all cases we assume $n_1 = n_2$.

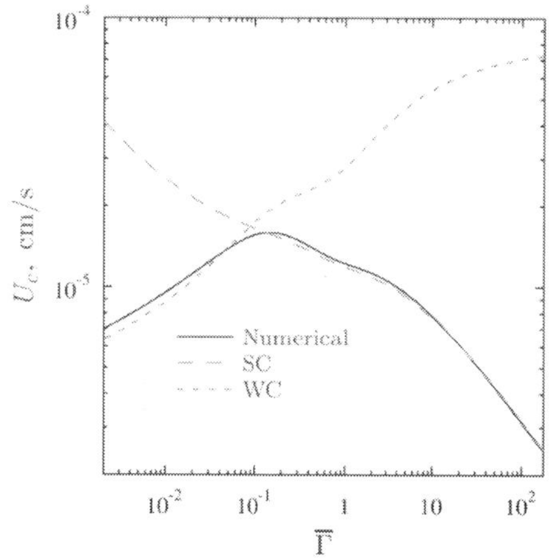

Figure 2: (Color online) Settling velocity of ^{12}C ions mixed with ^{4}He ($n_{\rm C}/n = 0.4$) in the outer neutron star envelope ($\rho = 10^6$ g cm^{-3}, $g = 2 \times 10^{14}$ cm s^{-2}) versus mean Coulomb coupling parameter $\overline{\Gamma} = (n_{\rm C}\Gamma_{\rm C} + n_{\rm He}\Gamma_{\rm He})/n$. See explanation in the text.

The Coulomb separation of ^{4}He, ^{12}C and ^{16}O ions can be important in isolated and accreting WDs. It affects chemical composition and, consequently, heat capacity, thermal conductivity, neutrino emission, nuclear reaction rates as well as chemical, thermal and nuclear evolution of WDs. Coulomb separation of ions may affect also asteroseismology of WDs.

Coulomb separation of ions with equal A/Z in NS envelopes is much stronger than in WDs. Figure 2 plots the sedimentation velocity $\boldsymbol{u}_c$ of ^{12}C ions mixed with ^{4}He in the outer NS envelope at $\rho = 10^6$ g cm^{-3} as a function of the mean Coulomb coupling parameter [16]. The solid line is a result of computation based on numerical calculation of $\mu_j^{(C)}$ [22] which is applicable for arbitrary coupling. SC stands for the approximation of strong coupling; this line represents velocity (16). WC represents the approximation of weak coupling. Details of general and weak coupling calculations will be discussed elsewhere. The envelope is non-isothermal and the temperature gradient can affect diffusion which we ignore for simplicity. For the densities of $\sim 10^5 - 10^7$ g cm^{-3} (from a few to a few tens of meters under the surface) the sedimentation velocity can reach several meters per year. The separation can affect nuclear evolution of the matter in the outer layers of accretting NSs. It will change the thermal conductivity of this matter, influence the relation between the surface and inner temperatures of NSs and affect cooling of isolated and accreting NSs (e.g., [6–8, 23–25]).

Acknowledgments
We thank A. I. Chugunov and A. Y. Potekhin for discussions. DGY acknowledges support from RFBR (grants 11-02-00253-a and 13-02-12017-ofi-M), RF Presidential Program NSh 4035.2012.2, and Ministry of Education and Science of Russian Federation (agreement No. 8409, 2012). The work of MB has also been partly supported by the Dynasty Foundation.

References

[1] Isern J, Hernanz M, Mochkovitch R and Garcia-Berro E 1991 *Astron. Astophys.* **241** L29–L32

[2] Bildsten L and Hall D M 2001 *Astrophys. J. Lett.* **549** L219–L223

[3] Deloye C J and Bildsten L 2002 *Astrophys. J.* **580** 1077–1090

[4] Althaus L G, García-Berro E, Renedo I, Isern J, Córsico A H and Rohrmann R D 2010 *Astrophys. J.* **719** 612–621

[5] García-Berro E, Torres S, Althaus L G, Renedo I, Lorén-Aguilar P, Córsico A H, Rohrmann R D, Salaris M and Isern J 2010 *Nature* **465** 194–196

[6] Chang P and Bildsten L 2003 *Astrophys. J.* **585** 464–474

[7] Chang P and Bildsten L 2004 *Astrophys. J.* **605** 830–839

[8] Chang P, Bildsten L and Arras P 2010 *Astrophys. J.* **723** 719–728

[9] Chapman S and Cowling T G 1952 *The Mathematical Theory of Non-Uniform Gases* (Cambridge Univ. Press, Cambridge)

[10] Hirschfelder J O, Curtiss C F and Bird R B 1954 *Molecular Theory of Gases and Liquids* (Wiley, New York)

[11] Hansen J P, McDonald I R and Pollock E L 1975 *Phys. Rev. A* **11** 1025–1039

[12] Hughto J, Schneider A S, Horowitz C J and Berry D K 2010 *Phys. Rev. E* **82** 066401

[13] Hughto J, Schneider A S, Horowitz C J and Berry D K 2011 *Phys. Rev. E* **84** 016401

[14] Donkó Z, Hartmann P and Kalman G J 2003 *Phys. Plasmas* **10** 1563–1568

[15] Beznogov M V and Yakovlev D G 2013 *Phys. Rev. Lett.* **111** 161101

[16] Haensel P, Potekhin A Y and Yakovlev D G 2007 *Neutron Stars. 1. Equation of State and Structure* (*Astrophysics and Space Science Library* vol 326) (Springer, New York)

[17] Landau L D and Lifshitz E M 1987 *Fluid Mechanics* (Butterworth-Heinemann, Oxford)

[18] Landau L D and Lifshitz E M 1993 *Statistical Physics, Part 1* (Pergamon, Oxford)

[19] Shapiro S L and Teukolsky S A 1983 *Black holes, white dwarfs, and neutron stars: The physics of compact objects* (Wiley-Interscience, New York)

[20] Paquette C, Pelletier C, Fontaine G and Michaud G 1986 *Astrophys. J. Suppl.* **61** 177–195

[21] Hansen B M S 1999 *Astrophys. J.* **520** 680–695

[22] Potekhin A Y and Chabrier G 2010 *Contrib. Plasma Phys.* **50** 82–87

[23] Potekhin A Y, Chabrier G and Yakovlev D G 1997 *Astron. Astrophys.* **323** 415–428

[24] Potekhin A Y and Yakovlev D G 2001 *Astron. Astrophys.* **374** 213–226

[25] Yakovlev D G and Pethick C J 2004 *Annu. Rev. Astron. Astrophys* **42** 169–210

16th Russian Youth Conference on Physics and Astronomy (PhysicA.SPb/2013) IOP Publishing
Journal of Physics: Conference Series **572** (2014) 012002 doi:10.1088/1742-6596/572/1/012002

Narrow-line Seyfert Galaxies. Connection between abundance and the large-scale structure

A.A. Ermash, B.V. Komberg

Astro Space Centre, P.N. Lebedev Physical Institute of Russian Academy of Sciences, Moscow, Russia

E-mail: `aermash@gmail.com`, `bkomberg@asc.rssi.ru`

Abstract. Utilizing methods, developed by the author the correlations between spatial concentrations of active nuclei (NLS and BLS) and concentration of galaxies of full uniform sample were obtained. Galaxies of this uniform sample trace the large-scale structure. We used SDSS DR 7 data. The correlations obtained are linear and the NLS/BLS ratio is constant. That leads to conclusion that amounts NLS and BLS are some fixed portion of all galaxies independent on the density of large-scale environment. In order to check validity of our results we also confirmed the well known result that fraction of red galaxies increases with density of environment. Also it was confirmed that this trend is more prominent for less massive galaxies.

1. Intoduction

According to the current classification the population of Seyfert galaxies is divided into two main types. Objects with broad permitted lines are classified as Sy1, objects without broad permitted lines as Sy2. This difference might be caused by the lack of BLR (Broad Line Region) or by obscuration of BLR by dusty torus. In former case such objects are called "true Sy2". Intermediate types (Sy 1.2 – Sy 1.8) in this classification are defined according to prominence of the broad components of permitted lines.

When we speak about Narrow Line Seyfert 1 galaxies we imply that their broad permitted lines are merely slightly wider than forbidden lines. But broad permitted lines in Narrow Line Seyfert galaxies are present! Various authors define the limit between NLS and BLS (Broad Line Seyfert Galaxies) differently, convenient value is in range 2000 – 2200 km/s.

There is a growing interest in this type of AGN in the recent years. Their host galaxies have on average later Hubble types, $\langle HT \rangle = 3.0$(Sb) and $\langle HT \rangle = 1.0$(Sa) for NLS and BLS respectively [2]. Bulges in host galaxies of NLS are always pseudobulges [3], their nuclear spirals show regular two-arm morphology, in contrast to BLS.

In recent years more and more authors had confirmed that NLS accrete with high Eddington ratios close to the Eddington limit and that their black holes are less massive than such of BLS [4, 5]. Black holes in NLS also seem to have high angular momenta [6, 3, 7, 8]. It is curious that NLS can produce relativistic jets in many ways similar to jets in blasars.

Many authors believe that NLS evolve through so-called secular processes in contrast with BLS. See, for example, [3]. Secular evolution implies slow evolution of the host galaxy via internal instabilities (internal secular evolution) or tidal interactions and minor mergings with small satellite galaxies (external secular evolution). For details see [9].

Relatively small attention was paid to the surroundings of NLS galaxies in the literature. In [10] the connection between environment and properties of host galaxies of LINERs[1] and TO[2] was studied. As a result, no connection was found. On the other hand, in [11] it was found that there is a significant difference between environments of Sy1 and Sy2. Sy2 have more close companions. Authors had also compared environments of control samples of galaxies of same morphological types as hosts of Sy1 and Sy2. No difference between host galaxies of AGN and normal galaxies of same morphological types was found. That led to conclusion that differences in environmental density of Sy1 and Sy2 are due to differences in morphological types of their host galaxies and not due to the presence of an active nucleus. Paper [12] was dedicated directly to the problem of environment of NLS. Authors came to conclusion that NLS tend to have close companions less frequently. Nevertheless, it should be noted that this paper is relatively old (year 2001) and all conclusions are based on a sample of 27 galaxies.

In our previous work [1] we have obtained the luminosity function of NLS in [OIII]λ5007Å. This forbidden line is emitted at large distance from the active nucleus in NLR, which has conical geometry, thus this line does not suffer from orientation effects and obscuration. Modified V/V_{max} method was used. For a complete magnitude limited sample it is very easy to calculate V_{max}, but the situation is quite difficult for spectral lines. In order to solve this problem we calculate the function of probability of observation of object for every luminosity bin $p(d_c)$, where d_c — comoving distance.

We also took into account variations of the density of the Universe due to the large-scale structure. The idea was as following. If we know the luminosity function of inactive galaxies with good accuracy, then having a uniform sample it is possible to calculate for considered volume N_{obs}/N_{calc}, where N_{obs} – the observed amount of galaxies, N_{calc} — calculated amount of galaxies from luminosity function. This approach allows us to normalize sample of considered objects to uniform complete sample of galaxies which traces variations of density of the Universe. For details see [1].

2. Data processing

In our work we have used SDSS DR 7 [13]. Luminosities of AGN were estimated using [OIII]λ5007Å line. For classification we used Hα line. We have also selected a magnitude limited sample of galaxies in r band taking extinction into account. This sample was used for normalization. In our work we used the following cosmological parameters: $\Omega_M = 0.279$, $\Omega_\Lambda = 0.721$, $h = 0.701$, as in SDSS DR 7.

Firstly we have split our sample of AGN into luminosity bins in [OIII]λ5007Å line (expressed in units $\lg(L_{[OIII]}/L_\odot)$ and set minimum and maximum redshifts of the sample. As in our previous work [1] we have used fixed step in comoving volume instead of comoving distance d_c or redshift z.

For every defined interval of redshift we calculated the normalization $\rho_{gal}/\langle\rho_{gal}\rangle$, where ρ_{gal} is the density of galaxies in the considered volume element, $\langle\rho_{gal}\rangle$ — average density of galaxies. Then for every luminosity bin we found amount of AGN objects N_{AGN}. The amount of AGN normalized to the average density of the Universe is:$N_{AGN,norm} = N_{AGN}(\rho_{gal}/\langle\rho_{gal}\rangle)^{-1}$ The ratio $\rho_{gal}/\langle\rho_{gal}\rangle$ was calculated the following way. Luminosity function is usually fitted with the Schechter function, for absolute magnitudes it takes the following form:

$$\Phi(M) = 0.4\lg(10)\Phi_* 10^{-0.4(M-M_*)(\alpha+1)} \exp\left(-10^{-0.4(M-M_*)}\right)$$

We took parameters for local luminosity function of galaxies from [14], where they are as following: $\Phi_* = 0.0090 \pm 0.0007$, $M_* - 5\lg_{10} h = -20.73 \pm 0.04$, $\alpha = -1.23 \pm 0.02$ in r band of

[1] Low-Ionization Nuclei Emission Region galaxies.

[2] Transition Objects.

SDSS survey. Knowing this parameters one can obtain total luminosity of galaxies per unit of volume:

$$L_{sch} = \int_{L_1}^{L_2} L'\phi(L')dL,$$

where L_1 and L_2 are the limits of integration.

Now we can obtain $\langle\rho_{gal}\rangle/\rho_{gal} = L_{sch}V/L_{obs}$, where L_{obs} — total observed luminosity of galaxies in considered volume element, V — volume. Luminosity of each galaxy is in range $L_1 < L < L_2$. It is crucial to stress that transition from luminosities to spatial densities is possible only if parameters of Schechter function α and L^* are constant. For the redshift interval considered this is correct. For details see [1]. Hence, normalized amount of AGN is $N_{AGN,norm} = N_{AGN}(L_{sch}V/L_{obs})$.

Having normalized sample of AGN, we can now calculate the probability function of observation of objects in considered luminosity bin $p(d_c)$, where d_c — comoving volume. This function has the following form: $p_{AGN}(d_c) = aexp(-b_1/d_c^2)exp(-b_2d_c^2)$. Details see in [1].

Then we have split considered volume of the Universe into elements limited by redshift $(z_i < z < z_{i+1})$, ascension and declination $(\alpha_{min,j} < \alpha < \alpha_{max,j}; \delta_{min,j} < \delta < \delta_{max,j})$. For every volume element the ratio $\rho_{gal}/\langle\rho_{gal}\rangle$ was calculated. Total count of NLS and BLS was obtained for the considered luminosity interval. It was then corrected using the probability function $p(d_c)$. Making the same calculations for each volume bin we obtained the dependence of N_{NLS}, N_{BLS} and N_{NLS}/N_{BLS} on $\rho_{gal}/\langle\rho_{gal}\rangle$, where N_{NLS} and N_{BLS} — spatial densities of NLS and BLS respectively.

3. Spatial density of AGN and the large-scale structure

Firstly, it is necessary to define volume bins of the SDSS survey. The grid used had the following parameters. Solid angle of each element was $\Omega = 0.011$sr or $36^{\square}$, minimum redshift $z_{min} = 0.022$, maximum redshift $z_{max} = 0.18$, amount of redshift bins was set to 15.

The AGN luminosity range was split in 4 intervals: 5.25 – 6.25, 6.25 – 7.0, 7.0 – 7.75, 7.75 – 9.0. Values are in units $\lg L_{[OIII]}/L_{\odot}$.

For each luminosity bin we have built three correlations between $N_{NLS}(Mpc^{-3})$, $N_{BLS}(Mpc^{-3})$,N_{NLS}/N_{BLS} and $\rho_{gal}/\langle\rho_{gal}\rangle$. We have built these correlations also for the whole luminosity interval of AGN ($L = 5.25 - 9.0$).

On figure 1 the following correlations are shown. On the upper row on panels a) – d) the correlations between N_{BLS} and $\rho_{gal}/\langle\rho_{gal}\rangle$ for four intervals of AGN luminosity are shown. The second row e) – h) displays the same for NLS. In the third row i – l) relations between N_{NLS}/N_{BLS} and $\rho_{gal}/\langle\rho_{gal}\rangle$ are shown. In the fourth row (m, n, o) three correlations between N_{BLS}, N_{NLS},N_{NLS}/N_{BLS} and $\rho_{gal}/\langle\rho_{gal}\rangle$ for the whole considered interval of AGN luminosity are shown.

The correlations between N_{NLS},N_{BLS} and $\rho_{gal}/\langle\rho_{gal}\rangle$ were fitted with linear function $ax + b$ and with exponential one in form $a \times exp(-c/x^2) + b$, where a, b and c — free parameters. The latter function was chosen because it grows with increasing x and asymptotically flattens at high x. The correlation between N_{NLS}/N_{BLS} and $\rho_{gal}/\langle\rho_{gal}\rangle$ was fitted with linear function and also with a cubic polynomial $ax^3 + bx^2 + cx + d$.

Let us consider the correlation between N_{BLS} and $\rho_{gal}/\langle\rho_{gal}\rangle$. For the first luminosity interval there is some evidence of deviation from the linear trend at high values of $\rho_{gal}/\langle\rho_{gal}\rangle$. Approximation by straight line gives $\chi^2 = 1.15$ and exponential function gives $\chi^2 = 1.01$. For the other three intervals in luminosity there are no signs of deviation from the linear trend. The same is true for the whole interval of nuclear luminosities. Chi-square gives 1.3 for linear and 1.27 for exponential functions respectively. For the relation between N_{NLS} and $\rho_{gal}/\langle\rho_{gal}\rangle$ there is some evidence of deviation from linear trend in the luminosity range $L_{AGN} = 7.0 - 7.75$ ($\chi^2 = 1.14$ and 1.02 for linear and exponential trends respectively).

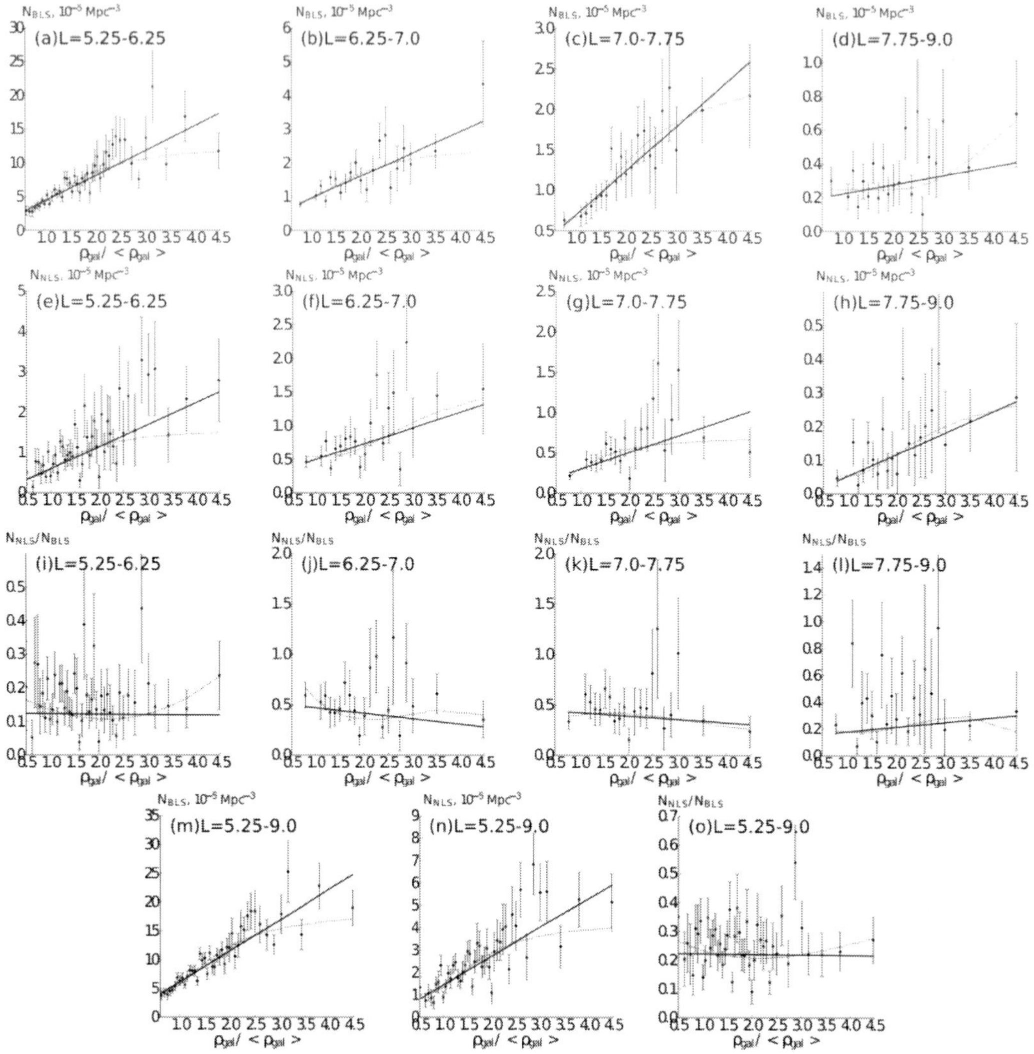

Figure 1. Upper row — correlations between N_{BLS} and $\rho_{gal}/\langle\rho_{gal}\rangle$. Four panels correspond to four intervals of AGN luminosity in units of $\lg L_{[OIII]}/L_{\odot}$. a) 5.25 – 6.25 b) 6.25 – 7.0 c) 7.0 – 7.75 d) 7.75 – 9.0. The second row displays correlations between N_{NLS} and $\rho_{gal}/\langle\rho_{gal}\rangle$. In the same way e), f), g), h) correspond to the same intervals of AGN luminosity. The third row displays the relations between N_{NLS}/N_{BLS} and $\rho_{gal}/\langle\rho_{gal}\rangle$. i), j), k) l) correspond to the same intervals of AGN luminosity. Lower row displays relations between m) N_{BLS}, n) N_{NLS}, o)N_{NLS}/N_{BLS} and $\rho_{gal}/\langle\rho_{gal}\rangle$ for the whole interval of the AGN luminosity, $L = 5.25 - 9.0$. Solid lines on all diagrams are the result of linear approximation. Dashed lines on all diagrams are the result of approximation with nonlinear function: a) – h), m), n) exponential function $a \times exp\left(-c/x^2\right) + b$;i) -l), o) cubic polynomial $ax^3 + bx^2 + cx + d$.

Relation between N_{NLS}/N_{BLS} and $\rho_{gal}/\langle\rho_{gal}\rangle$ either for all 4 intervals of luminosity and for the whole interval is well fitter by linear functions. The a coefficients are equal to zero within errors.

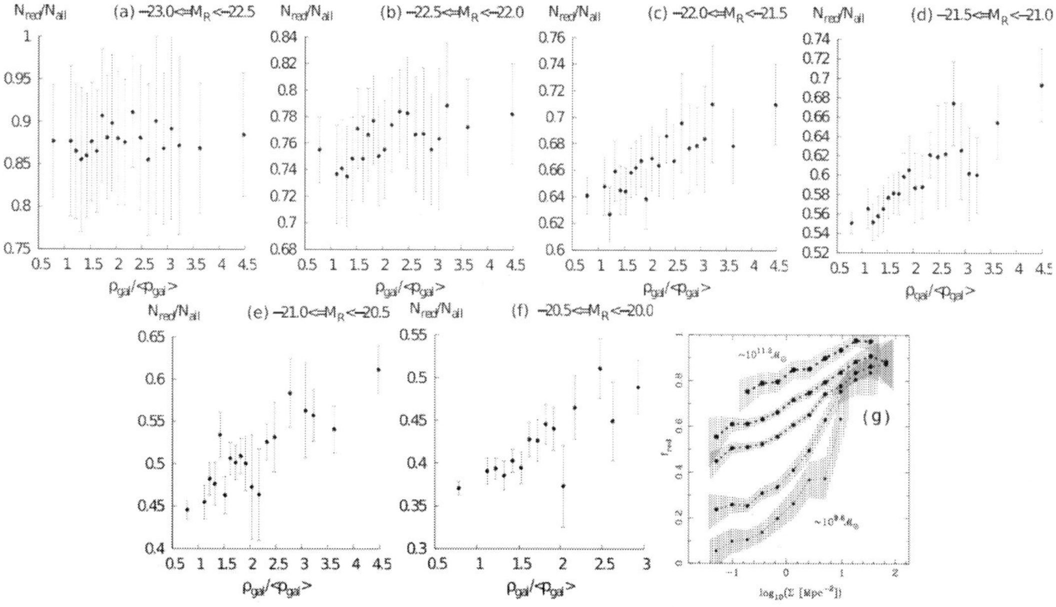

Figure 2. a - f) Correlation between fraction of red galaxies with $u - r > 2.2$ and $\rho_{gal}/\langle\rho_{gal}\rangle$ for six intervals of absolute magnitude M_r. g) Fraction of red galaxies vs $log_{10}(\Sigma[Mpc^{-3}])$ for different mass intervals. Figure was taken from [15].

Let us now discuss what do these results imply. If NLS and BLS are a fixed, albeit different fraction of all galaxies it is obvious that the relations considered will behave the following way. N_{NLS} and N_{BLS} will be linear and N_{NLS}/N_{BLS} will be constant. Deviation from such behavior will mean that such assumptions are not valid. *Deviation of N_{NLS}/N_{BLS} is not observed.*

For all the luminosity intervals considered correlations between N_{NLS}, N_{BLS} and $\rho_{gal}/\langle\rho_{gal}\rangle$ are linear. But there are two exceptions: $L = 5.25 - 6.25$(BLS) and $L = 7.0 - 7.75$(NLS). Evidence of deviation from linear trend at $\rho_{gal}/\rho_{gal} > 3$ is present.

Nevertheless, statistical significance is too small for any conclusions.

It seems that for all considered luminosity intervals the correlations between N_{NLS}, N_{BLS} and $\rho_{gal}/\langle\rho_{gal}\rangle$ are linear. NLS and BLS are some fraction of all galaxies which depends on nuclear luminosity but not on density of large-scale environment.

All this argue for the idea that NLS activity is launched through internal processes, not by interactions. It should be noted, however, that our approach allows to analyze connection of nuclear activity with variations of concentration of galaxies on scales of large-scale structure cells (i.e. with the position of AGN in large-scale structure), but it does not account at all for the presence or absence of close companions. Some authors state that interactions with close companions lead to formation of classical bulges in BLS (see, e.g., [3])

4. Dependence of the red fraction of galaxies on luminosity and density of environment

The fact that the fraction of early-type galaxies is higher in more dense regions is well-known. See, e.g., [15]. Also with increasing density of environment increases the fraction of red galaxies.

Paper [15] is dedicated to analysis of data provided by the Galaxy Zoo project. For today it is the largest catalog of visually classified galaxies. Right panel of fig. 12 in [15] shows the correlation between the fraction of red galaxies and $\lg_{10}(\Sigma[Mpc^-3])$. In our work this correlation is shown on fig. 2g. Authors of [15] defined the density of environment as $\Sigma_N = N(\pi d_N^2)$, where d_N — projected distance to the Nth companion brighter than $M_r = 20$. N was set to 4 or 5.

Let us check if our approach allows to independently confirm the result of dependence of red fraction on the density of environment. The correlations were obtained for 6 intervals in absolute magnitude. In order to classify galaxies we used $u - r$ colors. Galaxy was considered red if $u - r > 2.2$ according to [16]. The obtained correlations between red fraction N_{red}/N_{all} and $\rho_{gal}/\langle\rho_{gal}\rangle$ are shown on figure 2a-f.

Let us compare our result with the one from [15] shown on fig. 2g. We confirm not only that there is a tendency of increasing fraction of red galaxies with density of environment, but also the fact that this trend is more prominent for less massive galaxies.

Hence, confirmation of this result means that our approach is correct and results obtained about NLS and BLS are trustworthy.

Acknowledgments
The author would like to acknowledge O.V.Verkhodanov and S.V. Pilipenko and B.V.Komberg for interesting discussions and useful comments.

This work was partially supported by the Federal Targeted Program "Scientific and Science-Teaching Staff of Innovative Russia" for 2009–2013 (Agreement No. 8405), the Program of State Support for Leading Schools of the Russian Federation (grant NSh-2915.2012.2), and the Basic Research Program of the Presidium of the Russian Academy of Sciences P-21 "Non-stationary phenomena in objects in the Universe."

Funding for the SDSS and SDSS-II has been provided by the Alfred P. Sloan Foundation, the Participating Institutions, the National Science Foundation, the U.S. Department of Energy, the National Aeronautics and Space Administration, the Japanese Monbukagakusho, the Max Planck Society, and the Higher Education Funding Council for England. The SDSS Web Site is http://www.sdss.org/.

This research has made use of CosmoPY package for PYTHON http://roban.github.com/CosmoloPy/.

References

[1] Ermash A 2013 *Astron. rep.* **90**(5) 355
[2] Deo R P, Crenshaw D M and Kraemer S 2006 *Astron. J.* **132** 321
[3] Orban de Xivry G, Davies R, Schartmann M *et al.* 2011 *Monthly Not. Roy. Astron. Soc.* **417** 2721–2736
[4] Botte V, Ciroi S, Rafanellin P and Mille F D 2004 *Astron. J.* **127** 3168
[5] Xu D and Komossa S 2011 *Proceedings of the Workshop "Narrow-Line Seyfert 1 Galaxies and Their Place in the Universe", PoS(NLS1) 006* ed Foschini L, Colpi M, Gallo L *et al.* (Trieste, Italy: Proceedings of Science)
[6] Patrick A, Reeves J, Lobban A *et al.* 2011 *Monthly Not. Roy. Astron. Soc.* **416** 2725
[7] Fabian A, Kara E, Walton D *et al.* 2013 *Monthly Not. Roy. Astron. Soc.* **429** 2917
[8] Risaliti G, Harrison F A, Madsen K K *et al.* 2013 *NATURE* **494** 449
[9] Eliche-Moral M C, Balcells M, Aguerri J A L and González-García A C 2006 *Astron. Astrophys.* **457** 91
[10] Ho L C, Filippenko A V and Sargent W L W 2003 *Astrophys. J.* **583** 159–177
[11] Koulouridis E, Plionis M, Chavushyan V *et al.* 2006 *Astrophys. J.* **639** 37
[12] Krongold Y, Dultzin-Hacyan D and Marziani P 2001 *Astron. J.* **121** 702
[13] Abazajian K N, Adelman-McCarthy J K, Agüeros' M A *et al.* 2009 *Astrophys. J. Suppl. Ser.* **182** 543
[14] Montero-Dorta A D and Prada F 2009 *Monthly Not. Roy. Astron. Soc.* **399** 1106
[15] Bamford S P, Nichol R C, Baldry I K *et al.* 2009 *Monthly Not. Roy. Astron. Soc.* **393** 1324B
[16] Tempel E and Saar E 2011 *Astron. and Astrophys.* **529** 53

Maximal energies of the particles accelerated by the system of converging magnetohydrodynamic flows

P.E. Gladilin[1], **A.M. Bykov**[1] **and S.M. Osipov**[1]

[1]Ioffe Physical-Technical Institute of the Russian Academy of Sciences, Saint-Petersburg, Russia,

E-mail: peter.gladilin@gmail.com

Abstract. We have shown that maximal energies of the charged particles accelerated in the system of converging magnetohydrodynamic (MHD) flows can reach $\approx 10^{17}$ eV. The scheme of magnetic field amplification (MFA) applied to the previous non-linear model of particle acceleration in the colliding shocks allowed to make proper estimates for the value of turbulent magnetic field. The efficiency of the particle acceleration on the energy range larger than the "knee" in the cosmic rays spectrum ($\approx 10^{14} - 10^{15}$ eV) makes the systems of colliding MHD flows important contributors to the overall high-energy cosmic rays population in the Galaxy.

1. Introduction

Maximal energy of the charged particles attainable in supernova remnants (SNR) is one of the exciting questions of the astrophysics of cosmic rays (CR). SNR shocks are considered as the main CR accelerator up to the "knee" in the spectrum of galactic CR ($\approx 5 \cdot 10^{14}$ eV) and even further (up to $10^{16} - 10^{17}$ eV). Understanding the processes occuring on the maximal energies of the particles accelerated in the vicinity of the SNRs is extremely important [1].

Recently Ptuskin and Zirakashvili have obtained that in particular conditions the energy of protons accelerated on the shocks of different types of SNRs can reach $10^{15} - 10^{17}$ eV. Their calculations predict that for Fe-ions accelerated on IIb-type SNR the maximal energy can reach even $5 \cdot 10^{18}$ eV [1].

Recent observations and theoretical models, suggest that the maximal energy of the accelerated particle for the bulk population of SNRs is below 10^{16} eV (see, e.g., [2, 3]). The maximal energy of the proton accelerated at the SNR shock depends on shock velocity, the strength of the amplified turbulent magnetic field, the radius of SNR, and the density of the ambient medium.

The maximum energy of accelerated particles strongly differs for different types of supernovae (see e.g., [1]), it depends on the circumstellar medium around the supernova progenitor star. Moreover, core-collapsed supernovae produced by massive stars often occur in OB-star associations where the intense radiation of hot massive stars, powerful stellar winds and supernova shocks strongly modify the interstellar environment, producing large hot cavities of a few tens of parsec size, called superbubbles.

In the present paper we model a class of a few parsec size particle accelerators associated with collision of a young supernova shock with a fast stellar wind of a massive star. Some basic features of such systems were noticed in the previous papers [4, 5]. The modeled stage starts a

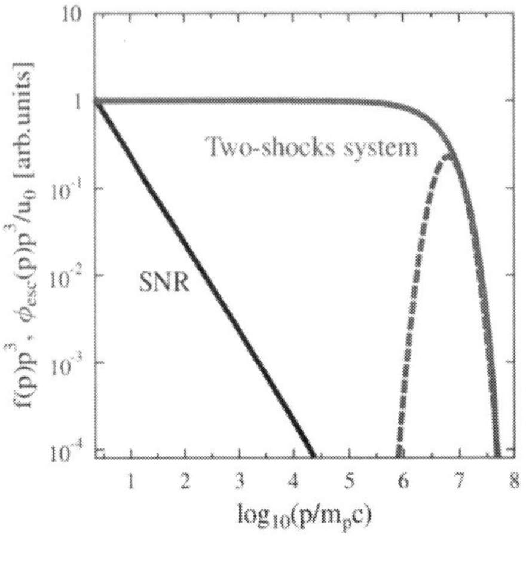

Figure 1. Spectral energy distribution of protons for the cases of acceleration on the isolated supernova shock (black line) and by the system of two colliding shocks (blue line). Dashed blue line corresponds to the spectrum of particles escaping from the system on the free escape boundary L_{FEB}.

few hundred years before the supernova shock collides with the wind termination shock. At this stage the maximal energy particles accelerated via diffusive shock acceleration mechanism (DSA) at the SNR shock reach the fast wind termination shock and are scattered back by magnetic fluctuations carried by the fast stellar wind. Therefore, the high energy particles that have mean free path $\Lambda(p)$ larger than the distance between the two shocks L_{12} start to be accelerated by the converging fast flows. This is the most favorable circumstance for the efficient Fermi acceleration. While the structure of the MHD flow in the vicinity of a supernova shell colliding with the stellar wind termination shock is rather complex, it is possible to consider a simplified plane-parallel model. When the magnetic field fluctuations are strong enough to provide the so-called Bohm diffusion regime with $\Lambda(p) = \xi r_g(p)$, where p is the particle momentum, $r_g(p)$ is the particle gyroradius in the mean magnetic field and $\xi \gtrsim 1$, the high energy particles bouncing between the converging flows are likely to have a spectrum harder than that produced at an individual shock and may contain a sizeable fraction of the total kinetic energy of the converging MHD flows.

In figure 1 we present simulated spectra for protons accelerated by the system of two colliding shocks compared with the spectrum derived for an isolated SNR. Spectra represent particle momentum distribution function $f(p)$ and particle escape flux $\phi_{esc}(p)$ multiplied by p^3. Spectral energy distribution (SED) and escape flux for the two-shocks system corresponds to the following model parameters: distance from the shock to the free escape boundary (FEB), where particles can freely escape from the system, $L_{FEB} = 3$ pc, flow velocity $u_0 = 5000$ km/s, the strength of the amplified magnetic field $B_{turb} = 275\,\mu G$ (see Section 2).

2. Maximal proton energy from colliding shocks system

Let now consider the case where SNR shock propagates into the interstellar medium (ISM) in the vicinity of the stellar wind of the nearest massive star. The efficiency of particle acceleration in the colliding shock system will increase comparing with the acceleration process on the ordinary shock. This system will characterized by hard spectral energy distribution of

Table 1. Maximal energy E_{max} of protons calculated for various model parameters, where L_{FEB} - the distance from the shock to the free escape boundary in the model [5], u_0 - the flows velocities, B_0 - the strength of the magnetic field in the vicinity of the shocks, B_{turb} - the strength of the amplified magnetic field, E_{max} - the maximal proton energy for the corresponding parameters.

L_{FEB}, pc	u_0, km/s	B_0, μG	E_{max}, eV	B_{turb}, μG	E_{max}, eV
	3000	3	$4.0 \cdot 10^{13}$		
		100	$1.3 \cdot 10^{15}$	228	$2.5 \cdot 10^{15}$
	5000	3	$6.2 \cdot 10^{13}$		
3		100	$1.9 \cdot 10^{15}$	275	$5.0 \cdot 10^{15}$
	10000	3	$2.0 \cdot 10^{14}$		
		100	$6.3 \cdot 10^{15}$	305	$1.3 \cdot 10^{16}$
	3000	3	$2.1 \cdot 10^{14}$		
		100	$6.3 \cdot 10^{15}$	239	$1.6 \cdot 10^{16}$
	5000	3	$3.2 \cdot 10^{14}$		
5		100	$1.3 \cdot 10^{16}$	281	$2.5 \cdot 10^{16}$
	10000	3	$6.4 \cdot 10^{14}$		
		100	$2.6 \cdot 10^{16}$	310	$8.0 \cdot 10^{16}$
	3000	3	$4.0 \cdot 10^{14}$		
		100	$1.3 \cdot 10^{16}$	265	$4.0 \cdot 10^{16}$
	5000	3	$5.1 \cdot 10^{14}$		
10		100	$2.1 \cdot 10^{16}$	296	$6.3 \cdot 10^{16}$
	10000	3	$1.3 \cdot 10^{15}$		
		100	$4.1 \cdot 10^{16}$	334	$1.6 \cdot 10^{17}$

accelerated particles, high efficiency of the acceleration, and increasing maximal energy of the particle for the time of efficient acceleration [5, 6].

Main features of these systems were investigated in the recent papers [4, 5, 6]. We have performed the non-linear model of particle acceleration in the two approaching magnetohydrodynamical (MHD) flows, having taken into account the flow modification by the pressure of the accelerated particles [5]. This model is based on the generalization of the semi-analytical model originally developed by Amato, Blasi and Caprioli [7, 8]. Their model was expanded for the case of two colliding shocks. We have concluded that due to the high efficiency of the particle acceleration on the colliding shocks system and extremely hard spectral energy distribution the maximal energy of the particles accelerated via such a system can be an order of magnitude (or even more) larger than it is for the isolated SNRs.

In the Table 1 we list the values of calculated maximal energy of protons accelerated by the two-shocks symmetric system. Numerical simulations were made via non-linear model of colliding shocks with free escape boundary [5]. The previous non-linear model was implemented by magnetic field amplification (MFA) procedure. It allowed us to calculate the values of turbulent magnetic field based on the calculated growth rates of plasma instabilities (see, e.g., [9, 10]). There was made the analysis for the dependence of the maximum possible proton energy E_{max} from different parameters, such as: L_{FEB} - the distance from the shock to the free escape boundary in the model [5], where particle escape from the acceleration area, u_0 - the flow velocity (we consider here symmetric model and velocities of the colliding flows accepted as equal), B_0 - the strength of the magnetic field in the vicinity of the shocks, B_{turb} - the strength of the amplified magnetic field. The density of the interstellar medium was taken as $n_0 = 1.0\ cm^{-3}$.

We consider E_{max} as the energy that corresponds to the maximum of the escape flux. For the model sets without MFA procedure the magnetic field B_0 was treated as free parameter instead of MFA setups where B_{turb} was calculated self-consistently. It is clearly seen that the maximal energy of proton grows up with the increase of the corresponding parameters. With included magnetic field amplification E_{max} confidently exceeds 10^{16} eV (see the last two columns in the Table 1).

It should be pointed out that the number of SNRs of IIb type, which accordingly to the research [1] can accelerate ions up to 10^{18} eV, is not very large and amounts only 1-2% from the full number of galactic SNRs [11]. Compare to these SNRs two-shocks systems have a higher efficiency of particle acceleration. Due to the hard spectrum of energetic particles the emissivity of these systems in a high energy range can by the order of magnitude exceed the emissivitiy of the isolated SNRs. Therefore these accelerators can be one of the important contributors to the cosmic rays population in the energy range $10^{15} - 10^{17}$ eV for the time of it's efficient acceleration (< 1000 yrs). Furthermore, these systems can be responsible for some "fluctuations" in the galactic CR spectrum in the energy range $10^{15} - 10^{17}$ eV [12, 13].

3. Discussion

It should be noted that the stage of the closely approaching SNR and stellar wind flows that is favorable for particle acceleration typically lasts for 200-1,000 years depending on the stellar wind velocity and the termination shock radius. Therefore, such sources can likely contribute to the high-energy galactic CRs population only for the time during the phase of the existence of necessary geometry (up to the moment when shocks collide and system morphology is destroyed).

The latest 2D-hydrodynamical simulations of circum-stellar medium around the rotating massive stars show that radii of their winds can reach 10 pc in size [14]. The average radii of the galactic SNRs lies around $5 - 6$ pc [11]. The efficient acceleration on two converging flows system begins only when the distance between the shocks L_{12} is less than the mean free path $\Lambda(p)$ of the energetic particle. Numerical simulations show that it occures only when $L_{12} \approx 0.1$ pc. Therefore, plane-parallel approach used in simulation is appropriate for modeling.

Maximal energies of accelerated protons attainable in the two-shocks systems allowed it to be the strong sources of high-energy gamma-emission. When the gas density of the interstellar medium in the starforming regions nearby SNR is large enough to provide efficient hadronuclear interactions of ions and protons accelerated in the system, nuclei being collided with ISM will produce π-mesons which in turn will produce gamma-photons and neutrinos. The energy E_ν of neutrino produced by p-p interactions is approximately $0.04 \cdot E_p$, where E_p - is the energy of the parent proton. It makes two-shocks systems possible sources of PeV-neutrinos detected by IceCube in 2013 [15]. For example, the system with $u_{01} = u_{02} = 5000\, km/s$, $X_{FEB} = 5$ pc, $B_{turb} = 281\, \mu G$ will produce proton with $E_p = 2.5 \cdot 10^{16}$ eV, that can be parent for the PeV neutrinos detected with IceCube. Detailed discussion of these systems as possible high-energy neutrino sources will be done in the following paper (A.M.Bykov, P.E.Gladilin, S.M.Osipov, 2014, in prep.).

Acknowledgments

The work was supported by the RFBR grant 14-02-31721-mol-a-2014. The numerical simulations were performed at SC at Ioffe Institute.

Journal of Physics: Conference Series **572** (2014) 012003 doi:10.1088/1742-6596/572/1/012003

References

[1] Ptuskin V, Zirakashvili V and Seo E S 2010 *ApJ* **718** 31–36 (*Preprint* 1006.0034)
[2] Bell A R, Schure K M, Reville B and Giacinti G 2013 *MNRAS* **431** 415–429 (*Preprint* 1301.7264)
[3] Helder E A, Vink J, Bykov A M, Ohira Y, Raymond J C and Terrier R 2012 *Space Sci. Rev.* **173** 369–431 (*Preprint* 1206.1593)
[4] Bykov A M, Gladilin P E and Osipov S M 2011 *Memories of Italian Astronomical Society* **82** 800 (*Preprint* 1111.2587)
[5] Bykov A M, Gladilin P E and Osipov S M 2013 *MNRAS* **429** 2755–2762 (*Preprint* 1212.1556)
[6] Bykov A M, Ellison D C, Gladilin P E and Osipov S M 2012 *American Institute of Physics Conference Series* (*American Institute of Physics Conference Series* vol 1505) ed Aharonian F A, Hofmann W and Rieger F M pp 46–55 (*Preprint* 1212.1985)
[7] Amato E and Blasi P 2005 *MNRAS* **364** L76–L80 (*Preprint* arXiv:astro-ph/0509673)
[8] Caprioli D, Blasi P, Amato E and Vietri M 2009 *MNRAS* **395** 895–906 (*Preprint* 0807.4261)
[9] Bykov A M, Brandenburg A, Malkov M A and Osipov S M 2013 *Space Sci. Rev.* (*Preprint* 1304.7081)
[10] Bykov A M, Osipov S M and Ellison D C 2011 *MNRAS* **410** 39–52 (*Preprint* 1010.0408)
[11] Green D A 2004 *Bulletin of the Astronomical Society of India* **32** 335–370 (*Preprint* arXiv:astro-ph/0411083)
[12] Hörandel J R 2013 *American Institute of Physics Conference Series* (*American Institute of Physics Conference Series* vol 1516) ed Ormes J F pp 185–194 (*Preprint* 1212.0739)
[13] Berezhnev S F, Besson D, Budnev N M, Chiavassa A, Chvalaev O A, Gress O A, Dyachok A N, Epimakhov S N, Haungs A, Karpov N I, Kalmykov N N, Konstantinov E N, Korobchenko A V, Korosteleva E E, Kozhin V A, Kuzmichev L A, Lubsandorzhiev B K, Lubsandorzhiev N B, Mirgazov R R, Panasyuk M I, Pankov L V, Popova E G, Prosin V V, Ptuskin V S, Semeney Y A, Shaibonov B A, Silaev A A, Silaev A A, Skurikhin A V, Snyder J, Spiering C, Schröder F G, Stockham M, Sveshnikova L G, Wischnewski R, Yashin I V and Zagorodnikov A V 2012 *Nuclear Instruments and Methods in Physics Research A* **692** 98–105 (*Preprint* 1201.2122)
[14] Georgy C, Walder R, Folini D, Bykov A, Marcowith A and Favre J M 2013 *ArXiv e-prints* (*Preprint* 1309.1360)
[15] Aartsen M G, Abbasi R, Abdou Y, Ackermann M, Adams J, Aguilar J A, Ahlers M, Altmann D, Auffenberg J and Bai X e a 2013 *Physical Review Letters* **111** 021103

The Properties of Solar Cycle 24 and Their Connection with Common Relationships of Cycles 19–23

P A Otkidychev
GAS GAO RAN, Kislovodsk, Russia
E-mail: otkid@list.ru

Abstract. A set of properties of solar activity cycle 24 is considered on the base of the data of Kislovodsk Mountain Astronomical Station. Some of those properties are in connection with common relationships of previous cycles.

1. Introduction

The Sun observations data set in white light at the Kislovodsk Station (GAS GAO RAN) embrace the period from 1954 until the present including complete cycles 19–23 and the beginning of cycle 24. The set of GAS GAO RAN observations is continuous [7] and in good correlation with data sets of other observatories [1]. The data archive is on the site www.solarstation.ru and represented by the daily summaries. The summaries contain following information: date and time of observations; number of each group by Kislovodsk classification; coordinates of each group (latitude, Carrington longitude, r/R); group area and area of maximal spot in group; quantity of sunspots in each group. Thus the Kislovodsk data set is to a certain extent unique, because among all the solar observatories it is the only one that contains simultaneously all the above-listed parameters.

The cycle terms in accordance with NGDC/NOAA data are following: 19: 04.1954–09.1964; 20: 10.1964–05.1976; 21: 06.1976–08.1986; 22: 09.1986–04.1996; 23: 05.1996–12.2008; 24: from 01.2009 up to 06.2013 inclusive.

2. The low activity of the cycle 24

Before the beginning of the current cycle there was dramatic discussion concerning its power. As is well known, the odd cycle 23 was weaker than previous one that is violation of the Gnevyshev – Ohl rule [3], [4], [5]. In this regard, the predictions of the power of cycle 24 had to confirm or disprove various propositions relative to the dynamo theory. There was opinions about both the high [2] and the low [9] activity of cycle 24. Now, almost half of the cycle, it became clear that the latter viewpoint is correct.

In [6] it was shown that the initial phase of cycle 24 has very low activity, the northern hemisphere is more active than the southern one. All these conclusions remain actual until the present. Mean area of sunspots for the first 4.5 years of cycle 24 is 602 millionth of solar hemisphere (msh), mean Wolf number is 57.5. The northern hemisphere surpasses the southern one at 1.59 times by the area and at 1.49 times by the Wolf number, mean spot latitude in the north is approximately two degrees less than in the south. Moreover, activity for the last about 2 years came, in fact, to a plateau (figures 1 & 2).

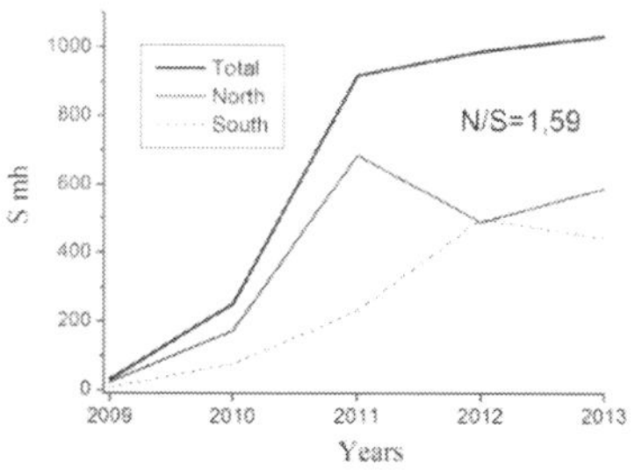

Figure 1. Mean annual sunspot areas of cycle 24.

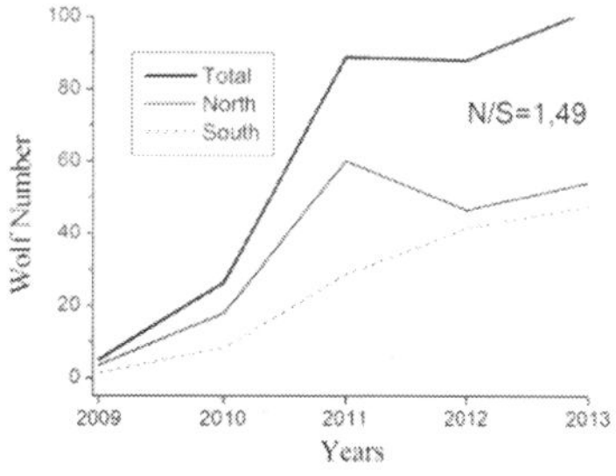

Figure 2. Mean annual Wolf number of cycle 24.

3. The low amount of very-large-scale sunspots

The current cycle is inferior to all previous ones by relative number of groups at which the area of the maximum spot exceeds 800 msh. It is remarkable that the leader by this parameter is not cycle 19 (the most powerful of all), but cycle 22. The relative quantity of such groups by cycles is given in Figure 3, the same dependence by years is in Figure 4 (mean annual areas of sunspot groups are given in overlapping by a dotted line).

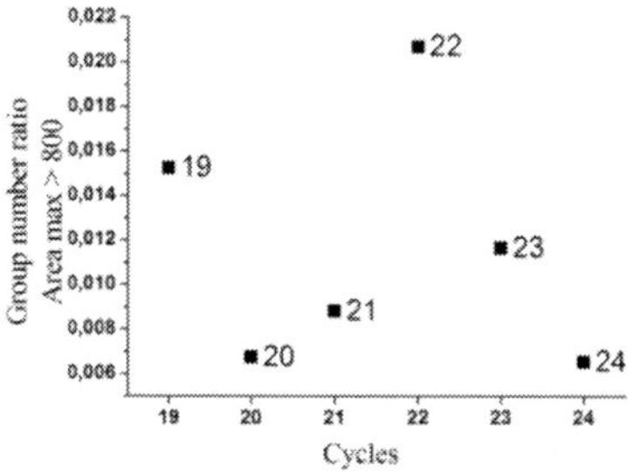

Figure 3. Mean relative amount of very-large-scale sunspots by cycles.

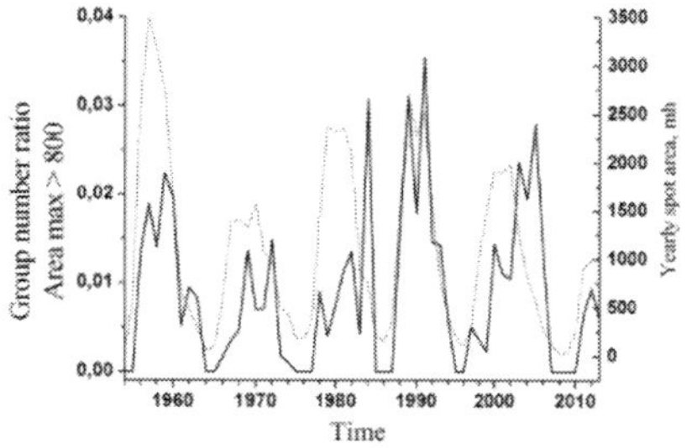

Figure 4. Mean relative yearly amount of very-large-scale sunspots.

4. The low duration of group life

The actual solar cycle has the least mean duration of group life among all the Kislovodsk cycles. This parameter is in good correlation with mean areas of sunspot groups (Figure 5) and in even more correlation with mean areas of maximal spots in groups (Figure 6). Cycle 22 is the leader by mean duration of group life, but it was cycle 22 that has maximal amount of very-large-scale sunspots as follows from the previous paragraph. This result is expected because the more the sunspots the more its life time.

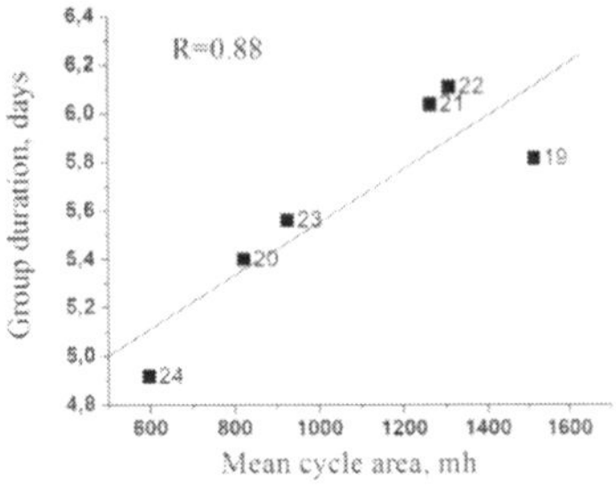

Figure 5. Mean relative duration of group life subject to mean areas of sunspots.

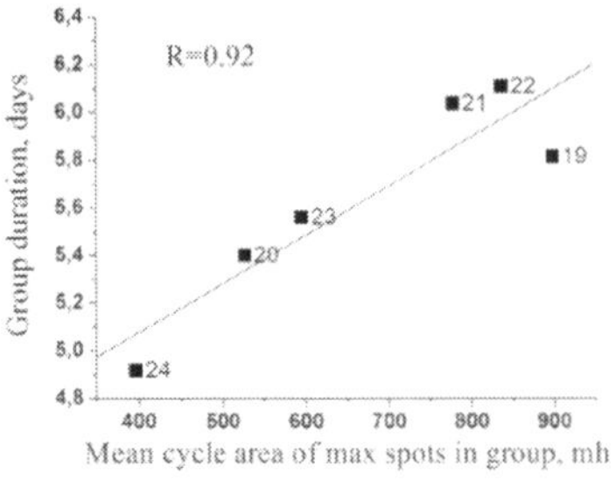

Figure 6. Mean relative duration of group life subject to mean areas of maximal spots in groups.

5. The great amount of short-lived groups and single sunspots

The present cycle is the leader both by the relative amount of short-lived groups (namely, the groups which has life time less than one day) and by the relative amount of single sunspots (i.e. groups which consist of the only sunspot). Figure 7 shows the relative amount of short-lived groups, Figure 8 presents the relative amount of single sunspots. Both dependencies has the identical view with pronounced growth trend from the beginning of cycle 22. In Figure 9 is given the growth trend of single sunspots by years. Both parameters are in good correlation (Figure 10). This can be explained by the fact that the short-lived groups are, as a rule, single sunspots (the opposite is generally incorrect).

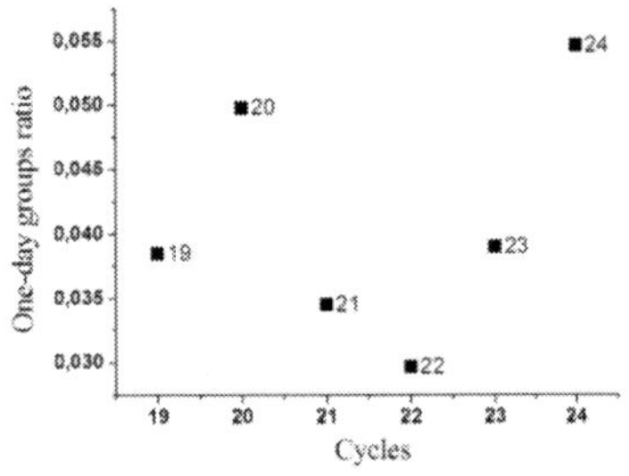

Figure 7. The relative amount of short-lived groups.

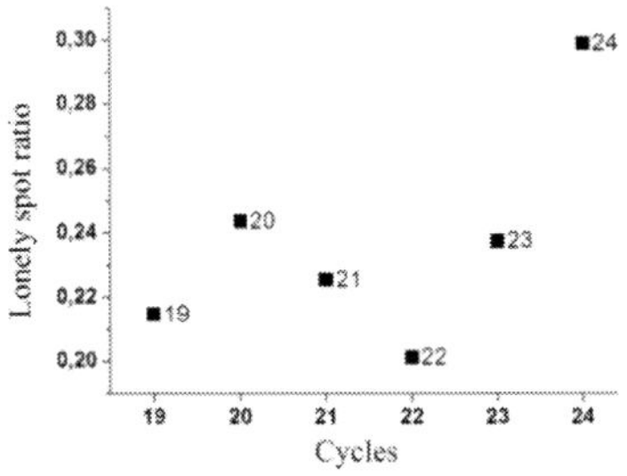

Figure 8. The relative amount of single sunspots.

6. Common relationships of sunspot groups for Kislovodsk cycles

Cycle 24 is the leader by a set of many parameters. Namely, it has: 1) the least mean amount of sunspots in a group; 2) the least mean amount of both sunspots and groups per day; 3) the least mean ratio of the Wolf number to the group amount; 4) the greatest mean ratio of the maximal spot area in group to the total group area (calculated as the sum of all maximal spots areas through the cycle divided by the total sum of sunspot areas). But in all these aspects cycle 24 continues the relationships that began from cycle 19. These relationships are given in Figures 11–18. Figure 11 shows the mean amount of sunspots in a group by years including trend line, in Figure 12 is the same dependence by cycles [8]. In Figures 13 and 14 are given the mean amounts of sunspots and groups per day respectively. Figure 15 presents the mean ratio

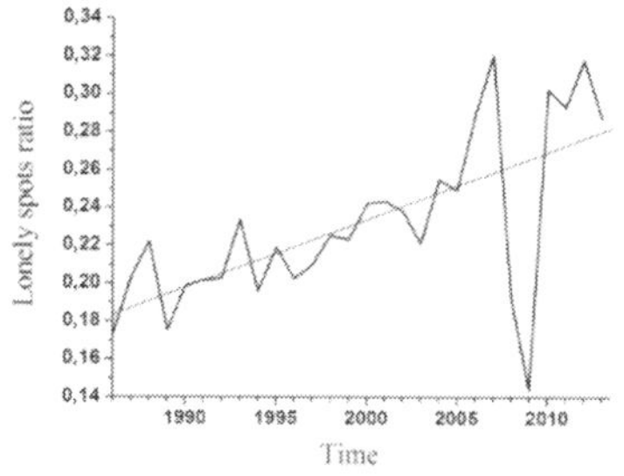

Figure 9. The growth trend of single sunspots from the beginning of cycle 22.

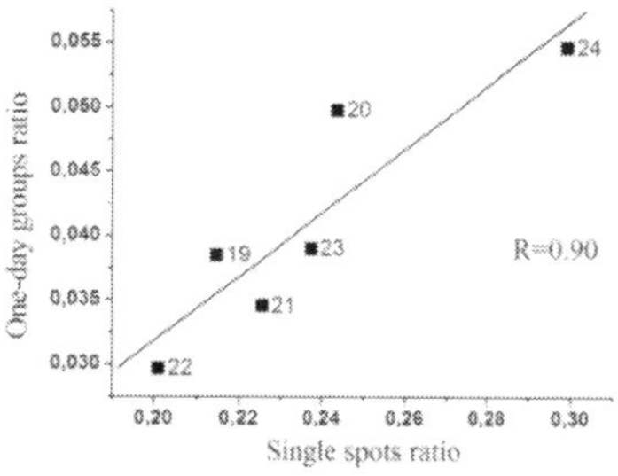

Figure 10. The correlation between single sunspots and short-lived groups.

of the Wolf number to the group amount. In Figure 16 is given the mean ratio of the maximal spot area in group to the total group area.

It is remarkable that the average size of an individual sunspot (calculated as the sum of all sunspot areas through the cycle divided by the total number of sunspots) is practically constant for each cycle except cycle 20 (Figure 17). On the other hand, the average size of an individual group (calculated by the same manner) is in direct dependence with cycle power (i.e. with mean areas of sunspot groups) (Figure 18).

The foregoing dependencies as well as the growth of short-lived groups and single sunspots demonstrate the tendency of sunspot groups to "break up". In other words, from cycle to cycle the relative number of groups with high amount of small spots decreases and the the relative number of groups with low amount of small spots (many of them are single and/or short-lived) increases.

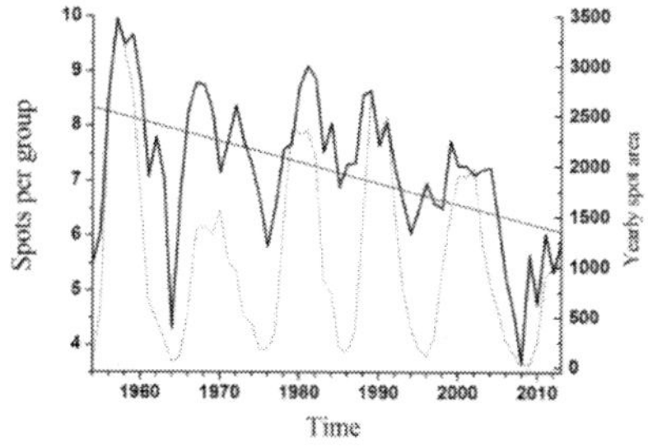

Figure 11. The decrease of mean amount of sunspots in a group through 1954–2013.

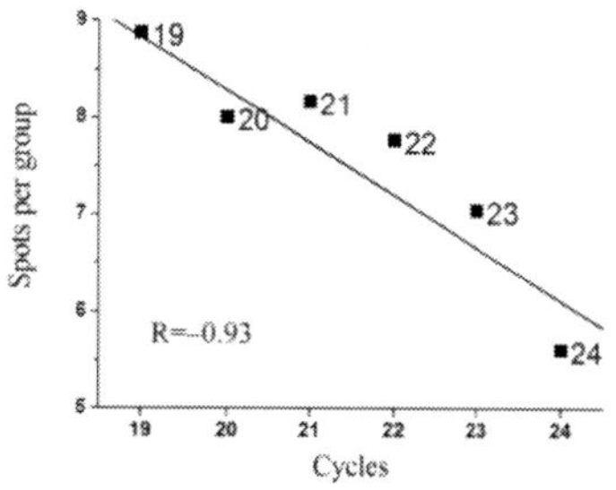

Figure 12. The decrease of mean amount of sunspots in a group through cycles 19–24.

7. Conclusions

Cycle 24 has the least relative amount of very-large-scale sunspots and the least mean group life time. On the other hand, it is the leader by the relative amount of short-lived groups and single sunspots. The mean amount of sunspots in a group and the average size of an individual group are the least among all the Kislovodsk cycles. Besides, it has the least mean amount of both

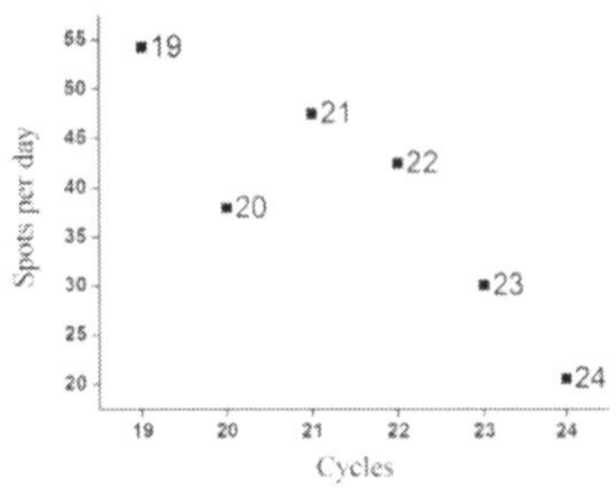

Figure 13. The mean amount of sunspots per day in cycles 19–24.

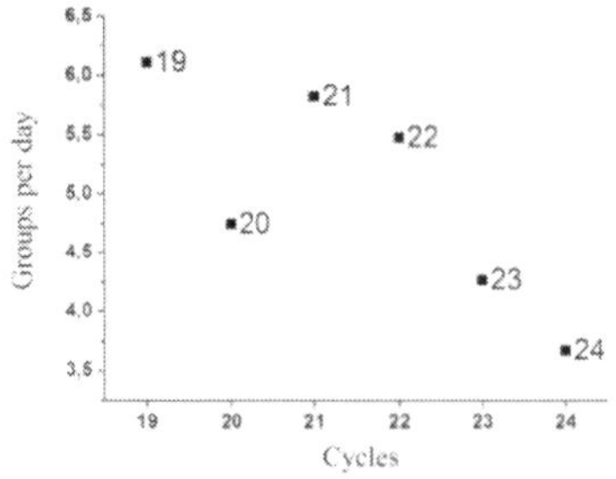

Figure 14. The mean amount of groups per day in cycles 19–24.

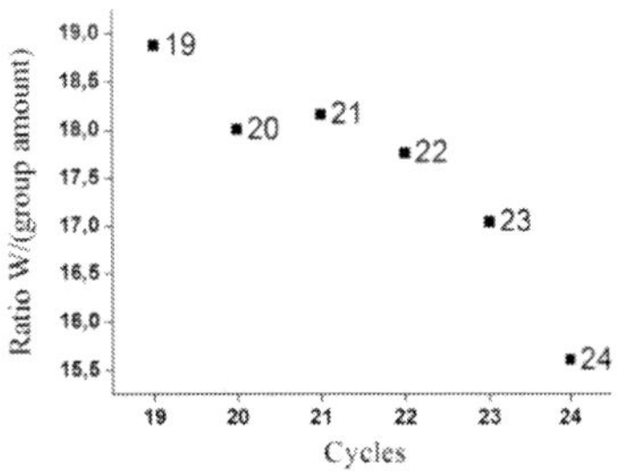

Figure 15. The mean ratio of the Wolf number to the group amount in cycles 19–24.

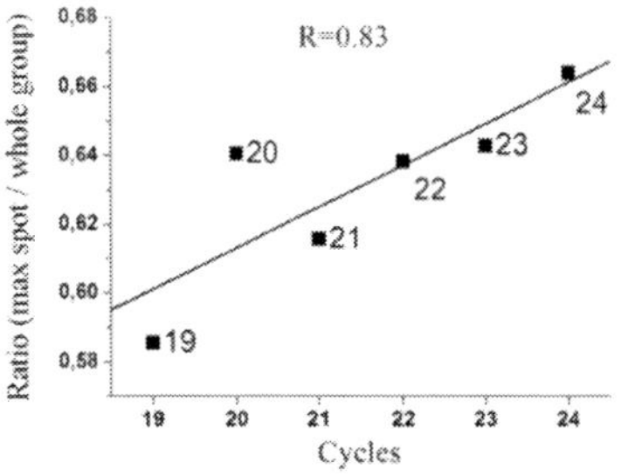

Figure 16. The mean ratio of the maximal spot area in group to the total group area in cycles 19–24.

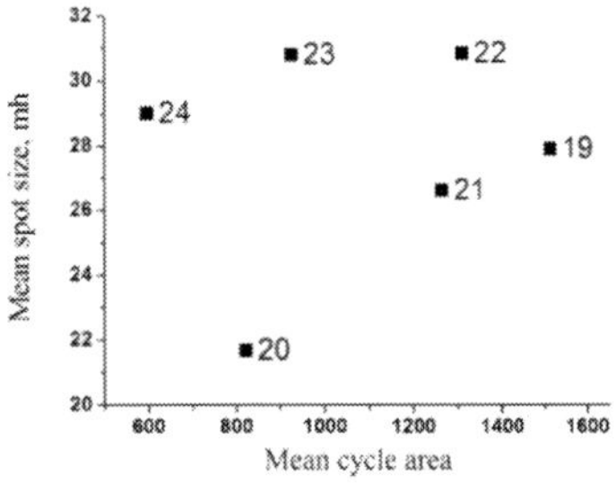

Figure 17. The average size of an individual sunspot through cycles 19–24.

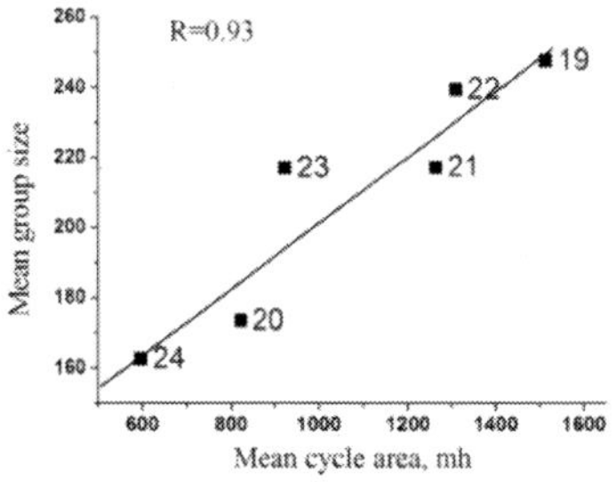

Figure 18. The average size of an individual group through cycles 19–24.

sunspots and groups per day. All these parameters are in correction with previous Kislovodsk cycles.

Cycles 19–24 are characterized by the tendency of decreasing the mean amount of sunspots in a group and increasing the relative amount of groups consisting of small number of sunspots. This phenomenon can be characterized as "crushing" of sunspot groups.

Acknowledgments
The work is supported by the grant of Russian Foundation for Basic Research 12-02-31128.

References

[1] Baranyi T, Király S and Coffey H E 2013 *MNRAS* **434** 1713
[2] Dikpati M, de Toma G and Gilman P A 2006 *GeoRL* **33** L05102
[3] Javaraiah J 2012 *SoPh* **281** 827
[4] Nagovitsyn Yu A, Nagovitsyna E Yu and Makarova V V 2009 *AstL*, **35** 564
[5] Nagovitsyn Yu A, Nagovitsyna E Yu and Makarova V V 2010 *AstL*, **36** 605
[6] Otkidychev P A and Skorbezh N N 2012 *Solar Activity Cycle: The Observal and Theoretical Aspects* (St. Petersburg: Pulkovo) pp 101–104
[7] Otkidychev P A and Tlatov A G 2012 *Solar Activity Cycle: The Observal and Theoretical Aspects* (St. Petersburg: Pulkovo) pp 105–108
[8] Tlatov A G 2012 *Solar Activity Cycle: The Observal and Theoretical Aspects* (St. Petersburg: Pulkovo) pp 133–136
[9] Svalgaard L, Cliver E W and Kamide Yo 2005 *GeoRL* **32** L01104

Relativistic solitons in pulsar wind nebulae

A E Petrov[1 2] **and A M Bykov**[2 1]

[1] Department of Cosmic Research, Saint-Petersburg State Polytechnical University, Polytechnicheskaya st. 29, St. Petersburg 195251, Russia
[2] Laboratory for High Energy Astrophysics, Ioffe Physical Technical Institute, Polytekhnicheskaya st. 26, St. Petersburg 194021, Russia

E-mail: alexey.e.petrov@gmail.com

Abstract. The conversion mechanisms responsible for the transformation of the relativistic stellar wind to the observed pulsar nebula electromagnetic radiation is one of the most interesting problems of the high energy astrophysics. In particular, the existence of a connection between the giant flares of the gamma-radiation from the Crab nebula (which were discovered in 2011 by Fermi and AGILE telescopes) and observed dynamical structures remains being a question. To address the origin of the observed dynamical structures in the shocked wind we consider here a model of a weak nonlinear perturbation propagating transverse to the mean quasi-stationary magnetic field in the pulsar nebula plasma. The plasma is supposed to be electron-positron and relativistic. It is shown that in the regime of the strong scattering of pairs by the stochastic magnetic field fluctuation the propagation of magnetic field perturbation can be described by well-known Korteweg — de Vries (KdV) equation. One of the KdV solutions is a soliton — a long lived solitary wave, propagating without changing of its shape. The width of plasma magnetic soliton in the presence of ultra-relativistic pairs is discussed.

1. Introduction

The relativistic pulsar wind pulsar nebulae are unique sources of electromagnetic radiation detected in all parts of the electromagnetic spectrum. The pulsar wind nebulae (PWN) are able to transform the energy of a rotating neutron star into electromagnetic energy of their emission with a high efficiency. The processes, responsible for this are now the hot issues of the high energy astrophysics. High angular resolution radio, optical and X-ray observations revealed the presence of dynamical structures in the Crab nebulae[1].

The study of the origin of the dynamical structures like jets and wisps may help to understand the physics of the energy conversion mechanism in the relativistic pair plasma of the pulsar nebula and the nature of the observed gamma-ray variability[1–3]. To interpret the observations one need a theoretical model for dynamical magnetic structures, propagating in the relativistic pair plasma of the PWNe.

Non-linear equations for magnetic perturbations in a collisionless pair plasma with infinite particle mean free path were derived in [4]. The authors come up with the Korteveg- de Vries type of equation. The model allows the soliton solutions - a long lived solitary wave, propagating without changing of its shape. The condition of collisionless might not be satisfied in the case of the PWN plasma where pair scattering by stochastic magnetic fluctuations may be important. Also, the energetic distribution of particles in the PWN plasma might be strongly non-equilibrium. The nonthermal case with specified above scatterings is considered.

2. The model

For our problem we use the kinetic approach. The kinetic equation:

$$\partial_t f + \mathbf{v}\partial_{\mathbf{r}} f + \dot{\mathbf{p}}\partial_{\mathbf{p}} f = I[f] \tag{1}$$

Here t, $\mathbf{r}$ and $\mathbf{p}$ are the time, the radius vector of the particle and the momentum of a particle (in the laboratory frame), respectively, $I[f]$ is a collisional operator, and $\dot{\mathbf{p}} = e\left(\mathbf{E} + \frac{1}{c}[\mathbf{vb}]\right)$ - the Lorentz force. The distribution function f is as usually sought in the form of $f = f_0 + \tilde{f}$, where f_0 is its isotropic part.

We suppose that the perturbation of the magnetic field propagates along x-axis, transverse to mean quasi-stationary magnetic field $\mathbf{B}_0||z$. Our perturbation could be considered as a linearly-polarized wave, whose electric field $\mathbf{E}||y$, and magnetic field $\mathbf{b}||z$.

As we say in the introduction, we consider a 'collisional' case, where we mean the regime of the strong scattering of pairs by the stochastic magnetic field fluctuation. As usual, a typical frequency of collisions ν can be introduced. We make an assumption that the typical frequencies of the considered processes related with propagation of the perturbation are small in comparison with the typical frequency of collisions ν. Also, we should introduce the gyrofrequency

$$\Omega = \frac{eB_0}{m_e c\gamma}$$

Here e - the elementary charge, m_e - the mass of a particle, B_0 - the induction of the mean quasi-stationary magnetic field, c - the light velocity, γ - the Lorentz-factor. It is known that the frequency of collisions might not be larger than the gyrofrequency. So, we can write:

$$\omega << \nu \leq \Omega \tag{2}$$

We use the relaxation time approximation for a collisional operator:

$$-\nu\left(f - f_0\right) \equiv -\nu\tilde{f} \tag{3}$$

Here f_0 is the isotropic part of the distribution function. The system of the kinetic equations takes the form:

$$\partial_t f_p + v_x\partial_x f_p - \Omega\cdot\hat{\mathbf{O}} f_p + e\left(E_y - \frac{1}{c}v_x b\right)\partial_{p_y} f_p + \frac{e}{c}v_y b\partial_{p_x} f_p = -\nu_1\tilde{f}_p + \nu_2\tilde{f}_e \tag{4}$$

$$\partial_t f_e + v_x\partial_x f_e + \Omega\cdot\hat{\mathbf{O}} f_e - e\left(E_y - \frac{1}{c}v_x b\right)\partial_{p_y} f_p - \frac{e}{c}v_y b\partial_{p_x} f_p = -\nu_2\tilde{f}_e + \nu_1\tilde{f}_p \tag{5}$$

Here $\hat{\mathbf{O}} = [\mathbf{p}\times\partial_{\mathbf{p}}]$ is the angular momentum operator. Subscripts 'p' and 'e' refer to positrons and electrons, respectively, and $\tilde{f}_p$, $\tilde{f}_e$ are the anisotropic parts of the distribution functions. Here we write the collisional operators in a symmetric to the sorts of the particles form.

3. The derivation of an evolution equation

To get an evolution equation we need to derive a current response at the field perturbation. The only non-zero component j_y is calculated as a moment of the distribution function:

$$j_y = \sum e_i \int v_y f_i d\mathbf{p} \tag{6}$$

Here the summation is done by the sorts of the particles. Thus, we need to derive the anisotropic part of the distribution function $\tilde{f}$. We seek $\tilde{f}$ as an expansion in the spherical functions.

3.1. Linear current response

Firstly, we derive a linear part of the current response. We leave in the kinetic equations only linear in the fields terms.

$$\partial_t f_p + v_x \partial_x f_p - \Omega_z \partial_\phi f_p + eE_y \partial_p f_0 \frac{p_y}{p} = -\nu_1 \tilde{f}_p + \nu_2 \tilde{f}_e \tag{7}$$

$$\partial_t f_e + v_x \partial_x f_e + \Omega_z \partial_\phi f_e - eE_y \partial_p f_0 \frac{p_y}{p} = \nu_1 \tilde{f}_p - \nu_2 \tilde{f}_e \tag{8}$$

Here the relation $\hat{O}_z = \partial_\phi$ is used. The distribution function is sought in the form of

$$\tilde{f} = f_x sin\theta cos\phi + f_y sin\theta sin\phi + f_z cos\theta$$

We take the Fourier transform, rewrite the kinetic equations (7),(8) in the spherical coordinates, multiply them at $sin\theta cos\phi$ or $sin\theta sin\phi$ and integrate over the angles. Thus, we get an algebraic system

$$\begin{gathered}(-i\omega + \nu_1) f_{px,k\omega} - \Omega f_{py,k\omega} - \nu_2 f_{ex,k\omega} = 0 \\ \Omega f_{px,k\omega} + (-i\omega + \nu_1) f_{py,k\omega} - \nu_2 f_{ey,k\omega} = -eE_{y,k\omega}\partial_p f_0 \\ -\nu_1 f_{px,k\omega} + (-i\omega + \nu_2) f_{ex,k\omega} + \Omega f_{ey,k\omega} = 0 \\ -\nu_1 f_{py,k\omega} - \Omega f_{ex,k\omega} + (-i\omega + \nu_2) f_{ey,k\omega} = eE_{y,k\omega}\partial_p f_0\end{gathered} \tag{9}$$

Only f_{py} gives a non-trivial contribution to the j_y. Substituting it to (6), we get

$$j_{y,k\omega} = \int \frac{2\left[\Omega^2 i\omega + (\nu_1+\nu_2)\omega^2 - i\omega^3\right] \cdot sin\theta sin\phi \cdot e^2 E_{y,k\omega}\partial_p f_0 v_y d^3p}{\omega^4 + 2(\nu_1+\nu_2)i\omega^3 - \left((\nu_1+\nu_2)^2 + 2\Omega^2\right)\omega^2 - 2(\nu_1+\nu_2)\Omega^2 i\omega + (\nu_1-\nu_2)^2\Omega^2 + \Omega^4} \tag{10}$$

Due to relations (2), the denominator can be expanded in the series in the small ratios of ω to the gyrofrequency and the frequency of collisions. Thus,

$$j_{y,k\omega} = \int \frac{2\left[\Omega^2 i\omega + (\nu_1+\nu_2)\omega^2 - i\omega^3\right]}{(\nu_1-\nu_2)^2\Omega^2 + \Omega^4}\left(1 + \frac{\left((\nu_1+\nu_2)^2 + 2\Omega^2\right)\omega^2 + 2(\nu_1+\nu_2)\Omega^2 i\omega}{(\nu_1-\nu_2)^2\Omega^2+\Omega^4}\right) sin\theta sin\phi \cdot e^2 E_{y,k\omega}\partial_p f_0 v_y d^3p \tag{11}$$

Then we write a Fourier transform for the Maxwell equation,

$$-ikb_{k\omega} = \frac{4\pi}{c} j_{y,k\omega} - \frac{1}{c} i\omega E_{y,k\omega} \tag{12}$$

and substitute here the result for $j_{y,k\omega}$. Thus, we can write a dispersion equation

$$\omega = Vk\left(1 - a_D^2 k^2\right) - i\chi k^2 \tag{13}$$

Here we introduce new designations: the phase velocity V, the dispersion length a_D and the damping χ, which are defined as

$$V = \frac{c}{\sqrt{1 + \omega_p^2 A_1}} \tag{14}$$

$$a_D^2 = \frac{V^4}{c^2}\omega_p^2 B_1 \tag{15}$$

$$\chi = \frac{V^4}{c^2}\omega_p^2 B_2 \tag{16}$$

And A_1, B_1, B_2 are the integrals over momentum

$$\omega_p^2 A_1 = \int \frac{2}{(\nu_1-\nu_2)^2 + \Omega^2} \cdot sin\theta sin\phi \cdot e^2 \partial_p f_0 v_y d^3p \tag{17}$$

$$\omega_p^2 B_1 = \int \left(-\frac{2}{((\nu_1-\nu_2)^2+\Omega^2)\Omega^2}\right) \cdot sin\theta sin\phi \cdot e^2 \partial_p f_0 v_y d^3p + \\ + \int \left(\frac{4(\nu_1+\nu_2)^2}{((\nu_1-\nu_2)^2+\Omega^2)^2\Omega^2} + \frac{2((\nu_1+\nu_2)^2+2\Omega^2)}{((\nu_1-\nu_2)^2+\Omega^2)^2\Omega^2}\right) \cdot sin\theta sin\phi \cdot e^2 \partial_p f_0 v_y d^3p \tag{18}$$

$$\omega_p^2 B_2 = -\int \left(\frac{2(\nu_1+\nu_2)}{((\nu_1-\nu_2)^2+\Omega^2)\Omega^2}\right) \cdot sin\theta sin\phi \cdot e^2 \partial_p f_0 v_y d^3p + \\ + \int \frac{4(\nu_1+\nu_2)}{((\nu_1-\nu_2)^2+\Omega^2)^2} \cdot sin\theta sin\phi \cdot e^2 \partial_p f_0 v_y d^3p \tag{19}$$

Here $\omega_p = \left(\frac{4\pi e^2 n_0}{m_e}\right)^{1/2}$ is the plasma frequency.

Making the inverse Fourier transform for $j_{y,k\omega}$, we can write for the linear current response in new designations:

$$j_y = \frac{1}{4\pi}\frac{c}{V}\left[\left(1-\frac{V^2}{c^2}\right)\partial_t b + V a_D^2 \partial_x^3 b - \chi \partial_x^2 b\right] \tag{20}$$

Here we used a useful relation $E_y = \frac{V}{c}b$, which could be easily got by taking the Fourier transform of the Maxwell equation $\partial_x E_y = -\frac{1}{c}\partial_t b$.

3.2. Nonlinear current response

We also have to derive the nonlinear current response. We are going to solve the kinetic equations (4), (5) in the coordinate space $(ct, \mathbf{r})$ taking into account the non-linear in the fields terms. We can use the approximation method for this part of the problem.

In the first approximation only linear in the fields terms should be left. We seek the first approximation of the anisotropic distribution function in the form of

$$f_{(p,e)1} = \sum C_{1m}^{(p,e)} Y_{1m}(\theta, \phi) \tag{21}$$

The kinetic equations (4),(5) are rewritten in the spherical coordinates, multiplied at $sin\theta cos\phi$ or $sin\theta sin\phi$ and integrated over the angles. Thus, we get a system of equations

$$\partial_t C_{1m}^p - i\Omega m C_{1m}^p + \nu_1 C_{1m}^p = im\sqrt{\frac{2\pi}{3}} e E_y \partial_p f_0 + \nu_2 C_{1m}^e \tag{22}$$

$$\partial_t C_{1m}^e + i\Omega m C_{1m}^e + \nu_2 C_{1m}^e = -im\sqrt{\frac{2\pi}{3}} e E_y \partial_p f_0 + \nu_1 C_{1m}^p \tag{23}$$

We can easily build a Green function for each of the equations (22),(23): it will take the form $G_{1m}^{(p,e)}(t-\tilde{t}) = e^{-\nu_{(1,2)}(t-\tilde{t}) \pm i\Omega m(t-\tilde{t})}\Theta(t-\tilde{t})$, where sign + and frequency ν_1 refer to positrons, and Θ - the Heaviside function. Then, according to the Green function definition, we can build a formal solution. For positrons:

$$C_{1m}^p = im\sqrt{\frac{2\pi}{3}}\partial_p f_0 \int_0^{+\infty} e E_y(t-\tilde{t}, x) e^{-\nu_1\tilde{t}+i\Omega m\tilde{t}} d\tilde{t} + \nu_2 \int_0^{+\infty} C_{1m}^e e^{-\nu_1\tilde{t}+i\Omega m\tilde{t}} d\tilde{t} \tag{24}$$

To get the solutions for electrons one should do the substitutions: $e \to -e$, $\Omega \to -\Omega$, $\nu_1 \to \nu_2$, $\nu_2 \to \nu_1$.

In the second approximation of the distribution function we have to consider nonlinear in the fields terms in the equations (4),(5). The second order of the distribution function is sought as

$$f_{(p,e)2} = \sum C_{2m}^{(p,e)} Y_{1m}(\theta, \phi) \tag{25}$$

After the same to the first order derivation transformations, we get for positrons:

$$\begin{aligned}C^{p}_{2m} = -\sqrt{\tfrac{2\pi}{3}}m^2\partial_p f_0 \tfrac{e\Omega}{B_0} \int_0^{+\infty} dt_1 b\left(t-t_1,x\right)e^{-\nu_1 t_1+i\Omega m t_1} \int_0^{+\infty} dt_2 E_y\left(t-t_1-t_2,x\right)e^{-\nu_1 t_2+i\Omega m t_2}+\\ +im\nu_2 \tfrac{\Omega}{B_0} \int_0^{+\infty} dt_1 b\left(t-t_1,x\right)e^{-\nu_1 t_1+i\Omega m t_1} \int_0^{+\infty} dt_2 C^{e}_{1m} e^{-\nu_1 t_2+i\Omega m t_2}+\\ +\nu_2 \int_0^{+\infty} dt_1 C^{e}_{2m} e^{-\nu_1 t_1+i\Omega m t_1}\end{aligned} \tag{26}$$

To get the solutions for electrons one again should do the substitutions: $e \to -e$, $\Omega \to -\Omega$, $\nu_1 \to \nu_2$, $\nu_2 \to \nu_1$.

The equations (24),(26) and the corresponding pair for electrons could be considered as a system of coupled integral equations. This system could be reduced to an algebraic system: one should rewrite the fields and the coefficients C as the inverse Fourier transformation integrals and rearrange the integrals (appearing fractions of the form $1/\left(\nu_a - i\left(\omega \pm m\Omega\right)\right)$ should be expanded in the series in the small ratio of ω to other frequencies).

Finally, the solution of the system gives us a nonlinear contribution in the current response:

$$j_y^{NL} = \frac{V^2}{c}\lambda_1 \frac{b}{B_0}\partial_x b \tag{27}$$

Here V - introduced earlier phase velocity, λ_1 — the coefficient, defined by integral over momentum.

Thus, we substitute the sum of (27) and (20) to the Maxwell equation and after simple transformations get the evolution equation:

$$\partial_t b + V\partial_x b + Va_D^2\partial_x^3 b + \frac{V^3}{c^2}\lambda_1 \frac{b}{B_0}\partial_x b = \chi\partial_x^2 b \tag{28}$$

Since this moment, we assume $\nu_1 = \nu_2 \equiv \nu$. We also take ν in the frequently used form of $\nu = a\Omega$. Further evaluations can show that if $a \ll 1$, the small damping $\chi\partial_x^2 b$ can be neglected in quite a wide range of the parameters. We introduce designations $\mu = Va_D^2$, $\lambda = \frac{V^3}{c^2}\lambda_1$. Thus, for the dimensionless function $h = b/B_0$ we can write

$$\partial_t h + V\partial_x h + \mu\partial_x^3 h + \lambda h\partial_x h = 0 \tag{29}$$

This is a well-known Korteweg—de Vries equation (KdV). One of the KdV solutions is a soliton — a solitary wave, propagating without changing of its shape.

4. Evaluations

A practically important problem is the evaluation of the soliton width w, which is defined as $w = \sqrt{\frac{12\mu}{\lambda h_0}}$ (here h_0 — the dimensionless soliton amplitude). The angular resolution of the Hubble Space Telescope (HST) lets us to resolve objects with the size larger than $\sim 10^{15}$ cm in the Crab nebula. The evolution equation, which was obtained under the conditions, relevant to the PWNe, has the soliton solutions in a wide range of the parameters. The propagating perturbation of the magnetic field in the Crab nebula can be observed as moving structures like wisps in the high resolution images of synchrotron emission. Thus, if the solitons can propagate in the conditions of the Crab nebula, and their width is $\sim 10^{15}$ cm or larger, they could be resolved by the HST.

In the evaluations the induction B_0 is taken, according to the observations, at $200\mu G$. The concentration is estimated from the pressure, which could be, from the one hand, evaluated as a distribution function moment $P = \frac{1}{3}\int pvfd^3p$, and, from another hand, estimated as a magnetic pressure $\frac{B_0^2}{8\pi}$. The evaluations are done for the temperatures from the lower part of the relativistic range. The frequency ν is taken in the form of $\nu = a\Omega$, where $a \ll 1$. The soliton amplitude is taken equal to 0.1, because we consider a weakly-nonlinear perturbation.

The soliton width is evaluated in both cases of the equilibrium Maxwell-Juttner (relativistic Maxwell) energy distribution, and of the non-equilibrium one. In the second case we use the Hybrid distribution, which is the matching of the Maxwell-Juttner distribution at relatively low energies and the power-law energy distribution with the index 2 and the cutoff Lorentz-factor 10^9 at high energies (these values are well corresponding to the observations). The results for

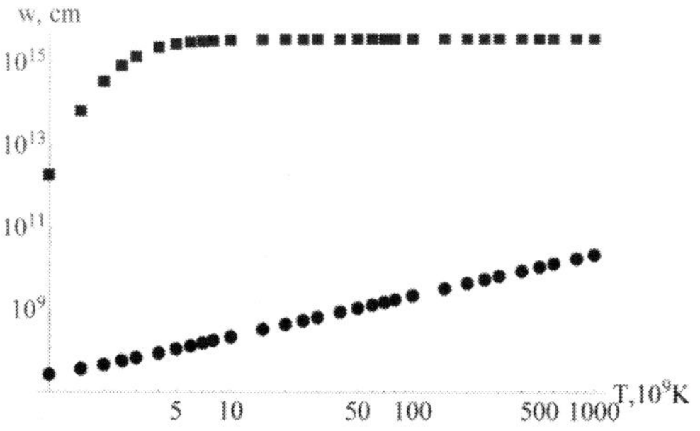

Figure 1. Soliton widths w for: ● - Maxwell-Juttner, ■ Hybrid distributions; $a = 0.1$

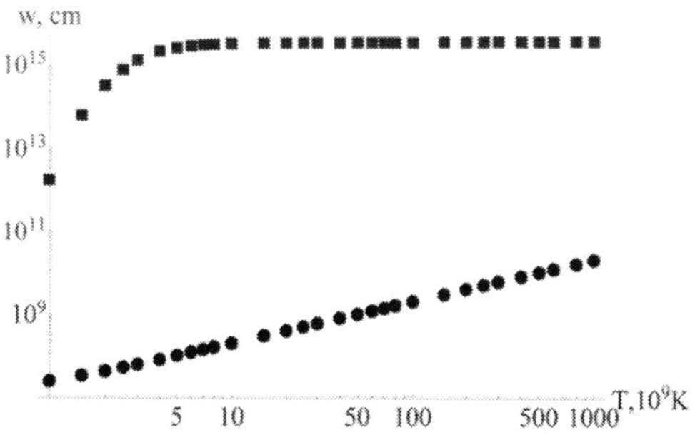

Figure 2. Soliton widths w for: ● - Maxwell-Juttner, ■ Hybrid distributions; $a = 10^{-3}$

different parameter values are presented in the Figures (1),(2) (they are not identical). We derived the widths $w(T)$ for various pair plasma temperatures T just to illustrate the range of the parameters space, the detailed analysis of this will be done elsewhere. The results show, that the soliton width w substantially depend on the energy distribution. In the case of quasi-equilibrium pair distribution w is in the range of 10^7—10^{10} cm, while in the presence of the power-law tails of pair distribution its values almost always exceed 10^{15} cm — which is about the resolution limit for the HST in the Crab nebula. Thus, the observed widths of the dynamical wisps in the Crab nebula may constrain the pair distribution function. This fact opens up a prospective of an observational study of the relativistic pairs gas properties (e.g., the pressure).

5. Conclusions

In this paper we discuss a theoretical model of a propagation of a weakly-nonlinear perturbation of the magnetic field transverse to the mean quasi-stationary magnetic field in the relativistic non-thermal electron-positron plasma of the pulsar wind nebula. By kinetic approach we derive an evolution equation, and show that it takes the form of well-known Korteweg—de Vries equation for the wide range of the parameters. This equation has a class of stable solutions — the solitons. The soliton widths are derived for quasi-equilibrium and for strongly non-thermal pair energy distributions. For the non-equlibrium case, when the pressure is dominated by the ultrarelativistic non-thermal synchrotron emitting pairs, the obtained soliton width exceeds the resolution limit of the Hubble Space Telescope for the Crab nebula. It means, that the solitons can be responsible for the moving structures like wisps, observed in the high resolution images of the synchrothron emission. Our results make it possible to constrain the pair distribution function (as a consequence, e.g., the relativistic pair gas pressure), by observational results.

References

[1] Hester J J 2008 *Annu. rev. astron. astrophys.* **46** 127-55
[2] Arons J 2012 *Space Sci Rev* **173** 341-67
[3] Bykov A M, Pavlov G G, Artemyev A V and Uvarov Yu A 2012 *MNRAS* **421** L67-71
[4] Mikhailovskii A B, Onishchenko O G and Tatarinov E G 1985 *Plasma Phys. Controlled Fusion* **27** 539-56

Orbital resonances in exoplanetary systems

E A Popova and I I Shevchenko

Pulkovo Observatory of the Russian Academy of Sciences,
Pulkovskoje ave. 65, Saint Petersburg 196140, Russia

E-mail: m02pea@gmail.com

Abstract. At present, more than 700 exoplanetary systems are known to have been discovered. They incorporate more than 130 multiplanet systems, i.e. those hosting two or more planets. The orbital resonance and near-resonance phenomena are ubiquitous in them. We present a statistical and dynamical analysis of the resonance structure of the multiplanet systems and planetary systems of binary stars. We have built distributions of the orbital period ratios, separately considering the cases of inner and outer location of the massive perturber. The histograms reveal apparent peaks close to the first order orbital resonances 2/1 and 3/2 in both cases; this confirms previous findings. We have performed analytical modelling of the histograms, and obtained exact positions of the peaks. Moreover, we have built the "period ratio – eccentricity" diagrams, with collision curves superimposed, so that to find anomalous systems.

1. Introduction

We analyze statistics of resonances in multiplanet systems, as well as in binary stellar systems with planets, using exoplanet database [1]. Orbital period ratios are calculated for each pair of objects in all discovered systems suitable for analysis. Differential distributions of the orbital period ratios are constructed separately for the following two cases: (a) the orbit of the dominating (in mass) body in the pair is inside the orbit of the smaller body in the pair (in the Solar system, this configuration is similar to the "Neptune – TNO (trans-Neptunian object)" configuration); (b) the opposite case (in the Solar system, this is similar to the "(main belt) asteroid – Jupiter" configuration). The dominating body in the pair is considered to be the "perturber" (analogously to the situation in the Solar system, where Neptune and Jupiter are the main perturbers for the TNOs and main-belt asteroids, respectively). In the both cases (a) and (b), the outer body is indexed with number 1 (e.g., its orbital period is T_1), and the inner body with number 2 (the orbital period T_2).

2. The period ratio distributions

We build the histograms of period ratios, using exoplanet database [1]. The obtained histograms are presented in figures 1 and 2 in the defined above cases (a) and (b), respectively. As one can see, peaks are prominent near the resonances 3/2, 2/1, 5/2, 3/1, and 4/1. However, some peaks seem to be shifted (mostly to the right) relative to their nominal positions both in (a) and (b) cases. This is a known phenomenon in exoplanetary systems, though discovered quite recently (see [2, 3] and references therein). Possible dynamical and cosmogonical mechanisms for such shifts were discussed in [3, 4, 5]. E.g., the shift can appear, if the planetary masses in the pair increase with time [3]. In [4, 5], the effect is explained by tidal dissipation.

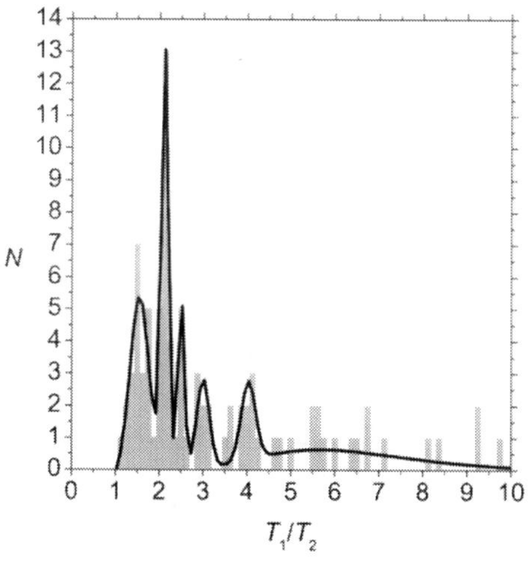

Figure 1. The period ratio histogram in case (a) (the inner perturber). The best-fit function (1) is shown by the solid curve.

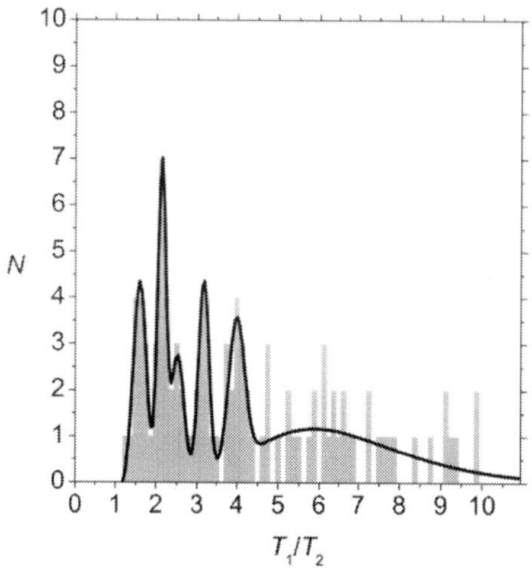

Figure 2. The same as in figure 1, but for case (b) (the outer perturber).

We fit the histograms in figures 1 and 2 with the curves described by the multi-peaked function

$$f\left(\frac{T_1}{T_2}\right) = \sum_{i=1}^{5} f_i\left(\frac{T_1}{T_2}\right) + \phi\left(\frac{T_1}{T_2}\right), \tag{1}$$

where

$$f_i\left(\frac{T_1}{T_2}\right) = \frac{A_i}{\sigma_i\sqrt{2\pi}} e^{-\frac{(T_1/T_2-\mu_i)^2}{2\sigma_i^2}}, \tag{2}$$

$$\phi\left(\frac{T_1}{T_2}\right) = B \cdot \left(\frac{T_1}{T_2} - C\right)^5 e^{-(T_1/T_2-C)/D}. \tag{3}$$

Every function f_i is nothing but the normal distribution of the period ratio values near resonance. The function ϕ describes the wide bulge and decay in the distribution tail. The values of parameters A_i, μ_i, σ_i, B, C, and D are obtained by the Levenberg–Marquardt method, using "Mathematica" software. The fits are shown in figures 1 and 2 by solid curves.

The best-fit positions of the peaks are given in Tables 1 and 2. It turns out that the peak shifts are much smaller in case (a) than in case (b). Moreover, almost no shift is present for resonances of order higher than 1 in case (a).

3. Collision curves and anomalous systems

In order to find anomalous (unstable or close to instability) systems, we have constructed the "period ratio – eccentricity" diagrams, using the same sample of objects. These diagrams are shown in figures 3 and 4 for cases (a) and (b), respectively.

The dots in both diagrams correspond to the planetary pairs in the sample. In figure 3, two squares correspond to the circumbinary systems Kepler-16 and Kepler-35, and the triangle

Table 1. Peak locations in case (a)

Resonance	Peak position
2/1	2.122 ± 0.007
3/2	1.547 ± 0.024
3/1	3.000 ± 0.039
4/1	4.033 ± 0.046
5/2	2.500 ± 0.018

Table 2. Peak locations in case (b)

Resonance	Peak position
2/1	2.131 ± 0.018
3/2	1.599 ± 0.025
3/1	3.168 ± 0.024
4/1	3.991 ± 0.039
5/2	2.517 ± 0.056

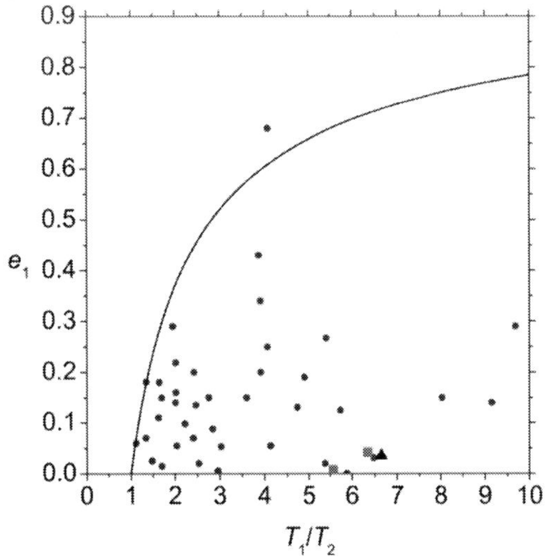

Figure 3. The "period ratio — eccentricity" diagram in case (a). The collision curve is superimposed (solid line).

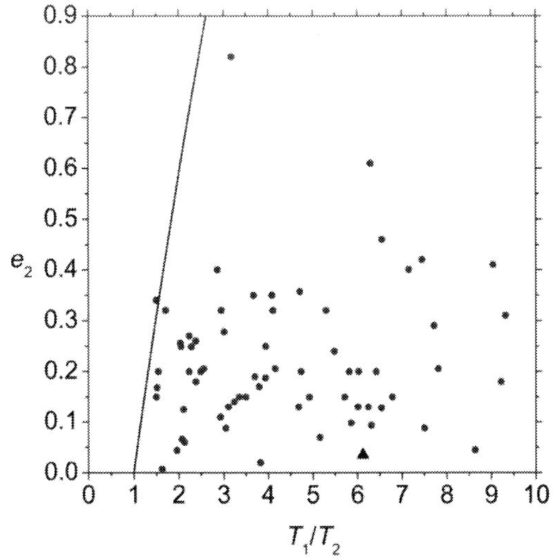

Figure 4. The same as in figure 3, but for case (b).

corresponds to the circumbinary system Kepler-47 (the b planet is taken into account solely). In figure 4, the triangle corresponds to the planetary pair b–c in the Kepler-47 system.

In order to identify anomalous systems, we have superimposed collision curves on the plots. These curves are given by the functions

$$e_1\left(\frac{T_1}{T_2}\right) = 1 - \left(\frac{T_1}{T_2}\right)^{-2/3} \tag{4}$$

and

$$e_2\left(\frac{T_1}{T_2}\right) = \left(\frac{T_1}{T_2}\right)^{2/3} - 1 \tag{5}$$

in cases (a) and (b), respectively. These formulas can be obtained easily from basic relations of the two-body problem theory [6].

Using these diagrams, we have identified anomalous systems, i.e. those with planets located outside the zones of conditional stability (zones limited by the collisional curves). The dots located higher than the collision curves betray anomalous systems that host planets with intersecting orbits. It turns out that two such systems are unambiguously identified: HD102272 (in case (a)) and HD215152 (in case (b)).

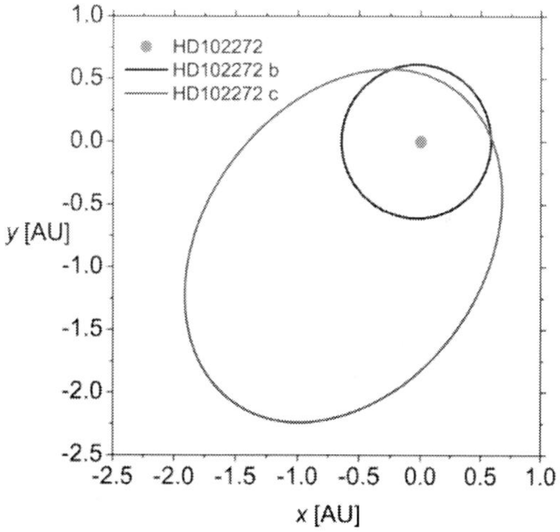

Figure 5. Configuration of the planetary orbits in the HD102272 system at zero time (when integration starts).

Table 3. Data on planets in the HD102272 system [9]

Parameters	b	c
Mass $m \sin i$ $[M_J]$	5.9 ± 0.2	2.6 ± 0.4
Semimajor axis a [AU]	0.614 ± 0.001	1.57 ± 0.05
Period P [d]	127.58 ± 0.3	520 ± 26
Eccentricity e	0.05 ± 0.04	0.68 ± 0.06
Argument of pericenter ω [°]	118 ± 58	320 ± 10

Unfortunately, the observational data available are not enough for modelling the second system, HD215152. In what concerns the HD102272 system, it is close to the mean motion resonance 4/1. We have integrated the system motion, using the software [7, 8], allowing one to compute the Lyapunov exponents of the motion. Figure 5 shows the initial orbital configuration. The planetary parameters and initial conditions, as given in [9], are collected in Table 3; the mass of the star is equal to $1.9 \pm 0.3 M_\odot$. Our numerical modelling shows that the system is stable on time intervals up to 10^6 yr, notwithstanding the fact of the intersection of orbits.

4. Conclusions

We have built distributions of the orbital period ratios, considering separately the cases of inner and outer location of the massive perturber. The histograms have prominent peaks close to the first order orbital resonances 2/1 and 3/2, as well as less significant peaks at resonances 5/2, 3/1 and 4/1. We have performed analytical modelling of the histograms, and obtained the best-fit positions of the peaks. It turns out that the peak shifts are much smaller in case (a) than in case (b). Moreover, almost no shift is present for resonances of order higher than 1 in case (a).

Moreover, we have constructed the "period ratio – eccentricity" diagrams, with collision curves superimposed. Using these diagrams, we have identified anomalous systems, i.e. those with planets located outside the zones of conditional stability (zones limited by the collisional curves).

References

[1] Schneider J 2013 http://exoplanet.eu
[2] Fabrycky D C *et al* 2012 *Preprint* arXiv:1202.6328
[3] Petrovich C, Malhotra R, and Tremaine S 2013 *Astrophys. J.* **770** 16
[4] Batygin K and Morbidelli A 2013 *Astron. J.* **145** 10
[5] Lithwick Y and Wu Y 2012 *Astrophys. J.* **756** L11
[6] Kholshevnikov K and Titov V 2002 *The Two-Body Problem* (St. Petersburg: St. Petersburg Univ.), in Russian
[7] Shevchenko I and Kouprianov V 2002 *Astron. Astrophys.* **394** 663
[8] Kouprianov V and Shevchenko I 2005 *Icarus* **176** 224
[9] Niedzielski A *et al* 2009 *Astrophys. J.* **693** 276

16th Russian Youth Conference on Physics and Astronomy (PhysicA.SPb/2013) IOP Publishing
Journal of Physics: Conference Series **572** (2014) 012007 doi:10.1088/1742-6596/572/1/012007

Numerical modeling of light-induced drift in a cloud of interstellar gas

A I Sobolev, D A Varshalovich

Department of Theoretical Astrophysics, Ioffe Physical Technical Institute, Saint Petersburg, Russia

E-mail: Asoboll@gmail.com

Abstract. We study velocity distribution of the gas in case of resonance interaction between directed radiation line and the gas in medium of non-resonating buffer gas. We have constructed a 6-parameter model numerically describing the effect of uniform quasi-stationary light-induced drift in a rarefied interstellar gas. Based on numerical calculations, we qualitatively determined the dependence of the light-induced drift velocity on the model parameters in the cases of optically thin and optically thick clouds. We checked that light-induced drift has resonance behavior and requires low collision rate of the gas. Also we showed and explained narrowing and shifting of emission line during propagation in cloud.

1. Introduction

The effect of light-induced drift (LID) is one of the most powerful effects of radiation impact on the translational motion of the particles. Theoretically predicted [1] and observed [2] in 1979. The essence of the effect is the appearance of macroscopic directed flow of particles absorbing radiation close to the resonance and being in mixture with the buffer particles. Kinetic energy and momentum of the flow of impurity gas is taken from thermal energy of the environment. In this case, reduction of the entropy of the gas is compensated by an increase in entropy of the scattered photons [3].

1.1. Problem statement

Consider interstellar gas cloud through which the radiation passes, which is close to the resonance frequency of the impurity gas components. Under the influence of radiation, a excitation of particles of impurity gas occurs. Moreover, due to the frequency shift of radiation to the exact resonance excited gas will have a specific asymmetric velocity distribution corresponding to its translational movement in one direction or another. This leads to the separation of components in the cloud.

1.2. Description of the gas

In this problem, velocity distribution of the impurity gas is most important. Since radiation has dedicated axis, we are interested in the distribution of the projection of the velocity on the axis. Define the distribution of impurity gas in the ground state as $f(v)$, and in the excited - as $g(v)$. In addition, one can assume that the resonating gas in the absence of radiation is distributed according to the Maxwell-Boltzmann:

$$f_0(v) = \frac{n_0}{\sqrt{2\pi v_T^2}} exp(-\frac{v^2}{2v_T^2}) \quad (1)$$

Where v_T - thermal (and turbulent) speed.

1.3. Description of the radiation

Consider radiation having spectral radiance $I_\nu(\nu)$ and propagating in a small solid angle Θ. For easy work with a spectral radiance we must take into account the correlation between the frequency of light and the speed of the molecules. If the frequency ν of the incident light differs from the exact resonant frequency of the impurity particles ν_0, it will resonate only with particles moving at the speed (2). Therefore, as the argument of the spectral intensity it is possible to use v instead of ν.

$$v = c\frac{\nu - \nu_0}{\nu_0} \quad (2)$$

1.4. Approach

Due to the fact that the task has many parameters, its analytical description is very rough. For a detailed study of this process i have built calculation scheme. The task is assumed stationary and quasi-homogeneous:

$$f/\frac{df}{dt} \propto f/\frac{df}{dx}\frac{1}{v_T} \propto T > \tau \quad (3)$$

Where T - typical time of the drift, and τ - relaxation time of velocity distribution of the gas that is determined by the slowest process. Stationary examination is justified by the big drift time in the cloud that much more than Maxwellization time and time of radiative transitions. And quasi-homogeneity condition actually identical to the condition of stationarity.

2. Model

2.1. Processes taken into account in model

Boltzmann equation, which describes the distribution of the gas in the ground and excited states:

$$\frac{\partial f(v)}{\partial t} + \frac{\partial f(v)}{\partial r}v + \frac{\partial f(v)}{\partial v}\frac{F}{m} = (\frac{\partial f(v)}{\partial t})_{int} \quad (4)$$

Problem stationary and quasi-uniform: $\frac{\partial f(v)}{\partial t} = 0$, $\frac{\partial f(v)}{\partial r} = 0$. Force affects the motion of all particles together. So, it does not participate in the effect of light-induced drift and we will assume that force is equal to zero: $F = 0$. Thus, the Boltzmann equation looks like:

$$0 = (\frac{\partial f(v)}{\partial t})_{int} = (\frac{\partial f(v)}{\partial t})_m + (\frac{\partial f(v)}{\partial t})_e + (\frac{\partial f(v)}{\partial t})_c \quad (5)$$

Where the terms are defined by the following processes:

- Maxwellization of the impurity gas in excited and ground states. I consider the relaxation time approximation and describe the resulting velocity distribution of impurity gas through the parameter Θ - the part of gas in the excited state. The equilibrium distribution of the resonating gas in the excited and ground state of the proportional concentration of the gas in this state: $n_g = \theta n_0$ for the excited state and $n_f = (1 - \theta)n_0$ for the ground state.

$$\theta = \frac{1}{n_0}\int g(v)dv \quad (6)$$

$$(\frac{\partial f(v)}{\partial t})_m = \frac{f_0(v)(1-\theta) - f(v)}{\tau_f} \tag{7}$$

$$(\frac{\partial g(v)}{\partial t})_m = \frac{\theta f_0(v) - g(v)}{\tau_g} \tag{8}$$

Where f_0 is Maxwell distribution of gas emission 1.

- Stimulated and spontaneous transitions between the ground and excited states:

$$(\frac{\partial f(v)}{\partial t})_e = -B\frac{\Omega I(v)}{c} f(v) + B\frac{\Omega I(v)}{c} g(v) + Ag(v) \tag{9}$$

$$(\frac{\partial g(v)}{\partial t})_e = B\frac{\Omega I(v)}{c} f(v) - B\frac{\Omega I(v)}{c} g(v) - Ag(v) \tag{10}$$

Where A and B - Einstein coefficients for this transition (for simplicity we assume that the statistical weights of the levels are equal: $g_u = g_d$).

- Transitions caused by collisions. Qualitatively, effect of light-induced drift exists without collisional transitions. However, the speed of drift may significantly increase.

$$(\frac{\partial f(v)}{\partial t})_c = Z\theta f_0(v) \tag{11}$$

$$(\frac{\partial g(v)}{\partial t})_c = -Zg(v) \tag{12}$$

$$Z = W^{coll}_{u \to d} \tag{13}$$

Substituting (7-12) into (5), we obtain the equations describing the balance of the gas in the excited and ground state:

$$\frac{f_0(v)(1-\theta) - f(v)}{\tau_f} - B\frac{\Omega I(v)}{c} f(v) + B\frac{\Omega I(v)}{c} g(v) + Ag(v) + Z\theta f_0(v) = 0 \tag{14}$$

$$\frac{\theta f_0(v) - g(v)}{\tau_g} + B\frac{\Omega I(v)}{c} f(v) - B\frac{\Omega I(v)}{c} g(v) - Ag(v) - Zg(v) = 0 \tag{15}$$

2.2. Parameters of the problem

For the numerical solution of this problem it is required to introduce dimensionless parameters. We will consider the emission with Gaussian spectral radiance, where v is related to the detuning of frequency according to (2).

$$I(v) = I_{max} exp(-\frac{(v - v_r)^2}{2\Delta v^2}) \tag{16}$$

It is characterized by three parameters: I_{max} - intensity in the maximum, v_r - position of the maximum, Δv - the width of the emission line. Thus, the task is determined by 10 parameters: n_0, v_T, v_r, Δv, τ_f, τ_g, A, B, Z, I_{max}, and B can be expressed in the following way:

$$B = A\frac{\pi^2 c^3}{h\omega_0^3} \tag{17}$$

It is possible to leave 6 dimensionless parameters affecting the solution:

$\beta = \dfrac{v_r}{v_T}$ – shift of the emission line relative to the absorption line

$\delta = \dfrac{\Delta v}{v_T}$ – relative width of the emission line

$\gamma = \tau_g A$ – relative time of Maxwellization of excited gas

$\eta = \dfrac{\tau_f}{\tau_g} = \dfrac{\sigma_g}{\sigma_f}$ – the ratio of the cross sections of elastic interactions in the gas in the excited and ground states

$\epsilon = \dfrac{B\Omega I_{max}}{Ac}$ – the ratio of probabilities of spontaneous and induced transitions

$\zeta = \dfrac{Z}{A}$ – the ratio of probabilities of collisional transitions and spontaneous radiative transitions

2.3. Solving method

Program was written in the package Wolfram Mathematica. The essence of the algorithm is as follows. The system of equations (14,15) is resolved with respect to $f(v)$ and $g(v)$. The solution has a linear dependence of the parameter Θ. Then, after the numerical integration of the coefficients under zero and first powers of Θ in $g(v)$, one can solve the equation (6) with respect to Θ, which finally determines the distribution of $f(v)$ and $g(v)$.

3. Results

3.1. Examples of the gas distribution

Qualitatively obtained graphs with distribution of resonating gas (1,2) are described as follows. Since the rate of Maxwellization of impurity gas in the ground state is small (compared with other processes), then gas accumulates in a state with low transitions activity. On the other hand, at velocities where transition activity is high, there is an alignment of the gas concentration in the excited and ground states.

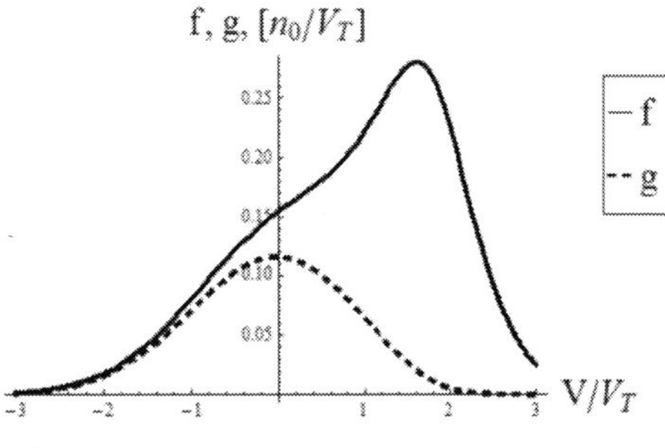

Figure 1. Gas distribution in the ground and excited states. Parameters $\beta = -1.5$, $\delta = 1$, $\gamma = 10$, $\eta = 10$, $\epsilon = 10$, $\zeta = 0.1$.

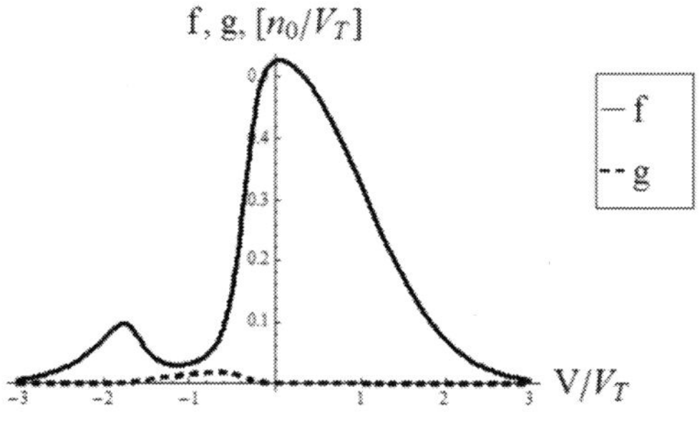

Figure 2. Gas distribution in the ground and excited states in case of narrow emission line. Parameters $\beta = -1$, $\delta = 1/4$, $\gamma = 10$, $\eta = 10$, $\epsilon = 1$, $\zeta = 0.1$.

3.2. The speed of light-induced drift

Here is some typical dependences of the average velocity of the gas on the parameters. The average speed of the impurity gas defined as follows:

$$\bar{v} = \frac{1}{n_0} \int (f(v) + g(v)) v dv \tag{18}$$

As we can see, drift velocity is comparable to thermal velocity. There is a maximum in dependence of drift velocity on the shift of the emission line relative to the absorption line β (figure 3), which is depends on the other parameters. Indeed, the dependence of the average velocity of the impurity gas from the shift of the emission line is antisymmetric. Therefore, when $\beta = 0$ there is no drift. Also, for large shift parameter, the activity of stimulated transitions is reduced by a Gaussian function, which is why the average speed of the impurity gas also tends to 0. Figure 4 also shows a maximum in dependence of drift velocity on the width of the emission line δ. Figure 5 shows that the drift occurs only when Maxwellization in ground state is slower than the radiative transitions. Figure 6 shows that it is also possible to change the direction of flow by varying the ratio of the cross section.

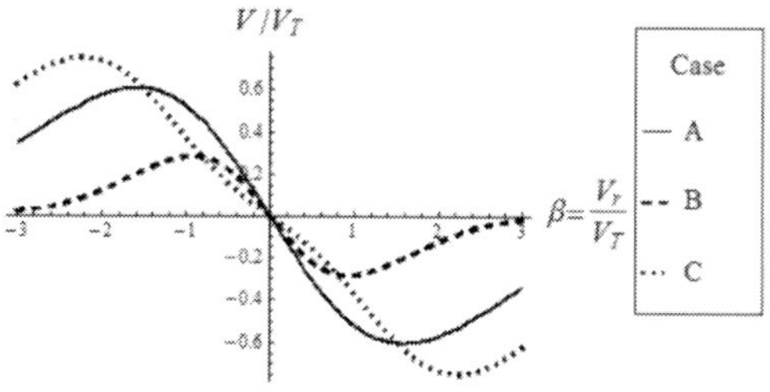

Figure 3. Graph of drift velocity of the impurity gas on the shift of the emission line relative to the absorption line.

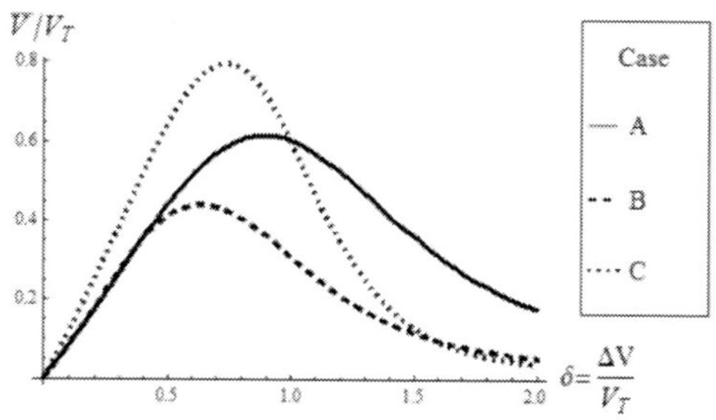

Figure 4. Graph of drift velocity of the gas on the emission line width.

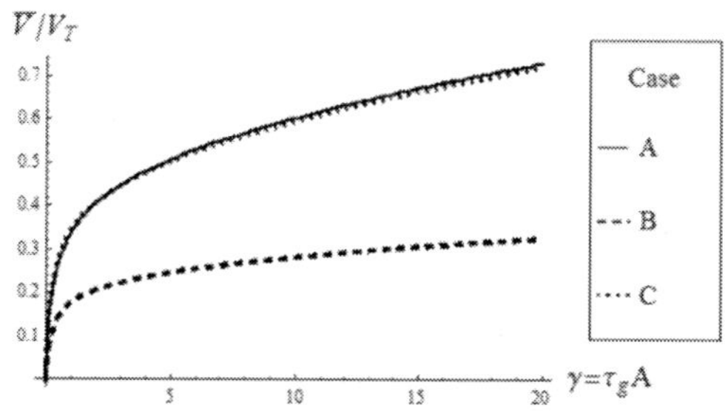

Figure 5. Graph of drift velocity of the gas on the relative Maxwellization time.

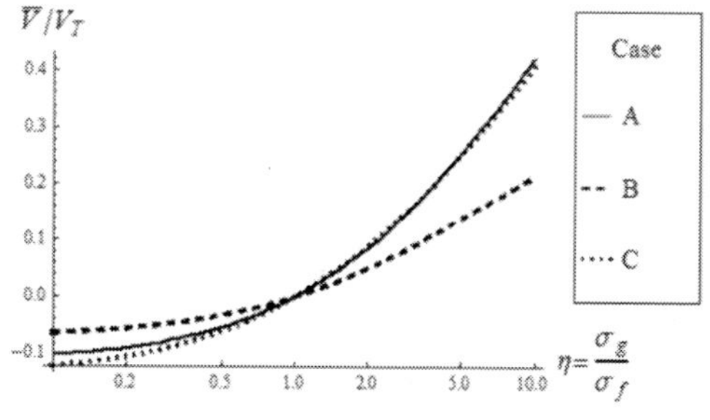

Figure 6. Graph of drift velocity of the gas on the ratio of the cross sections of elastic interactions.

All graphs here are for three sets of parameters:
A) the typical set of parameters: $\beta = -1.5$, $\delta = 1$, $\gamma = 10$, $\eta = 10$, $\epsilon = 1$, $\zeta = 0.1$
B) case of narrow emission line: $\beta = -0.5$, $\delta = 0.25$, $\gamma = 10$, $\eta = 10$, $\epsilon = 1$, $\zeta = 0.1$
C) case of strong radiation: $\beta = -1.5$, $\delta = 1$, $\gamma = 10$, $\eta = 10$, $\epsilon = 10$, $\zeta = 0.1$

3.3. The absorption of radiation in the optically thick cloud

The resulting solution for the distribution of the impurity gas can be generalized to the case of optically thick cloud, where there is a significant absorption of radiation. The derivative of the spectral intensity is determined according to:

$$\frac{dI(v)}{dx} = -hBI(v)(f(v) - g(v)) \tag{19}$$

Formula $x = z(\frac{v_T}{n_0 Bh}) = z(\frac{8\pi v_T}{A\lambda^3 n_0})$ for penetration depth is written under the assumption that the density of the gas n_0 is constant and can be generalized.

$$\int_0^x \frac{n_0(x')Bh}{v_T} dx' = \int_0^x \frac{n_0(x')\lambda^3 A}{8\pi v_T} dx' = z \tag{20}$$

Due to the peculiarities of the velocity distribution of the impurity gas, absorption of radiation is considerably decreased. Strong radiation forms a transparency window in gas distribution. Therefore, the absorption is faster at frequencies where the spectral emission intensity is smaller. Thus, emission line is narrowing and moving away from resonance frequency while spreading in cloud (figure 7).

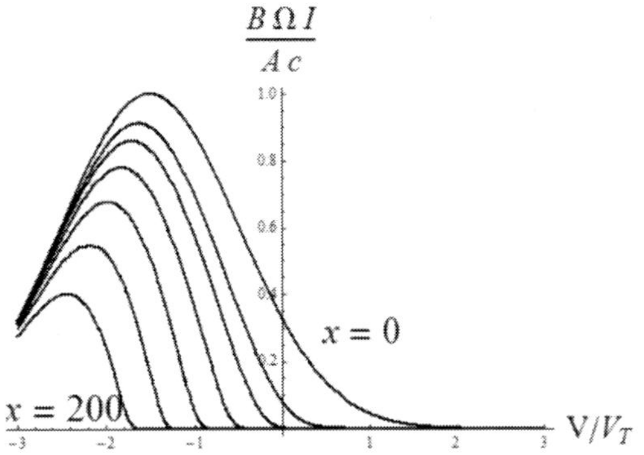

Figure 7. Dependence of the spectral radiance on the penetration depth in the cloud

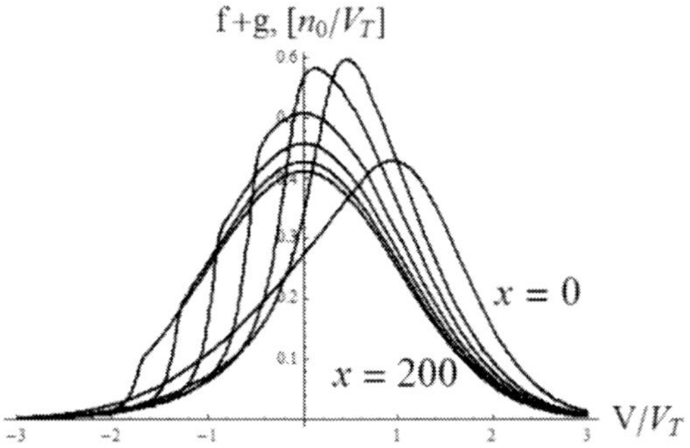

Figure 8. Dependence of the velocity distribution of the gas $f(v) + g(v)$ on the penetration depth in the cloud

Parameters $\beta = -1.5$, $\delta = 1$, $\gamma = 10$, $\eta = 10$, $\epsilon = 1$, $\zeta = 0.1$. Profiles are given for the depth of penetration $x = 0, 6, 12, 25, 50, 100, 200(\frac{v_T}{n_0 Bh}) = (\frac{8\pi v_T}{A\lambda^3 n_0})$

4. Conclusions

Based on calculation results of 6-parameter model we showed that velocity distribution of the gas can be significantly affected by radiation line close to its resonance. Then we analyzed dependence of the drift velocity on model parameters that shows resonance behavior of the LID effect. Also we found that Maxwellization of the gas should be slower than radiative transition for drift to occur. In case of optically thick cloud we got narrowing and shifting of the emission line spectral radiance in process of propagation in cloud.

References

[1] Gelmuhanov F H and Shalagin A M 1979 Light-induced diffusion of gases *JETP Letters* **29-12** 773-6.
[2] Antsygina V D, Atut S N, Gelmuhanov F H, Telegin G G and Shalagin A M 1979 *JETP Letters* **30-5** 262-5
[3] Chichigina O A 1999 *JETP* **116-1** 57-66.

Journal of Physics: Conference Series **572** (2014) 012008 doi:10.1088/1742-6596/572/1/012008

Anomalous behavior of neutron refraction index in a perfect crystal near the Bragg reflex

M.V. Lasitsa[1,2], Yu. P. Braginetz[1,2], E.O. Vezhlev[1,2], S. Yu. Semenikhin[1], I. A. Kuznetsov[1], V. V. Fedorov[1,2], V. V. Voronin[1,2].
[1)]Petersburg Nuclear Physics Institute, National Research Centre Kurchatov Institute, Gatchina, Leningrad region, 188300 Russia
[2)]St.Petersburg State Polytechical University 29, Polytechnicheskaya st., St.Petersburg, 195251 Russia.

E-mail: mishlas1@gmail.com

Abstract. Anomalous behavior of neutron refraction index in a perfect crystal near Bragg resonance was studied. This phenomenon is connected with the resonance behavior of potential of neutron interaction with crystal near the Bragg reflex. The amplitude of this resonance is equal to magnitude of g-harmonic of neutron interaction potential *Vg* and width is about the Bragg width of reflex. Recently, it was shown that for the case of noncentrosymmetric crystal this effect result in a large electric field acting on a neutron (value of the field can reach about 10^8V/cm) . This effect is planed using to search for the electric dipole moment of a neutron. If the degree of crystal imperfect is less than the Bragg reflection width (case of perfect crystal) the width of the reflex is determined by the own width of crystal reflex that is about 10^{-5} of the neutron energy. The value of g-harmonics of interaction of neutron with crystal *Vg* and optical potential of the interaction of neutron with crystal V_0 are usually about the same. Therefore the variation of neutron energy on a 10^{-5} of its value will change significantly a potential of neutron interaction with crystal.

1. Introduction.

Electric dipole moments (EDMs) of elementary particles belong to the most sensitive probes for CP violation beyond the Standard Model of Particle Physics. Constraining or detecting EDMs of different systems allows to gather experimental information about models for new physics that is complementary to high energy physics data.

The statistical sensitivity of any experiment to measure the nEDM is determined by the product $E\tau\sqrt{N}$, where E is the value of the electric field, τ the duration of the neutron's interaction with the field and N the number of counted neutrons.

Experiments with non-centrosymmetric crystals exploit the interplanar electric field. For quartz, this field was measured to be $E = 2\cdot 10^8$V/cm [4], several orders of magnitude higher than the electric field achievable in vacuum or liquid helium.

Time of neutron stay in the field τ is determined by the neutron velocity in crystal and abnormal refraction index can be the source of a systematic effects.

2. Neutron optics near the Bragg reflection

For the periodic crystal structure it is convenient to present the potential of neutron interaction with crystal as Fourier series over the reciprocal lattice vectors g:

$$V(r) = \sum_a V_a(\vec{r} - \vec{r}_a) = \sum_g V_g \exp(i\vec{g}\vec{r}) = V_0 + \sum_g 2v_g \cos(\vec{g}\vec{r} + \varphi_g) \quad (1)$$

where $V_a(\vec{r} - \vec{r}_a)$ is the potential of single atom, r_a is the atom position, $V_g = v_g \exp(i\varphi_g)$, v_g and φ_g are the amplitude and phase of g-harmonic V_g, $g = 2\pi / d$, d is the interplanar distance. Here we take into account $V_g = V^*_{-g}$, because we consider the real value potentials.

When the Bragg condition is not satisfied for any system of crystallographic planes the passage of neutron through the crystal is determined average potential V_0, otherwise all is determined by the g harmonic of the potential of the interaction of the neutron with the crystal V_g. Energy of the neutrons moving through the crystal close to the Bragg condition can be represented in the form:

$$\tilde{E}_k = E_k - V_0 + V_g \cdot \frac{1}{\Delta_B} \quad (2)$$

here $\Delta_B = (E_k - E_{kg}) / V_g = 2(E_k - E_B) / V_g$ – is parameter of the deviation from the Bragg condition for system of crystallographic planes g, $\vec{k}_g = \vec{k} + \vec{g}$ - is the wavevectors of the reflected wave. $E_k = \hbar^2 k^2 / 2m$, $E_{k_g} = \hbar^2 k_g^2 / 2m$ are the energies of a neutron in the $|k\rangle$ and $|k_g\rangle$ state, respectively. $E_B = \hbar^2 g^2 /(8m \sin^2 \theta_B)$ is the energy of the neutron corresponding to the exact Bragg condition.

According to Eq. (2), the neutron energy increases unlimitedly close to the exact the Bragg condition ($\Delta E \equiv E_k - E_B \to 0$). For this reason, perturbation theory is inapplicable at $\Delta E \cong |V_g| \equiv v_g$. For the exact Bragg condition ($\Delta E = 0$) two neutron states with the moments $\hbar\vec{k}$ and $\hbar\vec{k}_g$ have the same energy, that corresponds to the confluent state with the energy E_k. The amplitudes of these states are about the same. As a result, it is necessary to solve the two level problem

More accurate consideration of the last term in equation (2) results in:

$$\tilde{E}_k = E_k - V_0 + V_g \cdot \frac{\Delta_B}{\Delta_B^2 + 1} \quad (3)$$

One can see from the expression (3) that the last term quickly change on the scale of the Bragg reflex width Δ_B, i.e. for the neutron energy range $E_B - V_g < E_k < E_B + V_g$ ($V_g / E_B \cong 10^{-5}$).The magnitude of V_g and V_0 are about the same, therefore there is resonance like dependence of the potential of neutron interaction with the crystal on the neutron energy . For example, for the quartz crystal diffraction plane (110) $V_g = 4 \cdot 10^{-8} eV$, $V_0 = 10^{-7} eV$, $E_B = 3.2 \cdot 10^{-3} eV$ for the diffraction angle close to $\pi / 2$.

3. Method.

Setup allows us to change velocity of the crystal relatively to the laboratory coordinate system, that allows us to change it time the deviation from Bragg condition

$$\Delta_B(t) = \Delta_{B_0} + 4\frac{1}{\upsilon_n}\frac{E_B}{V_g}\upsilon(t) \qquad (4)$$

here υ_n is velocity of the neutron. The initial deviation from Bragg condition Δ_{B_0} was determined by the difference of the crystal interplanar distance, which was regulated by the temperature difference $T_{13} = T_1 - T_3$ of the crystals K_1 and K_3. The parameter deviation is defined as:

$$\Delta_{B_0} = 4\frac{E_B}{V_g}\chi T_{13} \qquad (5)$$

here χ is coefficient of thermal expansion of the crystal along the reciprocal lattice vector. Acceleration effect of the neutron is determined by the changing of potential of neutron interaction with the crystal during the time of neutron stay in crystal , and can be written as:

$$\Delta E(t_0) = \tilde{E}_k(t_2) - \tilde{E}_k(t_1) = \frac{1-\Delta_{B_0}^2}{(1+\Delta_{B_0}^2)^2}\frac{4E_B}{\upsilon_n}\Delta\upsilon(t_0) \quad (6)$$

here $\Delta\upsilon(t_0)$ - is the change of crystal velocity, t_0 is the time of neutron entrance the crystal.

4. Experiment.

For the experimental measurements of the refractive index of a neutron in a perfect crystal was made at the WWR-M reactor (PNPI, Gatchina).

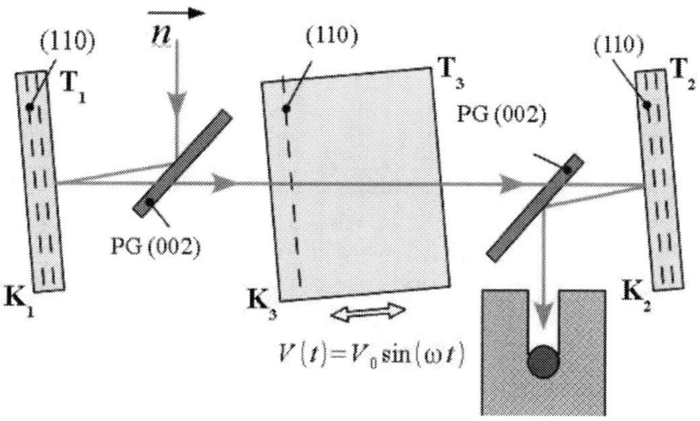

Figure 1. The experimental setup

Main crystal K3 was moved by the piezomotor with the oscillation frequency v_c = 4.5 kHz (t_c = 222 ms) , with an amplitude equal to 0.3 mm or 0.15 mm. The length of the crystal in the direction of the neutron velocity was L = 5 cm, the neutron wavelength λ_n = 4.9Å. Thus, the neutron time of flight through the crystal was approximately equal to a quarter of the period of oscillation of the crystal τ_n = 62 ms $\approx t_c$ / 4.

Crystals K1 and K2 as the double-crystal spectrometer. Where K1 is the monochromator and K2 is analyzer.: Neutron energy analysis after passage through the main crystal K3 was fulfilled by changing the temperature of crystal K2 and therefore the interpalnar distance and reflected neutron wavelength due to the temperature expansion

$$\lambda = 2d\sin(\theta_B) \qquad (7)$$

where d is the interpalnar distance.

5. Results.

The value of neutron acceleration in crystal is show in Figure 2. The two crystal temperature difference *T13* is directly connected with the deviation from Bragg condition, see (5), the calculated Bragg width of the reflex (110) of quartz crystal in a unit of *T13* is about 1K.

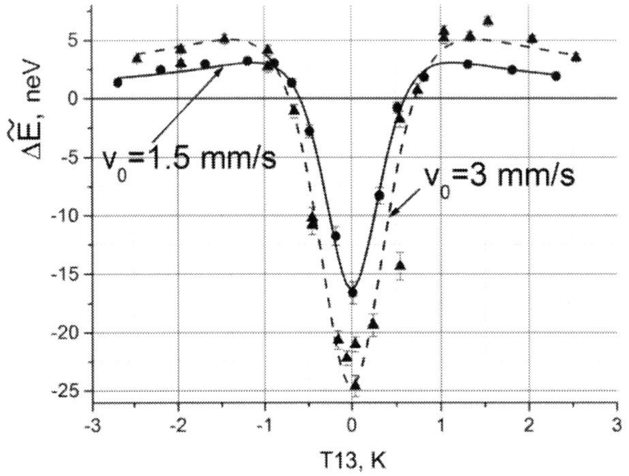

Figure 2. The change of energy of the neutron after the passage of the accelerated crystal.

One can see that effect of neutron accelerating reach about (20-30)neV, while the zero harmonics of neutron interaction with the crystal V_0 is about 100neV.

The value of neutron interaction with the crystal can be calculated from the effect of neutron accelerating (6) and (3). The result is shown in Figure 3. One can see that varying the incident neutron energy on a relative value a few 10^{-5} change the neutron interaction with crystal on about 40neV, that is about the zero harmonic V_0.

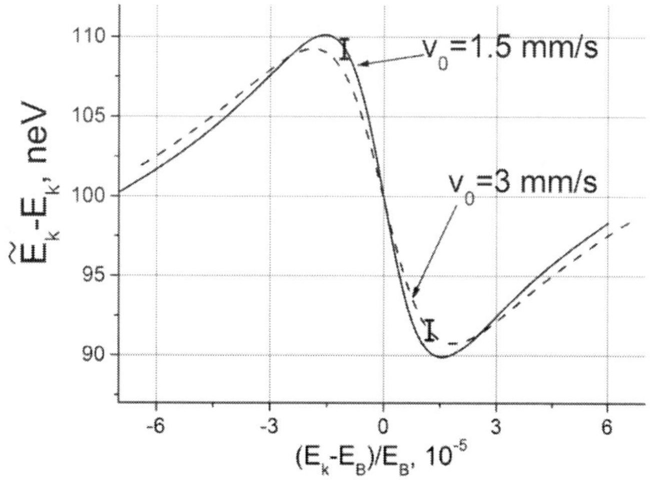

Figure 3. The potential interaction of neutrons with the crystal.

6. Conclusion:

We have studied the refraction index of a neutron in crystal near the Bragg reflection. It is shown that the dependence of neutron interaction potential with crystal on the deviation from Bragg condition is characterized by the resonance shape with the width of an order of the width of Bragg reflection (for thermal and cold neutrons $\Delta E / E \sim 10^{-5}$).

This fact opens up new prospects for the development of ultrahigh-resolution neutron optics experiment and new devices for managing the neutron beam.

A new phenomenon, the acceleration of the neutron near the Bragg resonance, was observed for the case of neutron passage through the crystal moving with variable speed.

This work is supported by the Russian Foundation for Basic Research (grant №. 12-02-00446-a) and The Ministry of education and science of The Russian Federation (agreement no. 8419 of program "Scientific and educational staff of innovation Russia").

7. References

[1] Fedorov V V, Voronin V V New possibility of a search for neutron EDM by polarization method in neutron diffraction by a crystal without center of symmetry, in Physics of Atomic Nuclei and Elementary Particles 1996 Proc. of the XXX PNPI Winter School 123

[2] Fedorov V V, Lapin E G, Semenikhin S Y and Voronin V V The effect of cold neutron spin rotation at passage through a noncentrosymmetric crystal 2002 Appl. Phys. **A74** s91- s93

[3] Fedorov V V, Jentschel M, Kuznetsov I A, Lapin E G, Lelièvre-Berna E, Nesvizhevsky V, Petoukhov A, Yu Semenikhin S, Soldner T, Voronin V V and Braginetz Yu P Measurement of the neutron electric dipole moment via spin rotation in a non-centrosymmetric crystal 2010 Physics Letters **B 694** 22-25

[4] Alexeev V L, Fedorov V V, Lapin E G and et al. 1989 Nucl. Instr. and Meth. A **284** 181 JETP **69** (1989) 1083.

Charge transfer in cold collisions of rubidium atoms with calcium and ytterbium ions

S A Yakovleva[1], **A K Belyaev**[1] **and A A Buchachenko**[2]

[1] Department of Theoretical Physics and Center for Advanced Study, Herzen University, St. Petersburg 191186, Russia

[2] Department of Chemistry, M.V. Lomonosov Moscow State University, Moscow 119991, Russia

Abstract. Low-energy collisions of the Ca and Yb cations with Rb atoms are investigated theoretically using accurate *ab initio* potential energy curves and coupling matrix elements to elucidate the dominant charge transfer mechanisms. The cross sections calculated at collision energies above 10^{-5} cm^{-1} exhibit the features typical to Langevin ion-atom collision regime, including a rich structure associated with the centrifugal barrier tunnelling (orbiting) resonances. It is shown that the dominant process in Yb^+ + Rb collisions is the radiative charge transfer, while in the case of Ca^+ + Rb collisions nonadiabatic transitions due to spin-orbit coupling dominate. Theoretical results are in a good agreement with available experimental data.

1. Introduction

With the development of the hybrid traps, which combine laser-cooled atomic ions stored in a radio-frequency trap and ultracold or statistically condensed neutral atoms in a magneto-optical trap, study of the ion-neutral reactions in the energy range between 1 mK and 1 K (usually termed as "cold" conditions) has recently become possible. Due to strong attractive induction interaction, cold ion-neutral collisions do not follow quantum regime being closer to "statistical" Langevin multiple partial wave picture. The effects of the orbiting resonances inherent to Langevin regime on charge transfer, a prototypical process for ion-molecular reaction, can only be understood theoretically since the micromotions of the trapped ions limit experimental collision energy resolution. The same is true for radiative-assisted processes, which are masked experimentally by the cooling laser field. Present work adresses these issues for the charge-transfer dynamics in cold collisions of the Yb^+ ions with Rb atom, studied recently using the hybrid trap technique [1], as well as the Ca^+ + Rb cold collisions [2], which, despite apparent similarity, exhibit quite distinct nonadiabatic charge transfer mechanism.

2. Interaction and dynamics

2.1. Ab initio data

Potential energy curves and coupling matrix elements used here were obtained *ab initio* under the scalar-relativistic approximation as described in great detail in [3, 4]. Those relevant to Yb^+ + Rb charge transfer are shown in Fig.1 (upper panel). The $A^1\Sigma^+$ state represents the initial channel, from which transition may occur to the ground $X^1\Sigma^+$ state. Nonadiabatic coupling matrix elements (NACMEs), responsible for nonradiative charge transfer mechanism, arise from the first- and second-order radial couplings and the second-order rotational couplings, presented in lower panel of the same figure.

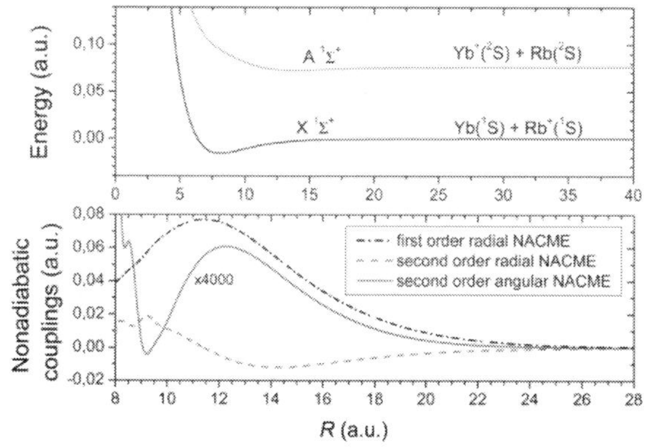

Figure 1. Potential energy curves and coupling matrix elements for $Yb^+(^2S)$ + $Rb(^2S)$ → $Yb(^1S)$ + $Rb^+(^1S)$ charge transfer process.

2.2. Nuclear dynamics

The nonadiabatic nuclear dynamics was studied within the formalism of the standard adiabatic (Born-Oppenheimer) approach described, for example, in Refs. [5, 6, 7, 8]. Due to the large energy splitting and small values of coupling matrix elements, the nonadiabatic transition probabilities are expected to be small at low collision energies, and the system of the coupled channel equations [7, 8] can be solved within the perturbation theory as shown in [5, 9].

The probabilities for a nonadiabatic transitions $i \to f$ represented by the unperturbed (elastic) scattering radial wave functions $F^{el}_{i(f)}(R)$ for the channel $i(f)$, respectively, are given by the following formulae:

$$P^{rad}_{if} = \frac{1}{4}\left|\int_0^\infty \langle\psi_{i\Lambda}|\frac{\partial}{\partial R}|\psi_{f\Lambda}\rangle\left[F^{el}_{i\Lambda}\frac{dF^{el}_{f\Lambda}}{dR} - F^{el}_{f\Lambda}\frac{dF^{el}_{i\Lambda}}{dR}\right]dR\right|^2 \tag{1}$$

for the radial couplings and

$$P^{ang}_{if} = \left|\int_0^\infty \frac{1}{R^2}\left[F^{el}_{i\Lambda}\langle\psi_{i\Lambda}|L_x^2 + L_y^2|\psi_{f\Lambda}\rangle F^{el}_{f\Lambda}\right]dR\right|^2, \tag{2}$$

for the angular coupling. In contrast to Yb^+ + Rb case, where the initial and final charge transfer channels are not coupled by spin-orbit (SO) interaction, it had to be considered for Ca^+ + Rb collisions. The corresponding probability is

$$P^{so}_{if} = \left|2\mu\int_0^\infty \left[F^{el}_{i\Lambda}\langle\psi_{i\Lambda}|\hat{H}^{so}|\psi_{f\Lambda'}\rangle F^{el}_{f\Lambda'}\right]dR\right|^2, \tag{3}$$

where μ is the reduced nuclear mass of the colliding partners.

The nonradiative charge transfer cross sections are computed as a sum over the total angular momentum quantum number J

$$\sigma_{if}(E) = \frac{\pi\hbar^2 p_i^{stat}}{2\mu E}\sum_{J=0}^{J_{max}} P_{if}(J,E)(2J+1), \tag{4}$$

where p_i^{stat} is the statistical probability for population of the initial channel i.

The dipole coupling between molecular states induces radiative charge transfer mechanism. An expression for the total radiative cross section follows from the optical potential approach [10, 11, 12, 13]. Being combined with the distorted wave approximation, it reads [10]

$$\sigma_{if}^{R}(E) = \frac{\pi}{2\mu E} p_i^{stat} \sum_J (2J+1)\{1 - \exp[-4\eta_J(E)]\} \tag{5}$$

with

$$\eta_J(E) = \frac{2\pi}{3}\sqrt{\frac{\mu}{2E}}\alpha^3 \int_0^\infty F_i^{J,E} d_{if}^2(R)[U_i - U_f]^3 F_i^{J,E} dR, \tag{6}$$

where $F_i^{J,E}$ is the unperturbed (elastic) scattering radial wave function, d_{if} is the transition dipole matrix element and $U_{i(f)}$ denotes adiabatic potential.

3. Results and conclusions

Cross sections for radiative and nonradiative charge transfer in Yb^+ + Rb collisions are shown in Fig.2. Evidently, the latter gives negligible contribution to the charge transfer due to large adiabatic splitting and weak interaction of the $A^1\Sigma^+$ and $X^1\Sigma^+$ states. At the same time, it is worthy of noting that all cross sections exhibit the same resonance structure. Partial wave analysis demonstrate that it entirely corresponds to the orbiting, strong narrow resonances – to the tunneling deep under the centrifugal barriers and weak broad features – to the capture slightly below or above the barriers. Detailed analysis of some resonances [14] clearly shows the occurrence of dynamical trapping of the colliding partners in the entrance channel, causing a correspondingly marked increase of the non-adiabatic transition probabilities.

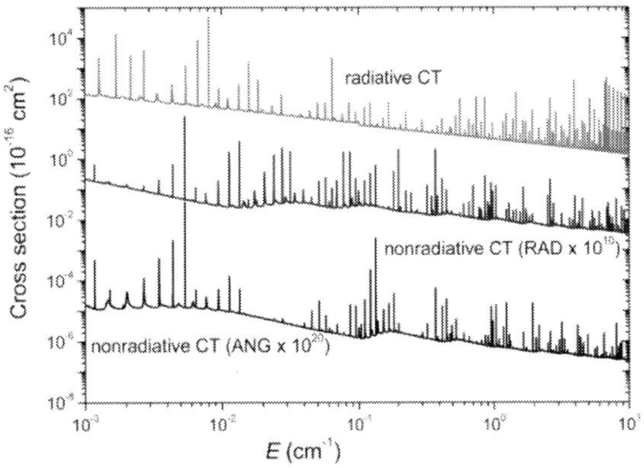

Figure 2. Energy-dependent cross sections for the $A^1\Sigma^+ \to X^1\Sigma^+$ process in Yb^+ + Rb collisions. The uppermost (red) trace shows the radiative cross section. The second from the top (blue) trace represents nonradiative cross section due to the radial NACMEs (multiplied by 10^{10}). The bottom (black) trace depicts the non-radiative cross section due to the angular NACME (multiplied by 10^{20}).

As far as the kinetic energy in a hybrid trap is controlled by ion micromotions, it is convenient to introduce energy-dependent rate coefficient as $R(E) = \sqrt{2E/\mu}\,\sigma_{AX}^{R}(E)$. It is shown in Fig.3 together with the experimental data. Note that the resonance component gives about 10% of the total rate. The mean theoretical rate coefficient is in good agreement with the measured one. It has been demonstrated that account for SO interaction that modifies the potential energy curve for initial channel further improves the coincidence with experiment [4].

In contrast, detailed investigation of Ca^+ + Rb collisions [3, 14] showed that the $Ca^+(^2S)$ + $Rb(^2S) \to Ca(^3P)$ + $Rb^+(^1S)$ charge transfer at very low collision energies proceeds by the non-radiative mechanism. The states involved, $A^1\Sigma$ and $b^3\Pi$, are coupled by the SO matrix element, whereas two crossings of the corresponding adiabatic potentials further facilitate the nonadiabatic charge transfer process. Its cross sections, shown in Fig.4, indicate that Langevin

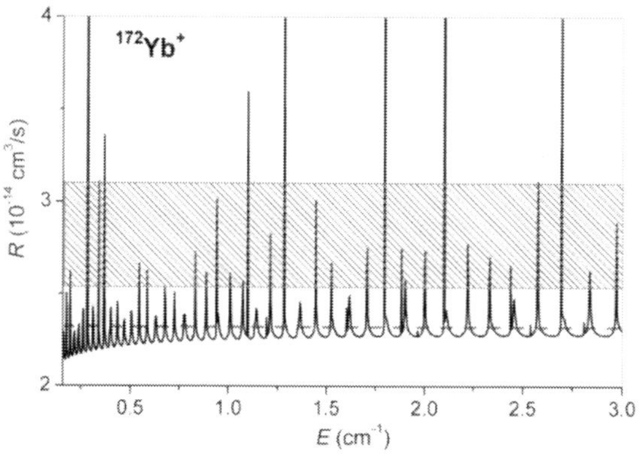

Figure 3. Energy-dependent rate coefficient for the radiative transfer in the $^{172}Yb^+$ + Rb collisions. Horizontal dashed line represents the mean value, shaded area – the experimental error bars [1].

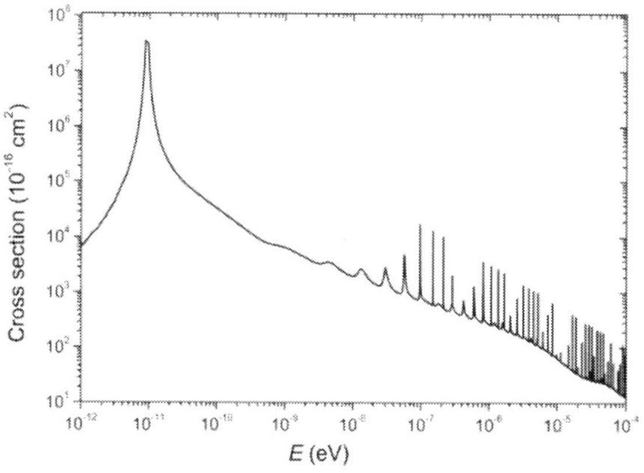

Figure 4. The calculated cross section for the $Ca^+(^2S)$ + $Rb(^2S) \rightarrow Ca(^3P) + Rb^+(^1S)$ nonradiative charge transfer process as a function of the collision energy.

regime onsets at collision energy above 10^{-5} cm^{-1}. The calculated rate coefficient for Ca^++Rb collisions is equal to 3.5×10^{-12} cm^3/s at the temperature $T = 2$ K and in good agreement with the experimental data 2×10^{-12} cm^3/s reported in [2].

To conclude, charge-transfer collisions at the energies typical to hybrid traps follow Langevin regime with its inherent rich orbiting resonance structure. This structure tend to enhance the rate of inelastic processes. The mechanism of charge transfer may vary from system to system being determined by the structure of adiabatic potential energy curves and magnitude of the couplings. The dominant mechanisms in the treated cold collisions are different: It is the radiative charge transfer in Yb^+ + Rb collisions, while the main mechanism in Ca^+ + Rb collisions corresponds to the nonradiative charge transfer. Spin-orbit interactions are essential for a quantitative description of the cold charge transfer dynamics.

Acknowledgments

We acknowledge financial support from the Russian Foundation for Basic Research (grants 11-03-00081 and 13-03-00163).

References

[1] Zipkes C, Palzer S, Ratschbacher L, Sias C and Köhl M 2010 *Phys. Rev. Lett.* **105** 133201
[2] Hall F H J, Eberle P, Hegi G, Willitsch S, Raoult M, Aymar M and Dulieu O 2013 *Mol. Phys.* **111** 2020–32
[3] Tacconi M, Gianturco F A and Belyaev A K 2011 *Phys. Chem. Chem. Phys.* **13** 19156
[4] Sayfutyarova E R, Buchachenko A A, Yakovleva S A and Belyaev A K 2013 *Phys. Rev.* A **87** 052717
[5] Mott N F and Massey H S W 1949 *The Theory of Atomic Collisions* (Oxford: Clarendon Press)

[6] Macias A and Riera A 1982 *Phys.Rep.* **90**, 299
[7] Grosser J 1986 *Z. Phys.* D **3** 39
[8] Belyaev A K, Egorova D, Grosser J and Menzel T 2001 *Phys. Rev.* A **64** 052701
[9] Belyaev A K 2007 *Eur. Phys. J.* D **44** 497
[10] Zygelman B and Dalgarno A 1988 *Phys. Rev.* A **38** 1877
[11] Cohen J S and Bardsley J N 1978 *Phys. Rev.* A **18** 1004
[12] Tellinghuisen J and Julienne P S 1984 *J. Chem. Phys.* **81** 5779
[13] Gustafsson M, Antipov S V, Franz J and Nyman G 2012 *J. Chem. Phys.* **137** 104301
[14] Belyaev A K, Yakovleva S A, Tacconi M and Gianturco F A 2012 *Phys. Rev.* A **85** 042716

Journal of Physics: Conference Series **572** (2014) 012010 doi:10.1088/1742-6596/572/1/012010

Inelastic cross sections for low-energy Mg + H collisions

D S Rodionov[1], A K Belyaev[1], M Guitou[2], A Spielfiedel[3], N Feautrier[3], and P S Barklem[4]

[1] Department of Theoretical Physics, Herzen University, St. Petersburg 191186, Russia
[2] Université Paris-Est, Laboratoire Modélisation et Simulation Multi-Echelle, UMR 8208 CNRS, 5 Bd Descartes, 77454 Marne-la-Vallée, France
[3] LERMA, Observatoire de Paris, 92195 Meudon Cedex, France
[4] Department of Physics and Astronomy, Uppsala University, Box 516, S-75120 Uppsala, Sweden

E-mail: `virtonoobne@bk.ru`

Abstract. Quantum calculations of cross sections for the inelastic processes in Mg + H collisions are improved. It is shown that the largest cross section among the endothermic processes with the value of approximately 80 $\AA^2$ corresponds to the process of the ion-pair formation: Mg($3s4s\,^1S$)+H $\rightarrow$ Mg$^+$ + H$^-$. The mechanism of the process is based on nonadiabatic transitions between the MgH($^2\Sigma^+$) molecular states, which provide the main mechanism for inelastic processes in Mg + H collisions. On the other hand, nonadiabatic transitions between MgH($^2\Pi$) states affect some cross sections rather significantly. For example, transitions between the MgH($^2\Pi$) states increase the cross section for the excitation process Mg($3s3p\,^1P$)+H$\rightarrow$ Mg($3s3d\,^1D$)+H almost by an order of magnitude as compared to the cross section obtained within the MgH($^2\Sigma^+$) symmetry.

1. Introduction

The measurement of abundances of chemical elements in stellar atmospheres, as interpreted from stellar spectra, is of fundamental importance in modern astrophysics, see, e.g., [1, 2, 3, 4]. Inelastic processes in collisions of different atoms with hydrogen atoms are important for the non-local thermodynamic equilibrium modeling of stellar atmospheres in determining spectra line shapes which is the main tool for relative and absolute chemical abundances, see, e.g., [2, 4]. Magnesium is an element of significant astrophysical importance, see [5, 6] and references therein. Thus, the need for investigation of inelastic collisions of hydrogen atoms with magnesium atoms is well justified.

In the earlier works, Ref. [7, 8], the quantum dynamical calculations have been performed for the three MgH($^2\Sigma^+$) plus two MgH($^2\Pi$) molecular states, as well as for the eight lowest-lying MgH($^2\Sigma^+$) states, respectively, based on the accurate *ab initio* quantum chemical data [9]. In Ref. [7] it was shown that among the low-lying states the transitions between $^2\Sigma^+$ molecular states dominate over transitions involving states of other symmetries. For this reason, the inelastic cross sections were calculated in Ref. [8] by taking into account only the transitions between the eight lowest-lying MgH($^2\Sigma^+$) states. In the present paper, we extend the earlier work in Ref. [8] through a calculation including nine $^2\Sigma^+$ states up to and including the ionic channel, and the five lowest $^2\Pi$ states based on the improved *ab initio* quantum-chemical data.

2. MgH interaction potentials, related couplings and cross sections

2.1. Ab initio quantum-chemical calculations

The adiabatic potentials and nonadiabatic couplings were calculated using large active spaces and basis sets since avoided crossings occur due to the $Mg^+ + H^-$ ionic configuration [9]. All $^2\Sigma^+$, $^2\Pi$, $^2\Delta$, $^4\Sigma^+$ and $^4\Pi$ electronic molecular states arising from Mg + H for energies up to about 6 eV above the lowest atomic asymptote Mg (^1S) + H (^2S_g) were calculated at the MRCI level, as well as the couplings between these states. The details are described in Refs. [7, 9] and will be described elsewhere.

The nine lowest MgH($^2\Sigma^+$) molecular states treated in the present work are collected in Table 1, including their asymptotic atomic limits. The corresponding potential energy curves

Table 1. The MgH molecular channels, the corresponding asymptotic atomic states and the calculated and experimental (NIST [10] weighted average values) asymptotic energies with respect to the ground state. Some calculated asymptotic values were adjusted to fit the experimental data.

j	Molecular states	Atomic asymptotic states	Asymptotic energies (eV) Calculation	Experiment
1	$1\,^2\Sigma^+$	Mg($3s^2\ ^1S$) + H	0.0	0.0
2	$2\,^2\Sigma^+$	Mg($3s3p\ ^3P$) + H	2.7142	2.7142
3	$3\,^2\Sigma^+$	Mg($3s3p\ ^1P$) + H	4.3894	4.3458
4	$4\,^2\Sigma^+$	Mg($3s4s\ ^3S$) + H	5.1342	5.1078
5	$5\,^2\Sigma^+$	Mg($3s4s\ ^1S$) + H	5.4237	5.3937
6	$6\,^2\Sigma^+$	Mg($3s3d\ ^1D$) + H	5.7532	5.7532
7	$7\,^2\Sigma^+$	Mg($3s4p\ ^3P$) + H	5.9321	5.9321
8	$8\,^2\Sigma^+$	Mg($3s3d\ ^3D$) + H	5.9460	5.9459
9	$9\,^2\Sigma^+$	Mg$^+$($3s\ ^2S$) + H$^-$	6.8916	6.8916

(PECs) are represented on Fig. 1 as a function of the internuclear distance. A series of avoided crossings for the $^2\Sigma^+$ states is clearly seen. The background of these avoided crossings is an interaction of covalent states with the Mg$^+$ + H$^-$ ionic state. For the high-lying molecular states the avoided crossings occur at large distances with small adiabatic splittings (for details see Fig. 2). These avoided crossings provide the main mechanism for inelastic processes in low-energy Mg + H collisions, that is, for the processes of excitation, de-excitation, ion-pair production, and mutual neutralization. The nonadiabatic couplings needed for performing quantum dynamical calculations are depicted in Fig. 3. They clearly confirm the presence of the nonadiabatic regions, where nonadiabatic transitions take place.

It is worth emphasizing that the five lowest MgH($^2\Sigma^+$) molecular states have practically the same adiabatic potentials and nonadiabatic couplings as in the previous calculations [7, 8, 9]. The improvement mainly concerns the higher-lying states, starting from $j = 6$. Thus, inelastic cross sections obtained with the present more accurate quantum-chemical data may be expected to deviate more substantially for partial processes with participation of high-lying states.

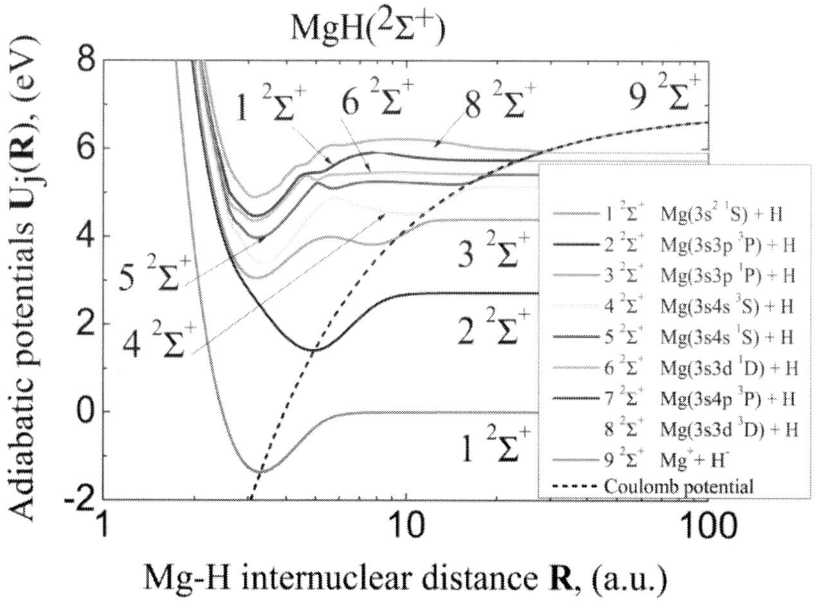

Figure 1. (Color online) Potential energy curves for the lowest nine $MgH(^2\Sigma^+)$ states.

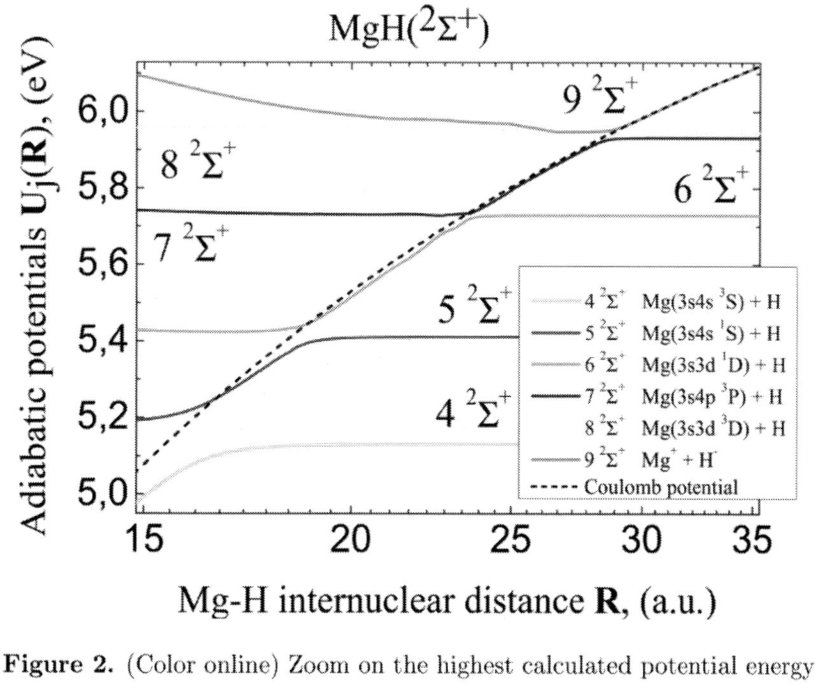

Figure 2. (Color online) Zoom on the highest calculated potential energy functions.

2.2. Nonadiabatic nuclear dynamics

The nonadiabatic nuclear dynamical calculations are carried out within the standard adiabatic Born-Oppenheimer approach by means of the multi-electron reprojection method [11] (see also

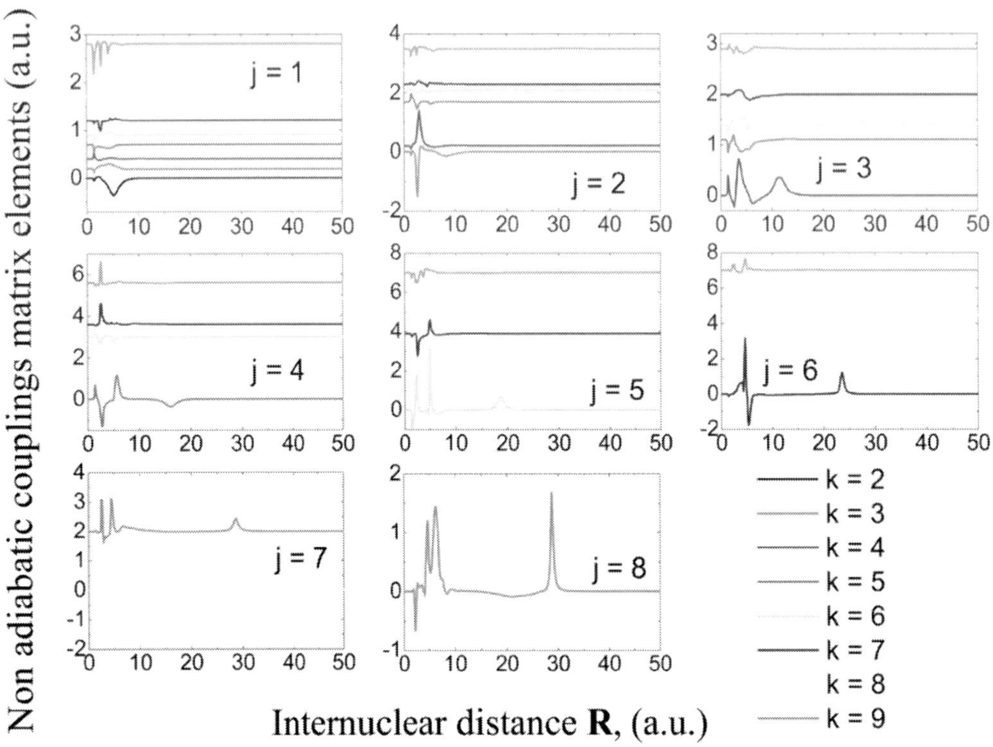

Figure 3. (Color online) Nonadiabatic radial coupling matrix elements $\left\langle j \left| \frac{\partial}{\partial R} \right| k \right\rangle$ plus offsets, between the nine $MgH(^2\Sigma^+)$ states used in the present study. The molecular state label j is indicated in each panel. The keys for the labels k are given in the right panel and are common to all panels. Offsets are used for better representation of the radial couplings.

references therein) to account for the so-called electron translation problem. The method takes into account non-vanishing asymptotic nonadiabatic matrix elements, provides the correct incoming and outgoing asymptotic total wave functions, and removes nonadiabatic transitions between atomic-state channels in the asymptotic region. First, the nuclear dynamics was performed for the three low-lying $^2\Sigma^+$ and the first two $^2\Pi$ states [7]. It was concluded that the main mechanism for nonadiabatic transitions between molecular states at low energies is due to the radial couplings associated with the avoided ionic crossings in the $^2\Sigma^+$ symmetry. The same mechanism, though in the $^1\Sigma^+$ symmetry, was found to be dominant in calculations for Li+H and Na+H low-energy collisions [12, 13, 14, 15, 16, 17]. Based on this conclusion, the nonadiabatic dynamics is then studied for the nine lowest molecular states in the $^2\Sigma^+$ symmetry including the ionic channel, as the astrophysical applications [6, 18, 19, 20] show the importance of ion-pair production processes.

The calculated cross sections for the endothermic processes are presented in Fig. 4 for each particular initial channel. The collision energy E is the kinetic energy in a given initial channel. The cross sections shown in Fig. 4 are the result of the calculation accomplished within the $^2\Sigma^+$ symmetry. The comparison of the present cross sections with those obtained in Ref. [8] shows that the calculations are stable, but some cross section deviate due to the new and more accurate quantum chemical data. It is seen that the process of the ion-pair formation: Mg($3s4s\,^1S$)+H $\rightarrow$ Mg^+ + H^- has the largest cross section of approximately 80 $\AA^2$ among the endothermic processes. The previously calculated cross section for the same process had a value of 100 $\AA^2$ [8], that is, the deviation is around 20%. This is within the expected accuracy.

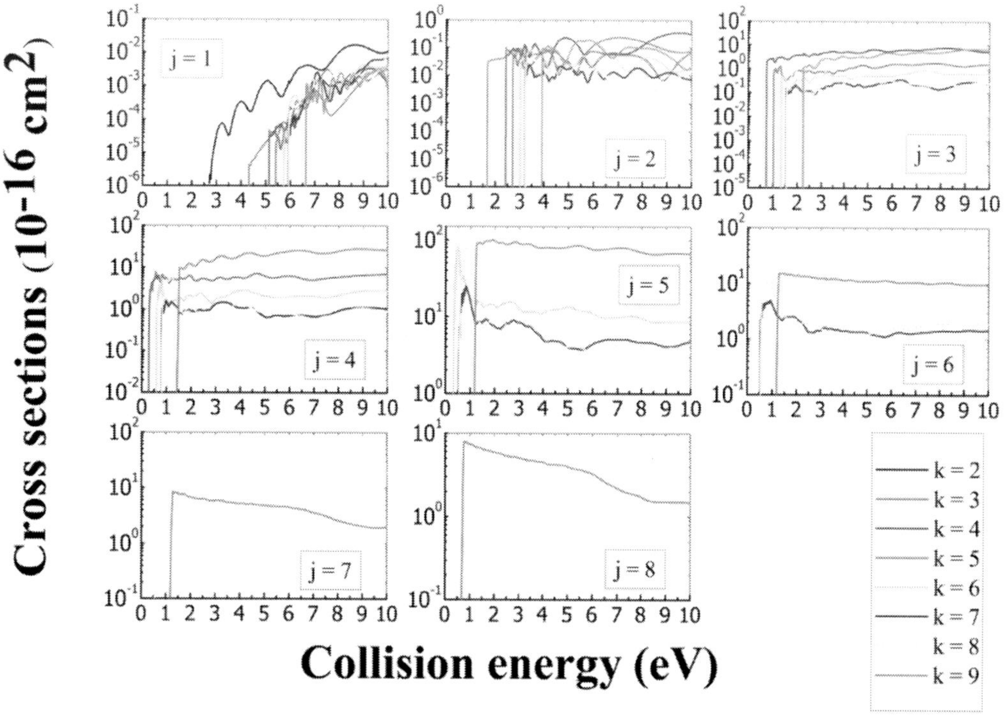

Figure 4. (Color online) The inelastic cross sections $\sigma_{jk}(E)$ for transitions $j \rightarrow k$ ($k > j$) in low-energy Mg + H collisions. The label j of the initial state from which transitions occur is indicated in each panel. The key for the final states k is given in the bottom right panel.

Nonadiabatic transitions between MgH($^2\Pi$) states may affect significantly the excitation (but not ion-pair formation nor mutual neutralization) cross sections due to transitions between highly excited states of magnesium atoms colliding with hydrogen (see Fig. 5). The labels for the processes within the MgH($^2\Pi$) is the same as in the MgH($^2\Sigma^+$) symmetry (see Table 1), but one should keep in mind that not all atomic channels collected in Table 1 can produce MgH($^2\Pi$) molecular channels. For example, in case of the excitation process Mg($3s3p\,^1P$)+H$\rightarrow$ Mg($3s3d\,^1D$)+H accounting transitions between the MgH($^2\Pi$) states increases the cross section almost by an order of magnitude as compared with the cross section obtained within the MgH($^2\Sigma^+$) states.

The physics for the inelastic processes in the treated collisions was carefully studied in Ref. [8] and several mechanisms were found. The present calculations confirm the previous conclusions. It is shown that some of the mechanisms are determined by interactions between adjacent molecular states due to interactions between ionic and covalent configurations. However, mechanisms at short distances, not restricted to adjacent states, are found to be important

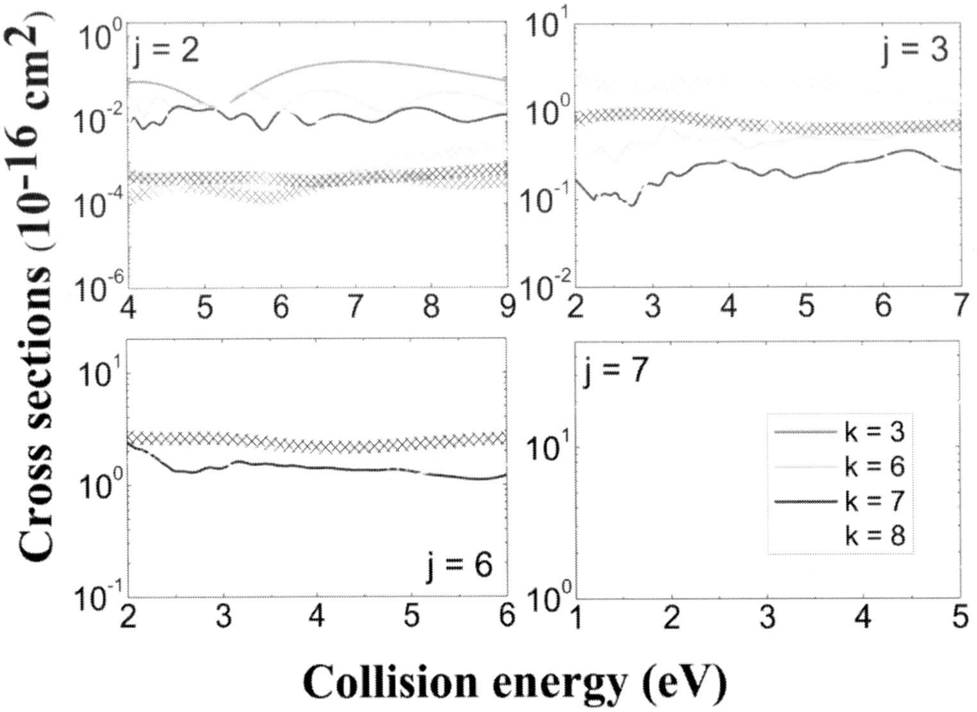

Figure 5. (Color online) The inelastic cross sections $\sigma_{jk}(E)$ for transitions $j \to k$ $(k > j)$ in low-energy Mg + H collisions calculated within the $^2\Sigma^+$ symmetry (solid lines) and within the $^2\Pi$ symmetry (lines plus symbols).

for some partial processes. These mechanisms explain the relatively large cross sections for excitation of highly excited states.

3. Concluding remarks

In the present study the cross sections for the inelastic processes in Mg+H collisions are calculated based on the improved quantum-chemical data and on estimates of accounting nonadiabatic transitions not only in the MgH($^2\Sigma^+$) symmetry, but also in the MgH($^2\Pi$) symmetry. It is shown that the process of the ion pair formation: Mg($3s4s\,^1S$)+H $\to$ Mg$^+$ + H$^-$ has the largest cross section of approximately 80 $\mathring{A}^2$ among the endothermic processes. This process occurs due to the nonadiabatic transitions between the $^2\Sigma^+$ states. On the other hand, nonadiabatic transitions between MgH($^2\Pi$) states affect significantly some excitation cross sections due to transitions between highly excited states of magnesium atoms colliding with hydrogen. As an example, accounting transitions between the MgH($^2\Pi$) states increases the cross section for the excitation process Mg($3s3p\,^1P$)+H $\to$ Mg($3s3d\,^1D$)+H almost by an order of magnitude as compared with the cross section obtained within the MgH($^2\Sigma^+$) symmetry.

Acknowledgements
The work partly was supported by the Russian Foundation for Basic Research (Grant No. 13-03-00163-a), by the Wenner-Gren Foundation (Sweden), by the French CNRS-PNPS (Programme National de Physique Stellaire) and by the GAIA program of Paris Observatory.

References
[1] Lambert D L 1993 *Phys. Scr.* T **47** 186
[2] Asplund M 2005 *A&A* **43** 481
[3] Barklem P S. 2007 *A&A* **466** 327
[4] Barklem P S 2012 *J. Phys.: Conf. Series* **397** 012049
[5] Arnone E, Ryan S G, Argast D, Norris J E and Beers T C 2005 *A&A* **430** 507
[6] Mashonkina L 2013 *A&A* **550** A28
[7] Guitou M, Belyaev A K, Barklem P S, Spielfiedel A and Feautrier N 2011 *J. Phys.* B **44** 035202
[8] Belyaev A K, Barklem P S, Spielfiedel A, Guitou M, Feautrier N, Rodionov D S and Vlasov D V 2012 *Phys. Rev.* A **85** 032704
[9] Guitou M, Spielfiedel A and Feautrier N 2010 *Chem. Phys. Lett.* **488** 145
[10] Ralchenko Y, Kramida A E and Reader J (NIST ASD Team), NIST Atomic Spectra Database (version 3.1.4), Gaithersburg, MD,USA: National Institute of Standards and Technology, 2008 [http://physics.nist.gov/asd3].
[11] Belyaev A K 2010 *Phys. Rev.* A **82** 060701(R)
[12] Belyaev A K, Grosser J, Hahne J and Menzel T 1999 *Phys. Rev.* A **60** 2151
[13] Croft H, Dickinson A S and Gadéa, F X 1999, *J. Phys.* B **32** 81
[14] Croft H, Dickinson A S and Gadéa F X 1999, *MNRAS* **304** 327
[15] Dickinson A S, Poteau R and Gadéa F X 1999 *J. Phys.* B **32** 5451
[16] Belyaev A K and Barklem P S 2003 *Phys. Rev.* A **68** 062703
[17] Belyaev A K, Barklem P S, Dickinson A S and Gadéa F X 2010 *Phys. Rev.* A **81** 032706
[18] Barklem P S, Belyaev A K and Asplund M 2003 *A&A* **409** L1
[19] Lind K, Asplund M and Barklem P S 2009 *A&A* *503* 541
[20] Lind K, Asplund M, Barklem P S and Belyaev A K 2011 *A&A* **528** A103

Properties of multicomponent system DNA – phtalocyanine metallocomplexes - surfactants

G. Alekseev and N. Kasyanenko
Laboratory of Molecular Biophysics, Department of Molecular Biophysics, St.Petersburg State University, Petergof, Ulyanovskaya st. 3, St.Petersburg, Russia

E-mail: komelbud@mail.ru

Abstract. DNA interaction with metallodiphtalocyanine with different metals (Cu, Ni, Co) and cationic azobenzene containing surfactant was investigated. The influence of concentrations of components and solution ionic strength on complex formation was regarded. The investigation shows the binding of metallodiphtalocyanines with DNA in a solution. The possibility of the formation of multicomponent system (DNA, surfactant and metallophtalocyanine) was examined.

Keywords: phtalocyanines, azobenzene containing surfactant, DNA packaging, multicomponent system

1. Introduction

The metallized diphtalocyanines (Pc) represent a new type of biologically active compounds. They have a wide range of applications from systems for gas analysis to microelectronic devices, where Pc are used as "building blocks" in MOSFET (metal-oxide-semiconductor-field-effect-transistor), and catalysts for redox reactions [1]. Some attempts were made to use them as cytotoxic and cytostatic drugs as well as photosensitizers for photodynamic therapy [2]. Such exotic applying as a catalytic treatment was also used [3]. Polycations and amphiphilic compounds can induce DNA packaging in a solution. Photosensitive surfactants in this case can produce the reversible DNA condensation [4, 5]. Such self-assembling systems can be used in nanotechnology. The introducing of such components as PC into gene vector may provide the additional opportunity for multicomponent systems in therapy and creation of new applications.

The aim of the study is to examine the properties of compounds under study, to analyze DNA-PC and DNA-surfactant interaction and to regard the opportunity to form the multicomponent system DNA-Pc-surfactant.

2. Experimental section.

2.1. Materials.

In our work we used the metallized homonuclear diphtalocyanines (PC) with Ni, Co and Cu as coordinating centers. PC compounds were preparing in Prof. A. Sengul group (Zonguldak Karaelmas University, Turkey). Cationic azobenzene containing surfactant was synthesized in group of Prof. S. Santer, Potsdam University. Calf thymus DNA with molecular mass 6,5, 10 and 10,5 MDa (Sigma), NaCl (chemically pure), DMSO (99%) were used.

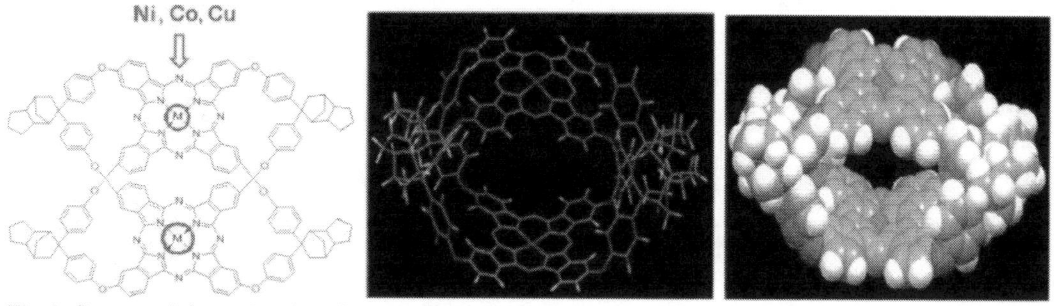

Fig.1. Structural formula of ball-type phtalocyanines.

The distance between the planes is 7,3 Å, the distance between Me-centres is 11,06 Å. The molecular mass and chemical formulas of these compounds are:
MM=2408 for $C_{152}H_{112}Ni_2N_{16}O_8$, MM=2408,48 for $C_{152}H_{112}Co_2N_{16}O_8$, and MM=2417,71, for $C_{152}H_{112}Cu_2N_{16}O_8$.

Chemical formula of surfactant AzoTMAB is $CH_3(CH_2)_3(H_4C_6N{=}NC_6H_4)O(CH_2)_6N^+(CH_3)_3Br^-$, molecular weight is M=476,59. The structure and scheme of photoinduced isomerization is shown on Fig. 2.

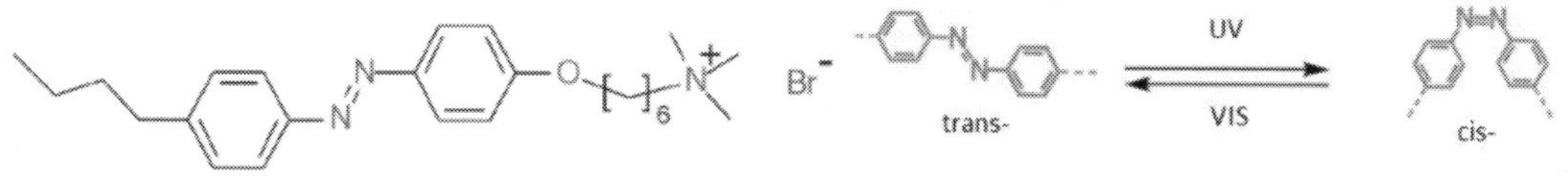

Fig. 2. The structure AzoTMAB surfactant (left) and photoinduced isomerisation of Azo-group (right).

2.2. Methods.
The systems were investigated by viscometry (low gradient Zimm-Crothers type viscometer; the specific viscosity η_r-1 of DNA solutions at constant DNA concentration was examined), spectrophotometry (spectrophotometer SF-56, Russia; the optical density of solutions D was examined), circular dichroism (autodichrograph Jobin Ivon Mark IV; the difference in extinction coefficient of DNA $\Delta\varepsilon$ for right and left light polarization was examined).

2.3. Preparation of solutions.
The phtalocyanines were dissolved in DMSO. Then solutions were filtered and gently mixed with water. The DNA solution in water was mixed with NaCl solution and filtered. DNA concentration was determined by spectral method [6].

3. Results and discussions.
The absorption spectra of PCs in DMSO and water solutions with different NaCl concentrations at λ >400 nm out of DNA absorption region are presented on Fig. 3. One can see two main bands with approximately similar positions of maximum (Fig. 3 a shows spectra normalized to the absorption at 619 nm). The difference in relative intensity of bands and noticeable shift of long-wave band for CoPC spectrum indicate that metals influence on the electron system of chromophores. The principal differences in PCs spectra remain in water solution with NaCl (see Fig 3 b, c): the short-wavelength band is playing a more prominent role for NiPC, the long-wavelength band has a higher intensity for CoPC. Both bands bring roughly equal contributions into spectrum for CuPC. NaCl concentration (from 0,005 M to 1 M) does not change significantly this picture (see Fig. 3d).

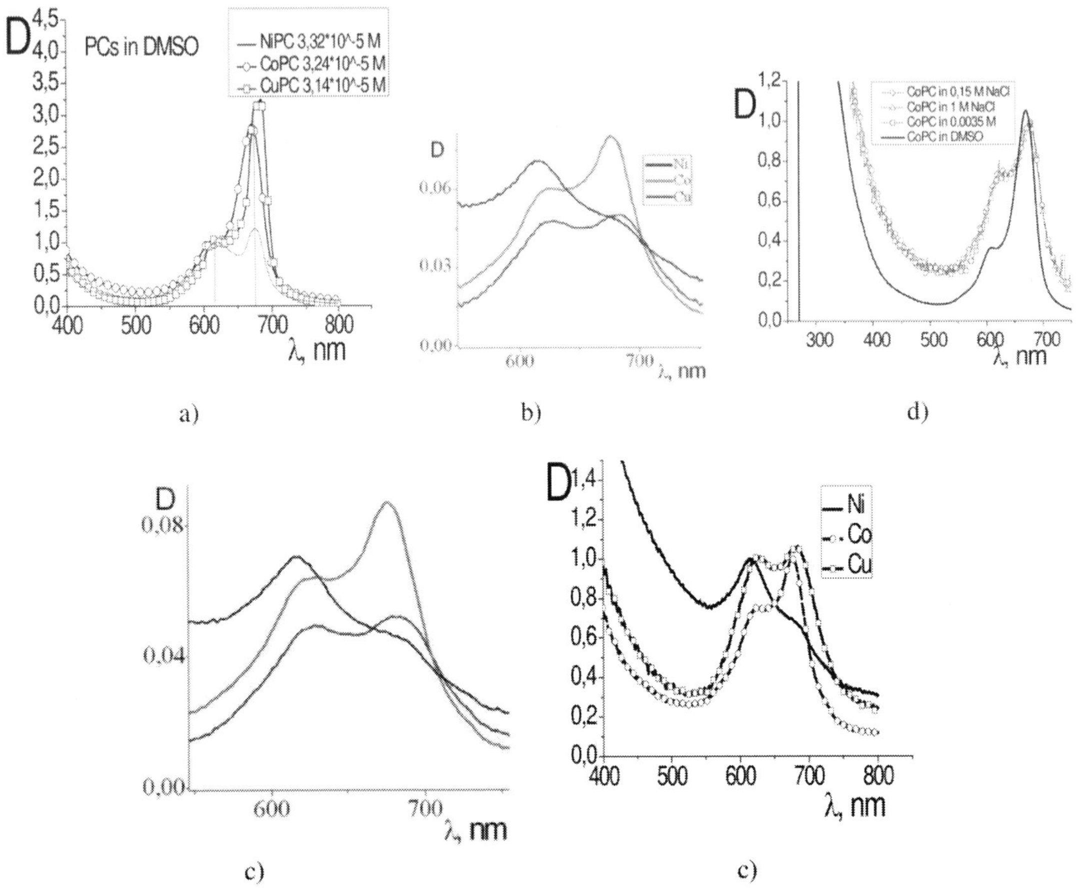

Fig. 3. MePC absorption spectra at DMSO (a), in water solution with 0.15 M NaCl (b), 1 M NaCl (c); CoPC spectra in DMSO and at different NaCl concentrations (d). Spectra (a) and (d) is normalized on absorption at definite λ.

After the mixing of DNA and PC solutions spectra of phtalocyanines have changed without great transforming of their shape (Fig. 4). The hypochromic effect can be explained with the formation of stack structure outside DNA after the binding of PC molecules via phosphate groups. Indeed, DNA secondary structure doesn't change during the interaction as it can be shown from unchanged CD spectra (Fig. 5)

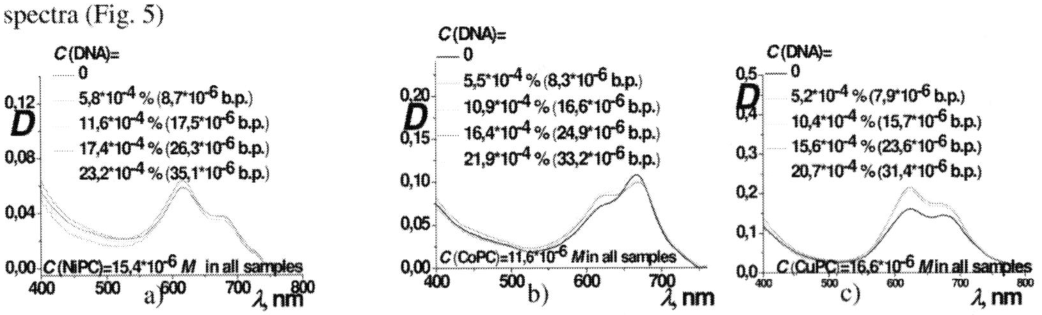

Fig. 4. Absorption spectra of PCs: NiPC (a), CoPC (b) and CuPC (c) in complexes with DNA.

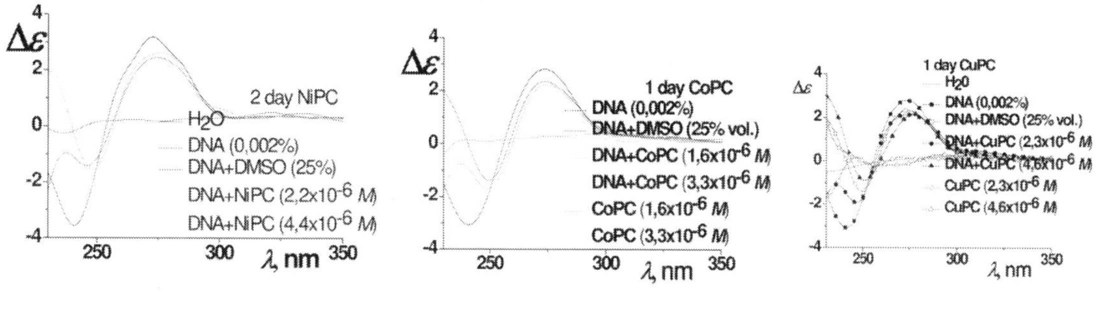

a) b) c)

Fig. 5. CD spectra of DNA in complexes with PCs: NiPC (a), CoPC (b) and CuPC (c)

The binding of PCs with DNA in a solution with low NaCl concentration causes DNA shrinkage (Fig. 6). We use viscometry for the investigation of change in the volume of DNA molecular coil (it is known that the change in specific viscosity of DNA solutions reflects the swelling or shrinking of molecular coil). It was shown that with the increase of PC concentration in DNA solution macromolecule decreases its volume for all compounds under study. We can conclude that the binding is realized just after the mixing of the solutions and does not change during time (in our experiments the systems were measured after the preparation and in 1 day storage at 4° C).

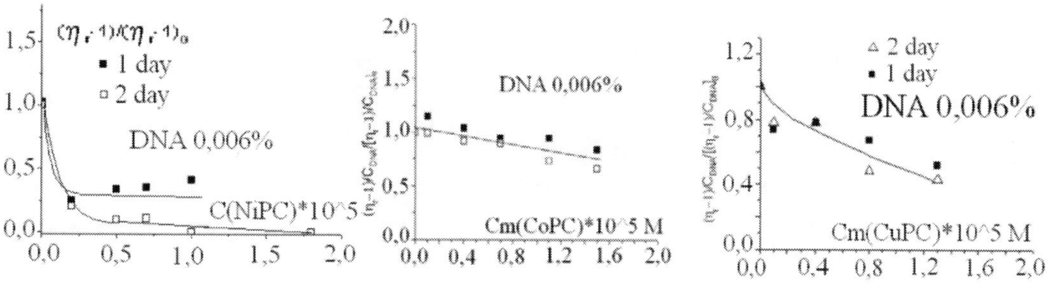

a) b) c)

Fig. 6. Dependence of relative change in specific viscosity of DNA solutions on PC concentration for NiPC (a), CoPC (b) and CuPC (c) measured just after the preparation of systems (dark signs) and in 1 day storage at 4°C (open signs).

We check the possibility of the formation of multicomponent system DNA +PC + surfactant in a solution. We use cationic azobenzene containing surfactant. It is well known that the reversible light-induced trans-cis isomerization of azobenzene containing compounds can be controlled via typical UV absorption spectra (Fig. 7). It is known that only azo- group is responsible for the spectral properties of surfactants under study. Trans isomer produces spectrum with maximum at 353 nm, same solution after UV-irradiation shows another spectrum with two bands at λ>280 nm typical for cis-isomer. Similar shape of spectra is registered for surfactant-DNA solution at λ>300 nm out of DNA adsorption band (Fig.8). Nevertheless, small blue shift of band maximum and disappearance of the "shoulder" at wavelength more than 400 nm during trans-isomer of surfactant binding to DNA are observed. UV irradiation of DNA-surfactant solutions leads to the typical spectra fixed for cis-isomer of free surfactant in a solution with same differences (red shift hyperchromic effect for the band at 350-400 nm). Moreover, same spectrum is observed if we use the preliminary irradiated surfactant (cis-isomer) for complex formation. It was shown [4] that DNA induces surfactant association on its chain with the formation of micelle-like state (pseudomicelles). The binding is not destroyed by trans-cis isomerization. After the investigation of DNA-surfactant solutions we prepare more complex systems containing DNA, surfactant and metal complexes.

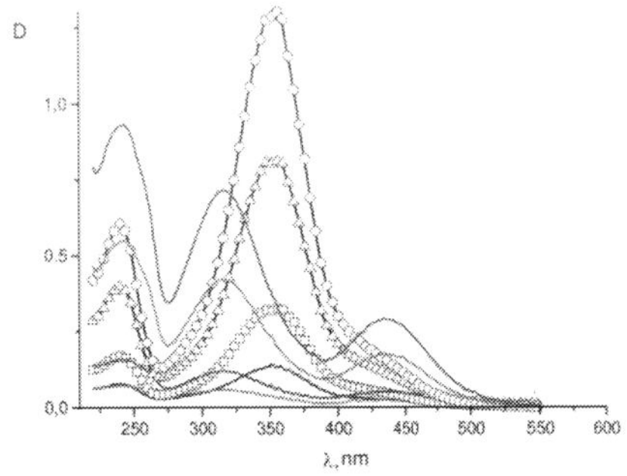

Fig. 7. The absorption spectra of the initial solutions of surfactants (trans-AzoTAB) in 0.005M NaCl and after UV irradiation at 365 nm for 2 minutes (cis-AzoTAB) (before irradiation - line geometric shapes).

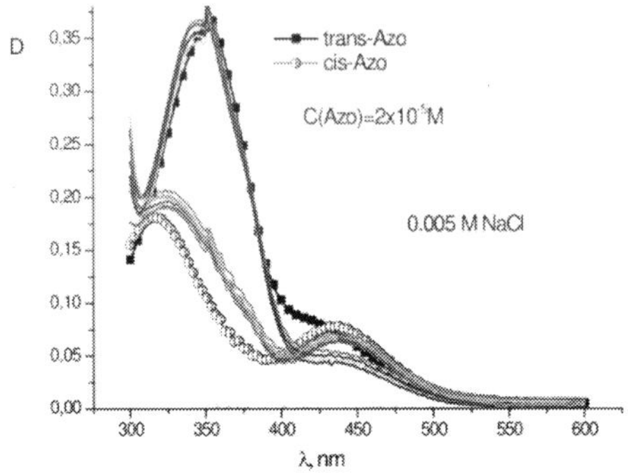

Fig. 8. Spectra of free surfactant (line with symbols) in a complex with DNA before and after 2 min UV irradiation (365 nm). C (DNA) = 0,003%, 0,006%, 0,009%, 0,012%, 0,015%, 0,018%; z = 0.44, 0.22, 0.15, 0.11, 0.09, 0.07.

We prepare solutions with DNA, surfactant and PCs in 0.005 M NaCl. We also check the influence of DMSO on spectra of compounds because the initial solution of PCs contains this solvent and finally our systems under investigations have 10% DMSO. It is clear from the experimental data that DMSO does not influence significantly on spectral properties of compounds. Fig. 9 demonstrates that the order of the addition of components into solution does not influence on the result. Indeed, absolutely similar spectra were observed for [(DNA+surfactant)+PC] and [(DNA+PC)+surfactant] systems for trans- and cis-isomers of surfactant.

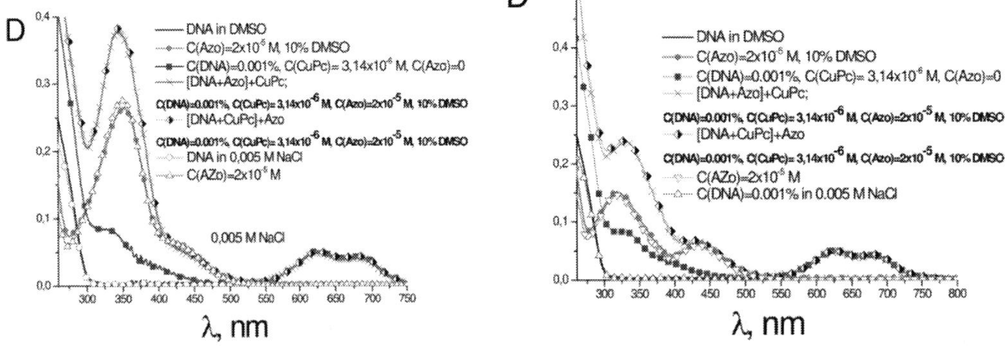

Fig. 9. Cis- and trans-Azo in complexes with DNA in a solution with PC and control bi-component systems. Concentrations of components and the way of system preparation indicate on Figures.

Finally, we can conclude that our experimental results show that the compounds CuPC, NiPC, CoPC bind to DNA in solution. The presence of DNA stabilizes PC solution and prevents precipitation of drugs. DNA-PC binding does not change the secondary structure of macromolecules. Drugs interact with DNA via the attraction to the phosphate groups outside of the helix. The state of PC compounds (chromophores) during the binding with DNA depends on the type of metal ion inside the PC rings, as it is evidenced by the variation in the absorption spectra in the long-wave area.

The interaction of DNA with photosensitive cationic surfactant in solution was regarded. It was shown that the binding of pre- UV-irradiated surfactant to DNA and the irradiation of DNA-trans-surfactant complexes lead to the same results. The preparation and investigation of multicomponent systems demonstrate that metal complexes bind to DNA in the presence of surfactants and DMSO. The formation of DNA-surfactant-metal complexes system does not depend on the order of addition of the components into solution. UV irradiation of complex system causes trans-cis transition of surfactant as well as in DNA-surfactant solution and in free surfactant in 0.005 M NaCl. Thus our experiments show the principle possibility to create the multicomponent systems with DNA-surfactant-metal complexes.

Acknowledgments.
Work was supported by a RFBR grant 13-03-01 192A and SPbU grant 11.38.644.2013.

References.

1. Şengül A, Doğan HZ, Altındal A, Özkaya AR, Salih B, Bekaroğlu Ö., Synthesis, interface (Au/M2Pc2/p-Si), electrochemical and electrocatalytic properties of novel ball-type phthalocyanines, Dalton Trans. 2012 Jul 7;41(25):7559-72.
2. HUANG Jinling, CHEN Naisheng, HUANG Jiandong, LIU Ersheng, XUE Jinping, YANG Suling, HUANG Ziqiang, SUN Jiancheng, Metal phthalocyanine as photosensitizer for photodynamic therapy (PDT) -- Preparation, characterization and anticancer activities of an amphiphilic phthalocyanine ZnPcS2P2, SCIENCE IN CHINA (Series B), April 2001, Vol. 44 No. 2, pages 113-122].
3. A. A. Chernonosov, L. I. Solov'eva, E. A. Luk'yanets, D. G. Knorre, O. S. Fedorova, Dimeric Fe-Co Phthalocyanine Complex as a Reagent for the Selective Damage of Nucleic Acids, Macroheterocycles, 2011, 4(2), pages 135-137].
4. PHYSICAL REVIEW E 84, 021909 (2011) DNA compaction by azobenzene-containing surfactant Yuriy Zakrevskyy, Alexey Kopyshev, Nino Lomadze, Elena Morozova, Ludmila Lysyakova, Nina Kasyanenko, and Svetlana Santer.
5. Anne-Laure M. Le Ny and C.Ted Lee. Jr., Photoreversible DNA Condensation Using Light-Responsive Surfactants, Journal of American Chemical Society, 2006, 128 (19), pp 6400-6408
6. Spirin A. S., Spectrophotometric determination of total amount of nucleic acids, Biochemistry (USSR), volume 23, issue 5, page 656, 1958

Microoptomechanical sensor for intracranial pressure monitoring

A V Andreeva, V V Luchinin, N A Lutetskiy, A N Sergushichev

Saint-Petersburg State Electrotechnical university, Saint–Petersburg 197376, Russia

Email: doutelet@gmail.com

Abstract. The main idea of this research is the development of microoptomechanical sensor for intracranial pressure monitoring. Currently, the authors studied the scientific and technical knowledge in this field, as well as develop and test a prototype of microoptomechanical sensor for intracranial pressure (ICP) monitoring.

1. Introduction

Treatment of patients with acute cerebral injuries almost always represents a serious problem. The main of the problem is the location of the system features in a closed hermetic cavity of a skull with the fixed volume. Moreover, the quantitative assessment of pressure values in a skull cavity for patients with cerebral pathology is very important. The mankind passed a certain way in methodological aspect, having developed different methods of the intracranial pressure (ICP) measurement. These methods are implemented by means of a wide range of various ICP sensors. These sensors are characterized by high invasiveness that is connected to their overall dimensions and to a method of ICP determination. Also, modern sensors do not allow receiving information on ICP status in a dynamic mode and have rather high cost.

The purpose of this work was development of the microoptomechanical sensor for ICP control eliminating listed above defects of existing sensors. The developed microoptomechanical sensor for control of intra cranial pressure has the following advantages:

- safe way of information obtaining:
  - the measurement scheme without electric signals impact on a human brain;
  - sensor membrane of Si_3N_4;
  - the semiconductor diode with the optical radiation wave length of 850 nanometers;
- reduction of the sensor geometrical sizes in section to 3 mm;
- increase of a sensitivity threshold;
- possibility of obtaining information in a dynamic mode.

2. ICP sensors

In the modern medicine wide variety of sensors for ICP monitoring having different advantages and disadvantages are used [1]. They share on the following types:

- intraventricular catheter: or the so-called outside ventricular drainage (OVD) connected to the outside pressure sensor by a tube, filled with liquid. It is the standard for testing other sensors, principal advantages of this sensor are:
  - rather low cost;

- except measurement of pressure it is possible to realize deduction of liquor with the medical purpose;
- it is possible to recalibrate for reducing a measurement error;

- intraparenchymal sensor: similar to an intraventricular catheter, but more expensive. In certain cases the measurement error about several;
- other less fine sensors:
  - the subarachnoidal screw ("bolt"), its main shortcoming is risk of infection 1% increasing after 3 days. In case of increase in ICP, the surface of a brain can superimpose a clearance that carries to the wrong indications;
  - epidural: it is possible to use a catheter filled with liquid.

Possible complications of ICP sensors installing (Table 1):

- Infection;
- hemorrhage: the overall incidence of 1.4% for all sensors. Risk of a large hematoma formation requiring removal ≈ 0,5%;
- malfunction or blockage: for devices running on the fluid pressure transmission, a higher incidence of occlusion is observed with an increase in intracranial pressure> 50 mm Hg. Art.;
- incorrect sensor position: intraventricular catheters in 3% of cases require prompt move to another position.

Table 1: The complication rate for different ICP sensors types.

Sensor type	Bacterial colonization	Hemorrhage	Sensor malfunction or blockage
Intraventricular catheter	Average: 10-17%, range: 0-40%	1,1%	6,3
Subarachnoidal screw ("bolt")	Average: 5%, range: 0-10%	0	16%

Typical system of outside ventricular drainage, as most often used for measurement of intracranial pressure is shown on Figure 1. The right side of human head, if there are no special reasons for use left (for example, blood bunches in the right side ventricle which can lock a catheter), is usually used.

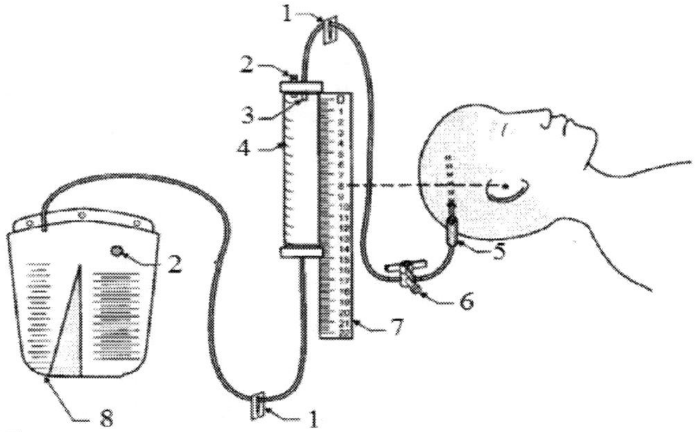

Figure 1. Typical system of outside ventricular drainage. 1 – The fixer, 2 - the air filter, 3 - a dropwise tip, 4 - the dropwise camera, 5 - the one-sided valve, 6 - port for the monitor, 7 - pressure scale, 8 - a bag for liquid collection [1].

The principle of operation of such ICP monitor is based on the intracranial pressure measurement transmitted by an air pole on a tube, which at the same time is a part of a measuring catheter located in a ventricle of a brain [1].

On the end of a catheter a latex barrel is located that is reporting on an air tube with a measuring pressure gage in the monitor. ICP is transmitted through a thin wall of a barrel, changing it volume and air pressure that register a pressure gage. After connecting a catheter to the monitor and switching on a power the compressor that built in monitor fills a barrel with air.

Example of this sensor type which is used in modern medicine, is the ICP measurement monitor Spiegelberg.

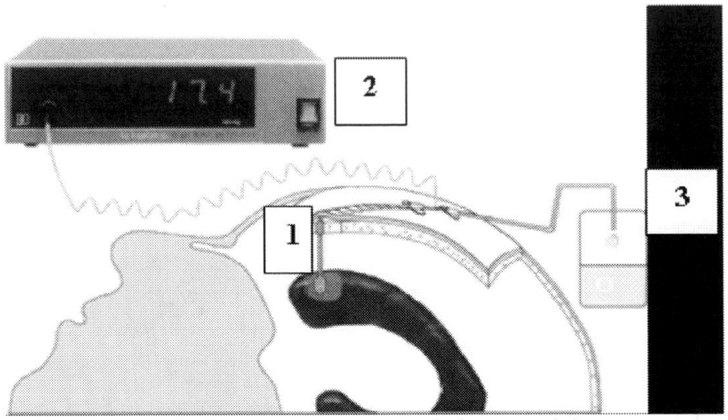

Figure 2. Intraventricular measurement of intracranial pressure monitor Spiegelberg scheme. Ventricular catheter with a balloon at the tip (1) mounted in front horn of the lateral ventricle and connected to a monitor (2). Additional channel catheter allows to drain spinal fluid cerebrovascular (3)

The operation principle of the monitor Spiegelberg is based on the ICP measurement transmitted through the air column in the tube, which is simultaneously part of the measuring catheter, which is set in the ventricle.

3. Amplitude sensors of pressure

To eliminate the disadvantages of existing ICP sensors the sensor based on the modern achievements of optoelectronics are developed.

The fiber-optic sensors (FOS) are now one of most dynamically developing areas of optoelectronics [2]. For the last 30 years there was a prompt transition from the simplest constructions of temperature and pressure fiber-optic sensors to creation of the wide nomenclature of physical quantities sensors. The important advantage of optical fiber sensors is introduction of new qualities into measuring systems, such as:

- small size;
- stability to uncontrollable and aggressive influences of environment and to electromagnetic interferences;
- high sensitivity;
- remoteness of measurements technological effectiveness of production.

Amplitude fiber-optic sensors are the simplest and convenient in maintenance constructions of FOS. The block-scheme of fiber-optic sensor is presented on Figure 2. Such sensors use the principle of intensity modulation of a laser light beam: the beam of light from laser source goes on an optical fiber and illuminates a diaphragm. In case of diaphragm oscillations the luminous flux is modulated (on intensity) and goes by the second optical fiber to the photodiode that transforms a signal to an alternating current, which value is shown on reader device.

According to such principle conversion of oscillations of a diaphragm directly in an electrical signal is not used. The diaphragm can be placed generally at distance of tens (hundreds) meters from a light source and the photodiode because of low transmission losses of a signal in an optical fiber (losses signal/noise make less than 3 dB on 1 km of an optical fiber). Such optical circuit does not cause any electromagnetic radiations. Also, it is insensitive to electromagnetic fields. The system can be placed in any hardly accessible place and can work in the strong magnetic or electrical fields because of its small sizes.

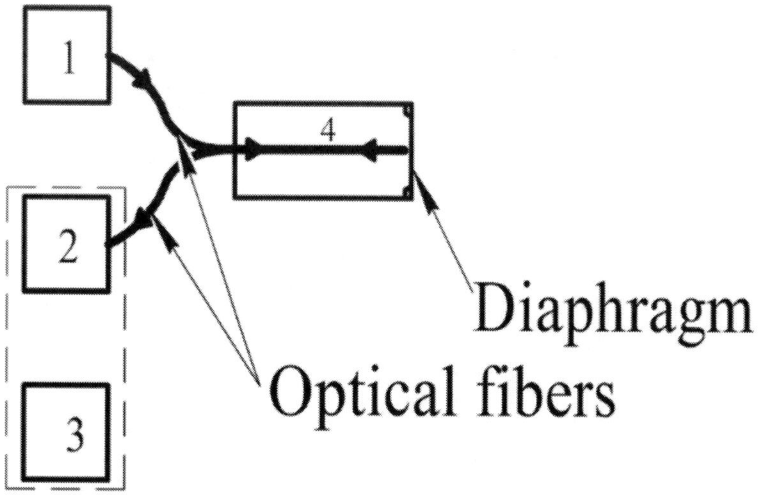

Figure 3. The block-scheme of fiber-optic sensor. 1 – light source, 2 – photodiode, 3 – reader device, 4 – optical fibers inside the sensor.

4. Diaphragm

One of the most important parts of construction of FOS is the diaphragm and many parameters of the sensor depend on its choice.

Hyperfine microsized diaphragms are a basis of many highly sensitive micromechanical, microheatphysical transformers, the active controlled microdevices and difficult analytics microsystems [3]. Physicomechanical properties of such diaphragms strongly depend on their composition, structure, properties of a surface and manufacturing techniques. Design of different transformers on the basis of hyperfine diaphragms requires the accounting of sized factors, and their manufacture assumes use of special technological operations.

Technological effectiveness and overall performance in composition with other constructional materials are the major criteria for choosing materials for implementation of constructions of membrane transformers. As a basic material of hyperfine diaphragms silicon nitride (Si_3N_4) – one of the main materials of microelectronics with well known technology of receiving and known properties is often used. Films of silicon nitride usually receive by silan amonolizing in the ammonia atmosphere at temperatures about 850 °C. The most important qualities of this material are mechanical and chemical stability, low porosity. It allows creating on its basis hyperfine, including nanosized films.
Diaphragms can be made by technology of volume and surface micromechanics depending from construction of membrane transformers.

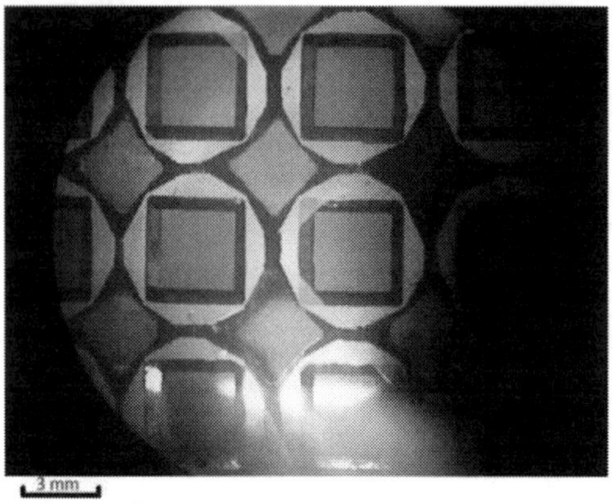

Figure 4. Photo of Si_3N_4 diaphragm matrix.

For the microoptomechanical sensor for monitoring of ICP implementation, we selected plane Si_3N_4 diaphragm. Photo of Si_3N_4 diaphragms is shown on Figure 4. This type of diaphragms possesses the following parameters:

- maximal withstood values of pressure up to 82 kPa;
- deflection at a pressure of 1 Pa is 0,9 nanometers;
- chemical inertness.

5. Construction and parameters of the ICP microoptomechanical sensor

Different options of the microoptomechanical sensor for ICP monitoring construction were developed. Figure 4 demonstrates the sensor with Si3N4 diaphragm fixed on a centralizer.

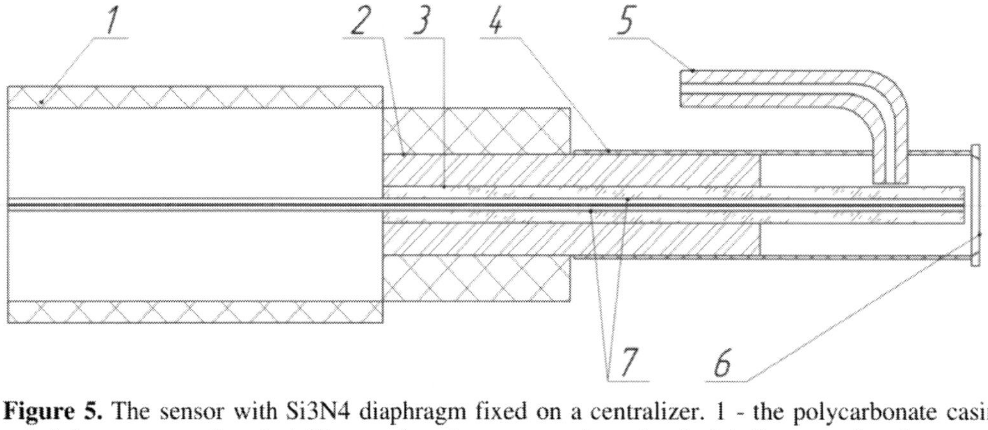

Figure 5. The sensor with Si3N4 diaphragm fixed on a centralizer. 1 - the polycarbonate casing, intended for support of optical fibers safety from external mechanical influences; 2 - the ceramic casing, one of the construction elements, providing centering of optical fibers on surface of a diaphragm; 3 - glass tube in which the optical fibers ground from an external covering are located; 4 - the centralizer also provides centering of optical fibers on surface of a diaphragm. The centralizer with the diaphragm strengthened on it (7) can be disconnected from the general construction; 5 - the hollow tube providing alignment of pressure in submembrane volume with external atmospheric pressure; 6 - Si3N4 diaphragm is the reflecting element providing modulation of light on intensity, as a result of external influence; 7 - optical fibers: receiving, the transferring and two separate fibers providing as much as possible the close layout of the main fibers. Diameter of a core of fibers of 62,5 microns

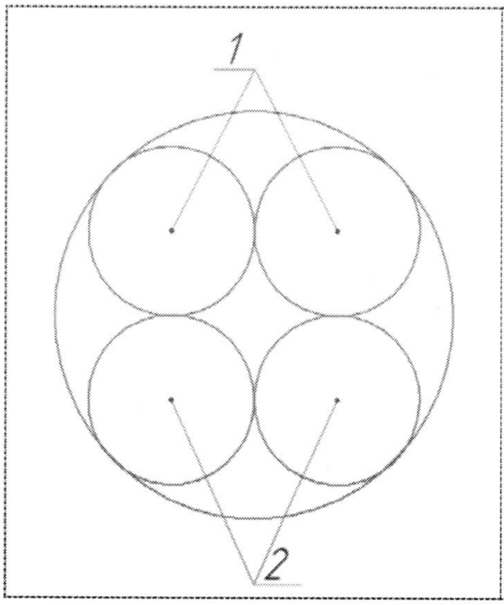

Figure 6. Construction of an end face of optical system. 1 – Transferring and receiving fibers, 2 – separate fibers.

Construction of an end face of optical system is shown on Figure 6. This option of a design was built and investigated with use of two types of optical fibers with diameters of core 62,5 and 100 microns.

For the structure shown in Figures 5 and 6, transfer function shown in Figure 7 was obtained.

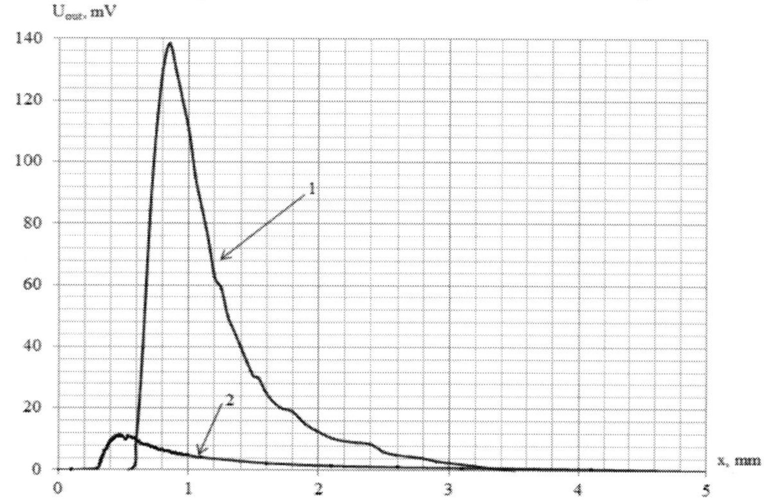

Figure 7. The output voltage as a function of distance between the ends of the optical fibers and a reflective surface for optical fibers of different types, 1 - 100/125 micron, 2 - 62.5 / 125 micron.

For obtaining the maximum sensitivity it is expedient to use optical fibers with a smaller diameter of a core. However on fibers with a larger diameter of a core it is possible to receive the maximum absolute value of output voltage. It is connected with fact that optical fiber with a 100 microns diameter of core possesses greater value of a numerical aperture. Larger range of measured pressure values at fibers by a 100 microns diameter is a result of wider initial quasilinear site of the transfer characteristic.

6. Conclusion

As a result of this work the level of scientific, technological and methodological knowledge in the field of human ICP measurement procedures was investigated as well as the existing ICP sensors, their advantages and disadvantages. Based on these studies microoptomechanical ICP monitoring sensor was developed, designed and investigated.

The analysis of the obtained data showed that the microoptomechanical sensor with a Si_3N_4 membrane that has been hermetically fixed on a centralizer, is the best solution for objectives. Also the dependence of the output signal and the sensor sensitivity from the optical fibers diameter was determined.

Results of this work can be used for further improvement of characteristics of sensors for control of intra cranial pressure, and also for development of the full-fledged medical system meeting all modern requirements.

References

[1] Greenberg M S 2010 *Handbook of neurosurgery* (Ontario: Webcom)

[2] Vetrov A A, Sergushichev A N, Shirshov A A 2010 *Fiber-optic vibroacoustic sensors* (Saint-Petersburg: SPBETU «LETI»)

[3] Korlyakov A V 2007 *J. of "NANO and MICROSYSTEM TECHNIQUE"* Ultrathin Membranes in Microsystem Technics **8(85)** 80

Journal of Physics: Conference Series **572** (2014) 012013 doi:10.1088/1742-6596/572/1/012013

Formation of intermolecular crosslinks by the actinocin derivatives with DNA in interaction under conditions of semidilute solution

D N Osinnikova and E B Moroshkina

St.Petersburg State University, St. Petersburg 199161, Russia

E-mail: osinnikovadasha@yandex.ru

Abstract. Interaction of native calf thymus DNA (ctDNA) with the actinocin derivatives containing protonated diethylamino groups, dimethylamino groups and unsubstituted amino groups and having different length of the alkyl chain have been studied by the method of viscometry. An anomalous hydrodynamic behavior of solutions of DNA with very low amount of ligands prepared under conditions of semidilute solution was revealed. We assumed that such an anomalous behavior of solutions of DNA complexes with actinocin derivatives associated with the formation of intermolecular crosslinks while the preparation of the complex was in terms of overlapping of macromolecular coils in solution. Comparative study of the hydrodynamic behavior of the DNA complexes with various actinocin structures lead us to the conclusion of the formation of crosslinks by the compounds containing protonated diethylamino groups.

1. Introduction

Currently, the problem of research of anticancer drugs is very relevant. The mechanisms of their action are various, one of them is disturbance of the DNA-dependent RNA synthesis. One of the most well-known antitumor antibiotics with a similar mechanism of action is actinomycin D. Actinomycin D was the first antitumor antibiotic that has been applied in the chemotherapeutic treatment of human tumors.

The most popular model of the interaction of actinomycin D with DNA is the intercalation binding model [1]. It is based on data from X-ray crystals diffraction of complex actinomycin D with deoxyguanosine [2]. Antibiotic chromophore is located between two GC pairs having the opposite orientation. The complex is formed by hydrogen bonds with the guanine residues on both sides of the chromophore. Intercalation occurs exclusively in the small groove.

In connection with the search of new effective antitumor antibiotics a large number of analogues of already known drugs has been synthesized. For example, actinocin derivatives containing dialkylaminoalkyl groups in the amide groups were synthesized. Previously it was shown that the actinocin derivatives of this group bind to DNA like actinomycin D by the way of intercalation of actinocin chromophore into DNA double helix [3].

These compounds possess three binding sites: a chromophore that can intercalate into DNA double helix and two protonated amino groups (figure 1).

Compounds with several binding sites, including the positively charged groups, can form intra- and intermolecular crosslinks. Intramolecular crosslinks are formed during a random approach of distant regions of DNA's chain, which are binding sites for specific groups of the ligand [4]. The molecules of histone H1 can form such crosslinks by interaction with a high-molecular DNA [5].

Intermolecular crosslinks can be formed under conditions of convergence of the elements of different molecules in solution. In the work [6] a method of molecular design, based on the "crosslinking" double-stranded nucleic acid molecules by polymeric chelate bridges, was proposed. Anthracyclines antibiotics which can form flat polymeric complexes were used to create the crosslinks. Favorable conditions for the formation of intermolecular crosslinks appear in the overlapping of macromolecular coils in semidilute polymer solutions [7][8].

In the present work we investigated complexes DNA with actinocin derivatives containing dialkylaminoalkyl groups in amide groups prepared in dilute and semidilute solutions.

2. Materials and methods

The interaction of DNA with actinocin derivatives containing various dialkylaminoalkyl groups in amide groups and having different length of the alkyl chain (figure 1) have been studied by the method of viscometry.

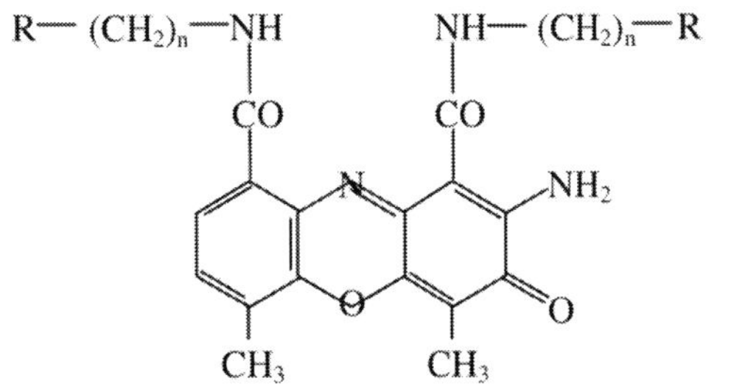

Compound 1 $R=N(C_2H_5)_2$, n=2
Compound 2 $R=N(C_2H_5)_2$, n=3
Compound 3 $R=N(CH_3)_2$, n=2
Compound 4 $R=N(CH_3)_2$, n=3
Compound 5 $R=NH_3$, n=2
Compound 6 $R=NH_3$, n=3

Figure 1. The structure of investigated compounds

The compounds were synthesized at St. Petersburg State Institute of Technology [9].

High-molecular-weight calf thymus DNA of firm "Sigma" (USA) has been used. Extinction coefficient ε_{260} = 6400–6700 $M^{-1}cm^{-1}$. The complexes were prepared by mixing the DNA and ligand solutions of the necessary concentrations. The number of the bound ligand molecules per pair of DNA bases (r) was 0,1 or 0,01.

Initial solutions were prepared in two ways: 1 - in a semidilute solution, when $[\eta] \cdot C_{DNA} > 1$, where $[\eta]$ - intrinsic viscosity of DNA ($C_{DNA} = 8 \cdot 10^{-3}\%$); 2 - in a dilute solution, when $[\eta] \cdot C_{DNA} < 1$ ($C_{DNA} = 4 \cdot 10^{-3}\%$). The ionic strength of the solution (μ) was 0.1 or 0.001 M and subsequently remained unchanged.

We measured the viscosity of solutions of pure DNA and DNA-ligand complexes. Viscosity of solutions has been measured with the magnetic rotational viscometer [10]. Measurements were carried out at the experimental setup to determine the viscosity of non-Newtonian fluids in a wide range of velocity gradient. Velocity gradient of viscometer g=(0.1 – 0.5) s^{-1}. Viscometer is a devise with two coaxially arranged cylinders. In the gap between them investigated solution is placed. The inner cylinder (rotor) is rotated by a rotated magnetic field. Rotor velocity depends on the frequency of magnetic field and viscosity of solution.

3. Results and discussion

The results of measurements of the dependence of the viscosity from concentration of DNA in solution at $r = 0.1$ are shown in figure 2, μ=0.001.

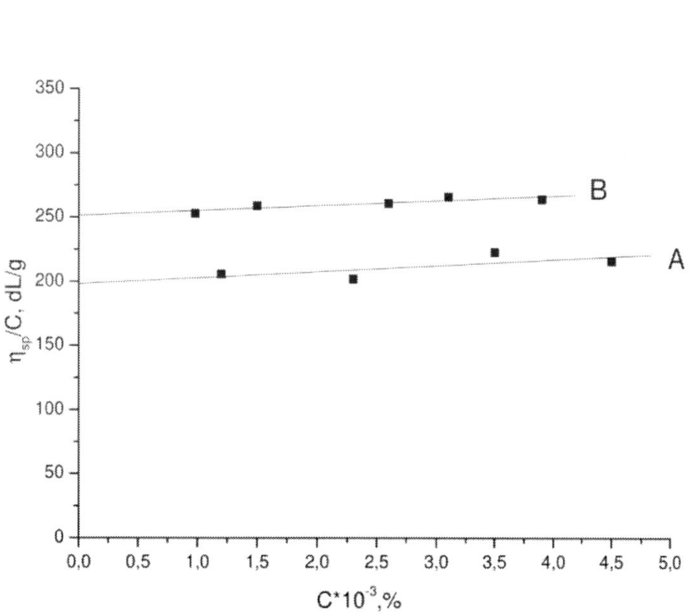

Figure 2. The dependence of the η_{sp}/C_{DNA} from concentration of DNA, μ=0.001, A - free DNA, B - complexes of compounds 1-6 with DNA at r=0.1

The concentration dependence of the viscosity of solutions of DNA and its complexes with ligands is rectilinear in dilute solutions regardless of the preparation conditions. This allows determine the intrinsic viscosity of DNA and its change in the process of interaction with the actinocin derivatives.

According to Flory formula [11], the intrinsic viscosity of the macromolecule is related to its parameters by the ratio:

$$[\eta]=F\frac{\left(\overline{h_0^2}\right)^{3/2}}{M}\alpha^3, \tag{1}$$

where $[\eta]$ - the intrinsic viscosity, F - Flory constant for a given polymer-solvent system, $\overline{h_0^2}$ – the mean-square distance between the ends of the chain, M – molecular weight, α – the coefficient of linear swelling.

For freely jointed chain:

$$h_0^2 = LA, \tag{2}$$

where L – contour length, A - length of statistical segment. According to a Flory formula [11], it is possible to write

$$\frac{[\eta]_r}{[\eta]_{DNA}}=\left(\frac{L_rA_r}{L_0A_0}\right)^{3/2}. \tag{3}$$

According to the intercalation model [1] when ligand bind with DNA $L_r=L_0(1+r)$.

For all investigated compounds at r=0.1 the relative change of $[\eta]$: $[\eta]_{0.1}/[\eta]_{DNA}$=1.25, which corresponds to intercalation mode of binding of ligand in these conditions [3].

The results of measurements of the viscosity dependence from concentration of DNA in solution at r = 0.01 are shown in figure 3.

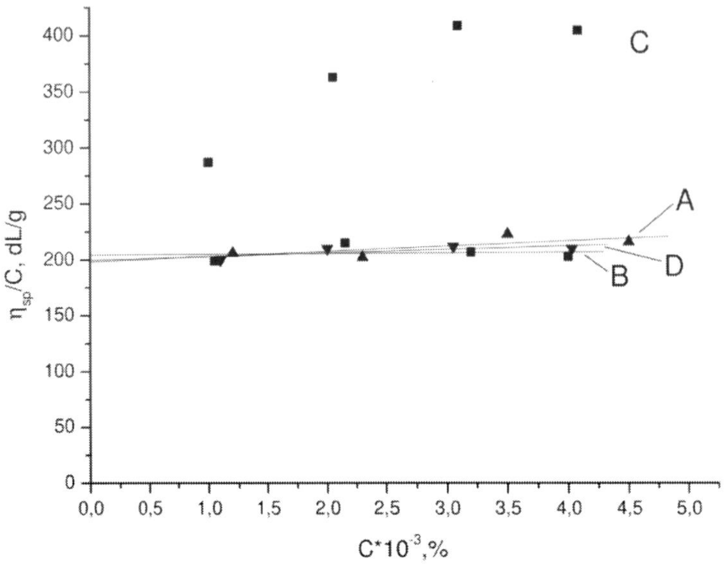

Figure 3. The dependence of the η_{sp}/C_{DNA} from concentration of DNA, μ=0.001, A - free DNA; B- complexes of compounds 1-6 with DNA, prepared under conditions of dilute solution, r=0.01; C- complexes of compounds 1,2 with DNA, prepared under conditions of semidilute solution, r=0.01; D- complexes of compounds 3-6 with DNA, prepared under conditions of semidilute solution, r=0.01

As figure 3 shows, there is no change between the viscosity of free DNA and viscosity of ligand-DNA complex while preparation of solutions of complexes of all compounds with DNA was under conditions of the dilute solution. However, while complexes were prepared under conditions of overlapping of macromolecular coils and then were diluted in 2 times an anomalous increase of the viscosity of the compounds 1, 2 was observed. In the case of compounds 3-6, the influence of the preparation conditions on the viscosity dependence of complexes from DNA concentration wasn't observed. Thus, we can see an anomalous increase in the viscosity of compounds containing diethylamino groups. Alkyl chain length doesn't affect the hydrodynamic behavior of the complex.

Similar studies of complexes at r = 0.01 were performed at ionic strength μ = 0.1. In this case we didn't observe any differences in viscosity of complexes compared to viscosity of free DNA, regardless of the preparation method of initial solutions of complexes.

Studies of viscosity dependence of complexes according to the velocity gradient showed that increasing gradient leads to disappearing of anomalous increase in viscosity.

We assumed that the observed anomalous behavior of solutions of DNA with actinocin derivatives related to the formation of intermolecular crosslinks in the process of preparation of the complex in terms of overlapping of macromolecular coils in solution. Saving these crosslinks upon dilution of the solution results in an increasing of specific viscosity due to the formation of large coiled aggregates composed of several DNA molecules. Their preservation in the measurement of viscosity due to low values of velocity gradients used in the gradient viscometer. Increasing of gradient to 0.5 s^{-1} leads to destruction of crosslinks and decreasing of viscosity.

Effect of ionic strength on the behavior of complex DNA with compounds 1, 2 shows that electrostatic interactions play an important role in the formation of crosslinks. We assumed that these interactions can exist between the protonated diethylamino groups of compound and phosphate groups of DNA.

At the same time, it was found that such crosslinks are formed in the case of compounds containing unsubstituted amino and dimethylamino groups. This points to the important role of hydrophobic interactions in the formation of intermolecular ligand-DNA crosslinks.

Difference in one CH_2 group (2 or 3) in the alkyl chain of the compound did not affect its ability to form a crosslink.

It was unexpected that complexes containing only very low amount of bound ligand ($r = 0.01$) have anomalous hydrodynamic behavior. While increasing this parameter to a value of 0.1 only intercalation complexes are formed. This indicates that the amount of ligand binding sites able to form crosslinks is very low, and the binding constant is high.

4.Conclusions

In this work an anomalous hydrodynamic behavior of DNA solutions with very low amount of actinocin derivatives, containing amide diethylamino groups and prepared in semidilute solution, was found. It has been suggested that this behavior is caused by the formation of intermolecular crosslinks. Study of the dependence of the hydrodynamic behavior from the ionic strength, structure of the compounds and velocity gradient has shown that this kind of interaction is electrostatic in nature, occurs with the participation of hydrophobic interactions and destroyed by increasing the velocity gradient.

References

[1] Lerman L S 1961 *J. Mol. Biol.* **3** 18
[2] Sobell H, Tsai C, Jain S and Gilbert S 1977 *J. Mol. Biol.* **114** 333
[3] Moroshkina E B, Krivtsova M A and Glibin E N 2002 *Biofizika(Rus)* **47** 444-448
[4] Frisman E V, Diakova E B, Golikova A I, Vorob'ev V I and Igamberdiev V M 1973 *Mol.Biol. (Rus)* **7** 745-752
[5] Birshtein T M, Elyashevich A M and Milenevsky A P 1973 *Biofizika(Rus)* **18** 804
[6] Evdokimov Yu M, Salyanov V I and Palumbo M 1999 *Sensornye sistemy (Rus)* **13** 82-91
[7] Mathukumar M 1997 *J. Chem. Phys.* **107** 2619-2635
[8] Seiffert S, Oppermann W 2008 *Polymer* **49** 4115-4126
[9] Popova E B, Plekhanova N G, Korshunova Z I and Glibin E N 1994 *Zhurnal Organicheskoi Khimii(Rus)* **30** 118-122
[10] Frisman E V, Shchagina L V and Vorob'ev V I 1965 *Kolloidn. J.* **27** 130
[11] Fox T and Flory P 1950 *J. Polymer Sci.* **6** 74

DNA interaction with cis- and trans- isomers of photosensitive surfactant

I N Unksov[1] and N A Kasyanenko[1]

[1] Faculty of Physics, St. Petersburg State University, Petrodvorets, Uluanovskaya,1 St. Petersburg, Russian Federation

E-mail: ivonijikt@gmail.com

Abstract. Interaction between DNA and photosensitive cationic surfactant in a solution is studied. Studies were conducted to examine the impact of the surfactant in its cis-conformation on the size of DNA molecule and also to investigate the phase behavior of the system depending on DNA and surfactant concentration. We conclude that trans- isomer of surfactant requires its smaller concentration to reach the DNA compaction compared with cis-isomer received by UV radiation of solutions. Studies of DNA-surfactant systems were performed by means of spectrophotometry and viscometry. Variation of surfactant concentration enables us to determine the precipitation zone on phase diagram. From the viscosity study it can be indicated that precipitation zone is narrower for UV-radiated surfactant and it shifts to higher surfactant concentration. Also we examine the reversibility of DNA compaction in systems with the surfactant in its trans- form.

1. Introduction

In general, cationic surfactants are believed to be effective in terms of DNA compaction because of their self-assembly into micellar-type aggregates on the "surface" of the DNA molecules. Since DNA is a negatively charged polyelectrolyte, electrostatic interactions between oppositely charged DNA phosphate groups and cationic "heads" of surfactants are very important for the formation of complexes [1, 2]. Complex formation also depends on the length of surfactant hydrophobic alkyl chain [2, 3, 4]. The process of the surfactant binding to DNA starts at the concentrations below the surfactant critical micelle concentration, CMC [1, 2]. The longer chain causes lower surfactant concentration for DNA packaging [4]. The changes of DNA molecule conformation can be controlled by adding compacting or decompacting agent into the solution. The reversible compaction can be achieved without direct interference in the solution contents. It is possible if the photosensitive or pH-sensitive cationic surfactant is used [1].

Interaction of photosensitive cationic azobenzene containing surfactant Azo(C6)TAB (figure 1) with DNA in a solution was studied. It is well-known that azobenzene containing surfactants can change conformation form trans- to cis- form under UV irradiation (353 nm) [1, 5]. Reversible cis-trans transition can be induced by visible light (453 nm) or can be realized as a relaxing in a dark. This conformational transition can be checked with UV-spectroscopy as cis- and trans- isomers have typical absorption spectra including bands out of DNA absorption region. The influence of surfactant conformation on the size of DNA molecular coil in a solution was investigated. The phase behavior of the system depending on DNA and surfactant concentration was also regarded.

2. Materials and methods

Azo(C6)TAB (chemical formula CH3(CH2)3-C6H5N=NC6H5-O(CH2)6N$^+$(CH3)3Br$^-$) surfactant we used was synthesized according to the procedure described in [6]. Calf thymus DNA (CT-DNA) with $M = 11 \times 10^6$ Da by Sigma was used. Studies of DNA-surfactant solutions in 0.005 NaCl systems were performed by means of spectrophotometry and viscometry. Absorption spectra for DNA-surfactant systems were registered in 2 mm or 5 mm cells depending on components concentration. During the spectral studies we used SF-56 UV/Vis spectrophotometer (LOMO, Russian Federation). Viscosity measurements were carried out on a low-gradient original construction rotary viscometer of Zimm-Crothers type [7].

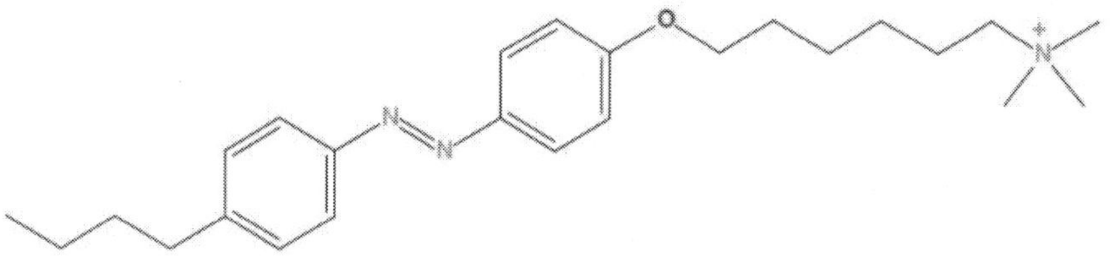

Figure 1. Structure of Azo(C6)TAB

3. Results and discussion

3.1. Viscosity studies

The relative viscosity of DNA solution $\eta_r = \eta / \eta_0$ (where η and η_0 are solution and solvent viscosities) at various surfactant concentrations C(Azo) was examined. Then the relative change of reduced viscosity ($\eta_r - 1$)/C(DNA) was plotted as a function of the ratio of surfactant concentration to DNA phosphate molar concentration z = C(Azo)/C(DNA). It is known that the change of reduced viscosity of DNA solution at constant C(DNA) reflects mainly change in the size of DNA molecular coil. Decrease in the viscosity of DNA solution in 0.005 M NaCl with z increase represents shrinkage of molecular coil induced by surfactant binding to DNA (at $z > 0.2$). DNA packaging with the viscosity of a solution similar to the solvent viscosity at a definite C(Azo) was observed at $z = 1$ (figure 2). DNA precipitation as visible flakes in a solution is realized at $z > 1$. Interestingly, at $z > 2$ the solution becomes transparent again and DNA nanoparticles formation was observed instead of DNA precipitation [1].

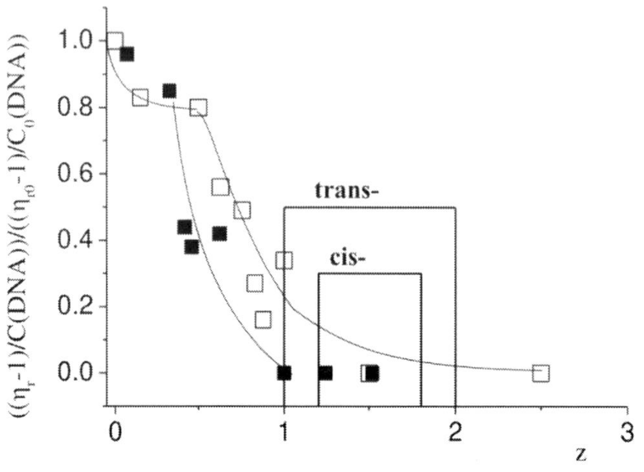

Figure 2. The relative change in DNA solution reduced viscosity with z value in 0,005 M NaCl for trans- (black signs) and cis- (open signs) isomers. The precipitation zones for trans- and cis- isomers are framed.

So the variation of surfactant concentration in a solution at constant DNA concentration enables us to determine phase separation in the system – the appearance of the precipitation zone (see figure 2). It can be indicated from the viscometric data that the precipitation zone becomes narrower for UV-irradiated surfactant (cis- isomer). This zone also shifts to higher surfactant concentration. We conclude on the base of experimental results that more hydrophobic trans- isomer of surfactant provides more effective DNA packaging [1, 8]. It causes DNA precipitation at smaller concentration (more precisely, z value) compared to cis- isomer (which was received by 10 minutes-UV irradiation of surfactant solution before mixing with DNA). For the same DNA and surfactant concentrations DNA precipitation was observed for cis- isomer at $z = 1.2$ up to $z = 1.8$ instead $z = 1$ and $z = 2$ for trans- isomer respectively.

3.2. Spectral studies

During the experiment we check light-induced cis-trans isomerization with UV absorption spectra (figure 3). One can see the difference in absorption spectra for cis- and trans- isomers of surfactant in 0.005 M NaCl. We change surfactant conformation by 10 minutes UV irradiation at 353 nm for trans-to-cis isomerization (figure 3 spectra a and c). After that we use visible light illumination for cis-to-trans isomerization (b and d spectra). Two different surfactant solutions in 0.005 M NaCl (with C(Azo) = 2×10^{-5} M below CMC (spectra a and b) and with C(Azo) = 3×10^{-4} M above CMC (spectra c and d) were used. The spectrum of trans-isomer consists of two resolved bands with the shoulder at $\lambda > 400$ nm. After trans-to-cis isomerization the main maximum shifts to the blue region with visible hypochromic effect, the short-wave band becomes wider and smaller and long-wave shoulder transforms into third band (spectra a and c). Visible light induces the returning of cis-conformation into trans- state again (spectra b and d). It is clear from figure 3 that at C(Azo) > CMC cis-to-trans transition also occurs, and micelle state for trans-isomers can be formed. Indeed, the most intensive band for trans-isomer has the tendency to the appearance of the second maximum at 345 nm typical for micelle with usual non-micellar maximum at 353 nm. This band for cis- isomer is not symmetric with the maximum at 325 nm. For C(Azo) < CMC the central band has one maximum at 353 nm for trans- and one maximum at 320 nm for cis- isomer. Nevertheless, surfactant state (micellar or non-micellar) in a solution does not influence significantly on trans-to-cis and cis-to-trans transitions.

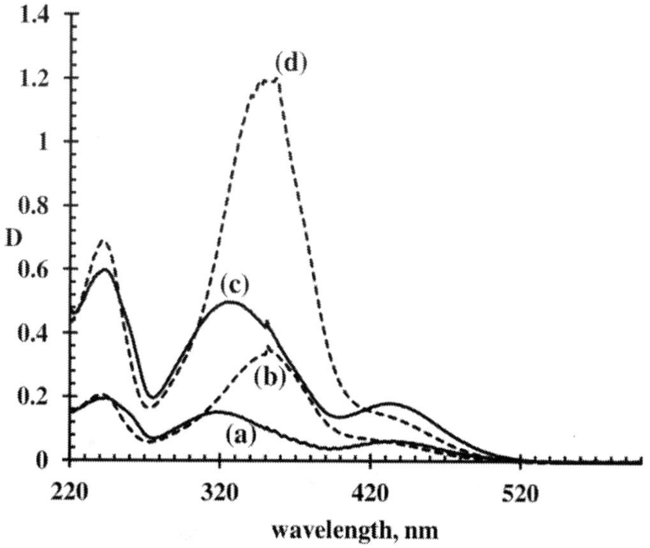

Figure 3. Surfactant spectra after UV (a, c) and further visible (b, d) illumination for C(Azo) = 2×10^{-5} M (a, b) and C(Azo) = 3×10^{-4} M (c, d).

After the mixing of DNA and surfactant solutions (for all systems C(Azo) < CMC) the spectrum of surfactant in trans- conformation out of DNA absorption band is similar to observed for micellar state. It was shown earlier that DNA as a template organizes micelle-like structure for bonded surfactants. After UV irradiation trans-to-cis transition is realized as it follows from spectral changes (figure 4a). The surfactant in these systems (spectra b', c', d') was in cis- form as the systems were UV light illuminated (dashed lines). The precipitation does not exist for these systems. Two of these systems with surfactant in its trans- and cis- forms and its concentration C(Azo) = 2×10^{-5} M compared to the corresponding DNA-free surfactant solutions of the same C(Azo) are shown in figure 4b.

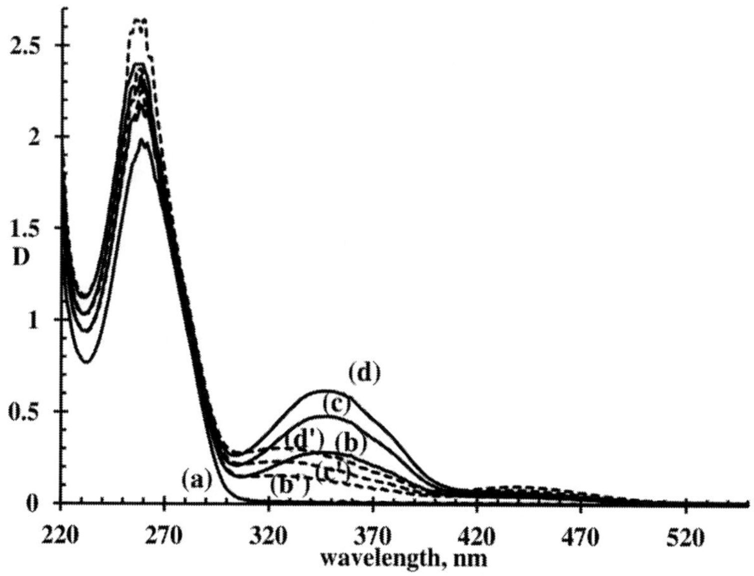

Figure 4a. Spectra for DNA-surfactant systems with Z < 0.5: DNA control (a), z = 0.09 C(Azo) = 2×10^{-5} M before UV illumination (the surfactant in trans- form) (b), z = 0.13 C(Azo) = 2.8×10^{-5} M before UV illumination (c), z = 0.19 C(Azo) = 4×10^{-5} M before UV illumination (d), z = 0.09 C(Azo) = 2×10^{-5} M after UV illumination (the surfactant in cis- form) (a'), z = 0.13 C(Azo) = 2.8×10^{-5} M after UV illumination (b'), z = 0.13 C(Azo) = 2.8×10^{-5} M after UV illumination (c'), z = 0.19 C(Azo) = 4×10^{-5} M after UV illumination (d').

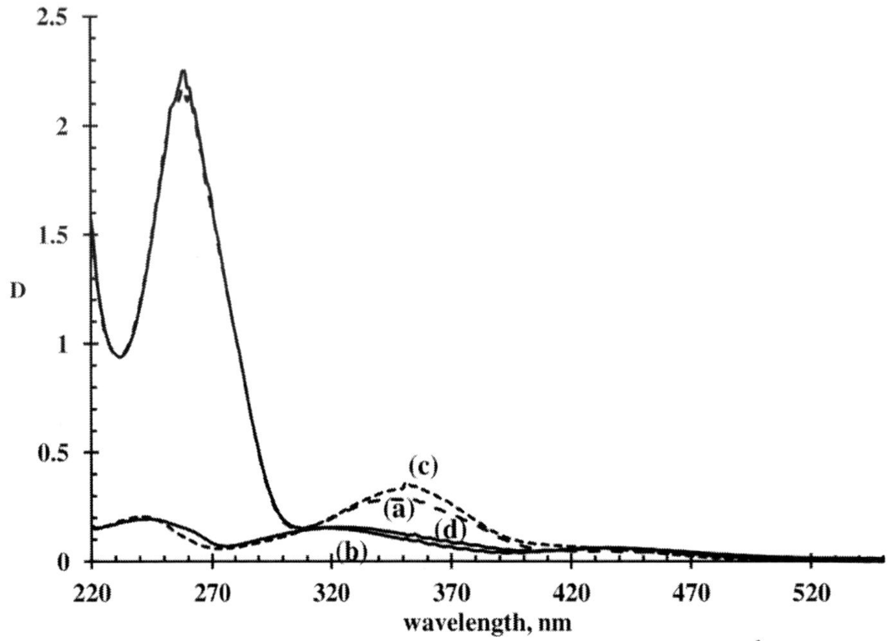

Figure 4b. Spectra for DNA-surfactant systems z = 0.09 C(Azo) = 2×10^{-5} M with the surfactant in trans- form (a, dashed line) and in cis- form (b) and for surfactant solutions C(Azo) = 2×10^{-5} M in trans- (c, dashed line) and cis- form (d).

3.3. Reversibility of transition

The reversibility of surfactant-induced DNA compaction in a solution also has been examined. The system with DNA C(DNA) = 0.01 % (phosphate molar concentration 3×10^{-4} M) after shrinkage and surfactant in trans- form at z = 0.8 was diluted with 0.005 M NaCl (scheme 1). The decompaction of DNA was indicated by the viscometric experiment (see table 1) via the comparison of examined system with the control solutions. Systems of z = 0.8 also can be compared with the system z = 0.4. It was shown earlier that the precipitation zone depends on C(DNA) [9] so the system with C(DNA) = 0.01 % at z = 0.8 is closed to the border of this zone without any turbidity in solution. The tendency to the reversibility can be manifested for some of the systems.

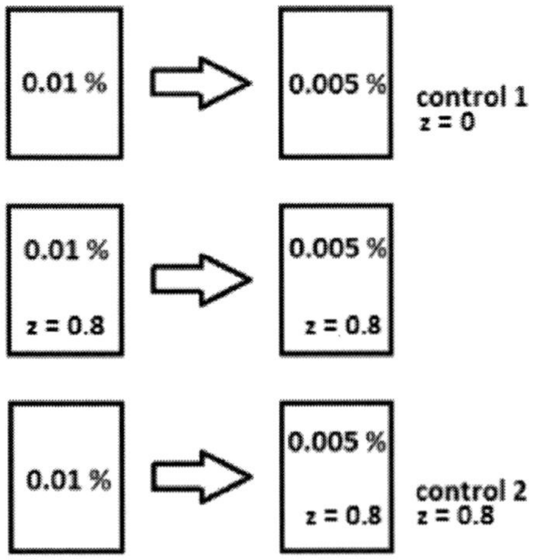

Scheme 1. The preparation of systems for the viscometric experiment. Stock DNA solution with C(DNA) = 0.01 % (3×10^{-4} M) was diluted with 0.005 M NaCl (control 1) and with surfactant (control 2). DNA-surfactant solution after DNA shrinkage (z = 0.8) was diluted with 0.005 M NaCl for the measurements and checking the DNA decompaction. System z = 0.4 was also measured for comparison.

Table 1. The results of viscometry for DNA-surfactant solutions according to scheme 1.

Syst. No.	C(DNA)	z	Definition	Reduced viscosity of DNA solutions (dl/g)
1	0.01 %	0	initial	105±5
2	0.005 %	0	control 1	90±5
3	0.01 %	0.8	DNA shrinkage	20±5
4	0.005 %	0.8	control 2	74±5
5	0.005 %	0.8	checked	76±5
6	0.005 %	0.4	for comparison	84±5

4. Conclusion

It can be concluded that cis- isomers cause DNA shrinkage and DNA precipitation at higher z value than trans- isomers. This transition is reversible according to change in z value. The formation of DNA-surfactant complexes does not prevent cis-trans-cis isomerization in a solution.

Acknowledgements
This work is partly supported by RFBR grant 13-03-01192 and Saint Petersburg State University financial support (grant 11.38.644.2013).

References

[1] Dias R S and Lindman B, editors 2008 *DNA interactions with polymers and surfactants* (Hoboken, NJ: John Wiley & Sons, Inc.)
[2] Dias R, Rosa M, Pais A C, Miguel M and Lindman B 2004 *J. Chin. Chem. Soc.* **51** 447–69
[3] Chatterjee R and Chattoraj D K 1979 *Biopolymers* **18** 147–66
[4] Hayakawa K, Santerre J P and Kwak J C T 1983 *Biophys. Chem.* **17** 175–81
[5] Le Ny A L M and Lee C T 2006 *J. Am. Chem. Soc.* **128** 6400–08
[6] Dumont D, Galstian T, Senkow S and Ritcey A 2002 *Mol. Cryst. Liq. Cryst.* **341** 375
[7] Frisman E V, Schagina L V and Vorobiev V I 1965 *Biorheology* **2** 189-94
[8] Diguet A, Mani N K, Geoffroy M, Sollogoub M and Baigl D 2010 *Chem. Eur. J.* **16** 11890–96
[9] Zakrevskyy Y, Kopyshev A, Lomadze N, Morozova E, Lysyakova L, Kasyanenko N and Santer S 2011 *Phys. Rev.* E **84** 021909

Journal of Physics: Conference Series **572** (2014) 012015 doi:10.1088/1742-6596/572/1/012015

Study of low-frequency dynamics of short peptides by Brillouin light scattering and Monte-Carlo global energy minimization

V E Vashchenkov[1,2], A I Fedoseev[1,2], M G Petukhov[3,4], V Kh Khavinson[5], E V Savvateeva-Popova[2,5] and S G Lushnikov[1,2]
[1] Ioffe Physical Technical Institute, St.Petersburg, Russia
[2] St.Petersburg State University, St.Petersburg, Russia
[3] St.Petersburg State Polytechnical University, St.Petersburg, Russia
[4] Petersburg Institute of Nuclear Physics, Gatchina, Leningrad district, Russia
[5] Pavlov Institute of Physiology , St.Petersburg, Russia
Email: vikvas6@ya.ru

Abstract. Brillouin light scattering studies of the behavior of hypersound velocity in solutions of short peptides Asp-Ser, Glu-Asp-Arg, Ala-Glu-Asp-Leu with widely varying concentrations at temperatures ranging from 295 to 360K are presented. As shown by analysis of experimental data and molecular simulation, the dipeptide Asp-Ser is characterized by formation of dimers which decompose at high temperatures. High-concentration solutions of tri- and tetrapeptides form aggregates at high temperatures which are preserved on cooling.

1. Introduction

The structure and dynamics of proteins and other biopolymers attract considerable attention. It has been shown that changes in structures and dynamics of proteins, DNA, and polysaccharides can lead to changes in their activity and function, which can result in different diseases. For example, it has been shown that one of the mechanisms responsible for neurodegenerative diseases is the conformational dynamics of proteins and/or its violation [1].

Phase transformations in biopolymers are intensively studied by Brillouin light scattering that gives information about dynamic processes at frequencies from 1 to 1000 GHz [2]. Brillouin scattering was successfully used for studies of thermal denaturation of lysozyme and DNA melting. On heating, the phase termed as the molten globula state, was found in lysozyme, and the subsequent temperature increase up to denaturation lead to formation of aggregates, organized in a gel-like structure [3, 4]. The conformational dynamics leading to melting of the DNA double helix was observed in Brillouin studies of DNA melting [5]. Thus, Brillouin light scattering is a highly informative technique for the investigations of phase transformations in biopolymers.

This paper deals with Brillouin light scattering studies of the low-frequency dynamics of short peptides. Peptides control the majority of processes in human organism, for instance, diverse functions in the central nervous system. The deficiency of peptides can accelerate ageing.

Peptides are natural or synthetic compounds build from amino acid residues connected by peptide (amide) bonds C(O)-NH. According to the number of amino acid residues they are classified as dipeptides, tripeptides, tetrapeptides, etc. The peptides containing up to 10 amino acids are called oligopeptides, more than 10 are called polypeptides. Though the short peptides used in our studies are very effective in clinics [6, 7], the mechanisms of their functioning, the structure and dynamics of peptides in solutions are still poorly studied.

2. Materials and Methods

Three peptides, i.e., dipeptide AD7 (Asp-Ser), tripeptide pinealon (Glu-Asp-Arg) and tetrapeptide bronchogen (Ala-Glu-Asp-Leu), were used. The peptides were synthesized at the Institute of Bioregulation and Gerontology (St.Petersburg, Russia) Peptide concentrations were analysed in the range from 0.05 mg/l to 10 mg/l. Peptides were dissolved in the sodium-phosphate buffer (pH = 7.0 for AD7, pH = 7.45 for bronchogen and pinealon).

Brillouin light scattering experiments were performed using a 3-pass Fabry-Perrot interferometer. The light source was an argon laser with wavelength λ=488 nm and power 100 mW. The laser beam was transmitted through the optical scheme and focused in the sample. The 180^0 light scattering geometry was employed. The distance between the interferometer plates was 7.5 mm corresponding to a free spectral interval (FSR) of 23 GHz. The time for spectrum recording was 300 s.

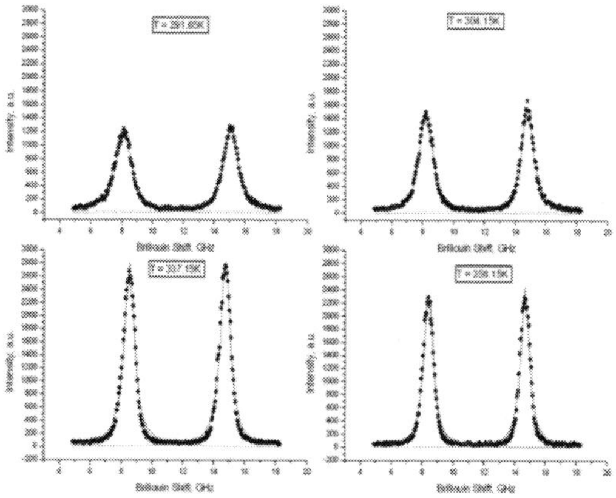

Figure 1. Brillouin spectra obtained for the solution of bronchogen (C = 5 mg/l) at different temperatures

Temperature of solutions was increased from 293 to 364 K with an accuracy of ±0.1°K. Examples of experimental Brillouin spectra obtained for the solution of bronchogen at different temperatures are presented in Fig.1. The dashed line is fitting. To describe the Brillouin doublet, the Lorentz function was used.

Models of spatial structures of DS dipeptide complexes were obtained by the Monte-Carlo global energy minimization as implemented in the ICM-Pro software package for molecular modeling of proteins and their complexes with ligands (Molsoft LLC, USA) [8]. All calculations were done in the ECEPP/3 force field with the all atom approximation [9]. All standard torsion angles (φ, ψ, ω and χ_i) of amino acids were allowed to vary during the Monte-Carlo energy minimization. The ICM default energy parameters including van-der-Waals, electrostatic, torsion energy interactions, hydrogen bonding and solvation energy term were used in the calculations. Free energy of peptide hydration was considered basing on calculations of peptide solvent accessible surface areas and continuum approximation model for protein solvent interactions [10].

3. Results and discussion

As follows from Fig.1, the position, i.e., frequency Δν, and the halfwidth at half-maximum (HWHM) of the Brillouin components change with increasing temperature. By using expression

$$\pm \Delta\omega/\omega = \pm\Omega/\omega = \pm\Delta\nu/\nu = 2n\left(\upsilon/c\right)\sin\left(\theta/2\right)$$

where $\upsilon = \omega/2\pi c$, θ is the scattering angle, and n is the refractive index, the hypersound velocity in the solution is found. We discuss the behavior of the Brillouin shift assuming that changes in the refractive index with varying temperature are small. The behavior of the Brillouin shift is proportional to the hypersound velocity.

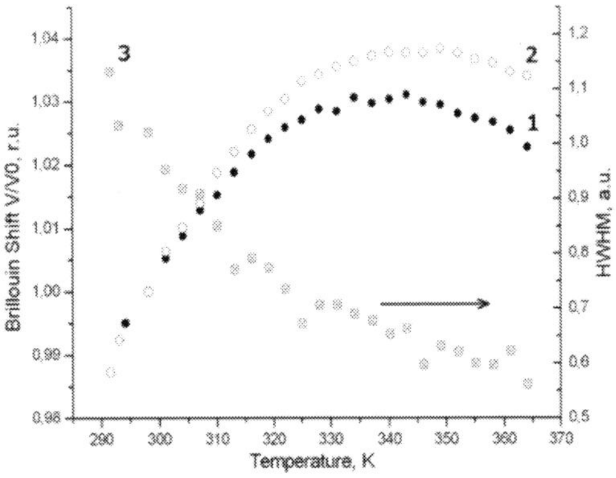

Figure 2. Temperature dependence of attenuation (3) and relative changes in the Brillouin shifts for bronchogen (2) and buffer (1)

The experiments and calculations yielded temperature dependences of the shift and HWHM of the Brillouin components for the buffer solution and for solutions of peptides with different concentrations. Fig.2 shows relative changes in the Brillouin shift for the sodium-phosphate buffer and for the solution of bronchogen with C=5 mg/l. As a basic for normalization the velocity at T=298.15 K was used.

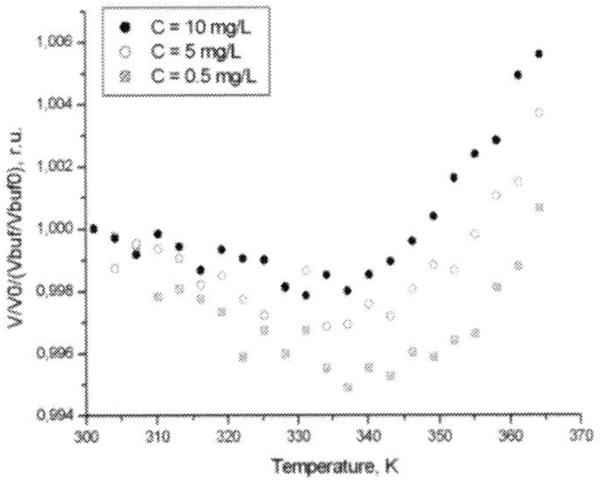

Figure 3. Temperature dependence of hypersound velocity for the solution of dipeptide AD7 normalized to the buffer

It is evident from the Fig.2 that the hypersound velocity in the sodium-phosphate buffer monotonically grows as temperature increases to 340 K. On further heating the hypersound velocity decreases (Fig.2). Interestingly, the relative change in the hypersound velocity in bronchogen (curve 2, Fig. 2) is similar to that of the buffer up to 310 K. Further temperature increase leads to a deviation of the curve 2 from the curve 1 which points to a peptide dynamics *per se* contribution to the temperature dependence of velocity. Similar behavior was observed for lysozyme solutions [3]. The temperature dependence of hypersound velocity in bronchogen has no anomalies, indicating that this peptide does not show phase transformations. A similar behavior – without anomalies and with a monotonic decrease with increasing temperature is demonstrated by the HWHM.

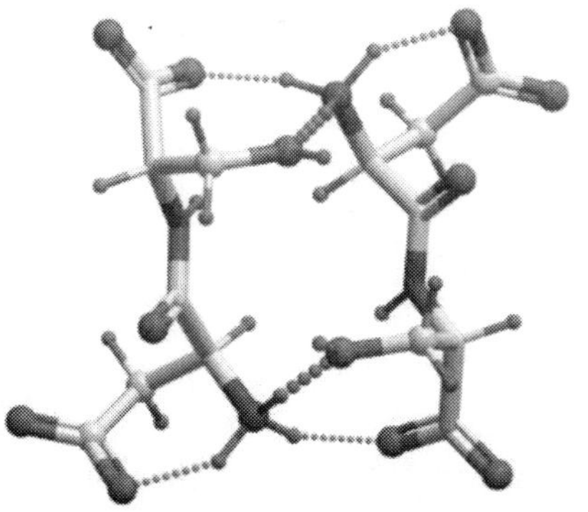

Figure 4. Dipeptide dimer.

To compare the behavior of hypersound velocity upon temperature variations in solutions with different peptide concentrations, the data were additionally processed. Each dependence was normalized to its magnitude at 305 K and then the normalized dependences were divided by the normalized dependence for the buffer (Fig.3, 5). As a result, we obtained the plot that showed the dynamics of given peptide. The normalized temperature dependence of hypersound velocity for the solution of dipeptide AD7 differs from that for the buffer (Fig.3): near 335-345 K there is a minimum in the velocity observed for all the concentrations studied. A further increase in temperature leads to a monotonic increase in the normalized hypersound velocity. Numerical analysis of possible configurations of dipeptides has shown that the energetically favorable situation is when AD7 exists in the solution in the form of dimers. Indeed, the lowest energy conformation obtained in the global energy minimization is a densely packed symmetrical complex with a head-to-legs arrangement of peptide main and side chains. In this conformation peptide alpha-amino-group forms multiple hydrogen bonds with C-terminal COO- group, as well as with hydroxyl group of Ser side chain. Hydrogen bond between the side chain of Asp- with peptide alpha-amino-group provides additional stabilization for the conformation of dipeptide complex. Nevertheless, other polar groups in the peptide main chain not involved in the intermolecular hydrogen bonding remain fully accessible for solvent hydration. All these observations make the lowest energy conformation a likely candidate for the DS dipeptide complex.

It should be noted that dimers are formed in the initial solution. It can be supposed that the anomalies observed in the hypersound velocity in the vicinity of 340 K are attributable to the decomposition of dimers.

The temperature dependences of the hypersound velocity normalized to the buffer for bronchogen and pinealon are similar. However, they markedly differ from that of AD7: the hypersound velocity for these peptides monotonically increases with increasing temperature (Fig. 5) at all concentrations. Calculations show that formation of the structures consisting of two and more molecules in these compounds is energetically unfavorable. The absence of anomalies on heating is in agreement with the calculations.

Let us consider the behavior of hypersound velocity on heating and cooling for different concentrations of bronchogen and pinealon. At high concentrations, the hypersound velocity behaves differently on heating and cooling (Fig. 6 A, C) below 350 K (Fig. 6).

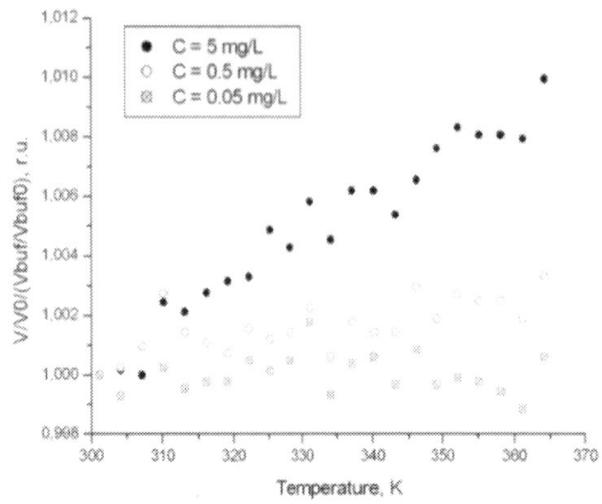

Figure 5. Temperature dependence of hypersound velocity for the solution of bronchogen normalized to the buffer

Suggestively, a structural change in the spatial organization of peptides occurs at 350 K. The differences in the behavior of hypersound velocity on heating and cooling for high concentrations of peptides can arise due to formation of aggregates between peptide molecules at decreasing temperature. For lower concentrations of pinealon and bronchogen, no differences in the behavior of hypersound velocity on heating and cooling are observed (Fig.6 B, D). This indicates a reversible character of conformational rearrangements of these peptides when no aggregates are formed.

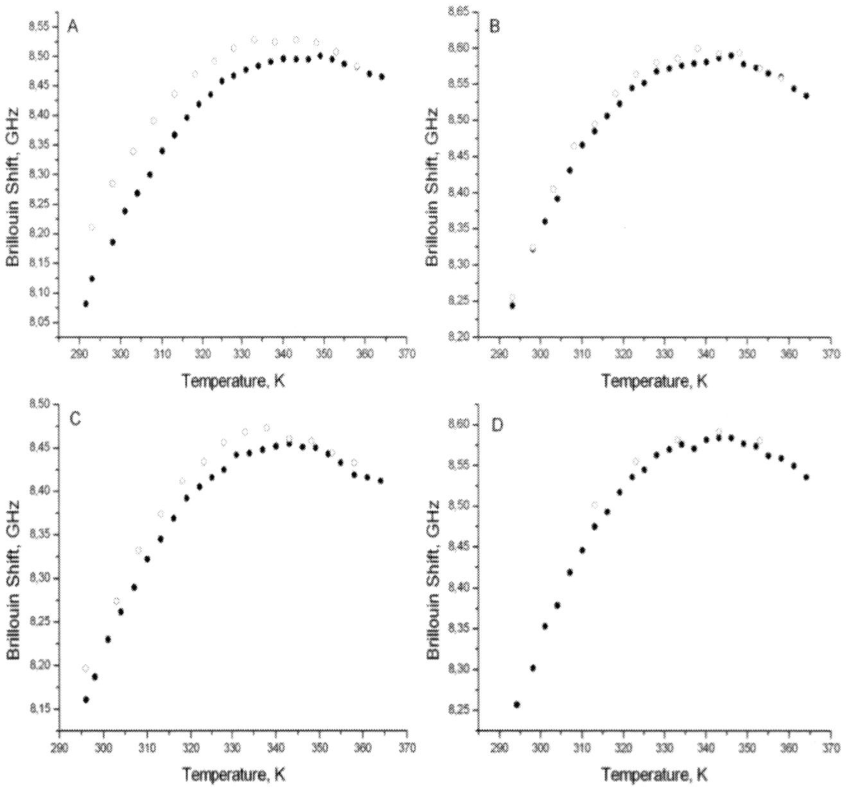

Figure 6. Temperature dependence of Brillouin shift for bronchogen in concentrations 5 mg/l (A) and 0.5 mg/l (B) and for pinealon in concentrations 5 mg/l (C) and 0.5 mg/l (D) at cooling.

Journal of Physics: Conference Series **572** (2014) 012015 doi:10.1088/1742-6596/572/1/012015

4. Conclusions

Studies of the hypersound velocity behavior upon varying temperature in solutions of short peptides with different concentrations demonstrated that the peptide dynamics depended both on their concentration and structure. Dipeptide AD7 in the solution at concentrations studied was capable of self- organization into dimeric structures which decomposed in the vicinity of 345 K. Tri- and tetrapeptides did not form sufficiently stable complexes on heating and their behavior was governed by their equilibrium dynamics. The molecules of pinealon and bronchogen in high-concentration solutions were likely to form aggregates n cooling down from the high-temperature region. This was not observed in low-concentration solutions of pinealon and bronchogen.

Acknowledgments

The work was supported by the RFBR grant (No 12-04-01737-a), Programs of Presidium of RAS (No 7) and Department of Scientific Research of St.Petersburg State University "Biomedicine and Human Health".

References

[1] Frost B and Diamond M I 2010 *Nature Reviews Neuroscience* **11** 155

[2] Vaughan J M 1989 *The Fabry-Perot Interferometer* (IOP Publishing Ltd)

[3] Svanidze A V, Lushnikov S G and Seiji Kojima 2009 *JETP Lett.* **90** 80

[4] Svanidze A V, Lushnikov S G, Romanov V P 2011 *JETP Lett.* **93** 409

[5] Lushnikov S G, A V Dmitriev, Fedoseev A I, Zakharov G A, Zhuravlev A V, Medvedeva A V, Schegolev B F, Savvateeva-Popova E V 2013 *JETP Lett.* **98** 735

[6] Khavinson V Kh, Lin'kova N S 2012 *Human Physiology* **38** 101

[7] Khavinson V Kh, Lin'kova N S, Dudkov A V, Polyakova V O, Kvetnoi I M 2012 *Bulletin of Experimental Biology and Medicine* **152** 615

[8] Abagyan R A and Totrov M M 1994 *J. Mol. Biol.* **235** 983

[9] Nemethy G, Pottle M S and Scheraga H A 1983 *J. Phys. Chem.* **87** 1883.

[10] Wesson L and Eisenberg D 1992 *Protein Sci.* **1** 227

Study of hepatocyte plasma membrane mechanical properties using optical trapping

A D Vedyaykin, N E Morozova , G E Pobegalov, A N Arseniev, M A Khodorkoskii and A V Sabantsev

St. Petersburg State Polytechnical University, Institute of Nanobiotechnologies, St. Petersburg, Russia

E-mail: god-fish@mail.ru

Abstract. In this paper we describe the use of membrane tether formation technique which is widely used to study mechanical properties of plasma membranes. This method was successfully used for the direct measurement of parameters characterizing membranes mechanical properties (static tether tension force and effective membrane viscosity) of human hepatocytes (HepG2 hepatocellular carcinoma line). These results allow using this method in future for diagnostics of the cell membrane, evaluating the influence on the mechanical parameters of various factors, including toxins and drugs.

1. Introduction

Mechanical properties of the cell membrane play an important role in the homeostasis of cell due to the influence on a number of membrane-bound enzymes and receptor systems [1]. Properties of the cell membrane of hepatocytes are closely related to the functioning of this type of cells [2].

Published data indicates that previous studies of hepatocyte membrane mechanics utilized bulk indirect methods as electron paramagnetic resonance and fluorescence depolarization [1, 2]. Only relative changes in membrane fluidity but not the absolute values of membrane mechanical characteristics can be obtained using these methods. Nevertheless, the literature describes relatively wide use of membrane tether formation technique for the study the mechanical properties of various cell lines[3-6] using optical trapping (OT) [7]. Thus we used this method to investigate hepatocytes membrane mechanical properties at the single cell level. The experiments were performed on human cell line HepG2.

The first membrane tether formation experiment using optical trapping was described in [6], where the mechanical properties of the membrane of neuronal growth cone were studied. Later the method described in this paper was used to investigate the properties of various eukaryotic and prokaryotic cells [8, 9].

The goal of our work was to measure parameters characterizing membrane mechanical properties of hepatocytes using membrane tether formation method.

2. Materials and methods

2.1. The experimental setup

Optical set-up was previously described elsewhere [10, 11]. Briefly, the optical trap is created by the 1064 nm infrared laser Spectra-Physics BL-106C, focused by the oil immersion objective with a numerical aperture of 1.46 "Plan-Apochromat" 100x/1.46 Oil DIC (Carl Zeiss). Piezo stage P-561 3DD (PI) is used to move the sample relative to the optical trap. Optical set-up is based on AxioImager.Z1 (Carl Zeiss) microscope. . Images are captured using EM-CCD cameras Cascade II 1024 (Photometrics) and iXon3 897 (Andor).

2.2. Cell culture preparation

Cells are grown on 18*18 mm coverslips coated with poly-L-lysine (Sigma). 25 μl of poly-L-lysine (0.01% w/v) water solution is applied to each two coverslips and incubated for 30 min at room temperature. After that, HepG2 culture, grown in 60 mm culture dish (Greiner), is harvested by trypsin addition, transferred to polyl-L-ysine coated coverslips and grown in DMEM medium supplemented with 10% fetal bovine serum (Gibco), 2 mM L-glutamine, and gentamicin at 20 μg/ml at 37 ° C in atmosphere of 5% CO_2.

2.3. The course of the experiment

The coverslips with cells are attached to microscope slides using double-sided tape (Scotch), forming a channel of about 10 mm in width and 130 μm in depth between the slide and the coverslip. Next, the channel is filled with polystyrene microspheres (3 μm in diameter, Bangs Laboratories) diluted in DMEM. During the experiment the sample temperature is maintained at 37 ° C by objective heating ring.

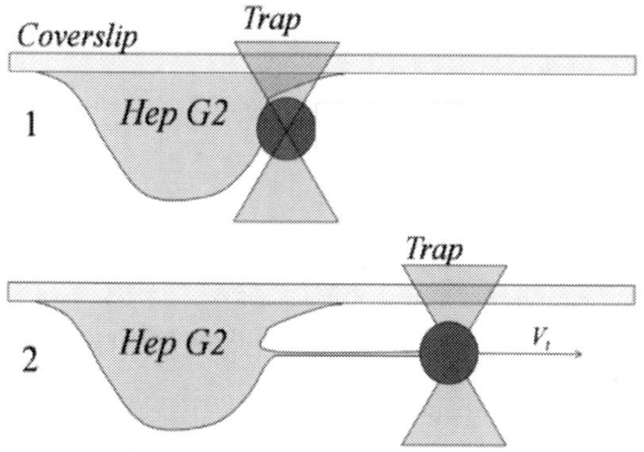

Figure 1. Schematic of the experiment on the membrane tether formation of cells HepG2: 1 – Microsphere fastening on the cell membrane. 2 - The formation of tether. V_t – tether elongation speed.

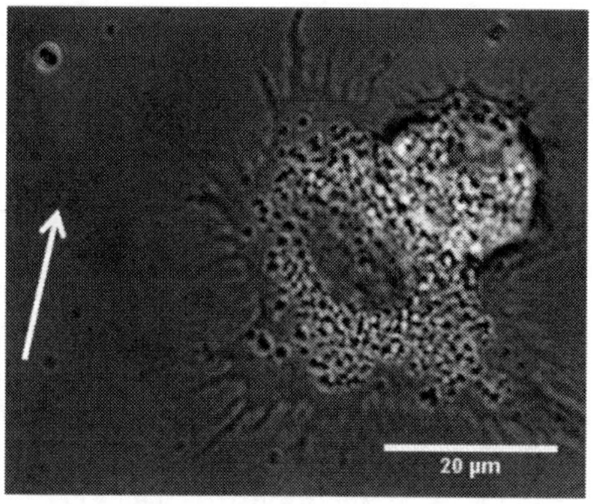

Figure 2. Image of membrane tubular structure ("tether") in the mode of differential interference contrast (DIC) (marked by a white arrow).

Microsphere is captured by the optical trap and is brought into contact with the cell membrane (see figure 1). Approximately 10 seconds later, the microsphere is moved at a constant speed in direction away from the cell surface using the piezo stage, as a result the membrane tether is formed. During this process, image recording is being performed using a CCD-camera. After predetermined displacement of stage, motion is stoped, but images are still being recorded (for detection of membrane relaxation processes). Finally, the OT is turned off and camera records microsphere retraction to the cell surface.

At the end of the experiment force calibration is performed using viscous drag force. For this purpose the sample (and water solution inside it) is repeatedly moved in opposite directions relative to trapped microsphere using the piezostage. The microsphere is displaced from the center of the trap to the point where force of viscous friction acting on the microsphere is compensated by the trapping force. Trajectory of microsphere is recorded using a CCD-camera, and the relationship between bead displacement and viscous drag force is fitted with a line to determine the trap stiffness.

2.4. Measured parameters and data processing

During the experiment tether tension force f is directly measured (by bead displacement from trap focus). It was shown [12] that in some cases (including our) the tension force of the membrane tether does not depend on its length:

$$f = f_0 + 2\pi\eta_{eff}v_t \tag{1}$$

where f_0 - stationary tension force of membrane tether; v_t- the rate of tether formation; η_{eff}– effective viscosity of membrane.

Sequential measurements of tension force at various speeds provides static tether tension force, that is connected to the membrane tension:

$$f_0 = 4\pi R_t(T + \gamma) \tag{2}$$

where T - in-plane tension of the membrane; γ– membrane-cytoskeleton adhesion term; R_t- radius of tether.

The measurement of tether radius R_t allows to get the membrane tension $(T + \gamma)$. However, radius measuring rather complicates the experiment, so we follow the authors of the earlier works - static tether tension force instead of membrane tension is indicated in our work.

The second parameter is effective membrane viscosity η_{eff} that consists of three terms:

$$\eta_{eff} = 2\eta_m + \eta_{si}h^2 \ln\frac{R_0}{R_t} + \eta_{sc}R_t^2 \ln\frac{R_0}{R_t} \tag{3}$$

where η_m - viscosity, determined by the motion of the membrane as a whole; η_{si} - viscosity, determined by the relative motion of monolayers; η_{sc} - viscosity due to the shift of the membrane relative to the cytoskeleton; h - thickness of the membrane; R_0 - the characteristic radius.

The experimental data were processed using the software package ImageJ [13] (Fiji distribution [14]). Microsphere positions are determined as centers-of-mass of inverted and high-pass filtered bead images.

3. Results and Discussion

Uncoated polystyrene microspheres 3 μm in diameter attached strongly to the cell membrane after 10 seconds and formed membrane tethers when pulled away from the cell surface by the optical trap (figure 2). Membrane tethers appear as lineal low-contrast objects when imaged by the microscope in differential-interference contrast mode. When the bead is released from the trap tether retracts and pulls the bead to the cell surface at up to 30 μm/sec.

During membrane tether extraction first the force acting on the bead grows rapidly until it reaches its peak value which corresponds to the moment of membrane tether formation (figure 3). After that it drops to the intermediate value (called plateau force) and remains roughly constant until the end of tether elongation. For the tether in figure 3 pulled at 1μm/sec peak force is about 60 pN and the plateau force is about 42 pN. As the elongation speed increases the plateau force increases too. Linear regression of this dependence yields effective membrane viscosity η_{eff} = 0,27 ± 0,07 pN·s/μm and static tether tension force f_0 = 40±3 pN [11]. Using published data on tether radius (about 200 nm) [12] we can estimate membrane tension $T + \gamma \sim 10$ pN/μm or 0.01 mN/m. Tether radius measurement should give us more precise value of membrane tension.

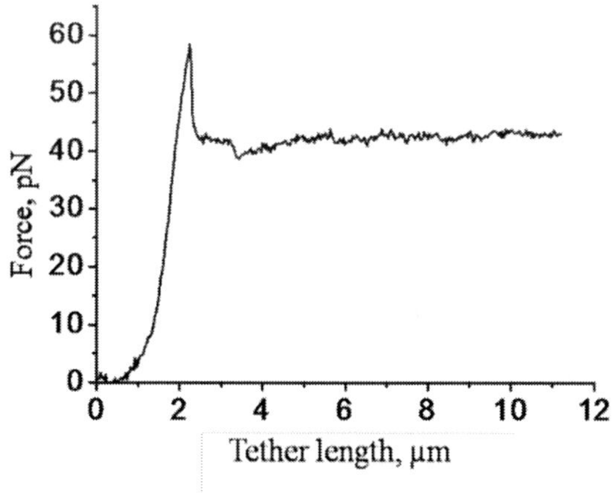

Figure 3. Tether tension force as a function of bead displacement from the cell surface when the bead is moved at a constant velocity of 1 μm/sec.

Experiments with actin-disrupting agents, such as cytochalasin D, will allow for the determination of cytoskeleton terms of effective membrane viscosity and membrane tension.

4. Conclusion
Membrane tether extraction is a powerful biophysical method that can be successfully used to study hepatocyte membrane mechanics. It allows for the determination of absolute membrane mechanical parameters values and can be used to assess cell-to-cell variability of these parameters. We are planning to analyze the cytoskeleton contribution to the measured parameters by disrupting F-actin. Also the effect of ethanol and other compounds on hepatocyte membrane mechanics will be tested.

5. References
[1] Schuller A, Solis-Herruzo J A, Moscat J, Fernandez-Checa J C and Municio A M 1986 The fluidity of liver plasma membranes from patients with different types of liver injury *Hepatology* **6** 714-7
[2] Schachter D 1984 Fluidity and function of hepatocyte plasma membranes *Hepatology* **4** 140-51
[3] Chen Y, Yao D K and Shao J Y 2010 The constitutive equation for membrane tether extraction *Ann Biomed Eng* **38** 3756-65
[4] Dai J and Sheetz M P 1999 Membrane tether formation from blebbing cells *Biophys J* **77** 3363-70
[5] Schumacher K R, Popel A S, Anvari B, Brownell W E and Spector A A 2009 Computational analysis of the tether-pulling experiment to probe plasma membrane-cytoskeleton interaction in cells *Phys Rev E Stat Nonlin Soft Matter Phys* **80** 041905
[6] Dai J and Sheetz M P 1995 Mechanical properties of neuronal growth cone membranes studied by tether formation with laser optical tweezers *Biophys J* **68** 988-96
[7] Ashkin A, Dziedzic J M, Bjorkholm J E and Chu S 1986 Observation of a single-beam gradient force optical trap for dielectric particles *Opt Lett* **11** 288

[8] Jauffred L, Callisen T H and Oddershede L B 2007 Visco-elastic membrane tethers extracted from Escherichia coli by optical tweezers *Biophys J* **93** 4068-75
[9] Li Z, Anvari B, Takashima M, Brecht P, Torres J H and Brownell W E 2002 Membrane tether formation from outer hair cells with optical tweezers *Biophys J* **82** 1386-95
[10] Pobegalov G E, Arseniev A N, Vedyaykin A D, Sokolova M L, Fedorova Y V and Sabantsev A V 2013 DNA molecule mechanical properties during interaction with YOYO-1 intercalating dye *St. Petersburg State Pol. Univ. J. Phys. and Math.* **1(165)** 136-43
[11] Morozova N E, Vedyaykin A D, Sabantsev A V, Pobegalov G E, Murashov S V and Khodorkovskiy M A 2013 Membrane tether formation from human hepatocytes *St. Petersburg State Pol. Univ. J. Phys. and Math.* **3(177)** 167-76
[12] Hochmuth F M, Shao J Y, Dai J and Sheetz M P 1996 Deformation and flow of membrane into tethers extracted from neuronal growth cones *Biophys J* **70** 358-69
[13] Schneider C A, Rasband W S and Eliceiri K W 2012 NIH Image to ImageJ: 25 years of image analysis *Nat Methods* **9** 671-5
[14] Schindelin J, Arganda-Carreras I, Frise E, Kaynig V, Longair M, Pietzsch T, Preibisch S, Rueden C, Saalfeld S, Schmid B, Tinevez J Y, White D J, Hartenstein V, Eliceiri K, Tomancak P and Cardona A 2012 Fiji: an open-source platform for biological-image analysis *Nat Methods* **9** 676-82

Hsp70 and ceramide release by diode laser-treated mouse skin cells in vivo

G S Sokolovskii[1], S B Onikienko[2], A V Zemlyanoi[2], K K Soboleva[3], N A Pikhtin[1], I S Tarasov[1], I V Guzova[4], B A Margulis[4]
[1] Ioffe Physical-Technical Institute, St Petersburg, Russia
[2] Military Medical Academy, St Petersburg, Russia
[3] St Petersburg State Polytechnical University, St Petersburg, Russia
[4] Institute of Cytology, St Petersburg, Russia

E-mail: gs@mail.ioffe.ru

Abstract. We report experimental study of generation of extracellular heat shock proteins (Hsp70) and ceramides under pulsed irradiation by quantum-well laser diodes. Our results are of great promise for applications in practical medicine such as protection against biopathogenes and abiotic stress factor challenges.

1. Introduction

Heat shock proteins (HSP) are functionally similar class of proteins whose expression is enhanced by heat shock [1, 2]. Increased expression of genes encoding heat shock proteins are regulated by transcription step. Heat shock proteins are named according to their molecular weights, for example Hsp70 belongs to the family of proteins having molecular weights of 70 kDa. When cells are exposed to elevated temperatures or other stress heat shock proteins expression is increased. Heat shock proteins are very important in protein-protein interactions, such as folding and assist in establishment of correct protein conformation and prevention of undesirable protein aggregation. Heat shock proteins stabilize partially folded proteins and facilitate their transport through the cell membrane. Heat shock proteins are expressed in all types of cells of all living organisms as playing a key role in the existence of proteins [3, 4].

Extracellular heat shock proteins (Hsp70) and ceramides are known to possess high adjuvant activity for cancer vaccines and low-immunogenic vaccines against dangerous infections [5-8]. Hsp70 and ceramides can activate receptors of the innate immunity (TLR4) and boost protective immune response reactions against abiotic stress factors and biopathogens [9, 10].

2. Experiment

In this work, we have used semiconductor lasers that offer many advantages over conventionally used gas and solid-state counterparts, which are more expensive, cumbersome and inefficient. Generation of laser pulses of sufficient amplitude was made possible by utilization epitaxially-stacked GaAs:Si/GaAs:C tunnel-junction asymmetric separate confinement quantum-well AlGaAs/GaAs/InGaAs laser structures [11]. Deep mesa stripe diode lasers with 150x12 μm aperture manufactured on these structures have demonstrated threshold current density of 96 A/cm^2 with internal optical loss of 0.82 cm^{-1} and series resistance 0.28 Ohm. Pumping of these structures by 100 ns pulses of 130 A amplitude has allowed generation of over 450 W peak optical power at 1.06 μm wavelength at 10 kHz repetition rate.

We have shown that pulsed laser-diode irradiation of mouse skin cause significant release of Hsp70 and ceramides from keratinocytes and dermal fibroblasts. These results are of great promise for medical applications such as development of new laser in-situ high-effective vaccine adjuvants and induction of stress-protective immune responses.

3. Results and discussion

In our experiments, we investigated the level of endogenous adjuvants (Hsp70 and ceramides) in the ear skin biopsy samples of the Balb/c mice (50 animals) one hour, 24 hours and 48 hours after laser treatment of the ear skin internal surface. Laser radiation was applied to the left ear of each mouse while the right ear was used for control.

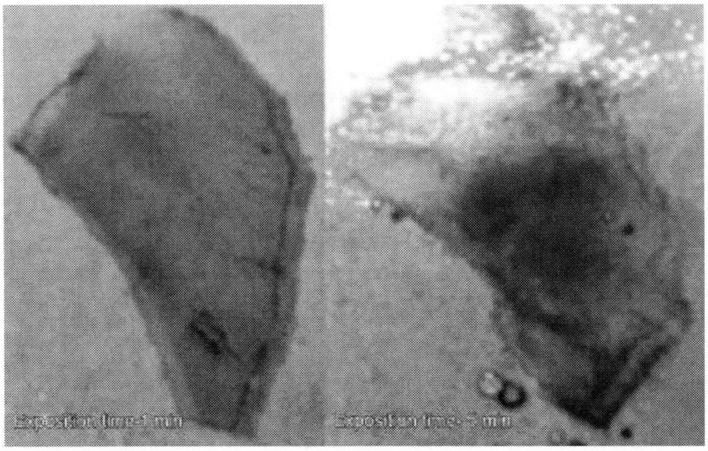

Figure 1. Laser-induced Hsp70 release in mouse ventral ear skin surface one hour after irradiation (immunohistochemistry).

Hsp70 and ceramide concentrations were also studied in the mice blood serum samples. Hsp70 levels were detemined by Western blotting, with anti-Hsp70 antibodies, while ceramide concentrations were measured with anti-ceramide antibodies and by liquid chromatography.

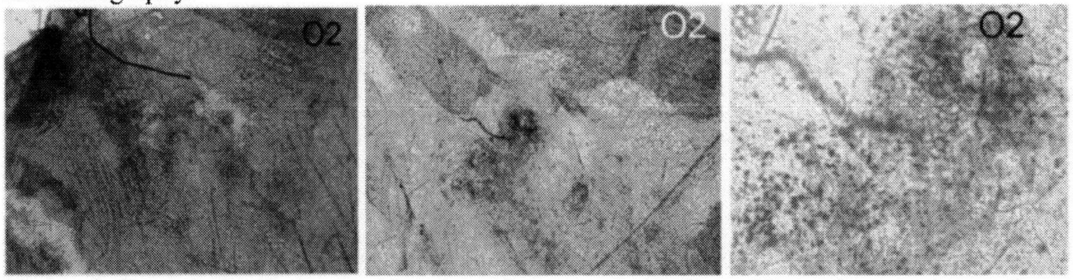

Figure 2. Dose-dependant ceramide release in mouse ventral ear skin surface one hour after 5 min laser irradiation at 0.5-0.75 W/cm^2 detected with immunohistochemistry.

One hour after laser skin irradiation the dramatic release of intracellular Hsp70 and ceramides was shown in the mice skin biopsy samples at the laser-treated areas as shown in figs. 1 and 2 compared with non-treated opposite ear (fig.2). Forty eight hours later Hsp70 content returned back to basal value. The effect was dose-dependent (figure 2). The ceramide level did not returned to the initial values 48 hours after laser treatment. As shown in the figure 3, serum Hsp70 level was increased 1.4 – 3.8 fold one hour after laser skin treatment, and after 24 hours it returned to the basal value. Serum ceramide level was increased 1.6- 4.5 fold and remained still elevated 48 hours after laser irradiation.

a)

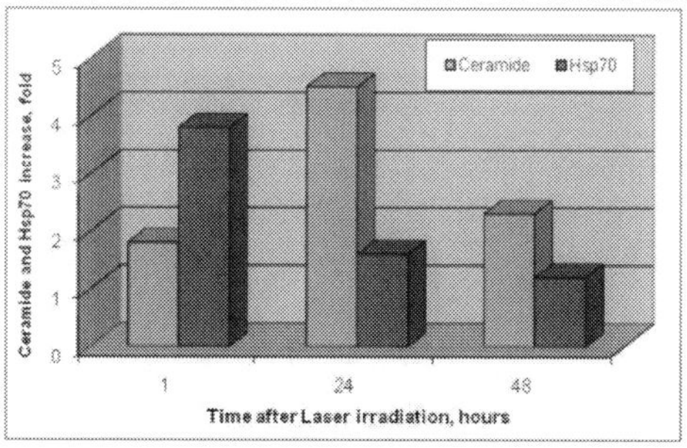

b)

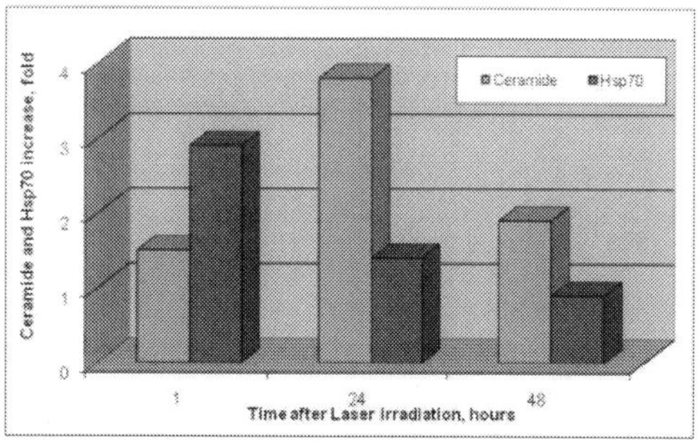

c)

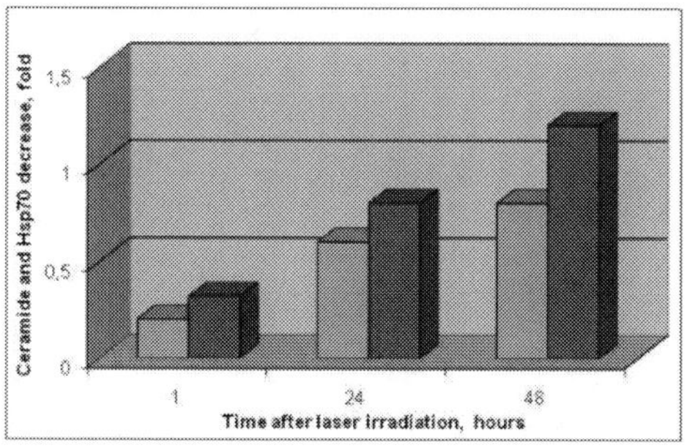

Figure 3. Laser-induced ceramide and Hsp70 increase in the blood serum after 5 min mouse ear skin ventral surface irradiation at 0.75 W/cm^2 (a) and 0.5 W/cm^2 (b). Laser-induced ceramide and Hsp70 release from mouse ear skin ventral surface slices after 5 min irradiation at 0.75 W/cm^2 (c).

It should be noted that laser-induced release of endogenous Hsp70 correlated positively ($P<0.01$) with increase of ceramide serum concentration with significant changes of Hsp70 levels observed at infrared radiation dose exceeding 3.5 J/cm^2.

4. Conclusion

In conclusion, our tests have proved that the gain-switched laser diode irradiation of mice ear skin can cause dramatic release of endogenous chaperones and stress-protectors – heat shock proteins and ceramides. Doses of applied semiconductor laser radiation are safe and do not cause irreversible cell damage. Therefore the compact high-intensity short-pulsed infrared semiconductor laser is a promising tool for applications in practical medicine for protection against biopathogenes and abiotic stress factor challenges.

References

[1] Lindquist S, Craig E A 1988 The heat shock proteins *Annu. Rev. Genet.* **22** 316
[2] Schlesinger M J 1990 Heat shock proteins *Jour. Bio. Chem.* **265**(21) 12111
[3] Craig E A, Gambill B D, Nelson R J 1993 Heat shock proteins: molecular chaperones of protein biogenesis *Microbiol Rev.* **57**(2) 402
[4] Richter K, Haslbeck M, Buchner J 2010 The heat shock response: life on the verge of death *Mol. Cell.* **40**(2) p 253
[5] Gong J, Zhang Y, Durfee J, Weng D, Liu C, Koido S, Song B, Apostolopoulos V, Calderwood S K 2010 A heat shock protein 70-based vaccine with enhanced immunogenicity for clinical use *J. Immunol.* **184**(1) 488

[6] Bolhassani A, Rafati S 2008 Heat-shock proteins as powerful weapons in vaccine development *Expert Rev. Vaccin.* **7**(8) 1185

[7] Kopecky-Bromberg S A, Fraser K A, Pica N, Carnero E, Moran T M, Franck R W, Tsuji M, Palese P 2009 Alpha-C-galactosylceramide as an adjuvant for a live attenuated influenza virus vaccine *Vaccine.* **27**(28) 3766

[8] Choi Y S, Hoory T, Monie A, Wu A, Connolly D, Hung C F 2008 α-Galactosylceramide enhances the protective and therapeutic effects of tumor cell based vaccines for ovarian tumors *Vaccine.* **26**(46) 5555

[9] Fischer H, Ellström P, Ekström K, Gustafsson L, Gustafsson M, Svanborg C 2007 Ceramide as a TLR4 agonist; a putative signalling intermediate between sphingolipid receptors for microbial ligands and TLR4 *Cell Microbiol.* **9**(5) 1239

[10] Wheeler D S, Chase M A, Senft A P, Poynter S E, Wong H R, Page K 2009 Extracellular Hsp72, an endogenous DAMP, is released by virally infected airway epithelial cells and activates neutrophils via Toll-like receptor (TLR)-4 *Respir Res.* **10**(1) 31

[11] Vinokurov D A, Konyaev V P, Ladugin M A, Lyutetskiy A V, Marmalyuk A A, Padalitsa A A, Petrunov A N, Pikhtin N A, Simakov V A, Slipchenko S O, Sukharev A V, Fetisova N V, Shamakhov V V, Tarasov I S 2010 A study of epitaxially stacked tunnel-junction semiconductor lasers grown by MOCVD *Semiconductors* **44**(2) 251

Defect formation in 4H-SiC single crystal grown on the prismatic seeds

A Yu Fadeev[1,2], A O Lebedev[1] and Yu M Tairov[1]

Micro- and Nanoelectronics Department, St. Petersburg Electrotechnical University, Prof. Popova str. 5, 197376 St. Petersburg, Russia

E-mail: [1]siclab-Tairov@yandex.ru, [2]alfadeus@gmail.com

Abstract. The defect structure of 4H silicon carbide single crystals grown by PVT method on three prismatic seeds (10-10), (11-20) and (8.3.-11.0) is considered. The only defects existing in the grown ingots are stacking faults and basal plane dislocations. The type of stacking fault is studied. The dependence of stacking fault morphology on the seed orientation is analyzed.

1. Introduction

Silicon carbide is known as a semiconductor material for producing high-power devices and systems operating under extreme conditions. SiC ingots grown by PVT method on basal plane contain well studied but extremely complex defect structure. Therefore, non-basal planes of SiC are of interest itself to design the microelectronics devices as well as for so-called RAF-process [1, 2] and further getting defect-free (0001)-grown ingots. Though there are the advantages of using non-basal planes of SiC (prismatic or rhombohedral) for designing the electron devices [3, 4] such substrates are still commercially not available. In spite of number of studies covering the growth of SiC on non-basal seeds [5-9] the dependence of defect structure of grown ingot on orientation of the seeds is still examined not enough. In this study the defect structure of the 4H-SiC crystals grown on (10-10), (11-20) and near to (8.3.-11.0) planes is considered. The last plane makes an angle of about 15° with (10-10) and (11-20).

2. Experiment

4H-SiC single crystals were grown by PVT method in a resistive heating reactor. The growth temperature was 2100–2200°C. The pressure during growth was 3-10 Torr. The temperature gradients were in the range of 10-30 K/cm. The duration of the growth was about 40 hours.

Initial 4H-SiC (10-10), (11-20) and (8.3.-11.0) substrates were cut from the single [000-1] grown crystal. The defect structure of the seeds (Figure 1) was characterized by several SFs (narrow linear etch figures) at the initial stage of growth, polytype inclusions (wide linear etch figures) and micropipes (triangle etch figures).

The SiC crystal growth was implemented on these three seeds in the same growth experiment simultaneously. The following wafers were cut from the ingot: (10-10), (11-20), (8.3.-11.0), i.e. normal to the growth direction and basal plane samples (0001) for every crystal. Defect analysis was held using optical and atomic force microscopy after KOH etching (type of defects was identified by etch pits form) as well as X-ray topography and transmission electron microscope.

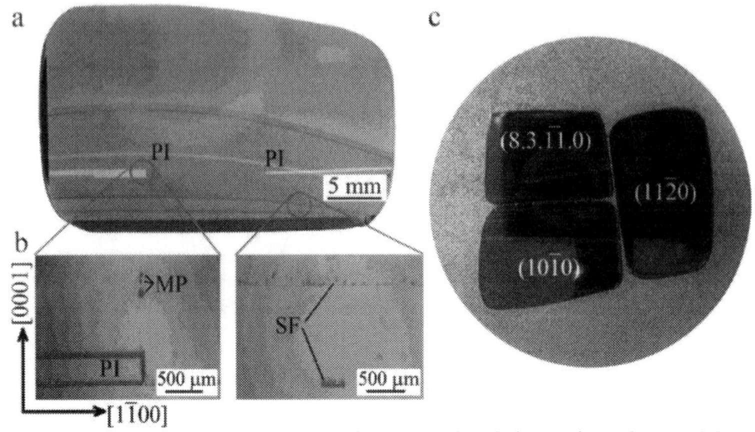

Figure 1. (a) 4H-SiC (11-20) seed and (b) its defect microphotos: PI — polytype inclusion, SF — stacking fault, MP — micropipes; (c) seed arrangement on the holder.

3. Results and discussion

Growth on prismatic seeds leads to stacking fault formation [8]. SF is planar defect parallel to (0001), usually observed [5] as linear etch figure. We have found that morphology of the SF depends on the orientation of the prismatic seed. (11-20)-cut of [11-20]-ingot revealed the SF etch figures with size along [1-100] tending to zero, while (8.3.-11.0)- and (10-10)-cuts of [8.3.-11.0]- and [10-10]-ingots, respectively, showed much longer etch figures along the direction normal to the [0001] in plane of growth. The former SF looked like 'non-opened' SFs (Figure 2a, b) contrary to the latter ones.

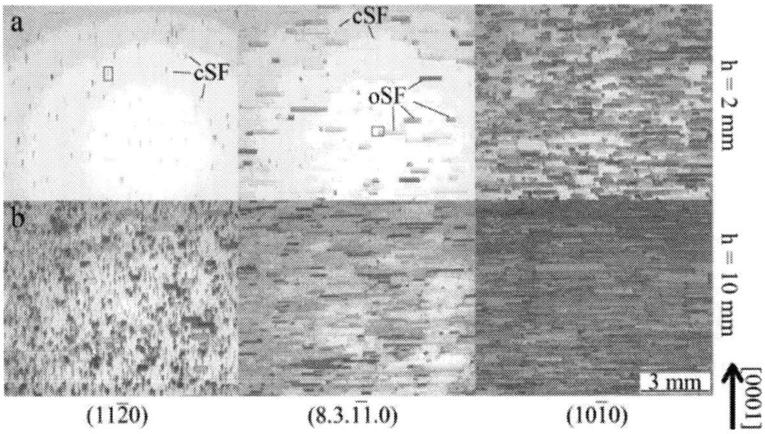

Figure 2. Defect structure of planes cut normal to the growth direction: (a) near the seed, (b) near the growth front. 'h' defines the distance between the seed and cut substrate. cSF, oSF — closed and opened stacking faults, respectively.

We called them 'closed' (cSF) and 'opened' (oSF) stacking faults. We revealed that oSF formation probability strongly depends on seed crystallographic orientation and gradually increases with seed inclination from (11-20)-plane to (10-10)-plane. For [11-20]-grown crystal it was also noticed that oSFs prevailed on the edges of crystal where convex growth front had maximal inclination from initial (11-20)-plane.

AFM study of the samples after etching in the molten KOH (T = 500°C, t = 15 min) revealed that stacking faults form etch hillocks contrary to the dislocations which form etch pits [10]. Fig. 3 shows 3D pictures of cSF and oSF (marked with frames on fig. 2) made by AFM. The cSF on the surface of (11-20)-plane (Figure 3,a) is about 100 nm height, 20 μm length along [0001] and 2 μm width along [1-100]. The inclination of the seed at an angle of 15° to (8.3.-11.0) leads to the 'opening' of SF (Figure 3,b). The 'opening' of SF obviously should be considered as gliding of partial dislocations bounding the SF. The height of oSF is about 100-200 nm, the size along [0001] is 2 μm, the size along [-34-10] can be 100 μm and more. The nature of such etch hillocks needs to be studied.

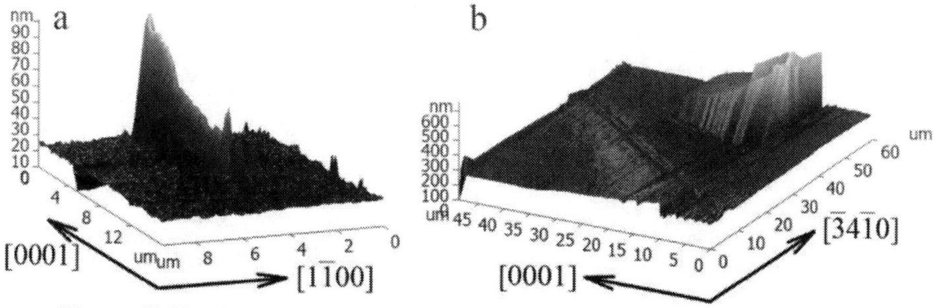

Figure 3. 3D pictures of cSF (a) and the edge of oSF (b) made by AFM.

The type of stacking fault was studied using high resolution transmission electron microscope Jeol Jem-2100F with an accelerating voltage of 200 kV. The high resolution surface area image is shown in Figure 4.

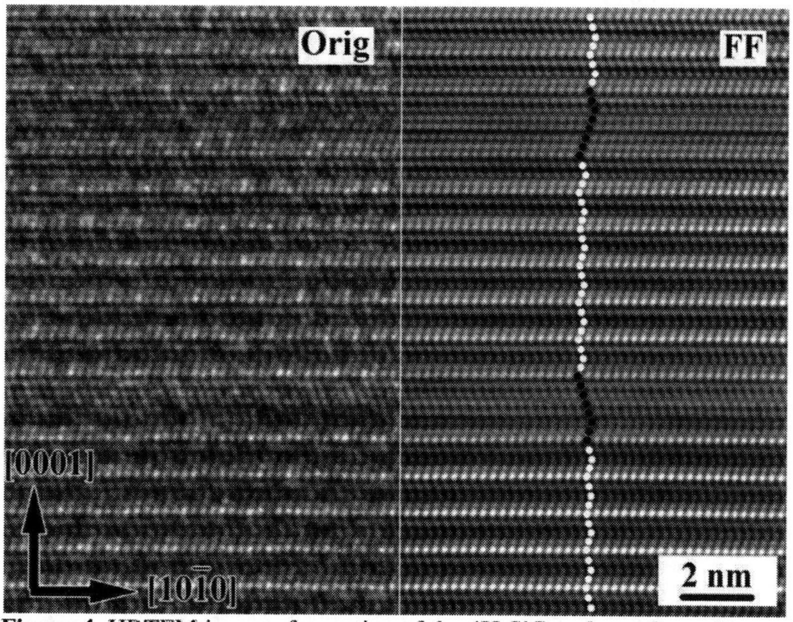

Figure 4. HRTEM image of a portion of the 4H-SiC wafer surface with two stacking faults (5,2). The left side (Orig) is the original image; the right side (FF) is the image after Fourier filtering.

We can see that visible stacking faults are internal Frank SFs and correspond to the formula (5,2) in Zhdanov notation. That means that stacking faults have growth nature of formation and cannot form due to dislocation splitting and partial gliding during after-growth cooling down.

Figure 5 shows (0001)-plane of grown crystals after KOH etching. This plane revealed basal plane dislocations (BPD) as the only visible defect on the surface. The density of BPD is decreased from $4{\cdot}10^3$ cm^{-2} for [11-20] to 10^3 cm^{-2} for [10-10]. It is also clear that the inclination of the seed from (11-20)-plane to (10-10)-plane leads to merging of single BPDs and formation of low-angle grain boundaries (LAGB).

Figure 5. Defect structure of basal plane (0001) cut out of ingots grown in [11-20]-, [8.3.-11.0]- and [10-10]-directions: (a) near the seed, (b) near the growth front.

Gliding of the BPD with LAGB formation and partials with oSF formation is possible under shear stress. One of the possible causes of induced stress can be the difference of thermal expansion coefficients of seed and holder materials. Using the common principles described in [11, 12], we calculated the shear stresses for the prismatic SiC seed 1 mm thick attached to the seed holder 5 mm thick made of extruded graphite (Figure 6) and found that shear stress gradually increases from (11-20) to (10-10). This shear stress lead to secondary gliding system {11-20}⟨0001⟩ activation and formation of LAGB as well as oSF.

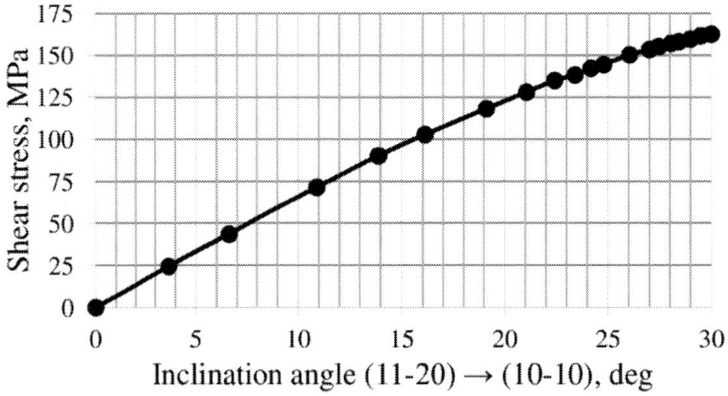

Figure 6. Calculated dependence of the shear stress in SiC seed on orientation of the seed.

4. Conclusion

The defect structure of the 4H-SiC single crystals grown on the prismatic seeds is characterized only by stacking faults and basal plane dislocations. We revealed that stacking faults form etch hillocks while dislocations form etch pits after KOH etching. For the first time it was shown the difference of SF morphology — 'closed' and 'opened' SF — depending on orientation of prismatic seed. The type of SF is internal Frank SF and correspond to the formula (5,2) in Zhdanov notation. That means that stacking faults have growth nature of formation. The 'opening' of SF at the growth plane as well as LAGB formation at the basal plane are the result of partials gliding under high shear stress which gradually increases from (11-20) to (10-10). We assume that the stress is induced in the grown ingot mainly as a result of difference of thermal expansion coefficients of seed and holder materials.

Acknowledgements
The authors are grateful to A.V. Zubkova for implemented AFM studies and V.N. Nevedomski for TEM studies.

References

[1] Nakamura D et al 2004 *Nature* **430** 1009
[2] Urakami Y et al 2012 *Mater. Sci. Forum* **717-720** 1069
[3] Yano H et al 1999 *IEEE Electron Device Lett.* **20** 611
[4] Senzaki J et al 2013 *Mater. Sci. Forum* **740-742** 621
[5] Takahashi J et al 1997 *J. Cryst. Growth* **181** 229
[6] Ohtani N et al 2003 *Jpn. J. Appl. Phys.* **42** L277
[7] Herro Z G et al 2005 *J. Cryst. Growth* **275** 496
[8] Rost H-J et al 2006 *J. Cryst. Growth* **290** 137
[9] Epelbaum B M et al 2008 *Phys. Stat. Sol. (b)* **245** 1257
[10] Amelinckx S et al 1960 *J. Appl. Phys.* **31** 1359
[11] Efimov A N and Lebedev A O 1995 *Surf. Sci.* **344** 276
[12] Efimov A N et al 1998 *J. Appl. Crystallogr.* **31** 461

High-frequency dielectric study of proustite crystals Ag_3AsS_3

V A Bordovsky, N Yu Gunia, R A Castro
Herzen State Pedagogical University of Russia,

naala4kaka@mail.ru

Abstract. The dielectric properties of the crystals proustite in the frequency of 10^6-10^9 Hz and a temperature range of 173 to 473 K were studied. The dispersion of the dielectric parameters indicates the existence of non-Debye relaxation mechanism correlates with structural changes in the phase transition region. The charge transfer is temperature activated with an activation energy of 2.40 ± 0.01 eV.

1. Introduction

Proustite monocrystal Ag_3AsS_3 is a ferroelectric material with a melting point 580 C. The crystal system is characterized by the high values of dielectric permittivity, which makes it challenging to use in nonlinear optic devices, electro-acoustic systems [1]. Dielectric study is a powerful tool in learning the effects of temperature and light factors on properties of wide class systems [1,2]. Currently, there are practically no data on the special features of polarization processes in proustite crystals in high frequency field.

Therefore, the aim of the work was to determine the special features of the dielectric relaxation mechanism in the proustite crystals Ag_3AsS_3 for frequencies $10^6 - 10^9$ Hz.

2. Experiment method

The measurements were taken with the spectrometer "Concept" 81 of the company NOVOCONTROL Technologies GmbH & Co. The samples consisted of monocrystal plate thickness of about 1 mm and an area of 1 cm^2, with a flat surface perpendicular to the triad axis. Measurements of dielectric parameters (components of the complex dielectric permittivity (ε', ε'') of proustite crystals samples were done in a temperature range 173 to 473 K. Temperature measurement accuracy was 0.5°C. The measuring voltage on the sample was 1.0V.

3. Results and discussion

The study found that the dielectric permittivity frequency dependence is characterized by the presence of dispersion (figure 1). At higher frequencies the value ε' practically does not depend on the alternating field frequency. The main reason of it is the electronic polarization. As it is known , electronic polarization is a flexible displacement and deformation of the electron shells of atoms and ions. Electronic polarization is observed in all types of dielectrics and not associated with loss of energy up to the resonance frequencies. The polarizability of the particles under the electronic polarization does not depend of temperature, and the dielectric permittivity ε decreases with rise in temperature due to thermal expansion of the dielectric and the reduction of the number of particles per unit volume.

When you reduce the frequency in the whole temperature range you observe the rise of ε', most likely due to the fact of dipole-relaxation polarization. It is known that if there are polar molecules in a dielectric and the relationship between them is small, they can relatively easily be rotated under the influence of the field. The thermal movement prevents from the orientation of the dipoles in the field. The result is a dipole polarization, which depends on the thermal movement. In the case of proustite, the silver ions can serve as dipoles and execute transition through the lattice, creating positively and negatively charged fields.

The dielectric permittivity temperature dependence reveals the existence of two zones: a monotonous dependence and a sharp increase of ε' under temperature increasing (figure 2). According to the literature the dependence mode change temperature T ≈ 420K (147°C) is the temperature at which the proustite executes of the first-order phase transition [2] due to the abrupt change of conductivity.

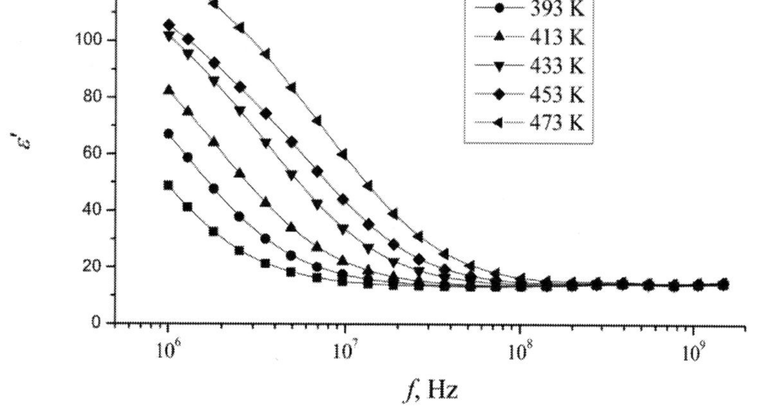

Figure 1. Frequency dependence of the dielectric permittivity ε' under different temperatures

Journal of Physics: Conference Series **572** (2014) 012019 doi:10.1088/1742-6596/572/1/012019

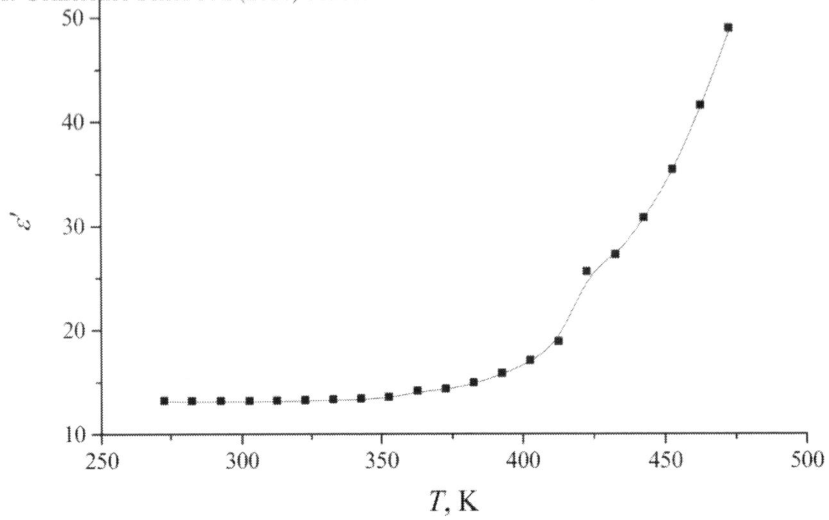

Figure 2. Temperature dependence of the dielectric permittivity under frequency f=1.37*10^7 Hz.

Absence of evident relaxation processes under low temperatures (figure 3) may be due to fact that up to a certain temperature, the silver ions are placed in different lattice sites. At higher temperatures the silver ions gain sufficient energy to execute transition between unoccupied sites in the structure. In this case, we can speak about an effect of relaxation loss mechanism. Relaxation losses arise from active components of polarization currents. They are typical for dielectrics with slower types of polarization, and show themselves under very high frequencies when the polarization is behind the field change.

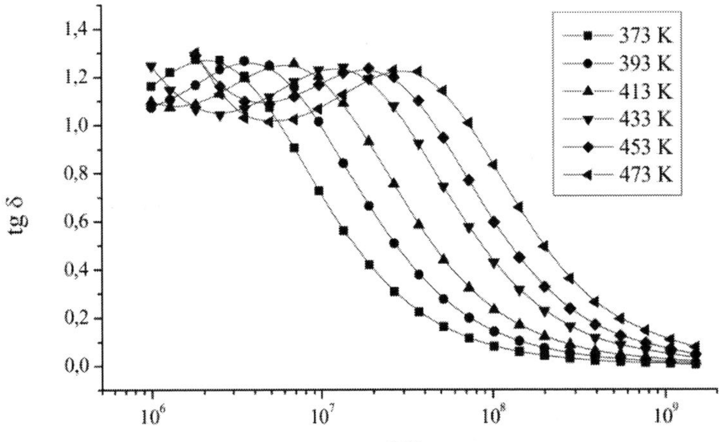

Figure 3. Frequency dependence of dielectric loss angle under different temperatures

In many dielectrics, relaxation processes are associated with the existence of not one relaxation time, but with a set of relaxation times. In this case we are dealing with the relaxation time distribution and consequently with the activation energies. This distribution can be attributed to a variety of relaxation processes according to their nature, or to distribution in accordance with concentration of dipoles in the structure. In the case of ionic jump processes, such as in proustite, it is assumed that after each jump the potential energy is changed and you need some time to return to a minimum of potential energy. Taking into account the deposition of large numbers of moving defects, then we get a set of relaxation times [3].

Deviation from the classical model of relaxation by Debye for the case of one time of relaxation can be studied by using the Cole-Cole or Cole-Davidson model representation for the case of symmetric and asymmetric distribution of relaxation times respectively. If you build a Cole-Cole chart (characteristic curve ε'=f(ε')), then the existence of distribution of relaxation times will appear as a deviation from hemispherical dependence with the radius of the hemisphere $(\varepsilon_s-\varepsilon_\infty)/2$.

The existence of Non-Dedye relaxation mechanism is confirmed by the estimation of parameter values in Havriliak-Negami two-parametric empirical function (HN) [4]:

$$\varepsilon^*(\omega) = \varepsilon_\infty + \frac{\Delta\varepsilon}{\left[1+(i\omega\tau)^{\alpha_{HN}}\right]^{\beta_{HN}}}$$

where ε_∞ - the high-frequency limit of the real part of the dielectric permittivity, $\Delta\varepsilon$ – dielectric increment (difference between the low-frequency and high-frequency limits), $\omega=2\pi f$, α_{HN} and β_{HN} – form parameters that describe respectively the symmetric (β = 1 - Cole-Cole distribution) and asymmetric (α = 1 - Cole-Davidson distribution) relaxation function extension [4]. Results of dielectric spectra measurements were approximated by using the software Novocontrol Winfit. Based on this approximation the positions of the dielectric loss peaks where emphasized and the parameters (HN) for the studied relaxation processes where defined:

Dielectric spectra of proustite samples have been measured in a wide temperature range, enabling the identification of the relaxation process existence in the range 90 to 200°C (figure 4). Uneven change of relaxation parameters with the temperature increasing was found. Their values are increased from a relatively constant value to higher values, against the background of the increasing intensity of the ions Ag^+ migration. Uneven change of relaxation parameters by temperature is attributed to the change in conductivity due to phase transition of the first kind and the second kind. Here the transfer jumping mechanism is replaced by the ion conductivity mechanism. The diagram of temperature against the most probable relaxation time τ_{max} (figure 5) can be divided into three sections. The first and third sections correspond to the status of the system before and after a phase transition, respectively (T=420 K). At the second section dependence becomes complicated (not a simple exponential), that corresponds to the transition field. Activation energy of the third section of the dielectric relaxation process is 0.44EV (calculated according to the positions of dielectric losses peaks for different temperatures (figure 4)). The activation energy of a similar relaxation process observed in crystals Ag_3SbS_3 is a value of 0.40 that is close to the value of activation energy in proustite and may indicate that jumping mobility is stipulated by the movement of silver ions, but is not associated with the SbS_3 or AsS_3 groups. Throughout the temperature range investigated the charge transfer is also thermally activated process, the activation energy is equal to 2.40 ± 0.01 eV.

Thus, in view of the approximation error we can conclude that in the investigated range of frequencies and temperatures in proustite crystals a Debye vibration process is observed with the distribution of relaxation times according to the Cole-Cole model.

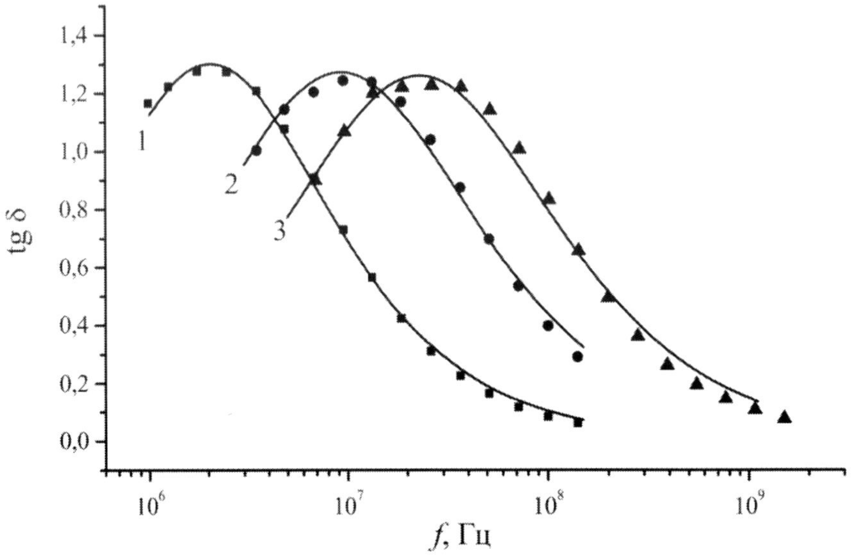

Figure 4. Frequency dependence of dielectric losses angle under different temperatures: 1 – 373 K, 2 – 423 K, 3 – 473 K. Solid lines – approximation by HN function.

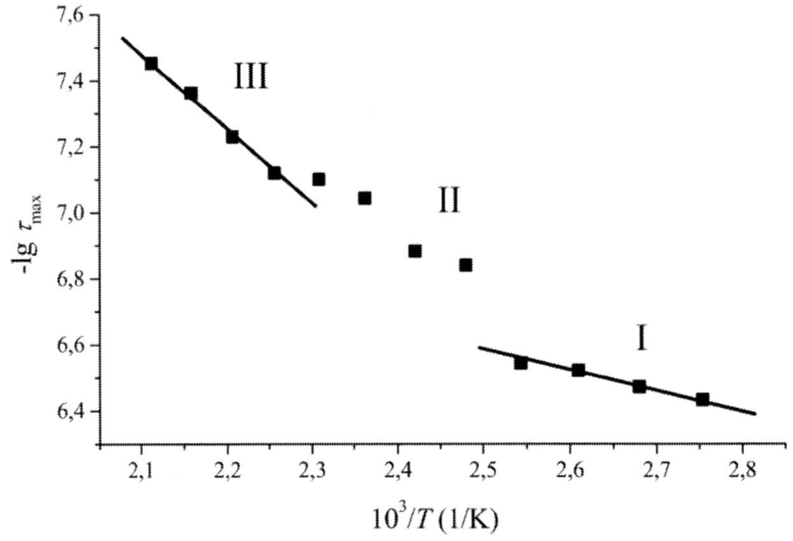

Figure 5. Temperature dependence of the most likely relaxation time for the high frequency field.

Journal of Physics: Conference Series **572** (2014) 012019 doi:10.1088/1742-6596/572/1/012019

4. Conclusion

Thus, in the course of the executed study the following main results where obtained:

1. We have found, that in the high frequency area (10^6 ...10^9 Hz) the dielectric permittivity ε' practically does not depend on the alternating field frequency, and if the frequency is decreased in the whole temperature interval we observe that the value ε' is increased, most likely due to the start of dipole-relaxation polarization along with electronic polarization.
2. The dielectric permittivity temperature dependence reveals the existence of two sections: monotonous dependence and sharp increase of the value ε' under temperature increasing. The dependence nature change temperature t≈420K (147°C) corresponds of the first-order phase transition, caused by a sharp change of the crystal conductivity.
3. The results of the dielectric spectra approximation by Havriliak-Negami function allow conclude, that at the studied field of high frequencies and temperatures Non-Debye relaxation process is observed in proustite crystals with symmetric distribution of relaxation times according to the Cole-Cole model.

5. Literature

[1] Yang S R, Taylor K N R 1991 Phase transitions in single crystals of proustite Ag_3AsS_3 *Phase Transitions* **36** 233-256

[2] Kristin A and Schonau 2002 High-temperature phase transitions, dielectric relaxation, and ionic mobility of proustite, Ag_3AsS_3, and pyrargyrite, Ag_3SbS_3 *Journal of Applied Physics* **92** 12 pp 7415-7424

[3] Kremer K and Schonhals A 2003 *Broadband dielectric spectroscopy*, Springer, Berlin Heidelberg, 729

[4] Gupta V, Bamzai K K, Kotru P N and Wanklyn B M 2006 Dielectric Properties, AC Conductivity and Thermal Behaviour of Flux Grown Cadmiumtitanate *Crystals Materials Science and Engineering* **130** 163-172

Dielectric relaxation in sillenite crystals $Bi_{12}SiO_{20}$:Fe

Castro R A [1], Nabiullina L A[1], Shadrin E B [2]
1 Herzen State Pedagogical University of Russia,
Saint Petersburg, Russia.
2 Ioffe Physical-technical Institute of Russia Academy of Sciences,
Saint Petersburg, Russia.

e-mail: lia-nabiullina@yandex.ru

Abstract
Within the range of low frequency and temperature between 293 and 373 K, temperature-frequency dependences of the dielectric permittivity and losses in iron-alloyed $Bi_{12}SiO_{20}$ crystal are investigated. Introduction of impurities into the crystal structure leads to greatest losses and change of the dielectric constant. The results are being discussed involving the concept of formation of the "impurity-vacancy"-type quasi dipoles with iron, resulting from the substitution of M-cations by Fe^{3+} ions at the centers of oxygen tetrahedra.

1. Introduction

Sillenite crystals of $Bi_{12}SiO_{20}$ (BSO) have body-centered cubic lattice of symmorphic space group *I23* (T^3) [1]. Currently more than 60 compounds like sillenite $Bi_2O_3M_xO_y$ with cations M^{n+} were synthesized, as well as their numerous solid solutions. Alloying of Fe sillenites is of particular interest [2], because ferrum (as well as chrome) is always present in these compounds as background impurity. Increased value of dielectric capacitance, minor losses and thermal stability make them advanced / perspective materials for microwave technology [3]. In particular, sillenite crystals doped with ferrum are successfully used as optical stoppers for eye protection or protection of light sensors in photosensitive devices against blinding effect of laser radiation of visible range.

The objective of this work is analysis of peculiarities of dielectric relaxation processes in BSO crystals: undoped crystals, as well as ones doped with ferrum ions, by means of analysis of their dielectric characteristics at applied field's temperature and frequency variation.

2. Experiment procedure

In this experiment, there were used samples of bismuth silicate grown by Czochralski method. Doping with ferrum ions was carried out at the synthesis stage by means of adding iron trioxide Fe_2O_3 into original load. Doping concentration of Fe was defined according to transmission spectrum and EPR [4]. The samples were prepared in form of discs with diameter 20.0 mm and thickness 1.0 mm. For electrical measurements a layer of silver paste was applied on both sides to provide secure contact. Doping concentration of ferrum ions was approximately $1{,}7 \cdot 10^{16}$ cm^{-3}.

Measurements of dielectric parameters were carried out in frequency range of $10^{-2} \div 10^{4}$ Hz and interval range from 293 to 473 K by means of a spectrometer «Concept-41» made by Novocontrol Technologies GmbH company. Voltage delivered to the sample was 1.0 B, temperature stabilization accuracy was ±0,3°C. Electrical quantity measurement error did not exceed 5%.

3. Results and discussion

According to figure 1, dielectric capacitance ε' drops dramatically with frequency increase, its significant reduction starts from 0,01 Hz up to 10 kHz for all chosen temperatures. Maximum value of ε' can be clearly observed at frequency 0,01 Hz for all analyzed temperatures. Thus, ε' depends heavily on temperature in low frequency range. High value of dielectric capacitance at low frequency in the crystals under study is probably associated with cavities, dislocations and other defects [7], as well as with participation of interfacial polarization [6, 8].

Figure 2, showing the temperature dependence ε', demonstrates that ε' is constantly rising at all frequencies. Although, maximum value of ε' turned out to be at 473 K for all analyzed frequencies. With temperature rise, the polarizability is rising at the expense of lattice defects development [5], which can be seen in increase of dielectric capacitance with increase of temperature. In this search sharp increase of ε' with temperature at all frequencies analyzed here can also be associated with polarization of space charge [6].

Figure 3 shows frequency dependence of dielectric dissipation factor, and this dependence demonstrates that the parameter is considerably reducing with increase of frequency from the first measurement to frequency of 10 Hz. In this frequency range, tanδ depend heavily on frequency, after which they slightly decrease with the increase of frequency to the value of 10 kHz, at which saturation phase comes, i.e. tanδ depend slightly on frequency in high frequency range.

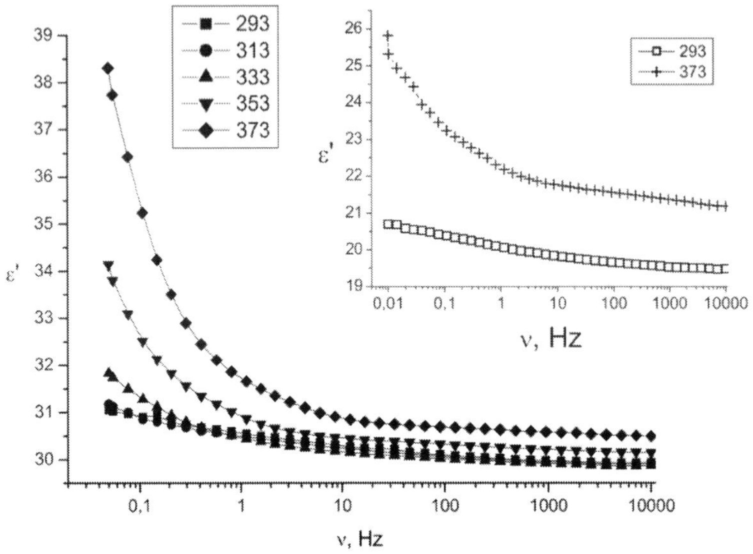

Figure 1. Frequency dependence of dielectric capacitance at following temperatures: 313, 353, 393, 433, 473 K.

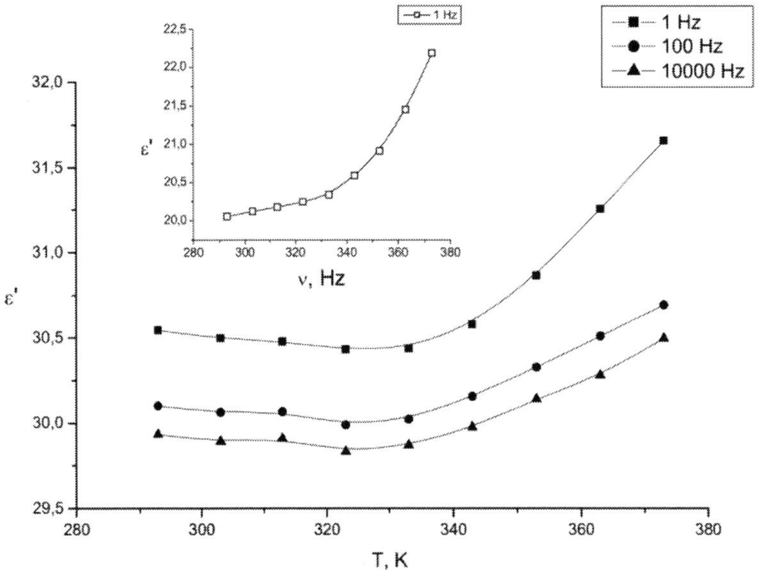

Figure 2. Temperature dependence of dielectric capacitance at following frequencies: 10^0, 10^2, 10^4 Hz.

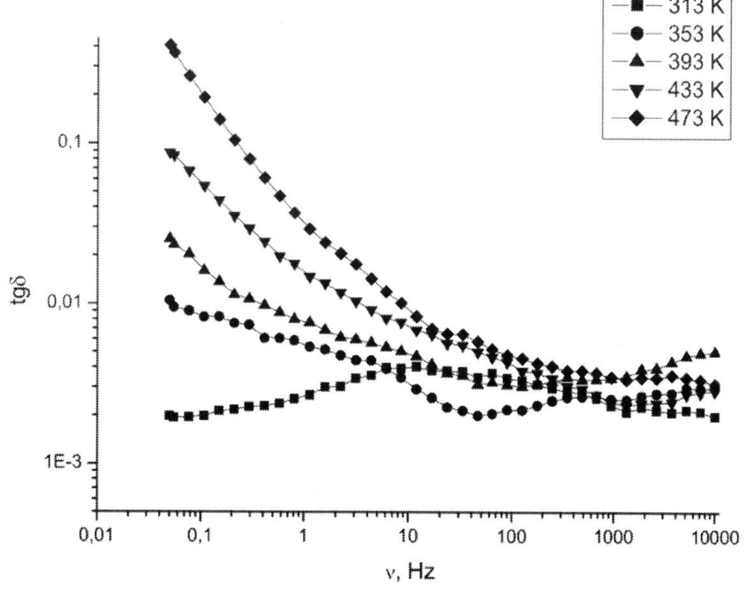

Figure 3. Frequency dependence of dielectric dissipation factor at following temperatures: 313, 353, 393, 433, 473 K.

Frequency dependence of complex dielectric capacitance ε" imaginary part of BSO samples is shown in Figure 4. Discovered maximum losses at room temperature suggest that there is relaxation process of non-Debye type. Existence of non-Debye type relaxation process is confirmed during parameter estimate in two-parameter empiric function of Havrilyak-Negami (HN) [9]:

$$\varepsilon^*(\omega) = \varepsilon_\infty + \frac{\Delta\varepsilon}{\left[1 + (i\omega\tau)^{\alpha_{HN}}\right]^{\beta_{HN}}} \qquad (2)$$

where ε_∞ - high frequency limit of dielectric capacitance's real part, $\Delta\varepsilon$ – dielectric increment (difference between low frequency and high frequency limits), $\omega=2\pi f$, α_{HN} and β_{HN} – form parameters, describing symmetrical (β=1 – Cole-Cole model) and asymmetrical (α=1 – Cole –Davidson model) expansion of relaxation function [9]. Values of the parameters are shown in Table 1.

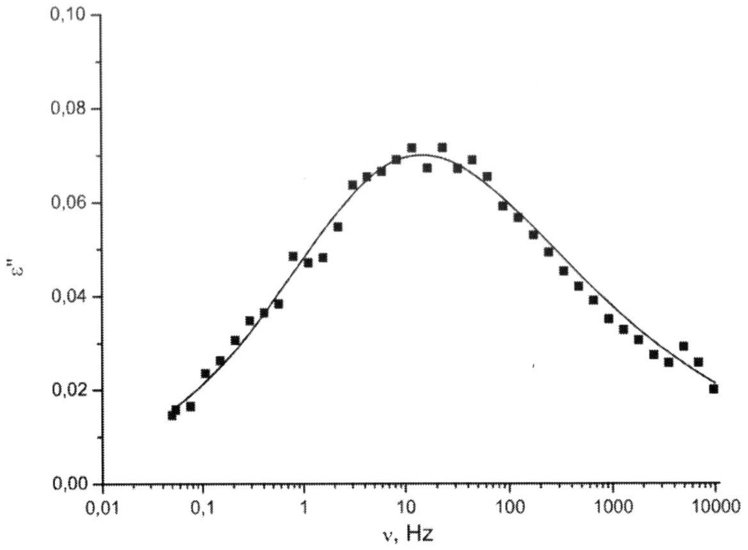

Figure. 4. Frequency dependence of imaginary part of dielectric capacitance at 293 K. Dots indicate experimental values, full line shows theoretical curve.

Table 1. Value of relaxation parameters according to Havrilyak-Negami approximation

τ_{max} [s]	$\Delta\varepsilon$	τ, s	ε_∞	α	β	Temp. [K]	AC Volt [Vrms]
0.01	0.43	0.03	1.0000	0.50	0.55	293	1.0000

The look of Cole-Cole diagram (figure 5) also indicates a wide spectrum of relaxation times in the sillenite crystals under study, caused by structural defects and presence of ferrum impurity. Quasidipoles of "impurity-vacancy" type with participation of ferrum can be relaxation oscillators. They appear if Fe^{3+} ions replace M-cations in the centers of oxygen tetrahedrons; and in one of the sites there should be an oxygen vacancy that captured an electron (according to charge compensation).

Migrations of this vacancy in the area of multiwell potential of oxygen tetrahedrons cause participation of quasidipoles in thermal orientation of polarization.

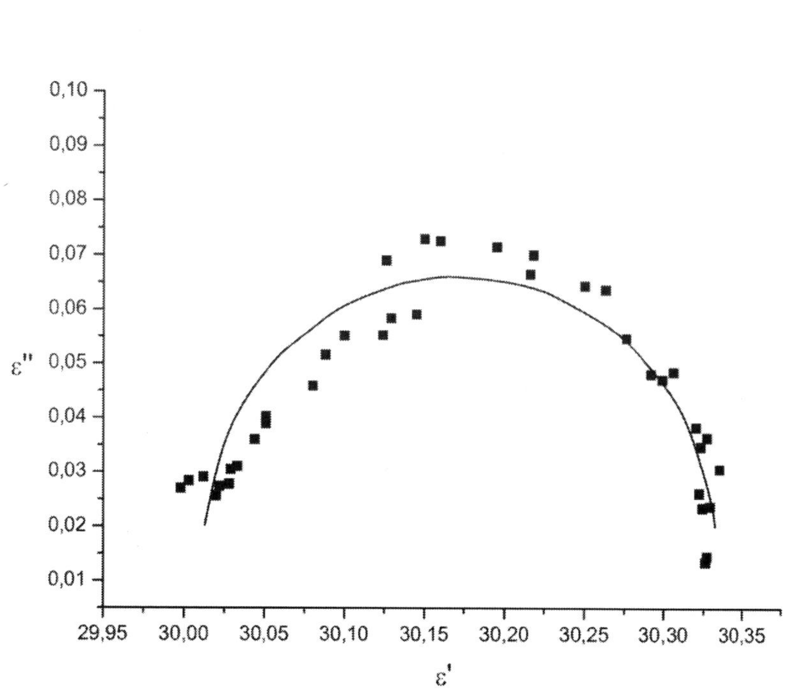

Figure. 5. Cole-Cole Diagram for T = 293 K.

4. Conclusion

Introduction of Iron impurity into the structure of sillenite crystals causes maximum losses and excursion of dielectric capacitance. The look of Cole-Cole diagram for alloy samples indicates a wide spectrum of relaxation times in the crystals under study, caused by structural defects and presence of ferrum impurity.

References

[1] Yu Hu and Derek C Sinclair 2013 Relaxor-like Dielectric Behavior in Stoichiometric Sillenite $Bi_{12}SiO_{20}$ *Chemistry of materials* **25** pp 48-54

[2] Briat B Grachev V G Malovichko G I Schirmer O F and Wohlecke M 2007 Defects in Inorganic Photorefractive Materials and Their Investigations *Photorefractive Materials and Their Applications* Springer Series in Optical Sciences **114** pp 9-49

[3] Nabiullina L A Shadrin E B 2012 Magnetic resonance properties of single crystals $Bi_{12}SiO_{20}$:Fe *Izvestia: Herzen University Jornal of Humanities & Sciences* **147** pp 63-72

[4] Panchenko T V Karpova L M Duda V M 2000 Dielectric relaxation in $Bi_{12}SiO_{20}$: Cr crystals *Physics of the Solid State* **42** 4 pp 689-693

[5] Amarendra Singh K Goel T C and Mendiratta R G 2002 Dielectric Properties of Mn-substituted Ni-Zn Ferrites *Journal of Applied Physics* **91** 10 pp 6626-6629

[6] Lal B Khosa S K Tickoo R Bamzai K K and Kotru P N 2004 Dielectric Characteristics of Melt Grown Doped $KMgF_3$ Crystals *Materials Chemistry and Physics* **83** pp 158-168

[7] Sirdeshmukh L Krishna Kumar K Laxaman S B Krishna A R and Sathaiah G 1998 Dielectric Properties and Electrical Conduction in Yttrium Iron Garnet (YIG) *Bulletin of Materials Science* **21** p 219
[8] Mahajan R P Patankar K K Kothale M B and Patil S A 2000 Conductivity, Dielectric Behaviour and Magneto-Electric Effect in Copper Ferrite-Barium Titanate Composites *Bulletin of Materials Science* **23** p 273
[9] Kremer K Schonhals A (Eds.) 2003 Broadband dielectric spectroscopy *Springer* Berlin Heidelberg p 729

Phase transitions study in $Ba_{(1-x)}Mn_xTiO_3$ and $BaMn_xTi_{(1-x)}O_3$ ceramics

E D Rumyantseva[1], V G Zalessky[2] and N V Zaitseva[2]
[1]Student, University ITMO (National Research University), Kronverksky prospect 49, Saint-Petersburg, 197101, Russia
[2]Ioffe Physical-Technical Institute RAS, Politekhnicheskaya 26, Saint-Petersburg, 194021, Russia

E-mail: ska-kotya@mail.ru

Abstract. Dielectric properties of the ceramics $Ba_{(1-x)}Mn_xTiO_3$ and $BaMn_xTi_{(1-x)}O_3$ of the compositions x=0.01, 0.03, and 0.1 have been studied at temperatures 77 – 450 K and at frequencies 10 Hz -100 kHz. The phase transitions diagram and parameters of the both system have been presented and compared. Dielectric relaxations in $BaMn_xTi_{(1-x)}O_3$ with x=0.1 have been shown.

1. Introduction

Barium titanate (BT) solid solutions, $BaTiO_3$ -ABO_3 (where A = Ba, Ca, Sr, Pb, Mn and B = Ti, Zr, Sn, Hf, Mn), have been intensively studied due to very interesting electric, piezoelectric, and optic properties [1]. Every solid solution is characterized by a phase diagram T - x, as a phase transition temperature versus second component concentration. The T - x diagrams of barium titanate solid solutions consist of the three lines phase transitions: O_h^1- C_{4v}^1 , C_{4v}^1– C_{2v}^{14} , and C_{2v}^{14}- C_{3v}^5. In $SrTiO_3$-$BaTiO_3$ [2] and $CoTiO_3$-$BaTiO_3$ [3] where ions Sr^{2+} and Co^{2+} substitute for Ba^{2+} (A-substitution) the three lines converge to a multiphase point. The multiphase points are also observed for solid solutions with B-substitution: $BaZrO_3$ [4, 5], $BaSnO_3$ [6], $Ba(Mg_{1/3}Nb_{2/3})O_3$, and $Ba(Mg_{1/3}Ta_{2/3})O_3$ [7], and both with A- and B-substitution: $KTaO_3$ [8], and $KNbO_3$ [6]. However, the T- x diagrams with non converged lines are known in the system $CaTiO_3$, $PbTiO_3$ [6], and $MnTiO_3$ [9]. All of them are the solid solutions with the A-substitution excepting $MnTiO_3$, where Mn ions can occupy the both A- and B- sites.

The studies of the system $Ba_{(1-x)}Mn_xTiO_3$ with concentration x<0.01 were reported in [10, 11]. However the T - x diagram of this solid solution still remains indefinite because technological difficulties of the sample preparation. It is also known that Mn^{2+} ions with concentration x>0.15 - 0.18 produce the hexagon phase in BT [10, 11]. We decided to fabricate the specimens of solid solutions both with A- and B- substitutions: $Ba_{(1-x)}Mn_xTiO_3$ and $BaMn_xTi_{(1-x)}O_3$ with x>0.01, study their phase transitions, and compare dielectric properties.

2. Experimental techniques and results

Samples of $Ba_{(1-x)}Mn_xTiO_3$ and $BaMn_xTi_{(1-x)}O_3$ solid solutions with x=0.01, 0.03, and 0.1 were prepared by standard ceramic technology [7, 8]. The starting materials were high purity $BaCO_3$, TiO_2, and MnO. There were the two step of the solid solution preparation. Primary, it was synthesized BT. After pressing, the mixture of the powdered $BaTiO_3$ and MnO was calcined at 1200 - 1400°C for 2-3 days.

The material was reground and the pellets were formed by pressing for 9 mm diameter discs at 50 MPa. After pressing, the samples were fired at 1370-1420°C for 1 hour. The phase content was monitored by X-ray diffraction study (CuK_α and CoK_α lines) and indicated the tetragonal perovskite single phase. Relative density of the samples was measured to be 90-93%. The samples were coated with silver-burned electrodes for dielectric measurements. The dielectric spectroscopy study was performed by impedance meter Goodwill LCR – 819 at the frequency range 12 Hz -100 kHz and amplitude of ac electric field 1 V. Temperature dependence of dielectric parameters measurement was performed by heating or cooling at a constant rate 1 K/min between 77 K and 450 K.

2.1. $Ba_{(1-x)}Mn_xTiO_3$ (A -substitution)

The temperature dependences of real $\varepsilon'(T)$ and imaginary $\varepsilon''(T)$ part of the dielectric permittivity in $Ba_{(1-x)}Mn_xTiO_3$ solid solution with $x = 0.01$, 0.03, and 0.1 are shown in figures 1 and 2. One can see that composition at the range from $x = 0$ to 0.03 changes the high temperature phase transition parameters such as: Curie temperature T_C, Curie-Weiss temperature T_{CW}, and Curie-Weiss constant C_{CW} (see table 1 and inset figure 1). The sequence of the low temperature phase transitions at T_1 and T_2 are also shifted down to lower values (figure 1). The $T - x$ diagram in figure 3 shows the three non converged lines.

Table 1. $Ba_{(1-x)}Mn_xTiO_3$ (A -substitution).

	T_C	T_{CW}	C_{CW}
$x = 0$	392 K	378 K	$1.2 \cdot 10^5$ K
$x = 0.01$	387 K	358 K	$1.3 \cdot 10^5$ K
$x = 0.03$	383 K	340 K	$9 \cdot 10^4$ K

The composition with the concentration $x = 0.1$ changes the temperature dependence $\varepsilon'(T)$ drastically. All the peaks disappear excepting smeared maximum at $T_2 = 186$ K, that is 4 K below the temperature upon the orthorhombic-rhombohedral phase transition in pure BT.

The temperature dependence $\varepsilon(T)$ in $Ba_{(1-x)}Mn_xTiO_3$ solid solutions doesn't show an evidence of a dielectric relaxation (figures 1 and 2). Increasing of real and imaginary parts of the dielectric permittivity in the case $x = 0.1$ is related to the thermal activated conductivity in the sample (Figure 2). The ac conductivity can be described using the relationship $\sigma = \sigma_0 \exp(E_a/kT)$, where E_a is the activation energy and k is the Boltzmann constant. The activation energy is estimated to be $E_a = 0.46$ eV. This value can be characterized the acceptors or donors charge carriers activated in the energy gap bound.

2.2. $BaMn_xTi_{(1-x)}O_3$ (B -substitution)

The temperature dependence of the real part of the dielectric permittivity $\varepsilon'(T)$ of $BaMn_xTi_{(1-x)}O_3$ with $x = 0.01$, 0.03, and 0.1 is shown in figure 4. One can see that the high temperature phase transition parameters are also changed by manganese doping at the range $x = 0$ - 0.01 (see table 2 and inset figure 4). The rest two phase transitions peaks at low temperatures T_1 and T_2 are also shifted down to lower value. The $T - x$ diagram in figure 3 shows the three non converged lines. Despite the diffuseness, this material does not exhibit the strong frequency dispersion of the dielectric permittivity as the evidence of the relaxor behavior [1].

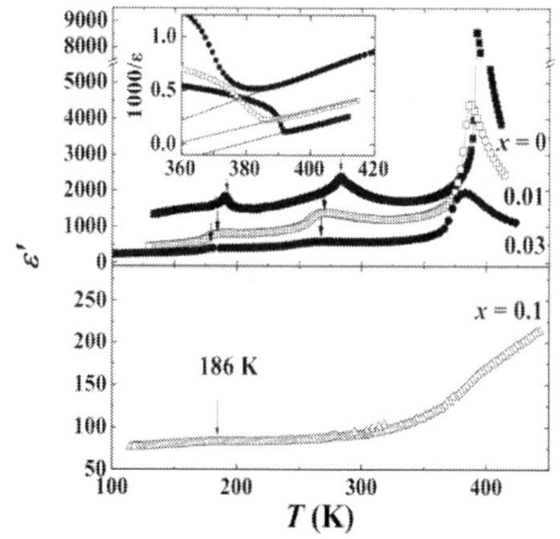

Figure 1. The temperature dependence of the real part of the dielectric permittivity $\varepsilon'(T)$ at 1kHz in $Ba_{(1-x)}Mn_xTiO_3$ solid solution with x=0.01, 0.03, and 0.1. Inset reciprocal permittivity fitted by the Curie-Weiss low.

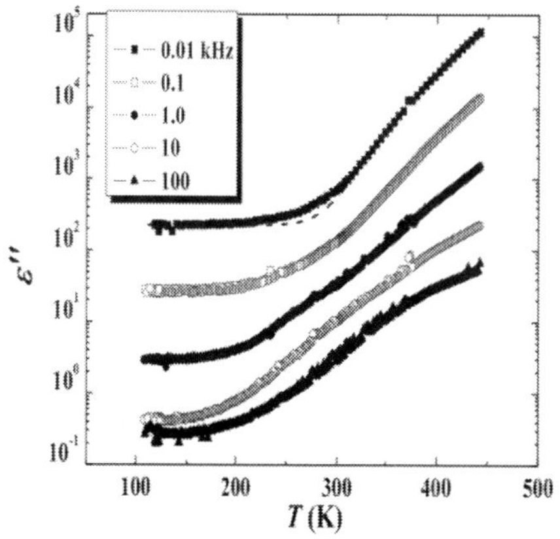

Figure 2. The temperature and frequency dependence of the imaginary part of the dielectric permittivity $\varepsilon''(T)$ in $Ba_{(1-x)}Mn_xTiO_3$ with x = 0.1. Dashed line is the imaginary permittivity fitted by the thermal activated conductivity.

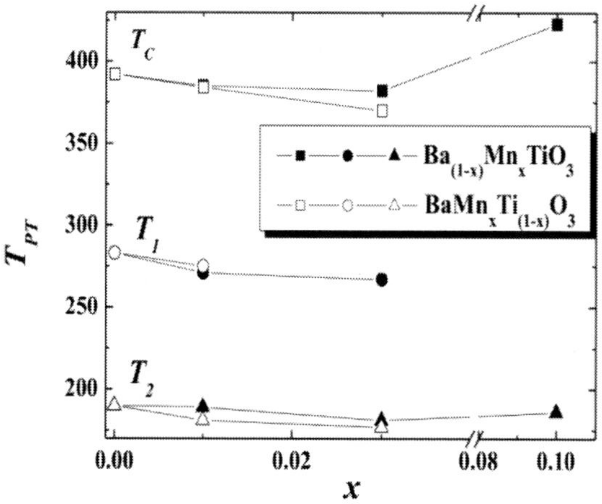

Figure 3. Phase transition temperatures vs. second component concentration x in $Ba_{(1-x)}Mn_xTiO_3$ and $BaMn_xTi_{(1-x)}O_3$ solid solutions (phase transition diagram).

Table 2. $BaMn_xTi_{(1-x)}O_3$ (B -substitution).

	T_C	T_{CW}	C_{CW}
$x = 0$	392 K	378 K	$1.2 \cdot 10^5$ K
$x = 0.01$	385 K	370 K	$7.1 \cdot 10^4$ K

In comparison to BT with A-substitution, the concentration $x = 0.03$ makes the significant changes in the temperature dependence $\varepsilon'(T)$ (figure 4). All the peaks disappear excepting smeared maximum T_2 that is 14 K below the temperature upon the orthorhombic-rhombohedral phase transition in pure BT.

For the composition with concentration $x = 0.1$ any maxima at the temperature dependence of the real permittivity $\varepsilon'(T)$ cannot be recognized. However, the temperature dependence of the imaginary part $\varepsilon''(T)$ (figure 5) shows an evidence of a dielectric relaxation. At the low temperatures range150-250 K it can be recognized the maximum $\varepsilon''(T)$ that shifts up to higher temperature with frequency increasing. In the other range 250-450 K one can also see the maximum upon monotonic increased background curve (figure 5). The maximum is also shifted to higher temperatures as the frequency increases. The relationship between the frequency f (or cycling frequency $\omega = 2\pi f$) and the temperature of maximum can be described using the Arrhenius relationship $\omega = \omega_0 \exp(-E_a/kT_m)$, where ω_0 is Debye frequency, T_m is the temperature of the imaginary dielectric maximum at the given frequency, E_a is the activation energy, and k is the Boltzmann constant. As a result, we estimate the values for both relaxations: for low temperatures $\omega_0 = 1.4 \cdot 10^{12}$ s^{-1} and $E_a = 0.3$ eV, and for high temperatures $\omega_0 = 4 \cdot 10^{15}$ s^{-1} and $E_a = 0.8$ eV.

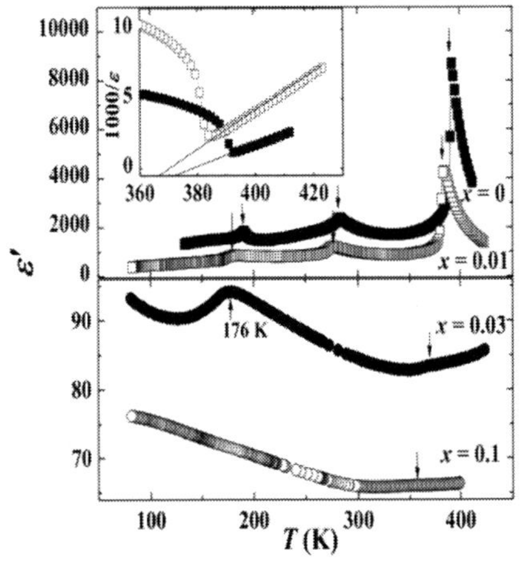

Figure 4. The temperature dependence of the real part of the dielectric permittivity $\varepsilon'(T)$ at 1 kHz in $BaMn_xTi_{(1-x)}O_3$ solid solution with x=0.01, 0.03, and 0.1. Inset reciprocal permittivity fitted by the Curie-Weiss low.

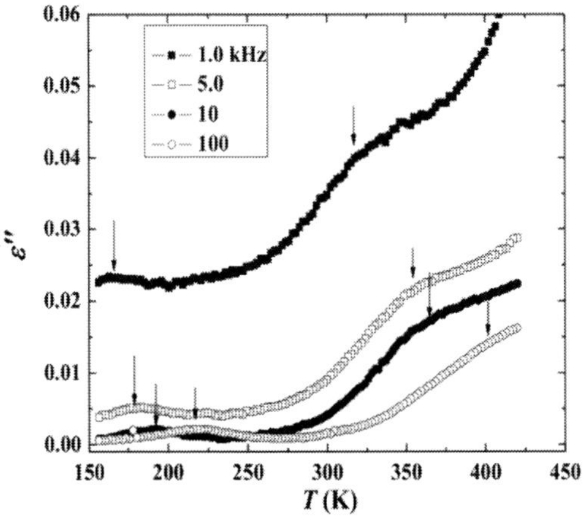

Figure 5. The temperature and frequency dependence of the imaginary part of the dielectric permittivity $\varepsilon''(T)$ in $BaMn_xTi_{(1-x)}O_3$ with x=0.1.

We attribute the low temperature relaxation to impurity centers {Mn^{2+} - Vo} in BT as well as in $SrTiO_3$ [12, 13]. In strontium titanate the Mn^{2+} impurity ion substitutes the Ti^{4+} ion and produces oxygen vacancy V_O in oxygen rhombohedra to compensate the charge. As a result, the dipole impurity centers {Mn^{2+} - Vo} is formed. It is known [12-14] that the thermally activated dipole reorientation via the vacancy jumping in $SrTiO_3$ is observed with $E_a \sim 0.1$ - 0.3 eV and $\tau_0 \sim 10^{-10}$ - 10^{-12} s. Therefore, the estimated parameters E_a = 0.3 eV and $\omega_0 = 1.4 \cdot 10^{12}$ s^{-1} evidence for the same relaxation mechanism in BT.

The high temperature relaxation can be related to electronic subsystem because the high value of Debye frequency $\omega_0 = 4 \cdot 10^{15}$ s^{-1}. The estimated activation energy E_a = 0.8 eV is of the same order as in $SrTiO_3$ E_a = 0.8 eV [14] and can be determined by energy of donor or acceptors in the band gap. On the other hand, it is well known [14, 15] that the thermal activated conductivity in ceramic grains and boundaries produces a strong dielectric relaxation. Therefore, we can associate the high temperature relaxation with the Maxwell-Wagner mechanism due to the heterogeneous nature of the ceramics with grain and boundary layers.

3. Conclusion

We studied the BT ceramic with A and B- substitution: $Ba_{(1-x)}Mn_xTiO_3$ and $BaMn_xTi_{(1-x)}O_3$ with manganese concentration x = 0.01, 0.03 and 0.1. The phase diagrams (T-x) was plotted according to the temperature dependence of the dielectric permittivity Compositions dependences of the phase transition parameters were obtained. Debye relaxation associated with impurity polar centers and relaxation Maxwell - Wagner due to grain and boundary layers in ceramic $BaMn_xTi_{(1-x)}O_3$ with x = 0.1 were investigated.

Acknowledge

The authors would like to acknowledge E. P. Smirnova and A.V. Sotnikov for helpful comments. They thank V. V. Krasovskaya and V. A. Yankovskaya for preparing samples.

References

[1] Smolenskii G, Bokov V,. Isupov V, Krainik N, Pasynkov R and Sokolov A 1984 *Ferroelectrics and Related Materials* (New York: Gordon and Breach Science Publishers)
[2] Lemanov V, Smirnova E, Syrnikov P and Tarakanov E 996 *Phys. Rev.* **B 52** 3151
[3] Benguigui L and Beaucamps Y 1981 *Phys. Rev.* **B23** 5866
[4] Simon A, Ravez J and Maglione M 2004 *J. Phys. Cond. Matt.* 16963
[5] Zhi Yu, Chen Ang, Ruyan Guo and Bhalla A 2002 *J. Appl. Phys.* **92** 1489
[6] Jaffe B, Cook W and Jaffe H 1971 *Piezoelectric Ceramics* (NewYork: Academic Press)
[7] Zalesskii V, Lemanov V, Smirnova E, Sotnikov A and Zaitseva N 2007 *Phys. of the Solid State* **49** 113
[8] Lemanov V, Zaitseva N, Smirnova E and Syrnikov P 1995 *Ferroelectric Letters*, **19** 7
[9] Wang X, Gu M, Yang B, Zhu S and Cao W 2003 *Microelectronic Engineering* **66** 855
[10] Langhammer H, Muller T, Felgner K and. Abicht H, 2000 *Mater. Lett.* **42** 21
[11] Wanga S, Hsu Y, Chu J and Cheng-Hui Wu 2006 *Appl. Phys. Lett.* **88** 042909
[12] Iguchi E and Lee K 1993 *J. Mater. Sci.* **28**, 5809
[13] Lemanov V, Sotnikov A, Smirnova E and Weinacht M 2002 *Phys. of the Solid State* **44** 2039
[14] Neuman H and Arlt G 1986 *Ferroelectrics* **69** 179
[15] O'Neill D, Bowman R and Gregg J 2000 *Appl. Phys. Lett.* **77** 1520-1522

Optical and electrical studies of ZnO thin films heavily implanted with silver ions

N M Lyadov[1*,2], A I Gumarov[2], V F Valeev[1], V I Nuzhdin[1], R I Khaibullin[1,2], and I A Faizrakhmanov[1,2]

[1] Zavoisky Physical-Technical Institute of Russian Academy of Sciences, Sibirsky Tract 10/7, Kazan 420029, Russia
[2] Department of Physics, Kazan Federal University, Kremlyovskaya st. 18, Kazan 420008, Russia

*E-mail: nik061287@mail.ru

Abstract. Thin films of zinc oxide (ZnO) with the thickness of 200 nm have been deposited on quartz substrates by using ion-beam sputtering technique. Then Ag+ ions with the energy of 30 keV have been implanted into as-deposited ZnO films to the fluences in the range of $(0.25-1.00)\times10^{17}$ ions/cm^2 to form ZnO:Ag composite layers with different concentrations of the silver impurity. The analysis of the microstructure has shown that the thickness of the ZnO film decreases, and the Ag dopant concentration tends to the saturation with increasing Ag implantation fluence. The ZnO:Ag composite layers reveal the optical selective absorption at the wavelength of the surface plasmon resonance that is typical for silver nanoparticles dispersed in the ZnO matrix. The red shift of the plasmon resonance peak from 480 to 500 nm is observed with the increase in the implantation fluence to 0.75×10^{17} Ag ions/cm^2. Then the absorption peak position starts the backward motion, and the absorption intensity decreases with the subsequent increase in the implantation fluence. The non-monotonic dependence of the absorption peak position on the implantation fluence has been analyzed within of Maxwell Garnet theory and taking into account the strong sputtering of ZnO films during implantation. The ZnO:Ag composite layers exhibit the p-type conductivity indicating that a part of Ag^+ ions is in the form of acceptor impurities implanted into the ZnO lattice.

1. Introduction

Zinc oxide (ZnO) is a direct-gap semiconductor with the band gap of 3.4 eV. This oxide semiconductor has many attractive properties, in particular the high exciton binding energy of ~ 60 meV at room temperature, which makes it promising for creating ultraviolet LEDs, lasers, photosensors, piezoelectric and transparent electrodes, etc. [1,2]. On the other hand, nanocomposite materials based on the dispersion of metal nanoparticles in optical transparent semiconductors are the objects of intensive studies for over the last 20 years. Depending on the kind of metal nanoparticles these nanocomposite materials have important magnetic, optical, non-linear optical or catalytic properties for applications [3]. A special place is occupied by dispersion systems with nanoparticles of noble metals such as Cu, Ag and Au, which exhibit selective surface plasmon resonance (SPR) absorption in the visible and near infrared region. These unique properties of the plasmon-active particles of noble metals cause a number of new physical effects in the composite semiconductors, such as section increase in Raman scattering, the strong increase in light emitting structures efficiency, etc. [4].

It is well known that usually semiconductor ZnO thin films and bulk crystals reveal the n-type conductivity which is caused either by oxygen vacancies or Zn interstitial atoms being electron donors in the ZnO lattice. However, one of the basic problems of semiconductor optoelectronics based on the ZnO material consists in the production of the ZnO material with the stable p-type conductivity and sufficiently high optical transparency. Silver or nitrogen is the most intensively studied acceptor impurity, using which many authors already obtained the p-type conductivity in ZnO using different methods [1, 5]. In this work we used high-fluence implantation with Ag^+ ions into ZnO films to form p-type ZnO:Ag composite layers with silver metal nanoparticles. The influence of the implantation fluence on the structural, optical and electric properties of the Ag-implanted ZnO films was studied by using scanning electron microscopy with elemental microanalysis, optical spectroscopy, as well as dc electric resistance, Hall and Seebeck effect measurements at room temperature.

2. Experimental

ZnO thin films were deposited on the quartz substrate (SiO_2) by ion-beam sputtering of the zinc metal target (purity: 99.99 %) in oxygen atmosphere. Chemically pure xenon (Xe) as a working gas was used in the sputtering procedure. The energy of Xe^+ ions was 1 keV, and the ion beam current density was 100 $\mu A/cm^2$ during sputtering of the zinc target for 30 min.. Then Ag^+ ions with the energy of 30 keV were implanted into as-deposited ZnO films to high fluences in the range of $(0.25 - 1.00)\times10^{17}$ ions/cm^2 to form composite ZnO layers with different concentrations of silver impurity. The Ag implantation was carried out in ZnO films at room temperature on an *ILU-3* ion-beam accelerator (Zavoisky Physical-Technical Institute, Kazan). Some as-prepared samples were subjected to thermal annealing for 15 min in vacuum (10^{-6} Torr) or in air at the temperatures ranging from 500 to 1000 K.

The measurements of the ZnO film thickness before and after ion implantation, as well the analysis of the surface morphology and elemental composition of samples under study were performed on an *EVO 50 XVP* scanning electron microscope (SEM) (*Carl Zeiss*) equipped with an *INCA Energy-350* elemental analyzer (*Oxford Instruments*). Optical spectra were measured on a *Hitachi-330* dual-beam spectrophotometer in the wavelength range of 200 – 700 nm. The standard van der Pauw geometry was used in the measurement of the electrical resistivity, Hall and Seebeek parameters of as-deposited and implanted ZnO films. All above physical measurements were performed at room temperature.

3. Results and discussion

Figure 1 shows cross-section SEM images of as-deposited ZnO film and the same film implanted with Ag^+ ions to the fluence of 1.0×10^{17} ions/cm^2. It is seen clearly that the initial film thickness is 215 nm, and it decreases to 165 nm due to strong sputtering of the ZnO film after high-fluence implantation. These values of the film thickness before and after silver ion implantation was used to estimate the sputtering coefficient as $S \cong 4$ atom/ion with allowance for the ZnO atomic density of 8.28×10^{22} at/cm^3 and the fluence value.

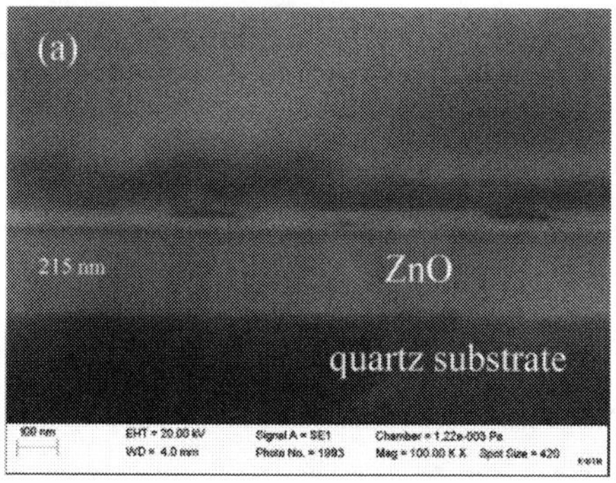

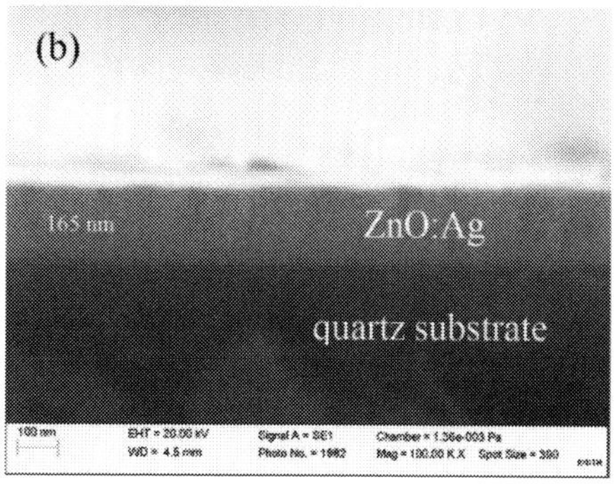

Figure 1. Cross section SEM images: (a) – as-deposited ZnO film, (b) – the same film implanted with Ag^+ ions to the fluence of 1.0×10^{17} ions/cm^2.

The analysis of in-plane SEM images of ZnO thin films shows that all samples have a homogeneous and smooth surface before and after ion implantation. Table 1 presents the elemental composition of the samples under study. It shows the presence of implanted silver, and zinc and oxygen, elements forming the ZnO structure. Since the probing depth of 20 keV electrons in the microanalysis is on the order of 1 μm, the high silicon and oxygen concentration is due to the quartz substrate (SiO_2), and the low Ag content is induced by the averaging over the full depth probing of the electron beam.

Table 1. Elemental composition of as-deposited and Ag-implanted ZnO thin films.

Sample	**Fluence (ion/cm^2)**	**Composition (at.%)**			
		O (K_α)	**Si (K_α)**	**Zn (K_α)**	**Ag (K_α)**
ZnO as-deposited		56.1	28.2	15.7	
ZnO:Ag(1)	0.25×10^{17}	55.3	29.0	15.3	0.4
ZnO:Ag(2)	0.50×10^{17}	54.7	29.4	15.1	0.8
ZnO:Ag(3)	0.75×10^{17}	54.1	31.2	14.9	0.8
ZnO:Ag(4)	1.00×10^{17}	53.6	31.4	14.2	0.8

Depth profiles of the silver concentration were calculated using the *SRIM-2008* simulation code [6]. We used the formula 11 from [7] to take into account the surface sputtering of ZnO with the above coefficient S = 4 atom/ion during implantation. The calculated profiles of the silver concentration in ZnO films implanted to different fluences are shown in figure 2. According to SRIM calculations, the mean range (R_p) of 30 keV Ag ions is about 12.1 nm with a straggling (ΔR_p) of 4.8 nm in the Gaussian-like depth distribution. Thus, the real thickness of the modified ZnO layer amounts to (Rp +2ΔRp) ~ 20-25 nm. The shape of the implantation profile initially changes with the fluence due to surface sputtering. The profile shape becomes smooth and the peak of Ag concentration shifts to the surface with increasing fluence. Our calculations and EDS elemental microanalysis show that the silver concentration increases with increasing fluence, and it tends to the saturation nominally reaching 25 at % in the peak of the depth profile for the highest fluence of 1.0×10^{17} ion/cm^2. Such high concentration of the doping impurity leads to its precipitation in the form of Ag nanoparticles.

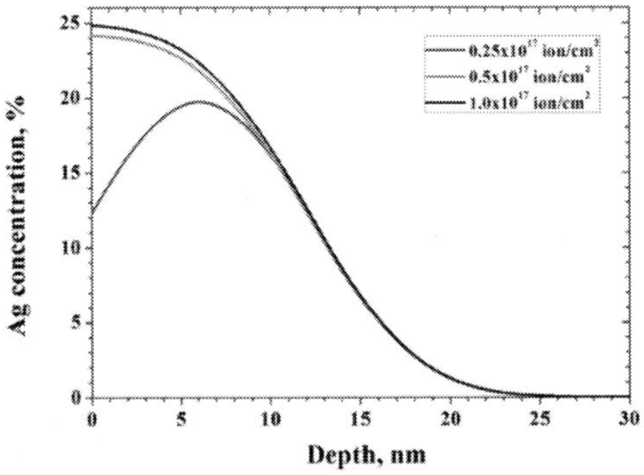

Figure 2. The depth profiles of silver concentration in ZnO films implanted with Ag^+ ions to different fluences.

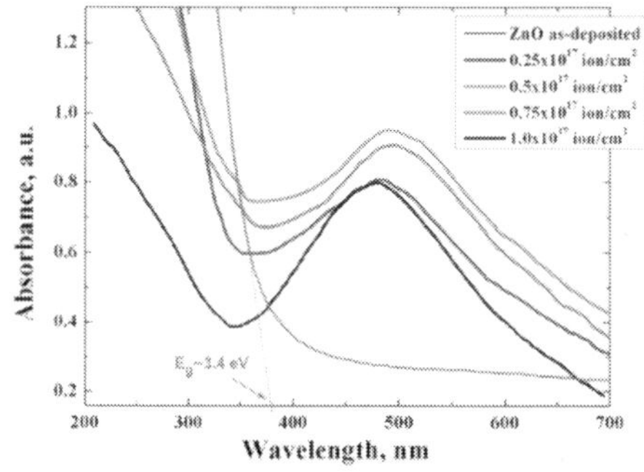

Figure 3. Optical absorption spectra of as-deposited ZnO film and ZnO films implanted with Ag^+ ions to different fluences.

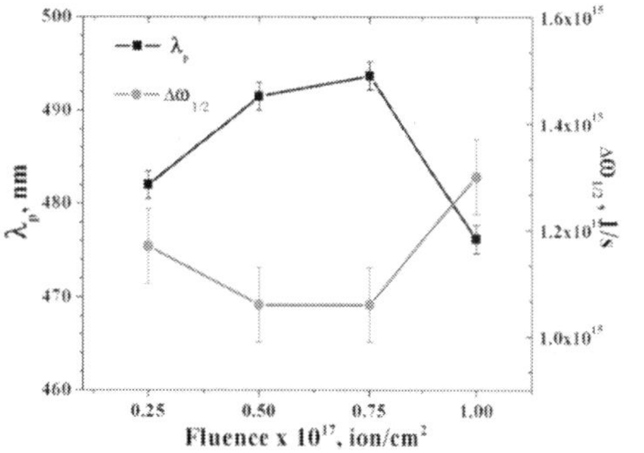

Figure 4. Fluence dependencies of peak position (λ_p) and half-width ($\Delta\omega_{1/2}$) of plasmon resonance in Ag-implanted ZnO films.

The formation of silver nanoparticles in Ag-implanted ZnO films was confirmed by optical absorption spectroscopy. Figure 3 shows the changes in optical absorption spectra of ZnO films after Ag ion implantation at different fluences. It can be seen that the selective absorption is observed at the wavelength $\lambda_p \cong 482$ nm in the sample implanted at low fluence 0.25×10^{17} ions/cm^2. The absorption increases in intensity and shifts toward longer wavelengths for large fluence values. Then the absorption peak starts the backward motion, and the absorption intensity decreases in the ZnO film implanted to the highest fluence of 1.0×10^{17} ions/cm^2. The observed selective absorption is attributed to the surface plasmon resonance (SPR) in the small silver particles because the dispersed isolated Ag ions or atoms do not exhibit such resonance absorption.

The dependences of the peak position (λ_p) and half-width ($\Delta\omega_{1/2}$) of plasmon resonance on the implantation fluence in Ag-implanted ZnO films are shown in figure 4. The experimental dependences are not monotonic and have a maximum for the peak position and a minimum for the half-width at the fluence of 0.75×10^{17} ion/cm^2. Note that similar fluence dependences were observed earlier in the polymers heavily implanted with 30 keV Ag^+ ions [8, 9]. It was shown that non-monotonic behavior of the SPR peak position versus the implantation fluence originates from the transformation size of silver nanoparticles and their agglomeration at high fluences. In fact, the peak position and absorption bandwidth depend on the size of silver nanoparticles according to both well-known Mie and Maxwell Garnett theories with allowance for the size-dependent collision frequency in metal particles [10, 11]. In particular, we used the bandwidth at half-maximum absorption (half-width) to estimate silver particle dimensions in the implanted ZnO films by using formula [10]:

$$R = \frac{\upsilon_F}{\Delta\omega_{1/2}} \qquad (1),$$

where R is the radius of spherical particles formed in the Ag-implanted ZnO film, $\Delta\omega_{1/2}$ are experimental values of the absorption half-width, and $\upsilon_F = 1.4\times10^6$ m/s is the Fermi velocity for the electron in silver. According to our calculations, the decrease in the half-width indicate the increase in the mean size of nanoparticles from 2.4 to 2.7 nm in the fluence range (0.25 to 0.75)$\times10^{17}$ with the subsequent decrease to 2.2 nm as the fluence is increased to 1.0×10^{17} ions/cm^2. This non-monotonic change of the nanoparticle size with increasing fluence can be understood by taking into account the non-uniform depth distribution of the implanted silver and strong sputtering of ZnO films during implantation. The Gaussian-like shapes of Ag concentration curves (fig.2) allow an assumption that there is a wide size distribution of the forming particle over the depth of the implanted region. This means that the high concentration of the Ag impurity in the peak of the distribution stimulates the growth of silver nanoparticles with larger sizes in the near-surface region, whereas the lower density of Ag atoms in the tail of the concentration curve contributes to the nucleation of smaller size clusters that dominate at the depth of the ZnO film. At the first stage, the sizes of large and small silver particles increase with increasing fluence to 0.75×10^{17} ions/cm^2 as it is seen from our calculations. These silver nanoparticles move steadily toward the surface due to sputtering of the ZnO film. At fluences above 0.75×10^{17} ions/cm^2, the nanoparticles with large sizes come to the surface and disappear due to the apparent loss of Ag from the sample surface during its sputtering. The latter results in the decrease in the mean size of silver nanoparticles and the overall optical absorption decreases, as it takes place in the sample implanted to the highest fluence of 1.0×10^{17} ions/cm^2.

According to calculations [11] obtained using the Mie theory, the position of the maximum of the extinction coefficient is almost independent of the size of silver particles in radius range of 1-10 nm, while in the experiment we observe a noticeable shift of the absorption maximum (fig. 4). However, the Mie theory is valid only for low concentrations of metal particles in dielectric matrix. In our case, the Maxwell Garnett theory [12] is more appropriate, according to which the peak position of the maximum absorption is determined principally by the filling factor (Q) of metal nanoparticles in ZnO matrix as:

$$|\varepsilon_1(\lambda_p)| = \frac{(2+Q)}{(1-Q)} n(\lambda_p)_s^2 \qquad (2),$$

where $\varepsilon_I(\lambda_p)$ is real part of the complex dielectric function of the silver at the wavelength λ_p corresponding to the maximum of the plasmon resonance absorption at a given implantation fluence, and n_s is refractive index of ZnO film (here $n_s \cong 1.98$ at plasmon resonance wavelength). According to formula (2), the experimental red shift of the absorption maximum is induced by the increase in the filling factor from f≅0.05 to 0.1 with increasing implantation fluence of silver. The backshift in short-wave region for highest fluence of 1.0×10^{17} ions/cm^2 is probably due either a decrease in the refractive index of the implanted ZnO film or the apparent loss of Ag metal from the sample during its sputtering as it was above described.

Table 2 presents the results of electrical measurements in as-deposited ZnO films and the same films implanted with Ag^+ ions to the fluence of 0.25×10^{17} ions/cm^2. As expected, initial deposited ZnO films reveal the n-type conductivity and have the low specific resistance (~10^{-3} Ohm·cm) due to the high value of the free electron concentration (~10^{20} cm^{-3}) in the ZnO conduction band. The high carrier concentration in semiconductor oxides, such as ZnO, CuO, TiO_2 etc., usually caused by the low oxygen content in the stoichiometric composition of oxide materials. In fact, subsequent annealing of as-deposited ZnO films in the reducing environment of air at 1000 K leads to the recovery of stoichiometry and the huge decrease in electric conductivity. The annealed ZnO films exhibit the high-resistance behavior at room temperature in the absence of the doping impurity.

Table 2. Electrical parameters of as-deposited ZnO film and the same film implanted with Ag^+ ions to the fluence of 0.25×10^{17} ions/cm^2.

Sample	Type of conductivity	Carrier mobility (cm^2/V·s)	Carrier concentration (cm^{-3})	Specific resistance (Ohm·cm)
ZnO as-deposited	n	9.43	$2.5\cdot10^{20}$	$2.65\cdot10^{-3}$
ZnO as-deposited, *after annealing in air at 1000 K*	-	-	-	high-resistance
ZnO:Ag (1), implanted	p	1.2	$2.8\cdot10^{16}$	220
ZnO:Ag (1) implanted *after annealing in air at 1000 K*	-	-	-	high-resistance
ZnO:Ag (1), implanted *after one year of storage in air*	p	1.2	$2.8\cdot10^{16}$	220

High-fluence implantation with silver ions into as-deposited ZnO film also results in the significant increase in their specific resistance to the value about 10^2 Ohm·cm. It is well known that ion implantation induces a large number of structural defects in the crystal lattice of the implanted material. Thus, the observed decrease in the electrical conductivity in Ag-implanted ZnO film can be caused by the formation of deep electron traps in the ZnO band gap. But more important is the fact that the type of conductivity of the ZnO film changes from n-type (by electrons) to p-type (by holes) after Ag ion implantation as it shows our Hall and Seebeek effect measurements. This means that a part of the Ag^+ ions is implanted into the ZnO lattice in the form of acceptor impurities. Note that the carrier mobility and concentration listed in Table 2 for Ag-implanted ZnO sample are very similar to the ones in p-type ZnO films doped homogeneously with Ag impurity [13]. Moreover, the measurements of the electrical properties with the time interval of 12 months showed the stable p-type conductivity in Ag-implanted ZnO films. It should be noted that the carrier concentration and the mobility of charge carriers in the sample were also unchanged. Annealing of Ag-implanted samples in air at temperatures above 500 K results in the high-resistance state of ZnO films. On the other hand, annealing in high vacuum at the same temperatures causes the complete evaporation of as-deposited and Ag-implanted films.

4. Conclusions

Our structural and optical studies show that high-fluence implantation with Ag^+ ions make it possible to synthesize silver nanoparticles in the surface region of the semiconductor ZnO matrix. The ion-synthesized silver nanoparticles reveal selective absorption at the wavelength of surface plasmon resonance in optical spectra of Ag-implanted ZnO films. It is established that the intensity and spectral position of surface plasmon resonance are strongly dependent on the implantation fluence. ZnO:Ag layers with silver metal nanoparticles exhibit the stable p-type electric conductivity that is important for practical applications of these composite materials.

Acknowledgments

This study was supported by Russian Youth Scientific-Innovation Program "U.M.N.I.K.".

References

[1] Ozgur U, Alivov Ya I, Liu C, Teke A, Reshchikov M A, Dogan S, Avrutin V, Cho S.-J., Morkoc H 2005 *J. Appl. Phys.* **98** 041301
[2] Gorla C R, Emanetoglu N W, Liang S, Mayo W E, Lu Y et al. 1999 *J. Appl. Phys.* **85**(5) 2595
[3] Stepanov A L, Khaibullin I B 2005 *Rev. Adv. Mater. Sci.* № 9 109
[4] Maier S A 2007 *Plasmonics: Fundamentals and Applications, Springer Science* **XXV** 223 p.
[5] Sun F, Shan C X, Li B H, Zhang Z Z, Shen D Z, Zhang Z Y, and Fan D 2011 *Optics Letters* **36**(4) 499
[6] Ziegler J F, Biersack J P, Littmark U 1985 *The Stopping and Range of Ions in Solids* (Pergamon Press, New York) (*SRIM-2008* software at http://www.srim.org/)
[7] Achkeev A A, Khaibullin R I, Tagirov L R, Mackova A, Hnatowicz V, and Cherkashin N 2011 *Physics of Solid State* **53**(3) 543
[8] Khaibullin R I, Osin Y N, Stepanov A L, Khaibullin I B 1998 *Vacuum* **51** (2) 289
[9] Khaibullin R I, Osin Y N, Stepanov A L, Khaibullin I B 1999 *Nucl. Instr. and Meth. in Phys. Res.* **B 148** 1023
[10] Arnold G W 1975 *J. Appl. Phys.* **46**(10) 4466
[11] Arnold G W, Borders J A 1977 *J. Appl. Phys.* **48**(4) 1488
[12] Maxwell Garnett J.C., 1906 *Philos. Trans. R. Soc. Lond* **205** 237
[13] Duan L., Zhang W., Yu X., Wang P., Jiang Z., Luan L., Chen Y., Li D., 2013 *Solid State Commun.* **157** 45.

Journal of Physics: Conference Series **572** (2014) 012023 doi:10.1088/1742-6596/572/1/012023

Sol-gel synthesis of Pb-free thin-film nanomaterials for electrocaloric devices

E V Abrashova, I E Kononova, V A Moshnikov, S S Nalimova

Saint-Petersburg Electrotechnical University "LETI", Pr. Popova 5, St. Petersburg, 197376 Russia

E-mail: katabr@mail.ru

Abstract. Lead-free $Ba_xCa_{(1-x)}TiO_3$, $BaSn_yTi_{(1-y)}O_3$ and composite $Ba_xCa_{(1-x)}TiO_3$ / $BaSn_yTi_{(1-y)}O_3$ thin films were synthesized by a sol-gel technique. A large electrocaloric effect is expected in the obtained films. Atomic force microscopy studies revealed that the thin films consists of ellipsoidal objects enclosed by an amorphous matrix. The electrical properties of synthesized nanomaterials were investigated by impedance spectroscopy.

1. Introduction.

Recently the development of nanoscale electronic devices has posed a problem of heat removal from very small surface area. There is a great interest in the design of miniature solid-state cooling systems having low power consumption and maximum temperature drop. To date, devices based on thermoelectric effects, in particular, the electrocaloric [1-3] are the most promising for the study and creation.

The electrocaloric effect is described as a reversible temperature change of the material under applied electric field at adiabatic conditions. The physical reason for the effect is the reorientation of the electric dipoles in the presence of an electric field and a decrease in entropy of the sample and increasing its temperature under adiabatic conditions. When removing the electric field dipoles acquire random orientation, entropy increases and the temperature decreases [2]. Thus, the presence of electric dipoles in the material is a prerequisite for obtaining the electrocaloric effect.

Perovskite-like lead-containing materials are generally used to produce cooling devices based on electrocaloric effect [1]. Lead is a heavy metal, well known to be extremely toxic to all life forms.

It is known that electrocaloric effect in ferroelectrics is strong near the Curie temperature. Therefore, the development of materials with phase transition at operating temperature is an important task. Table 1 shows phase transition temperatures and maximum values of temperature change in some materials under adiabatic conditions driven by the change in the electric field. [2].

Another problem in the production of miniature cooling devices is to achieve high electric field intensity while minimizing the applied voltage. This effect can be realized by the reduction in thickness of the active material. Thin-film nanomaterials best meet these requirements. Sol-gel process is one of the most suitable techniques to obtain such materials because of its economical efficiency, ease equipment, the ability to easily control the composition of resulting products and a wide range of precursors [4-6]. Sol-gel process has already become a well-established technique for the synthesis of functional layers used as gas sensors, inorganic adsorbents, catalysts and catalyst supports, synthetic zeolites, ceramics with special thermal, optical, magnetic and electrical properties, glass ceramics, transparent conductive coatings etc. [7-9].

Table 1. Maximum values of electrocaloric cooling (ΔT) and respective electric field changes (ΔE). T_{ph} - phase transition temperatures.

Material	T_{ph}, K	ΔT, K	ΔE, V/cm
$Pb_{0,99}Nb_{0,02}[(ZrSn_{0,27})_{0,93}Ti_{0,07})]_{0,98}O_3$ (ceramics, 250 nm)	420	2	20
$PbZr_{0,95}Ti_{0,05}O_3$ (film, 300 nm)	500	12	480
$PbSc_{1/2}Ta_{1/2}O_3$	291	1,8	25
$PbSc_{1/2}Ta_{1/2}O_3$ (16-layers ceramic film, 16×100 μm)	273	2,4	138
$0,85PbMg_{1/3}Nb_{2/3}O_3 - 0,15PbTiO_3$ (ceramics, 100 μm)	291	1,6	16
$0,9PbMg_{1/3}Nb_{2/3}O_3 - 0,1PbTiO_3$ (film, 260 nm)	350	5	895

The objective of this work was to develop an alternative lead-free material having large electrocaloric effect. The composites of $Ba_xCa_{(1-x)}TiO_3$ and $BaSn_yTi_{(1-y)}O_3$ at various ratios near the phase transition point were chosen for the study. The development of such complex materials requires the strict control of synthesis parameters having considerable influence on the structure and properties of composites. The choice of these composite materials is determined by their physical-chemical properties such as phase transition temperature, crystal lattice structure, the possibility to obtain electrocaloric effect [3].

2. Experiment.

Barium acetate, calcium acetate monohydrate, tin acetate and titanium isopropoxide were used as precursors in the synthesis process. Titanium isopropoxide was chosen as a gelation agent. Ethylene glycol monomethyl ether and acetic acid were used as a solvent and a reaction activator. Additional reagents provided the complete dissolution, the desired viscosity of solutions and structure of thin films. The solutions with different precursor concentrations were prepared and cast onto glass and silicon substrates. The resulting layers were spin-coated (3000 rpm) during 15 s and then annealed at different temperatures in the range 100 °C-600 °C.

The surface morphology of thin-film nanomaterials was studied by tapping mode atomic force microscopy (AFM) using a scanning probe nanolaboratory NTegra Therma (NT-MDT, Russia). Composite thin films were characterized by impedance spectroscopy (Z-500P, Elins, Russia). Measurements were done in the frequency range from 1 Hz to 500 MHz at 25 °C.

3. Results and Discussion.

AFM analysis showed that sphere-like aggregates form as a result of nucleofilic growth during sol ageing. Processes of aggregation and growth depend on precursor ratio of solutions. The grains have an ellipsoidal shape. The estimated sizes of the long and short diameters of grains are 160 nm and 130 nm, respectively (Fig. 1a, 1b). From a macroscopic point of view, the films are dense and uniform. It was observed that the layer completely covers the surface substrates. Thus, the films have good adhesion to glass and silicon which are commonly used to produce electronic devices.

Fractal dimension of the films was calculated by decomposition method using a free program Gwyddion for characterization of the surface. Value of fractal dimension was 2.57, 2.51, and 2.56, for Figures 1a, 1b, 1c respectively.

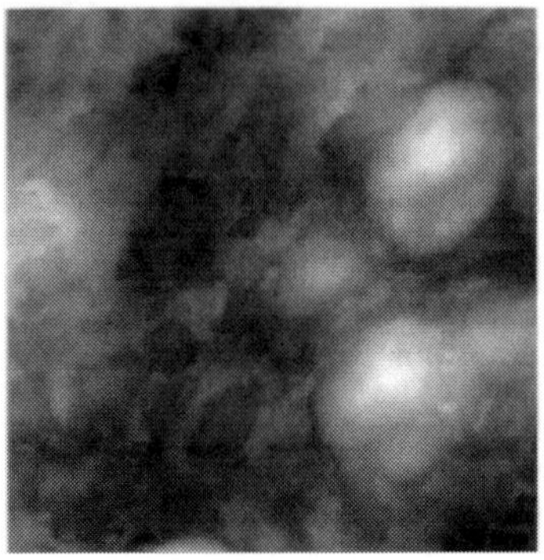

Figure 1a. Typical AFM image of $Ba_xCa_{(1-x)}TiO_3$ / $BaSn_yTi_{(1-y)}O_3$ thin film (the scanning area size is 0,5×0,5 μm).

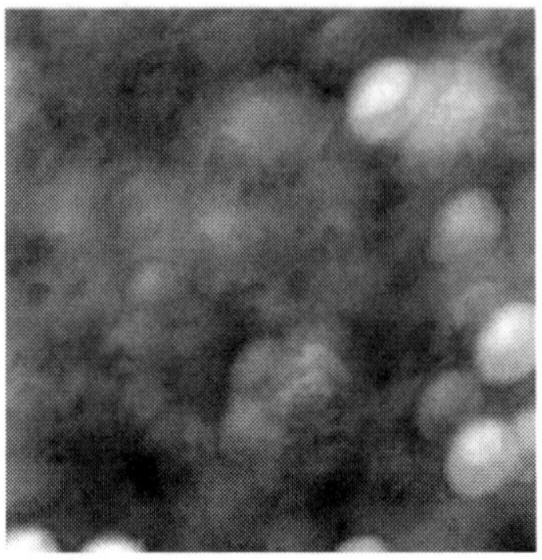

Figure 1b. Typical AFM image of $Ba_xCa_{(1-x)}TiO_3$ / $BaSn_yTi_{(1-y)}O_3$ thin film (the scanning area size is 1×1 μm).

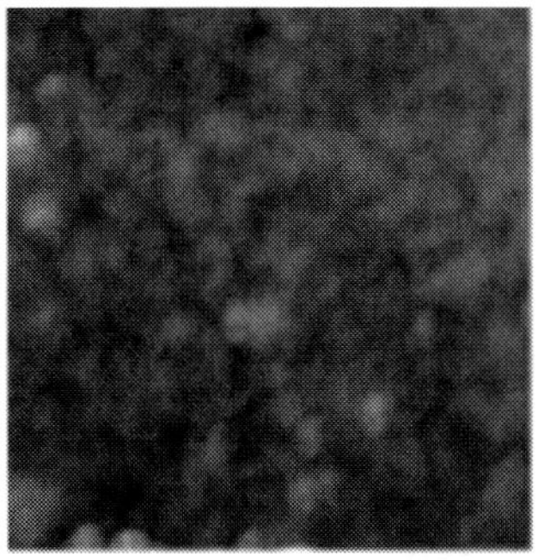

Figure 1c. Typical AFM image of $Ba_xCa_{(1-x)}TiO_3$ / $BaSn_yTi_{(1-y)}O_3$ thin film (the scanning area size is 2×2 μm).

The typical impedance spectrum of composite thin film presented as Nyquist plot is shown in fig. 4. Two regions corresponding to different relaxation processes are observed in Nyquist plot.

Impedance spectrum in the low-frequency range (region I – from 1 Hz to 10 kHz) can be approximated by a straight line and can be fitted to an equivalent circuit containing constant phase element (CPE). The impedance of a constant phase element is described by the equation [13]:

$$Z = 1 / A(j\omega)^n,$$

where A is a frequency-independent constant, n is an exponential index.

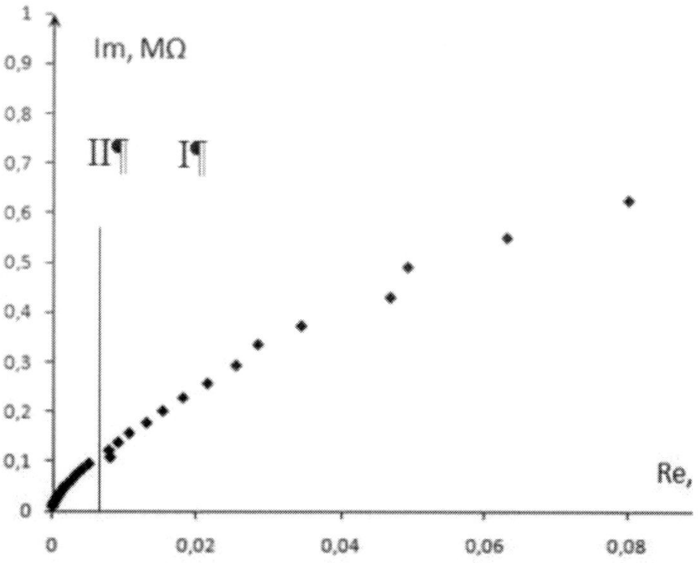

Figure 2. Impedance hodograph of $Ba_xCa_{(1-x)}TiO_3$ / $BaSn_yTi_{(1-y)}O_3$.

In the high frequency range (region II – from 10 kHz to 500 kHz) can be approximated by a circle segment with the center below the x axis and equivalent circuit of composite film consists of a CPE connected in parallel with a resistor. The changing view of the curve indicates that electrical barriers are exist in films. They are responsible an additional mechanism of dielectric polarization and the presence of polar regions [13]. In high-frequency range the processes of dipole relaxation and elastic dipole polarization contribute to electric properties [14, 15]. Thus, dipole structures that can cause electrocaloric effect are observed in the composite films.

Our further research will be aimed at experimental confirmation of electrocaloric effects in the obtained composite films.

4. Conclusion

In this study sol-gel technique was developed to synthesize lead-free thin-films of the composition $Ba_xCa_{(1-x)}TiO_3$ / $BaSn_yTi_{(1-y)}O_3$. A high electrocaloric effect is expected in such structures. It was found that the structure consists of sphere-like aggregates in an amorphous matrix. It was also shown that the processes of dipole relaxation and elastic dipole polarization occur in the obtained films that make them promising for electrocaloric devices.

5. References

[1] B.Peng, H.Fan, Q.Zhang // A Giant Electrocaloric Effect in Nanoscale Antiferroelectric and Ferroelectric Phases Coexisting in a Relaxor $Pb_{0.8}$ $Ba_{0.2}$ ZrO_3 Thin Film at Room Temperature / Adv. Funct. Mater. 2013. P. 1-6.A reference

[2] L.P. Bulat // Solid State Cooling Systems / Science & Technology. 2008. V 8. P. 10-17

[3] G. Suchaneck, A. Kleiner, G. Gerlach // Electrocaloric multilayer and radial cooling device concepts / SPP Focus Meeting B, Braunschweig, October 1st, 2013

[4] C.J.Brinker, G.W.Scherer // Sol-Gel Science. The Physics and Chemistry of Sol-Gel Processing / SanDiego: Academic Press. 1990. 908 p.

[5] S.Sakka // Handbook of sol-gel science and technology: processing, characterization, and applications / Ed.. New York, 2004.–V.1-3
[6] R.Corriu, T.A.Nguyen // Molecular Chemistry of Sol-Gel Derived Nanomaterials. John Wiley & Sons, 2009.
[7] I. E. Gracheva, V. A. Moshnikov, S. S. Karpova, E. V. Maraeva // Net-like structured materials for gas sensors / Journal of Physics: Conference Series. 2011. V. 291, № 1. P. 012017.
[8] E.V.Abrashova., M.V.Baranovskiy // Preparation and Analysis of spectral characteristics of nanocomposites based on wide band gap semiconductors metal oxides of ZnO-SnO_2-SiO_2/ News of the St. Petersburg Electrotechnical University, 2013. V. 5. p. 16-21.
[9] E.V.Abrashova, I.E.Gracheva, V.A.Moshnikov // Functional nanomaterials rased on metal oxides with hiererhical structure / Journal of Physics: Conference Series. 2013. T. 461. C. 012019.
[10] E.Barsoukov, J.Ross Macdonald // Impedance Spectroscopy. Theory, Experiment, and Applications. Second Edition.–Willey: Interscience, 2005.–595 p
[11] M.E.Orazem, T.Bernard // Electrochemical Impedance Spectroscopy. The Electrochemical Society Series.–Wiley: Interscience, 2008.–524 p
[12] A.S. Bozhinova, N.V. Kaneva, I.E. Kononova et al..// Photocatalite study and sensor properties of nanocomposite ZnO/SiO_2 layers. Semiconductors. V. 47, N 12. P. 1662-1666. 2013
[13] E.A.Ukshe, N.G.Bukun // Development of a model of adsorption relaxation in the double layer super ionic semiconductors / Eelectrochemistry, 1990, V. 26, N.2, p. 1373- 1381
[14] N.M. Olehnovich, Yu.V.Rad'ush, A.V.Pushkarev // Mechanisms of dielectric polarization of perovskite relaxor ferroelectrics ceramics $(1-x)(NaBi)_{1/2}\,TiO_3 - xBi(ZnTi)_{1/2}\,O_3$ / Solid State Physics, 2012, V. 54, N. 11
[15] N.M.Olehnovich, I.I.Moroz, A.V.Pushkarev et al. // Thermal impedance spectroscopy of solid solutions $(1\text{-}x)Na_{0.5}\,Bi_{0.5}\,TiO_3 - x\,LaMg_{0.5}\,Ti_{0.5}\,O_3$ / Solid State Physics, 2008, V.50, N3, p 472-478

Carbon encapsulation of magnetic metal nanoparticles: correlation between nanoscale structure of carbon matrix and electromagnetic properties

M S Chekulaev[1,3], S V Kozyrev[2], V I Ivanov-Omskii[3], S G Yastrebov[3], T K Zvonareva[3] and A V Siklitskaya[3]

[1] Saint-Petersburg National Research University of Information Technologies, Mechanics and Optics, Saint-Petersburg, Russia
[2] Center for Advanced Studies, Saint-Petersburg, Russia
[3] A.F.Ioffe Physico-technical Institute RAS, Saint-Petersburg, Russia

E-mail: mchs89@gmail.com

Abstract. A comparative analysis of modification of Raman spectra of hydrogenated amorphous carbon films containing encapsulated cobalt nanocrystals, their EM radiation losses and distribution function of nanosized fragments of graphene flatness constituting skeleton of amorphous carbon is performed. For this purpose, a model of confinement of graphene phonons by its fragments is developed, for two types of phonons contributing to the D and G peaks of Raman spectra, as a function of Co content.

1. Introduction

Metal nanoclusters are promising materials for creation on their basis magnetic media for ultra-high density recording and electromagnetic (EM) radiation absorbers. The ensembles of encapsulated clusters show emerging quantum physical effects. Methods of introduction of metal clusters in carbon matrix, also termed as "encapsulation" methods, are of two types. In the first method, metallic clusters are encapsulated into carbon cells. In the second method nanoclusters are embedded into a matrix of amorphous carbon. This technique is less expensive, simpler and more effective than the first one and compatible with thin film technologies for microelectronics.

2. Experimental

Films of amorphous carbon modified with Co (a-C:H(Co)) were produced by simultaneous sputtering of graphite and cobalt targets by Argon (80%) –Hydrogen (20%) plasma in DC magnetron. The films were deposited on Si substrates (for Raman experiments) and aramide tissue (for microwave absorption experiments). Cobalt content was varied by a change of surface of sputteing Co target. Its concentration in the produced films was controlled by Rutherford backscattering method as published elsewhere [7].

Intensity of Raman scattering (figure.1) was measured by SPEX RAMALOG spectrometer using unpolarised light with 488 nm wavelength operating in backscattering geometry.

Microwave absorption experiments were performed using technique published elsewhere [6].

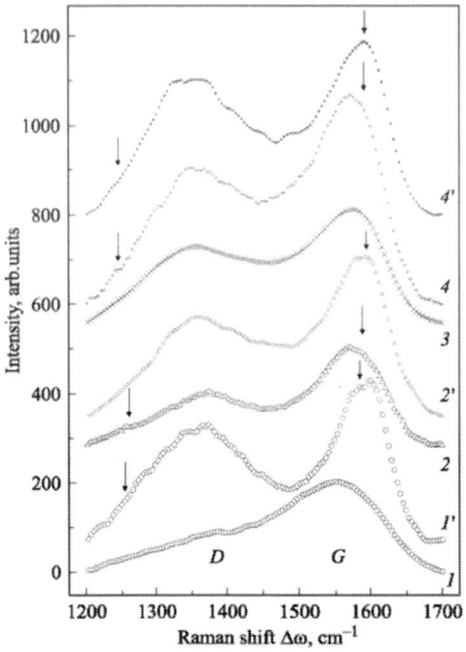

Figure.1. Raman spectra of a-C:H films (1,1') and a-C:H<Co> (2, 2',3,4,4') with different Co content: (1-4) – before thermal annealing and (1',2',4') – after thermal annealing: [Co/C]: 1, 1'—0; 2, 2':0.35; 3—0.5; 4, 4'—0.73

3. Results and discussion

We used data for microwave absorption published elsewhere [6] (figure.2) for analysis. The dependence of losses for microwaves measured at 10 GHz on Co content is presented in figure.3. It is seen that the dependence attains a maximal value for a certain Co content, for approximately 33 at.%.

These data for microwave absorption were obtained for the samples whose Raman spectra were investigated simultaneously and partially published in paper [7] and presented in figure.1.

Typical Raman spectrum of a-C:H(Co) film is presented in figure.4. The overall shape of the contours was not changed drastically with variation of Co concentration while positions of peaks D and G bands and their ratio were changed.

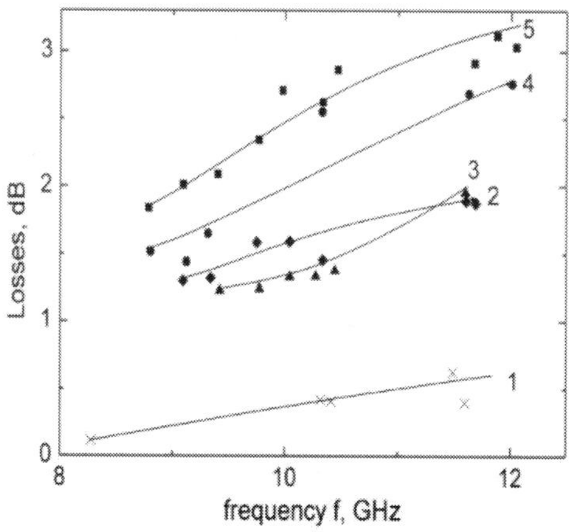

Figure.2. RF losses L of electromagnetic waves in four-layer a-C:H Co covers on aramide tissues. The a-C:H Co film thickness on aramide filaments and the Co content are: 1— four layers of the aramide tissue ; 2— 1.05 mm, 32-at.% Co; 3— 0.95 mm, 39-at.% Co; 4— 0.95 mm, 45-at.% Co; 5— 2.1 mm, 39-at.% Co;

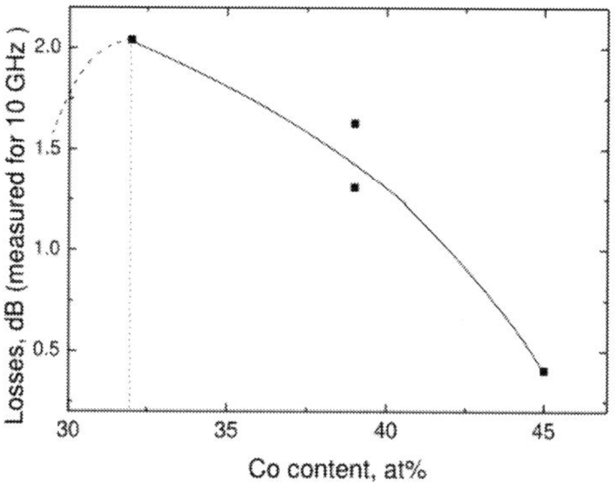

Figure.3. Black squares stand for the dependence of losses of microwaves measured at 10 GHz on Co content for different a-C:H(Co) samples. The curve shows the trend.

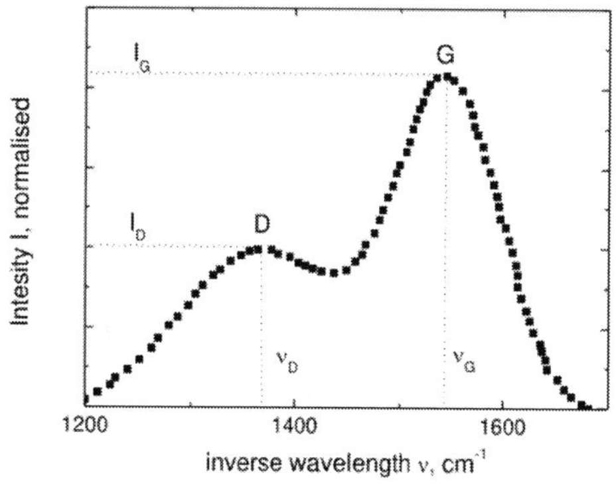

Figure.4. Typical Raman spectrum of a-C:H(Co) spectrum. Intensities of D and G bands and their spectral positions are marked as I_D, I_G and ν_D, ν_G correspondingly.

A quite natural to associate the effect of the intensity changes with variation of the mean size of graphene fragments constituting skeleton of carbon matrix. This variation may occur as a result of presence of Co atoms that may be scattered in the matrix. For the estimation of the mean size we used results presented in paper [8] where a correlation between size of graphene fragment and I_D/I_G ratio was investigated. The result is presented in figure.5.

The experimental dependence of I_D/I_G ratio obtained from figure.2 is shown in figure.6 together with restored diameter of the mean graphene fragments constituting matrix of a-C:H(Co) matrix. The restoration was performed using I_D/I_G ratio and dependence presenting in figure.4. As is easily seen, the mean size of graphene fragment slightly increases with increase of concentration of Co, while the overall trend attains saturation at higher Co values. The Co/C ratio where the trend attains saturation is about 0.5 (that corresponds to almost 30 at.% of Co). This concentration corresponds to a stechiometric formula CoC_2 for the considering substance. At the same time, as is seen in figure.6, the mean diameter of graphene fragment equals to approximately 11 Å.

This value correlates with concentration of Co where losses of microwaves are maximal, as figure.3 shows. As one might expect a molecular cluster with approximate CoC_2 formula might be responsible for absorption of microwaves and might be associated with Co-intercalated fragments of graphene either single or layered fragments of graphene.

We have linear dependence between the number of the rings and the approximate diameter of fragment (figure.7), so the mean number of benzene rings, M, that cover the compact graphene cluster with mean size of 11 Å, might be restored using that dependence.

It is seen that M=20 for the case.

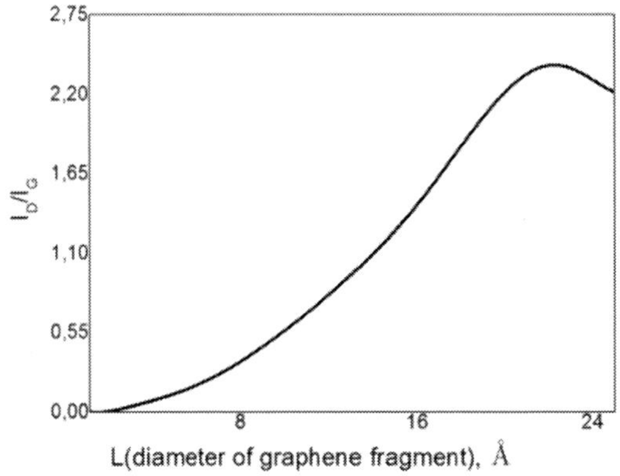

Figure.5. The dependence of I_D/ I_G ratio on the size of graphene fragment.

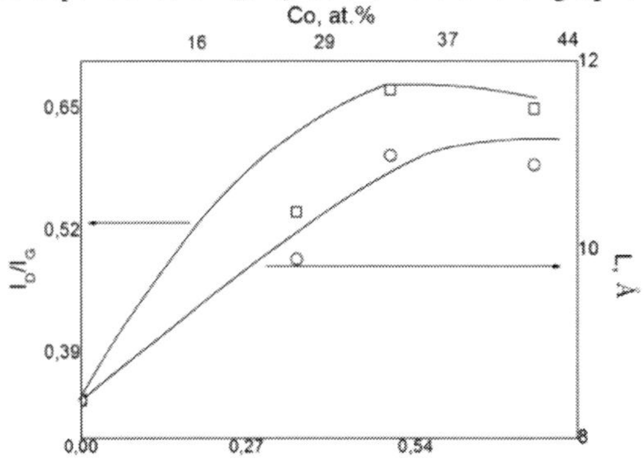

Figure.6 Open squares portrait experimental values for ration of intensities of D and G bands for a-C:H(Co) samples v.s. concentration of cobalt. Open circles show calculated values of the mean diameter of graphene fragments calculated using open square data and the function presenting in Figure 5. Full lines demonstrate corresponding trends.

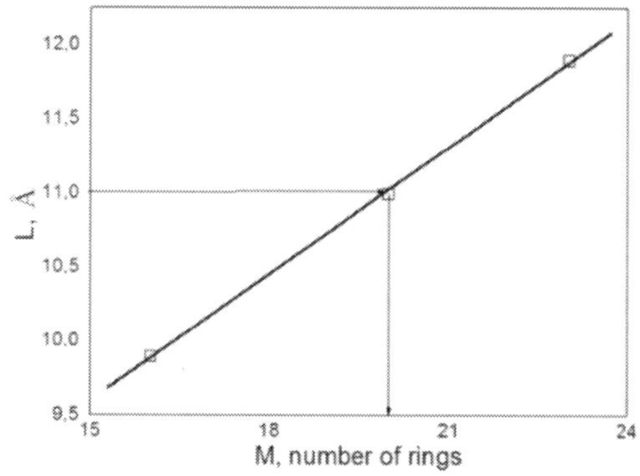

Figure.7 Dependence of mean size of graphene fragment on the number of six-member rings constituting it (results are collected from [7]).

A model of graphene fragment intercalated with Co is shown in figure. 9

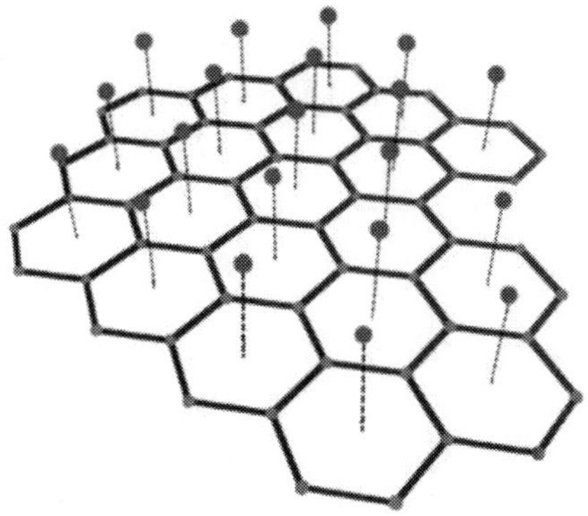

Figure. 9 A model of graphene fragment intercalated with Co.

4. Conclusions

Correlation in between microwave absorption and Raman spectra of a-C:H(Co) was shown. Maximum absorbtion is reached for the concentration of Cobalt approximately 30% that corresponds to the mean diameter of graphene fragment approximately 11 Å.

Fragments of Co-intercalated graphene might be responsible for this effect along with massive cobalt nanoclusters.

5. References

[1] Hayashi T, Hirono S, Tomito M and Umemura S 1996 *Nature* **381** 772.

[2] Alexeev A, Shtager E and Kozyrev S 2007 *Physical Foundation of Stealth Technology*, (VVM Ltd Publishing, Saint-Petersburg).

[3] Banhart F, Hernandez E and Terrones M 2003 *Phys. Rev. Lett.* **90** 185502.

[4] I.vanov-Omskii V I, Krivorotov I N and Yastrebov S G 1995 *Tech. Phys.* **40** 930-937.

[5] Ivanov-Omskii V I, Kolobov A V, Lodygin A B and Yastrebov S G 2004 *Semiconductors* **38** 1416.

[6] Lutsev L V, Yakovlev S V, Zvonareva T K, Alexeyev A G, Starostin A P and Kozyrev S V 2005 *JAP* **97** 104327.

[7] Zvonareva T K, Ivanova E I, Novak I I, Ivanov-Omskii V I 2003 *Phys. Solid State* **45** 9 1658-1668

[8] Ferrari A C and Robertson J 2000 *PRB* **61** (20) 14095, 14107

The dielectric properties of the diamond-like films grown by ion-plasma method

A N Brozdnichenko[1], **D M Dolgintsev**[1], **R A Castro**[1]
Herzen State Pedagogical University of Russia, 48 Moyka river Emb., St. Petersburg, Russian Federation
E-mail: dem.87@mail.ru

Abstract: Diamond-like films was deposited by ion-plasma method in impulse mode. The dielectric properties of the diamond-like films in the frequency of 10^{-1} – 10^{6} Hz at room temperature were studied. The dispersion of the dielectric parameters indicates the existence of non-Debye relaxation mechanism correlates with structural changes. The charge transfer is temperature activated hopping process.

1. Introduction

Diamond-like thick films are one of the first coating used for hardening the cutting tool runs at high speed without forced. Also the mechanical properties are of interest structural, optical and electrical characteristics of diamond-like films on various substrates. In connection with the possible practical application of the grown layers is especially important to study their dielectric properties. The aim of this study was to determine the characteristics of the dielectric relaxation processes in the diamond-like films by dielectric spectroscopy. This method allows us to investigate not only the dielectric properties, but their relationship with the structure of the systems [1].

Diamond-like films was deposited by ion-plasma method on device UVNIPA-1-001 in impulse mode. The substrate was a tantalum foil with a thickness of 0.5 mm. The thickness of the film was $d \sim 3.7$ μm [2]. The samples in the deposition chamber are rotated, the resulting films are deposited evenly and repeat the relief of the substrate (figure 1).

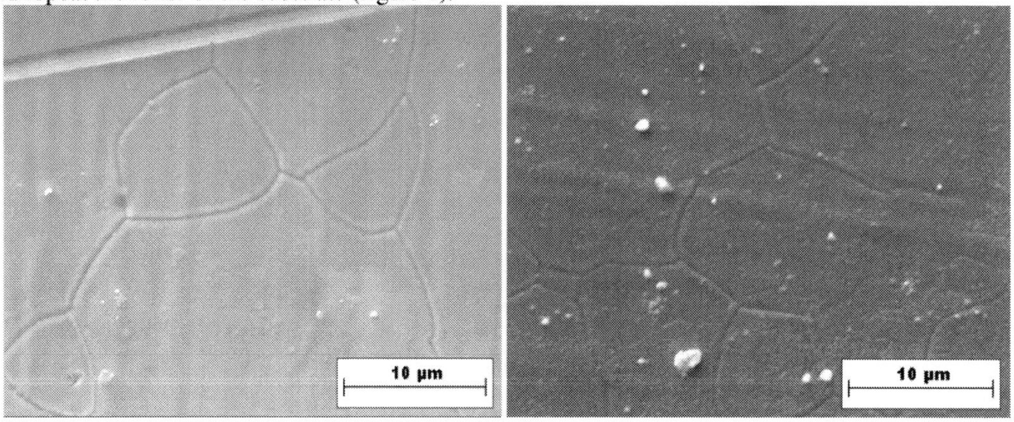

Figure 1. SEM image of the surface polished tantalum substrate (left) and diamond-like thin film thickness d = 3.7 μm on tantalum substrate (right)

2. Experimental

Measurements of the frequency dependence of the complex dielectric permittivity (dielectric constant ε ' and loss factor ε ") and complex conductivity were conducted on the dielectric spectrometer "Concept 41" of NOVOCONTROL Technologies GmbH & Co (Germany), in a wide range of frequencies ($f = 10^{-1}$ - 10^{6} Hz) at room temperature [3]. The device consists of a frequency analyzer, impedance measuring cell, temperature control system, automatic data collection with evaporation system and the supply of nitrogen. The spectrometer comprises an alternating voltage generator, which is fed to the sample analyzer and a current value measuring current and its phase shift relative to the voltage generator.

The complex impedance is given by:

$$Z^*(\omega) = R + \frac{1}{i\omega C} = Z' + iZ'' = \frac{U_0}{I^*(\omega)} \tag{1}$$

"Concept 41" is connected to a personal computer with software for data acquisition and processing. As the experimental data collected values of the imaginary and the real part of the impedance measured with the sample cell. The spectra of the complex permittivity and complex conductivity and impedance spectra were calculated from the following formulas:

$$\varepsilon^* = \varepsilon' - i\varepsilon'' = \frac{-i}{\omega Z^*(\omega)} \frac{1}{C_0} \tag{2}$$

where $C_0 = \frac{\varepsilon_0 S}{d}$ - capacity of the empty cell, $C_0 = \frac{\varepsilon\varepsilon_0 S}{d}$ - capacity of the sample cell.

Complex conductivity is expressed through the impedance:

$$\sigma^* = \sigma' - i\sigma'' = \frac{-i}{\omega Z^*(\omega)} \frac{S}{d} \tag{3}$$

The experimental curves are approximated by a function of Havriliak-Negami [4]:

$$\varepsilon^*(\omega) = \varepsilon_\infty + \frac{\Delta\varepsilon}{[1 + (i\omega\tau)^{\alpha_{HN}}]^{\beta_{HN}}} \tag{4}$$

where ε_∞ - high-frequency limit of the real part of the dielectric constant, $\Delta\varepsilon$ - dielectric increment (the difference between the low-frequency and high-frequency limits), $\omega = 2\pi f$, α_{HN} and β_{HN} - shape parameters that describe respectively the symmetric ($\beta = 1$ - the distribution of the Cole-Cole) and asymmetric ($\alpha = 1$ - the distribution of the Cole-Davidson) extension of the relaxation function.

3. Experimental results

The obtained values of the relaxation parameters α and β (Table 1) suggest that in the studied range of frequencies observed non-Debye relaxation process with an asymmetric distribution of relaxation according to the Cole-Davidson model. This fact is confirmed by the views of the Cole-Cole diagram (Figure 2)

Table 1. The values of the relaxation function parameters Havriliak-Negami.

τ_{max},	$\Delta\varepsilon$	α	β
$2{,}48 \cdot 10^{-6}$	2,29	0,97	0,89

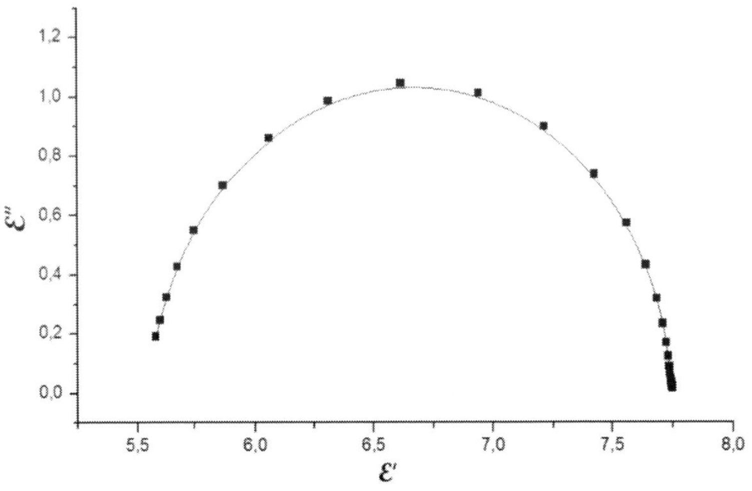

Figure 2. Cole-Cole diagram for the diamond-like film at room temperature

The observed maximum of ε'' (figure 3) and a sharp increase in ε' (figure 4) to the steady-state value in the midrange, indicate the existence of a mechanism of dipole relaxation and interfacial polarization. Dipole relaxation polarization caused by jumps of charge carriers between the centers with different energy to produce quasi dipoles. The charge carriers may be various impurities, absorbed on the surface after the film growth process. The interfacial polarization is related to the structural characteristics of the films, namely the existence of amorphous carbon clusters, the size of which increases with increasing film thickness.

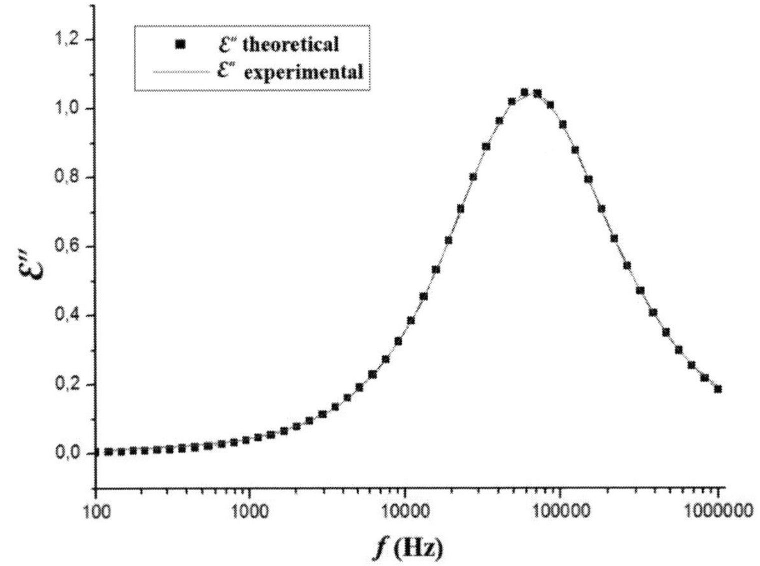

Figure 3. The frequency dependence of the imaginary part of the complex dielectric permittivity at room temperature

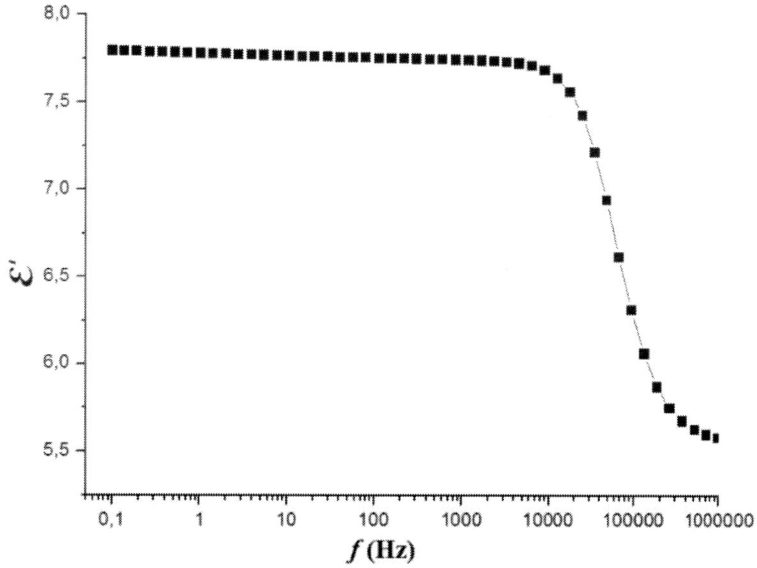

Figure 4. The frequency dependence of the real part of the complex dielectric permittivity at room temperature

Over the whole range of frequencies, for the real part of the complex conductivity, revealed the existence of dependence of the form $\sigma'(\omega) \sim \omega^s$, which is a sign of manifestation of hopping conduction mechanism. In this area there are two variance σ': first - high for $f > 10^3$ Hz, the second - the region of lower frequencies (figure 5). In both areas, the exponent is a frequency-dependent parameter and takes values $s = 0.65 \ldots 0.95$.

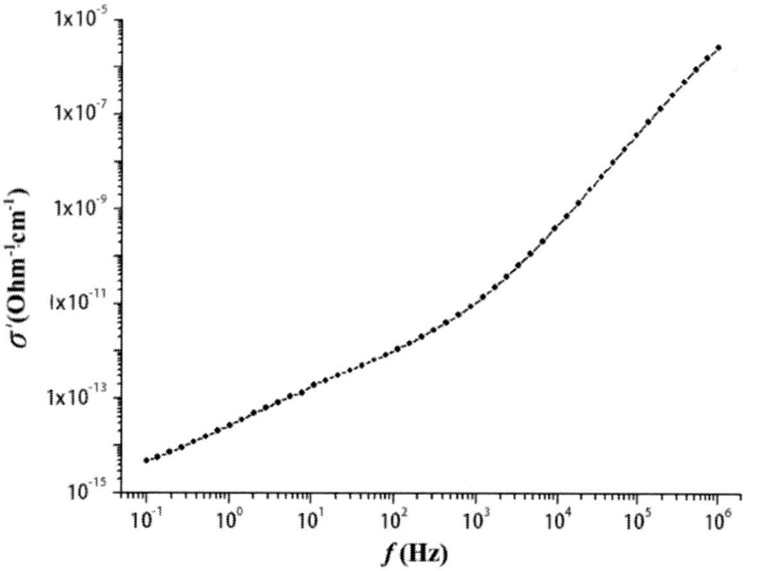

Figure 5. The frequency dependence of the real part of the complex conductivity at room temperature

Conclusions

The dispersion of the dielectric parameters indicates the existence of non-Debye relaxation mechanism correlates with structural changes. The charge transfer is temperature activated hopping process. The mechanism of the dipole relaxation and interfacial polarization due to the presence of impurities and the different phases in the film. The dielectric properties of diamond-like films was found that the relative permittivity of samples $\varepsilon \sim 11$.

References

[1] Castro R, Bordovsky V, Anisimova N, Grabko G 2009 Spectra of charged defects in glassy $Ge_{0.285}Pb_{0.15}S_{0.565}$ thin layers *Semiconductors* **43 № 3** pp 365-367.
[2] Brozdnichenko A, Dolgintsev D, Stozharov V 2012 Diamond-like films grown by ion-plasma method: structure and properties *Izvestia: Herzen University Journal of Humanities and Sciences* **147** pp 81-86.
[3] Castro R, Bordovskii V, Grabko G 2010 Dispersion of dielectric parameters in modified arsenic triselenide layers *Technical Physics Letters* **36** (9) pp. 783-785.
[4] Kremer K, Schonhals A (Eds.) 2003 Broadband dielectric spectroscopy *Springer* (Berlin: Heidelberg) 729.

The technique of field emission parameters research for nanostructured materials improvement

S V Filippov[1,2], E O Popov[1,2], A G Kolosko[1,3] and P A Romanov[1,2]

[1]A.F. Ioffe Physico-Technical Institute of Russian Academy of Sciences, ul. Polytechnitscheskaya 26, St.-Petersburg, 194021, Russia
[2]St. Petersburg State Polytechnical University, ul. Polytechnitscheskaya 29, St.-Petersburg, 195251, Russia
[3]The Bonch-Bruevich Saint-Petersburg State University of Telecommunications, pr. Bolshevikov 22, St.-Petersburg, 193232, Russia

E-mail: s.filippov@mail.ioffe.ru

Abstract. A new technique was developed to record and in situ process the data on the evolution of multiwall carbon nanotubes polymer (MWCNT-polymer) nanocomposite field emitters. It includes multichannel acquisition system of current voltage characteristic (IVC), vacuum level, temperature and mass-spectrum data with computer on-line processing of these data based on the LabVIEW program. We obtained the time dependence of the field enhancement factor, the number of emission sites and kinetics of main volatile products during field emission process.

1. Introduction

Carbon nanotube (CNT) polymeric nanocomposites as field emitters are now being evaluated for a wide range of vacuum electronic applications. Due to excellent field emission (FE) properties of CNTs, they can be used as unheated cathodes for X-ray sources, field emission displays, luminescent mercury-free tubes and vacuum triodes [1-5]. In spite of significant success in practical implementation of CNT for FE, many important problems have not been solved yet. Now it is very difficult to make a reliable conclusion concerning the physical processes occurring on the surface of nanosized emission sites by measuring macroscopical current and voltage values. Besides, additional difficulties arise in the study of field emission dynamics, when a change of the studied object or its environment may occur.

It is known that the investigation of carbon emitter stability is a multiple parameter problem, with a change of vacuum conditions and emitter surface temperature to be taken into account. Therefore the question of particular interest is registration of the phenomena accompanying functioning of flat multi-tip cold emitters, such as detecting an emitter radiation, visualization of emission sites distribution, field electronic microscopy and registration of volatile products by mass spectrometer.

Today, recording and processing of current-voltage characteristics (IVC) constitute a basic tool in the study of the materials promising as efficient field emitters. The computerized data acquisition systems used for these purposes are very effective. This allows better understanding of IVC dynamics and their correlation with the other emission phenomena.

The direct (on-line) processing of emission characteristics throughout the experiment was rarely used [6, 7]. For example, the work [8] describes the application custom LabVIEW program to the study of carbon nanosheets field emitters. The authors used the mass spectrometer for measuring partial pressures of reaction products. In addition, the similar method was used in [9]. The authors designed vacuum system in a cathode-anode configuration in line-of-sight to a residual gas analyzer. The data were recorded under LabView control.

We succeeded in developing the technique of multichannel recording and on-line processing of data on the field emitter evolution. On-line processing of signals has been performed on the basis of our program written by LabVIEW 2012.

2. Experimental details

As an object of research we used the nanocomposite polymer - multiwall carbon nanotube emitter (MWCNT-polymer) that has already shown remarkable properties: high values of emission currents, stability to the vacuum condition and sufficient stability of emission parameters [10]. This nanocomposite emitter easily enough forms the emitting surface of a large area with a good adhesion to the conducting substrate.

To make the samples we used a MWCNT suspension. As nanotubes we use certificated Graphistrength® MWCNT by Arkema firm (according to SEM, length of tubes is up to five microns, diameter is up to 15 nm) and as a polymer - polystyrene (PS), -polymethylmethacrylate (PMMA) and nitrocellulose (NC).

Suspension of MWCNT in a polymeric matrix was prepared by mixing of nanotubes suspension and 1 % polymer solution in the same solvent. Weight concentration of tubes in a polymeric matrix was taken equal to 10 %. Mixing in an ultrasonic bath for 8 hours was applied to prepare suspensions. Drawing of an emission surface on the flat substrates of stainless steel was made by spin coating technique. The field emission experiments were carried out in standard construction with flat metal electrodes.

To obtain IVC of the sample the high-voltage power unit with frequency 50 Hz was used. Electronic rectifier made positive impulses with duration 10 ms and consequently one such half-sinusoidal impulse gave one IVC. Other researchers in work [11] for the hardware analysis of IVC-FN evolution used a close idea.

Measurements were carried out at a cathode-anode gap d = 300 μm and various amplitudes of sinusoidal voltage impulse (thereby the range of amplitudes of corresponding current impulses was from 1 μA to 10 mA). Management of a high level voltage and change of interelectrode gap was carried out by means of the laboratory autotransformer and the micrometric screw. The system was evacuated to a base pressure of $5 \cdot 10^{-7}$ Torr.

To obtain mass spectrum signal we used reflectron type time-of-flight mass spectrometer (TOF-MS). The neutral molecules formed upon decomposition of the polymer (for PMMA and PS these are mainly the respective monomers) are ionized by the electron impact. For electrons of 70 eV the mass-spectra are tabulated. The TOF-MS is equipped with oil free pumping system and microchannel plate detector.

The measurement technique includes multichannel gathering system that records and sends to the personal computer values of emission current, anode-cathode voltage, cathode temperature, pressure and full mass spectrum signal (Figure 1). For this registration a multichannel plates DAQ by National Instruments NI PCIe-6351 are used.

The cathode temperature at high voltage is registered by multimeter UNI-T UT60G with optocoupling. For the purpose of computer plug-in-card protection against the vacuum breakdowns, measured and operating signals were directed through the multichannel block of protection and amplification. The synchronization was made by digital trigger formation scheme depending on the outlet supply phase. Thereby huge stream of IVC realizations that were in each period of scanning by the high voltage impulses was reached.

The computer analysis and recording of IVC evolution in time was carried out by own program written by LabView 2012 (figure 2). Program includes blocks that calculate and allow to observe a form of voltage and current impulses - *U(t)* and *I(t)* in real time mode, current-voltage dependence - *I(U)*, dependence of current density from electric field intensity - *J(E)*, IVC curve in Fowler–Nordheim coordinates - *Y(X)* and its diagram in the slope-intercept coordinates - *S(K)* (Seppen-Katamuki analysis [11,12]). Also it allows to control both the composition and kinetics of main volatile products.

Journal of Physics: Conference Series **572** (2014) 012026 doi:10.1088/1742-6596/572/1/012026

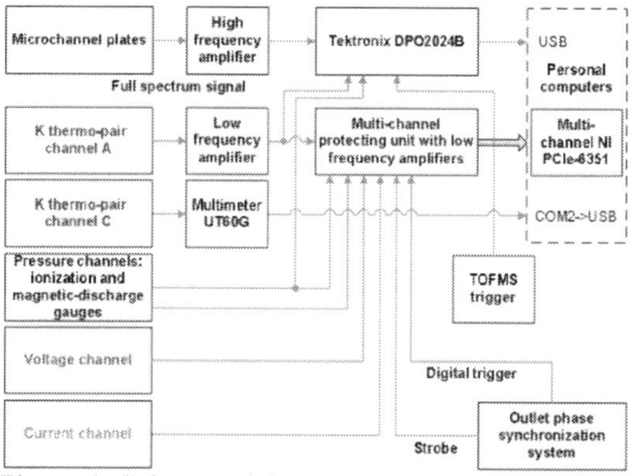

Figure 1. Scheme of the measurement system

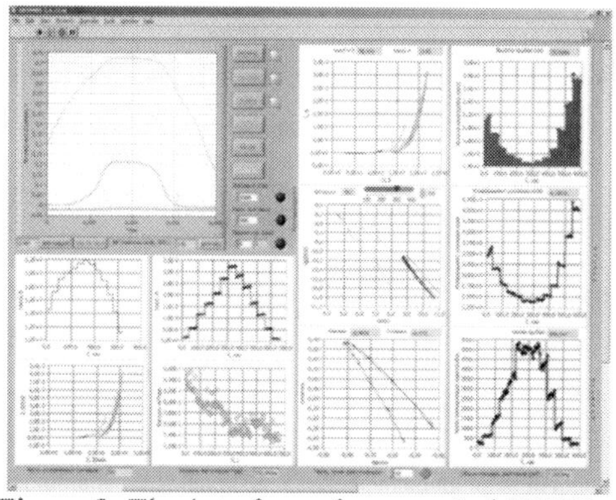
Figure 2. The interface of custom LabVIEW program

Additionally, data for representation in a convenient for the analysis form were processed: time dependence of the emitter power, time dependence of an emission interval active resistance and current loading on one MWCNT. The visualization of emission sites effective height and their number also was carried out.

The calculations were made by formulas from the known monograph with the same constants and values dimension [13]. The obtained data were deduced on plots of time dependences: $\beta(t)$, $N(t)$ and $h(t)$. For emitting site number evaluation we used the emitting area of the individual MWCNT which was offered in work [14]: S_{CNT} =1.8 nm^2 for individual MWCNT with diameter 15 nm. In our work we apply MWCNT of the same diameter. As shows the further reasoning this estimation gives quite adequate values of number of the emitting sites.

In our technique, we tried to work at high levels of current when field emission current much more than capacitive component. At small emission currents (several μA) a record and subtraction of parasitic capacitor "support" from the signal was used. Besides for increase in accuracy of measurement our program made accumulation and averaging profiles of current and voltage impulses in set time interval (about 100 impulses were averaged for 5 seconds).

The described apparatus has allowed us to obtain a number of promising results for field emitters. Some of them are shown below.

3. Results and discussion

Figure 3 demonstrates the characteristics recorded for MWCNT-NC emitter: time realizations of applied voltage, emission current and the main volatile products intensity registered by TOF-MS. In our experiments thermal instability of the sample was observed in mode with large emission currents (more than 10 mA). Intensity of main volatile product in the mass spectrum increases gradually: see kinetic of the peak h_{30} (fragment of nitrocellulose monomer) in the mass spectrum in figure 3. This increase points out that the gradual destruction of the polymer matrix occurs.

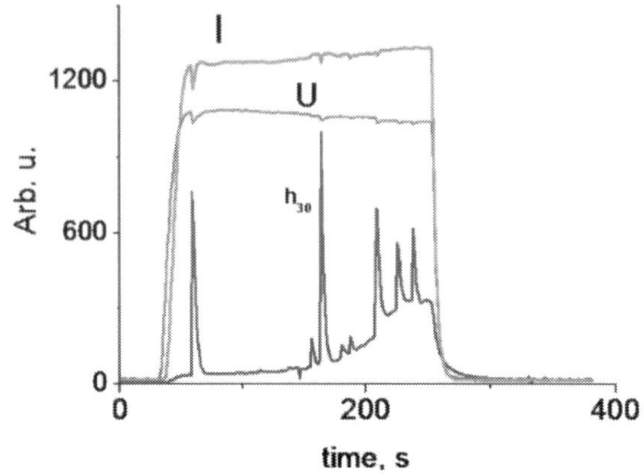

Figure 3. The characteristics record of MWCNT-NC emitter: integral (U, I) time realizations and the main volatile products registration

Figure 4 shows the SEM images of a MWCNT-NC emitter surface before and after FE experiment. One can see that the thickness of the polymer surrounding nanotubes has reduced significantly from ~74 to ~58 nm. It indicates that polymer destruction occurs right around the nanotubes. We suggest that this process is caused by the Joule heating of emitting sites.

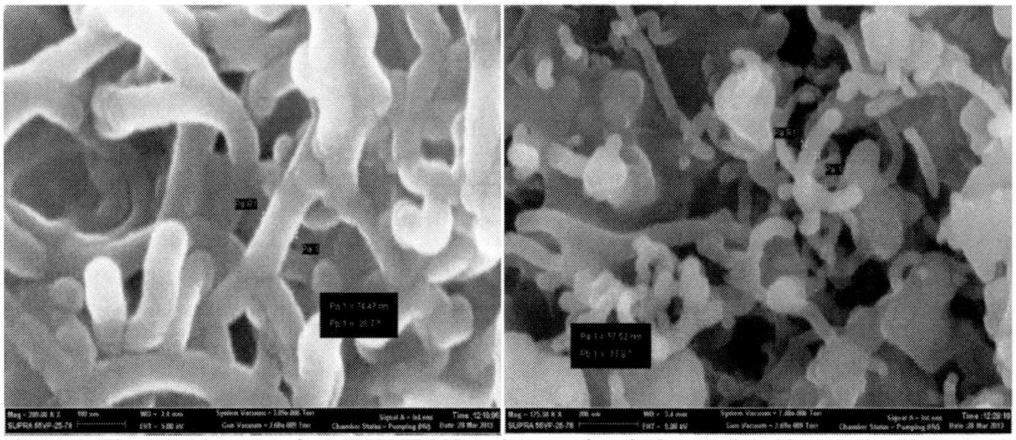

Figure 4. The SEM images of a MWCNT-NC emitter surface before (left half) and after (right half) FE operation

Aside from slow partial pressure change during the emission process, we also observed short-term ejections and accompanying sharp changes in the emission current and vacuum (figure 3). It is well known that this effect is connected with vacuum discharge and emission site degradation.

With the built-in video camera we observed the bright flashes that appear on the surface of the adjacent electrode. This glow can appear if something conductive protrusions are formed onto the surface of anode and electrons emitting from cathode impact them. We suggest that those protrusions are formed due to transfer of nanotubes from composite-covered electrode to adjacent electrode. Transfer of nanotubes is caused by "field decomposition" - influence of high electric fields on the emission sites.

The formation of carbon nanostructures (whiskers) on the anode as a result of mass transfer is of particular interest because it greatly restricts the emitter work. The application of high voltages leads to overheating and burning whiskers, accompanied by a vacuum discharge and breakdown of the system. In addition, the whiskers do not allow the use of this cathode system as the diode because they themselves are emission centers in reverse polarity mode.

For registration of the whiskers formation, we installed a clear adjacent electrode and applied a voltage of reverse polarity to the system. Thus we can register "reverse" emission current produced by whiskers on the anode.

Figure 5 demonstrates the result of our experiment. With stepwise voltage increases the current stepwise increases too with short sequence of discharges. And the time realization of the pressure in the vacuum chamber has accompanying sharp peaks.

The SEM image of the adjacent electrode received after the experiment (figure 6) shows that not only nanotubes, but clusters of nanotubes and polymer material were transferred to the anode.

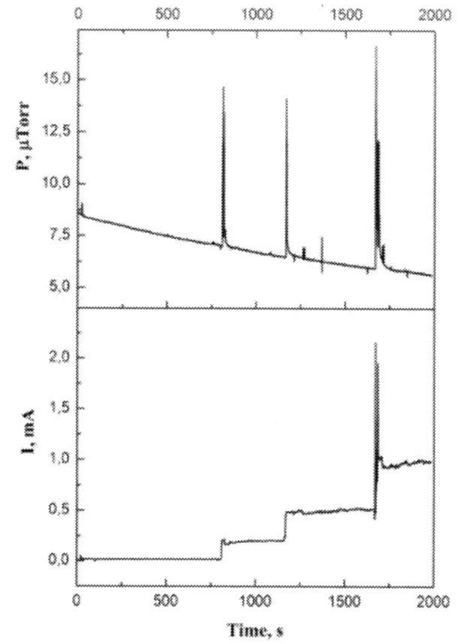

Figure 5. The time dependences of emission current and base pressure

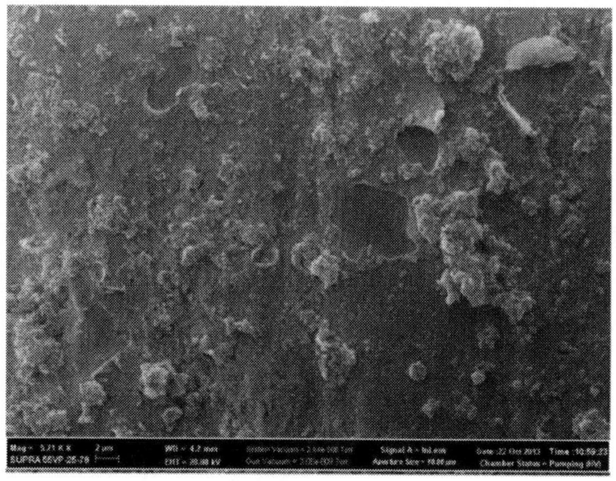

Figure 6. The SEM image of adjacent electrode (anode) after FE experiments

4. Conclusion

We have developed a method of multichannel acquisition and online computer processing of data on the evolution of working characteristics of flat multipoint field electron emitters. The jointed use of MS and FE techniques is suggested. In situ studies open new possibilities of FE research:

- the opportunity of observation the “activation” of the samples upon the initial stages of the FE;
- the opportunity of the research of the material transfer to the counter electrode;
- the long term study of the emitter working regime;
- the thermal decomposition of the electrode materials.

By using TOF-MS we detected the presence of monomers in the volatile products emitted from the sample surface of nanocomposite emitters at the large emission current. Moreover SEM analysis was shown that the gradual polymer destruction occurred right around the nanotubes.

We registered the transfer of the large clusters from nanocomposite material of cathode to the adjacent electrode. For registration of whiskers formation, we used the reverse polarity in the system. Thereby we observed electrical evolution of the whiskers on the anode during their growth.

Acknowledgments
The work was carried out with the partial support of the Foundation for Assistance to Small Innovative Enterprises in Science and Technology (№ 0002235).

References
[1] Chung D S, Park S H, Lee H W, Choi J H and Cha S N 2002 *Appl. Phys. Lett.* **80** 4045
[2] Sung H H, Aamir I, and Sung O C 2007 *Appl. Phys. Lett.* **90** 183109
[3] Haga A, Senda S, Sakai Y, Mizuta Y and Kita S 2004 *Appl. Phys. Lett.* **84** 2208
[4] Bonard J M, Stöckli T, Noury O and Châtelain A 2001 *Appl. Phys. Lett.* **78** 2775
[5] Jeon S, Kim G, Kim J, Shin D, Shin D H, Sun Y and Lee C J 2012 *Int.Vacuum Nanoelectronics* Conf. (Jeju: IEEE) p104
[6] Popov E O, Popov S O and Korovin O P 1999 *5th Int. Conf. on Computational Physics* (Kanazawa) p1-04
[7] Bormashov V S, Nikolski K N, Baturin A S and Sheshin E P 2003 *Appl. Surf. Sci.* **215** 178
[8] Bagge H M, Outlaw R A and Zhu M Y 2009 *J. Vac. Sci. Technol.* B **27** 2413–19
[9] Murray P T, Back T C, Cahay M M 2013 *Appl. Phys. Lett.* **103** 053113
[10] Popov E O 2010 *52nd Int. Field Emission Symp.* (Sydney) p54
[11] Kawasaki M, He Z, Gotoh Y, Tsuji H and Ishikawa J 2010 *J. Vac. Sci. Technol.* B **28** C2A77
[12] Popov E O and Ershov M V 2011 *Int.Vacuum Nanoelectronics Conf.* (Wuppertal: IEEE) p158
[13] Elinson M I 1974 *Unheated cathodes* (Moscow, Sov. Radio)
[14] Liu H, Kato S and Saito Y 2009 *J. Vac. Sci. Technol.* B **27** 2435

Influence of constant magnetic field on aggregation processes in magnetite colloids

K G Gareev, I E Kononova, V S Levitckii, V A Moshnikov and S S Nalimova

Saint-Petersburg Electrotechnical University LETI, 5 Popova Str., Saint-Petersburg, 197376, Russia

kggareev@yandex.ru, lev-vladimir@yandex.ru

Abstract. Colloids of Fe_3O_4/SiO_2 nanoparticles derived by a two-stage technique on the base of tetraetoxisilane soles were studied. Phase composition of the particles was analyzed by X-ray diffraction and Raman spectroscopy. It was found that phase transition of Fe_3O_4 into α-Fe_2O_3 occurs at a laser power exceeding a threshold value. The aggregation of the particles to linear structures in a constant magnetic field was studied by atomic force microscopy. The data on the change in resistance of Fe_3O_4/SiO_2 colloids under an applied magnetic field were obtained.

1. Introduction

Magnetic colloids have found many applications in medicine and electronics. A variety of techniques for the synthesis of magnetic nanoparticles were developed, that can be classified into two main groups: mechanical milling of bulk material and chemical synthesis [1, 2]. Particle coating within a shell of organic and inorganic material enhances the stability of magnetic colloids. An inert coating on the surface of magnetic nanoparticles also makes them suitable for biomedical application.

Magnetic nanoparticles are very promising as contrast agents for magnetic resonance imaging (MRI). The use of magnetite-based contrast agents even at low concentrations can enhance the quality of MRI because of the high efficiency of spin-spin relaxation [3]. Besides, magnetite is nontoxic to biological tissues and can be used for targeted drug delivery [4].

Physical properties of magnetic colloids are of great interest to researchers. Magnetohydrodynamics problems and the particle aggregation under an external and induced magnetic field (dipole-dipole interaction) were considered in one of the early publications [5]. Self-assembly of nanoparticles into chain-like structures, properties of the resultant aggregates and their dependence on the particle-size distribution were described in detail in [6-9].

Magnetic colloids can be characterized by thermal conductivity and electrical conductivity measurements under an external magnetic field [10, 11]. Raman spectroscopy can be used for the analysis of chemical composition of magnetic nanoparticles. This technique enables one to locally observe coexisting phases and phase transitions between magnetite (Fe_3O_4), maghemite (γ-Fe_2O_3) and hematite (α-Fe_2O_3) [12].

2. Experimental methods

Magnetic colloids were synthesized by a two-stage technique. First, silica nanoparticles were prepared by sol-gel technique in alcohol solution of tetraethoxysilane (TEOS) [13-16]. The surface morphology of SiO_2 powders was analyzed by nitrogen thermal desorption. The estimated mean particle size is 50-100 nm (the specific surface area is about 100 m^2/g). In the second stage, magnetite nanoparticles were precipitated from aqueous solution of Fe (II) and Fe (III) salts and superfine silica according to the reaction [17]. Fe_3O_4 nanoparticles are assumed to coat the surface of silica spheres. Thus, the formation of core-shell structures occurs. This assumption was confirmed experimentally. High conductivity was found in Fe_3O_4/SiO_2 film on dielectric substrate (magnetite is half-metal with $E_g \approx 0.2$ eV).

The aggregation of colloidal magnetic nanoparticles is intensified under a static magnetic field. Chain-like structures with different width were found to form because of the dipole-dipole interaction [6-9]. Fe_3O_4/SiO_2 layers were obtained by deposition of the colloidal solutions at different static magnetic field strength (0-150 kA/m) in order to observe these phenomena. Specimens were dried at the room temperature for 24 hours. Magnetic field was produced by Nd-Fe-B magnet that was influencing on objects for the whole time of drying. The surface morphology of layers was examined by atomic force microscopy (AFM). X-ray diffraction (XRD) and Raman spectroscopy were used to confirm the formation of magnetite phase by the reaction described in [17].

Raman spectra were recorded at room temperature using a LabRam HR800 spectrometer. A Nd:YAG laser operating at a wavelength of 532 nm was used as an excitation source. The diameter of the focused laser beam was ~ 1 μm. The laser beam was focused on the surface of the sample and the scattered radiation was collected by an X100 objective. The measurements were performed at different laser power values.

The change in resistance of magnetic colloid was measured under an alternating electric field (1 kHz) and under a constant magnetic field.

3. Experimental results

The XRD pattern of magnetite-silica powder (figure 1) indicated the presence of crystalline magnetite. The halo background at low angles is due to the amorphous phases.

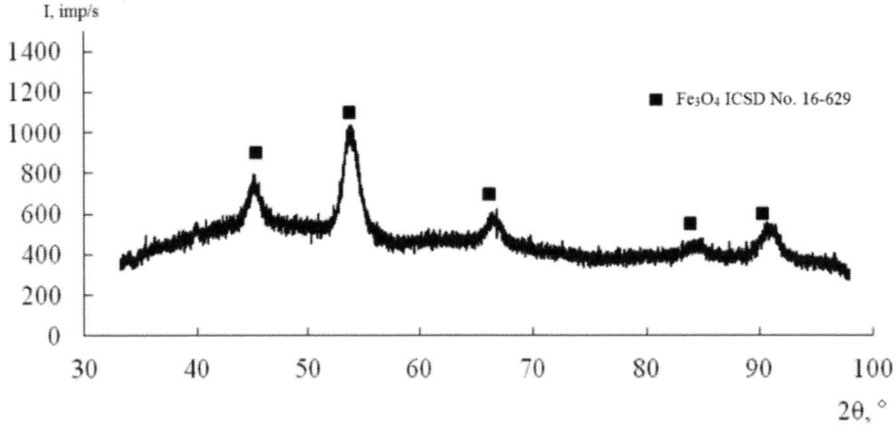

Figure 1. XRD pattern of powder obtained by drying 0.3 mM Fe_3O_4/SiO_2 solution with 10 vol. % TEOS in the initial sol

Raman spectra of Fe_3O_4/SiO_2 layer recorded at different laser power values are shown in figure 2. The most intense magnetite band corresponding to the A_{1g} mode vibration is observed at 670 cm^{-1} [12]. The bands at ~220 cm^{-1} and ~280 cm^{-1} are usually assigned to α-Fe_2O_3 lattice vibrations [12]. The analysis of Raman spectra (figure 2) revealed laser induced phase transition of magnetite to hematite via an intermediate maghemite phase at a laser power exceeding a threshold value (~1 mW). As it was stated early in works phase transition of magnetite to maghemite accomplishes at approximately 220°C [18,19] and maghemite transforms to hematite at 500°C [19]. So local heating over 500°C caused by 1 μm mean diameter laser spot may be supposed that can be proved by microphotographs containing SiO_2 melted areas. The characteristic band of maghemite is observed at

390 cm^{-1} [12]. When a laser power exceeds a threshold value, the intensities of bands at 390 cm^{-1} (maghemite) and 670 cm^{-1} (magnetite) decrease.

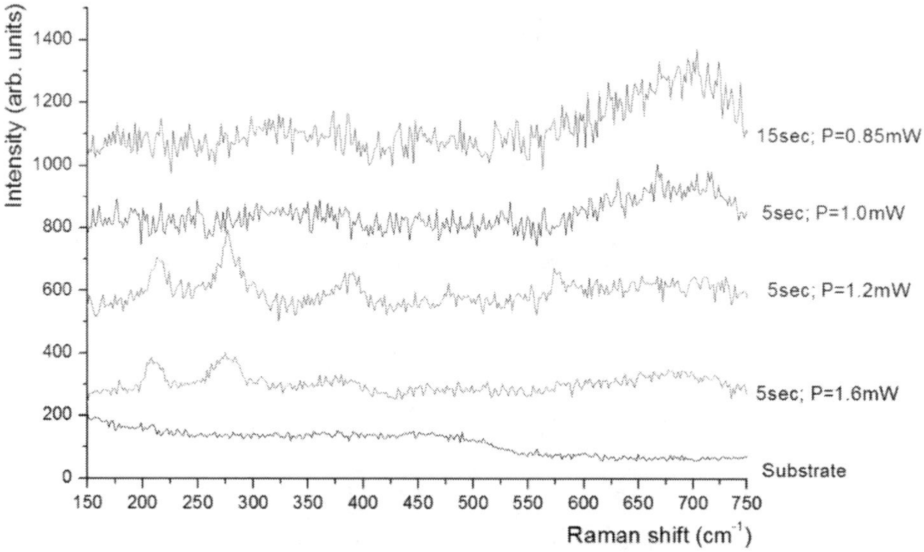

Figure 2. Raman spectra of the layer obtained from the 0.3 mM Fe_3O_4/SiO_2 solution with 50 vol. % TEOS in the initial sol

AFM studies of the layers deposited at different external magnetic field strength values revealed that nanoparticles begin to form linear structures at 1 kA/m. Isolated particles together with the linear structures are observed in the AFM images of the layers obtained at larger magnetic field strength values (figure 3-5). The layers deposited without a magnetic field consist of isotropic aggregates.

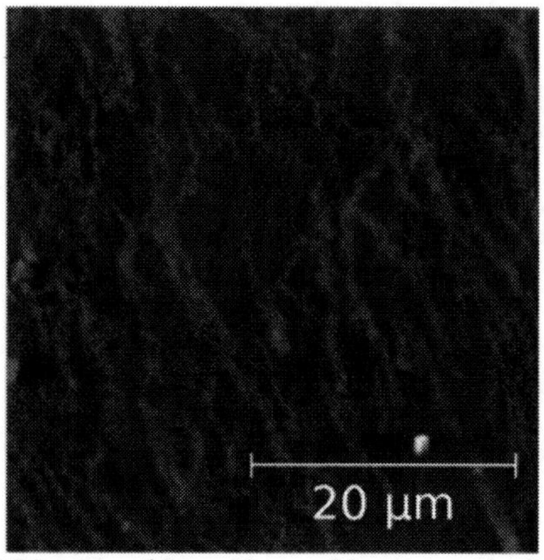

Figure 3. AFM image of layer formed in magnetic field 150 kA/m

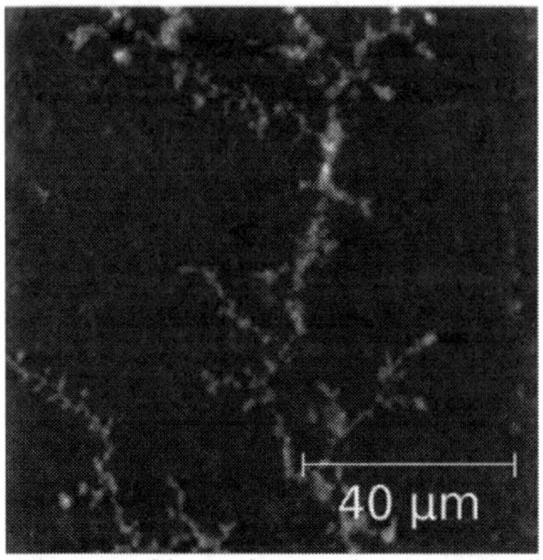

Figure 4. AFM image of layer formed in magnetic field 1.5 kA/m

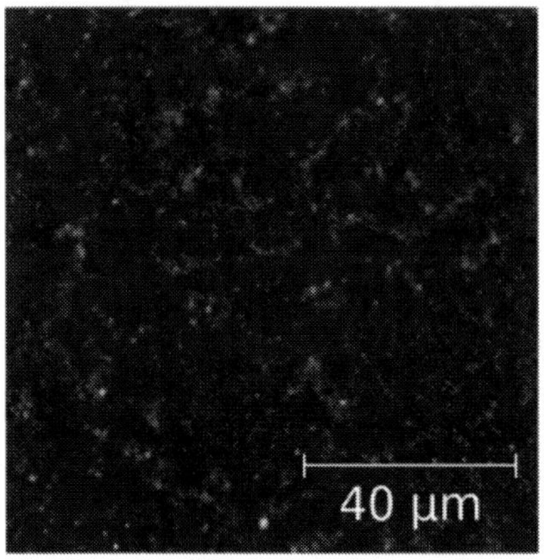

Figure 5. AFM image of layer formed in zero magnetic field

It was shown in [19] that the product of the reaction of magnetite formation is also ammonium chloride. The resulting colloid is known to exhibit ionic conduction. Therefore, resistance should be measured under an alternative electric field in order to avoid the adsorption of ions at electrodes. Furthermore, the assembly of Fe_3O_4/SiO_2 nanoparticles into linear chains leads to a decrease in resistance of magnetic colloid. The response time of the system to a static magnetic field is several seconds (figure 6). Thus, the effect of an alternative magnetic field on the electric properties of magnetic colloids was not studied. Besides, it was found that the change in resistance (after 180 s exposure) is maximal under a static magnetic field at 450 A/m and small at 150 A/m. This result is in good agreement with the observed stability of magnetic colloids in the earth magnetic field (about 40 A/m).

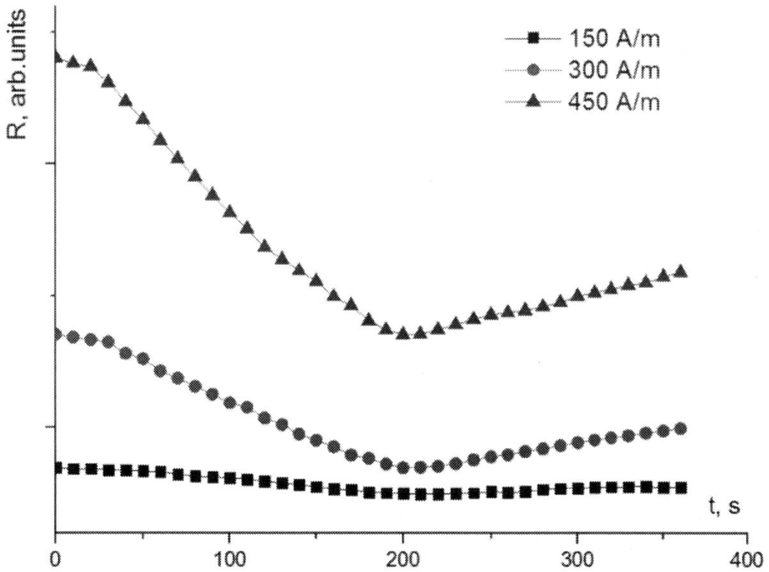

Figure 6. Resistance versus time dependence at various strength of magnetic field for 0.3 mM Fe_3O_4/SiO_2 solution with 60 vol. % TEOS in the initial sol

To determine the influence of static magnetic field on the relative change in resistance, the samples were exposed to a magnetic field with different strength for the same periods of time (180 s). As can be seen in figure 7, this dependence is linear.

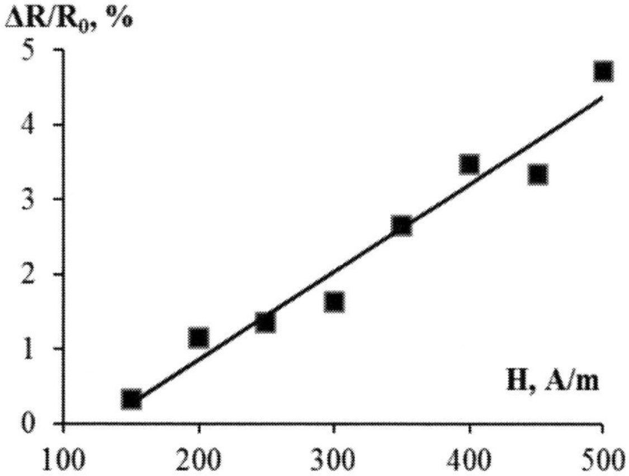

Figure 7. Relative resistance change versus magnetic field strength dependence for 0.3 mM Fe_3O_4/SiO_2 solution with 60 vol. % TEOS in the initial sol

4. Conclusions

XRD analysis of Fe_3O_4/SiO_2 layers obtained by deposition of the magnetic colloids revealed the formation of crystalline magnetite phase. Raman spectroscopy studies showed laser induced phase transition of magnetite to hematite via an intermediate maghemite phase at a laser power exceeding a threshold value (~1 mW).

AFM results indicated that the linear structures form in magnetic colloids under an external magnetic field. This structural modification leads to the change in the conductance of magnetic colloids under a magnetic field with strength larger than 100-200 A/m.

Thus, destruction of the stability of magnetic colloids under the earth magnetic field can be caused by a variation of synthesis conditions.

References

[1] Lu A-H, Salabas E L and Schuth F 2007 Angew. Chem. Int. Ed. **46** 1222-1244
[2] Hyeon T 2003 Chem. Commun **8** 927-934
[3] Rink P A Magnetic resonance in medicine 1993 Blackwell Scientific publication 246 p
[4] Nikolaev B P, Marchenko Ya Yu, Yakovleva L Yu Zimina T M, Soloviev A V, Luchinin V V, Petrov A V, Scharafutdinova T A and Dobrodumov A V 2013 IEEE Trans. on magn. **49** 1 Part 2 429-435
[5] Shliomis M I 1974 Sov. Phys. Usp. **17** 2 153-169
[6] Morozov K I and Shliomis M I 2004 J. Phys.: Condens. Matter. **16** 3807-3818
[7] Sorokina O N, Kovarski A L, Lagutina M A, Dubrovskii S A and Dzheparov F S 2012 Appl. Sci. **2** 342-350
[8] Kantorovich S and Ivanov A O 2002 J. of Magn. and Magn. Mat. **252** 244–246
[9] Ivanov A O and Zubarev A Yu 1998 Physica A **251** 348-367
[10] Gavili A, Zabihi F, Isfahani T D and Sabbaghzadeh J 2012 Exp. Therm. and Fluid Sci. **41** 94-98
[11] Racuciu M, Creanga D E and Calugaru Gh 2005 J. Optoelectr. and Adv. Mat. **7** 6 2859-2864
[12] Slavov L, Abrashev M V, Merodiiska T, Ghelev Ch, Vandenberghe R, Markova-Deneva I and Nedkov I 2010 J. of Magn. Magn. Mat. **322** 14 1904-1911
[13] Gracheva I E, Olchowik G, Gareev K G Moshnikov V A, Kuznetsov V V and Olchowik J M 2013 J. of Phys. and Chem. of Solids. **74** 5. 656-663
[14] Moshnikov V A, Gracheva I E, An'chkov M G 2011 Glass Phys. and Chem. **37** 5 485-495
[15] Abrashova E V, Gracheva I E and Moshnikov V A 2013 J. of Phys.: Conf. Ser. **461** 012019
[16] Gubin S P, Koksharov Yu A, Khomutov G B and Yurkov G Yu. 2005 Rus. Chem. Rev. **74** 6 489-520
[17] Massart R 1981 IEEE Trans. on Magn. **17** 2 1247-1248
[18] Haneda K and Morrish A H 1977 J. de Phys. **38** 4 C1-321 – C1-323
[19] Sidhu P S 1988 Clays and Clay Minerals **36** 1 31-38.

Journal of Physics: Conference Series **572** (2014) 012028 doi:10.1088/1742-6596/572/1/012028

Synthesis of cadmium selenide colloidal quantum dots in aquatic medium

D S Mazing, L B Matyushkin, O A Aleksandrova, I I Mikhailov, V A Moshnikov and S A Tarasov

Saint Petersburg Electrotechnical University "LETI", ul. Prof. Popova, 5, 197376 Russia

E-mail: dmazing@yandex.ru, leva.matyushkin@gmail.com

Abstract. Cadmium selenide nanocrystals were prepared in water phase through facile wet chemistry technique with thioglycolic acid (TGA) acting as capping agent. Structures were characterized using X-ray diffraction (XRD), UV-vis absorption and photoluminescence spectroscopies. Depending on synthesis conditions nanoparticles exhibit photoluminescence with maximum in the region of 580 – 680 nm. Influence of technological parameters and component concentrations on nanocrystals average size and properties was studied.

1. Introduction

Semiconductor quantum dots (QDs) have recently attracted much attention because of their application prospects in the variety of optoelectronic devices such as light-emitting diodes, photodetectors, solar cells and lasers. By virtue of technological procedure facility QDs are suitable material for flexible electronics, particularly for conductive polymers based structures. Another area where QDs may be widely used is biological and medical detecting [1,2]. In comparison with organic fluorophores semiconductor nanocrystals are much more stable to photodegradation. Absorption band wideness and variety of available emission colors make it possible to excite plurality of QDs with different sizes by means of the same radiation source, thus allowing detecting signals from different targets simultaneously. Cadmium selenide nanocrystals exhibit strong quantum confinement and by regulating their size one can obtain emission in almost entire visible range.

The hot-injection method proposed in early nineties [3] had instigated activity in the field of metal chalcogenide nanoparticles synthesis. It allowed to obtain highly luminescent crystals with good monodispersity but has certain disadvantages connected with expensive and hazardous pyrophoric reactives being used. Plenty of adaptations of original technology have been developed since then [4,5]. One of the natural choices of medium for QDs preparation might be polar solvents, for example water. Despite poorer crystallinity and therefore generally inferior optical properties as compared to nanoparticles synthesized in non-polar media these methods are considered to be more environmentally benign and provide nanocrystals compatibility with biological tissues eliminating necessity for ligand exchange operations. Suitable precursors in this case may include water-soluble metal and chalcogenide salts as sources of metal and chalcogenide ions respectively while thiols might be used for surface passivation.

2. Materials and methods

Commercially purchased $NaBH_4$, $CdCl_2{\cdot}2.5H_2O$, elemental Se (powder), thioglycolic acid (TGA, 80%), KOH (6M), propanol-2 (anhydrous) were used without any further purification. Distilled water was used as synthesis medium.

Absorption spectra were acquired on a PE-5400UV UV–vis spectrophotometer (LLC "Ekohim"). Photoluminescence measurements were carried out on equipment based on fast scanning spectrometer Ocean Optics USB-4000 using laser excitation on $\lambda = 445$ nm. XRD analysis was performed with diffractometer DRN "Farad" using CuKα radiation.

Average size of nanocrystals might be estimated via polynomial expression from the work [6]:

$$a=\left(1.6122\cdot10^{-9}\right)\lambda^4-\left(2.6575\cdot10^{-6}\right)\lambda^3+\left(1.6242\cdot10^{-3}\right)\lambda^2-0.4277\lambda+41.57, \quad (1)$$

where a is average size of particles, λ is a wavelength of the first excitonic peak. This equation was derived from various experimental data acquired through TEM studies and supposedly gives more reliable values for relatively small particles than effective mass approximation model.

3. Experiment and results

Typical synthesis procedure was adapted from reference [7] with several modifications and included following steps. Sodium hydroselenide solution was prepared by dissolving $NaBH_4$ and Se powder in distilled water in a glass vial. It resulted in vigorous reaction accompanied with intensive hydrogen gassing. The reaction might be described as follows:

$$4NaBH_4 + 2Se + 7H_2O \rightarrow 2NaHSe + Na_2B_4O_7 + 14H_2$$

To obtain cadmium precursor solution cadmium chloride hydrate was dissolved in distilled water and certain amount of TGA was added so that the mixture became turbid. By dropwise addition of KOH (6M) acidity of the solution was brought up to approximate value pH = 11 and it cleared. It is suggested that moderately alkaline environment facilitates formation of Cd^{2+}-TGA complexes which act as immediate reactant [8].

In one implementation component molar ratio was adjusted to $[Cd^{2+}]:[Se^{2-}]:[TGA]=1:0.5:2.4$. Selenium precursor solution was swiftly injected in solution containing cadmium ions under vigorous stirring at room temperature, followed by rapid heating. The first aliquot was withdrawn from the reaction volume immediately. The rest of samples were taken at 35 °C, 50 °C and 70 °C respectively. Absorption and photoluminescence spectra of this series are represented in the figures 1 and 2.

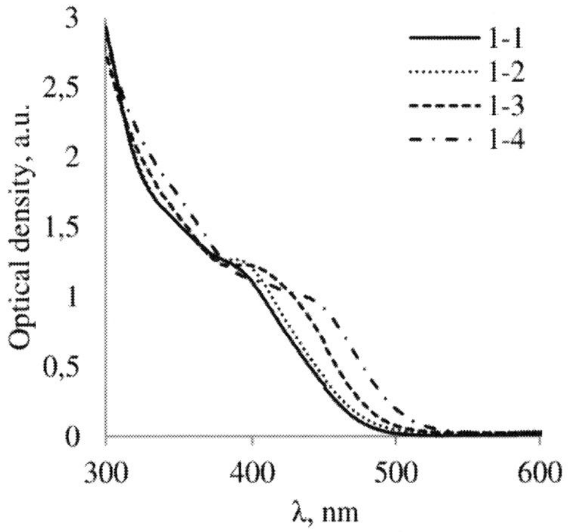

Figure 1. Absorption spectra of CdSe nanoparticles prepared with molar ratio $[Cd^{2+}]:[Se^{2-}]:[TGA]=1:0.5:2.4$.

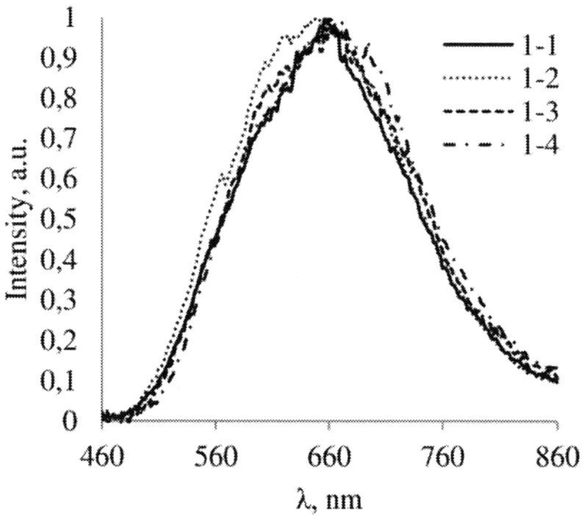

Figure 2. Photoluminescence spectra of CdSe nanoparticles prepared with molar ratio $[Cd^{2+}]:[Se^{2-}]:[TGA]=1:0.5:2.4$.

As might be seen from the figure 1 absorption onset shifts towards long wavelengths in the course of the reaction which points at nanocrystals growth. At the same time the first excitonic peak becomes lower and less distinct. Supposedly it indicates that the particle monodispersity somewhat worsens. Photoluminescence spectra almost don't change their position and feature broad peaks with large Stokes shift. It probably means that radiative recombination occurs via trap states on the surface of nanocrystals due to insufficient surface passivation.

In another implementation molar ratio $[Cd^{2+}]:[Se^{2-}]:[TGA]=1:0.25:1.2$ was used. Precursor solutions were prepared as described above and NaHSe solution was injected at 45 °C. Temperature was raised gradually throughout experiment and aliquots were taken from the reaction flask in the following order: 2-1 – 0.5 min (50 °C), 2-2 – 6 min (51 °C), 2-3 – 22 min (58 °C), 2-4 – 60 min (63 °C), 2-5 – 90 min (64 °C), 2-6 – 125 min (95 °C). Resulting spectra are shown in the figures 3 and 4.

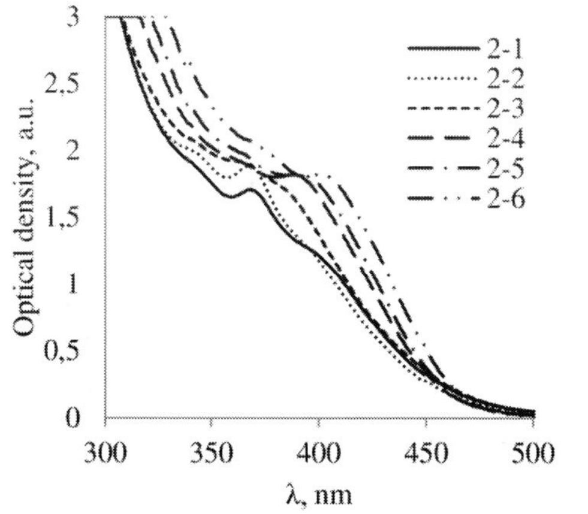

Figure 3. Absorption spectra of CdSe nanoparticles prepared with molar ratio $[Cd^{2+}]:[Se^{2-}]:[TGA]=1:0.25:1.2$.

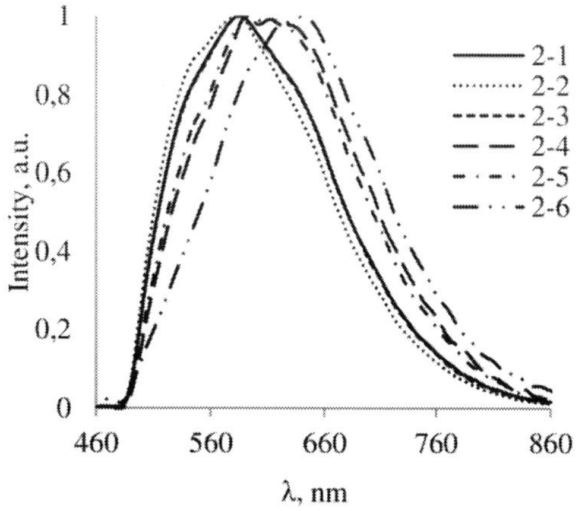

Figure 4. Photoluminescence spectra of CdSe nanoparticles prepared with molar ratio $[Cd^{2+}]:[Se^{2-}]:[TGA]=1:0.25:1.2$.

It is notable that in this case smaller nanoparticles were formed so that according to equation (1) approximate average size for the last sample was less than 1.6 nm. Absoption curves retain first excitonic transition maximum prompting uniform growth and relatively stable monodispersity in the course of the reaction. Corresponding emission band peak shifts from 580 to 640 nm.

In order to reveal influence of higher injection temperature precursors were mixed at 97 °C and temperature was maintained constant. Samples were consequently acquired in 0.5; 6; 25; 60 and 90 minutes after start of the reaction. Molar ratio was $[Cd^{2+}]:[Se^{2-}]:[TGA]=1:0.6:1.4$.

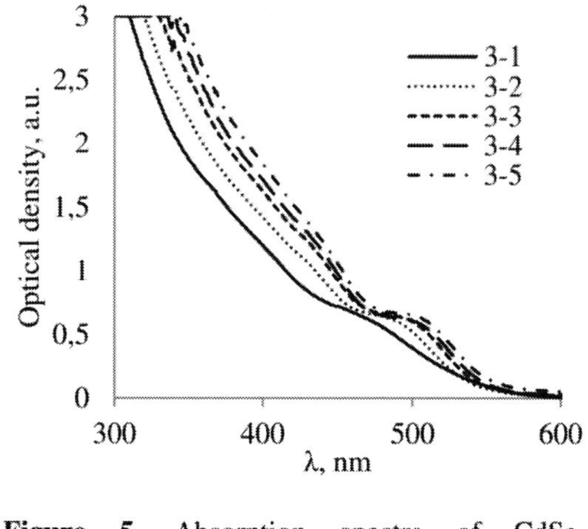

Figure 5. Absorption spectra of CdSe nanoparticles prepared at 97 °C.

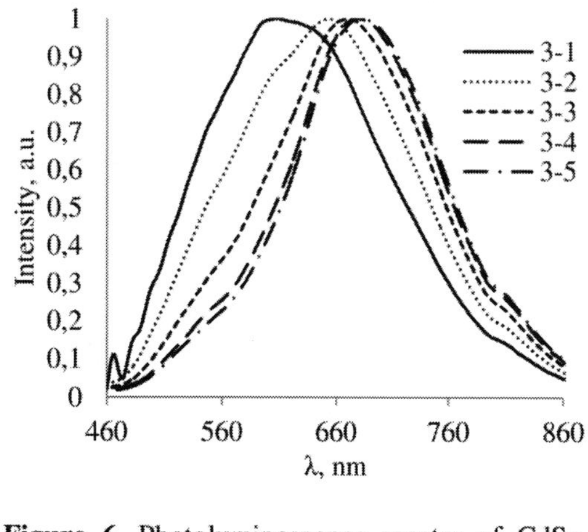

Figure 6. Photoluminescence spectra of CdSe nanoparticles prepared at 97 °C.

As can be seen from absorption spectra in the figure 5 larger nanocrystals were formed at nucleation stage. In 25 minutes crystal growth noticeably slowed down which is reflected on photoluminescence spectra either. Photoluminescence peak maximum (figure 6) shifts from 615 to 683 nm. The mean size of thus obtained nanoparticles was estimated less than 2.2 nm (for the last sample).

In order to confirm crystalinity of nanoparticles X – ray diffraction analysis was performed on the powder sample. For this fourth sample of the second series (2-4) was precipitated by means of propanol – 2 addition with subsequent centrifugation. X-ray analysis showed that nanocrystals are characterised by hexagonal wurzite structure rather than zinc blend which is not typical for CdSe nanoparticles synthesized in aqueous solutions [7–10]. Nevertheless possibility of two phase coexistence cannot be excluded particularly considering such a small average size of the QDs.

4. Conclusion

As a result of this work several series of CdSe QDs stabilized with TGA were synthesised in water phase. Depending on temperature regime and molar ratio of the components current method yields nanoparticles with mean size less than 2.2 nm. Particles exhibit photoluminescence in the range of 580 – 680 nm. Their emission that is characterized by wide peaks and large Stokes shift should be accounted for recombination via surface trap states. It was showed that higher precursor injection temperatures allow obtaining larger nanoparticles on nucleation stage while raising temperature along the course of the reaction results in higher growth rates. Application of higher molar ratio $[Cd^{2+}]:[Se^{2-}]$ facilitated formation of comparatively small nanocrystals (average size less than 1.6 nm). XRD analysis shows that thus prepared crystallites pertain to hexagonal wurzite structure.

Acknowledgements

This work was supported by grant of Russian Science Fund (project №14-15-00324).

References

[1] Musikhin S F, Aleksandrova O A, Luchinin V V, Maksimov A I, Moshnikov V A *Biotekhnosfera* 2012 **5(6)** 40–48

[2] Musikhin S F, Aleksandrova O A, Luchinin V V, Maksimov A I, Matyushkin L B, Moshnikov V A *Biotekhnosfera* 2013 **2(26)** 2–16

[3] Murray C B, Norris D J and Bawendi M G 1993 *J. Am. Chem. Soc.* **115** 8706–15

[4] Aleksandrova O A, Gordushenkov O E, Matyushkin L B, Menkovich E A, Mikhailov I I, Moshnikov V A, Romanovskiy D S and Tarasov S A 2013 *21st Int. Symp. "Nanostructures: Physics and Technology"* 298–299

[5] Matyushkin L B, Aleksandrova O A, Maksimov A I, Moshnikov V A, Musikhin S F *Biotekhnosfera* 2013 **2(26)** 27–32

[6] Yu W W, Qu L H, Guo W Z and Peng X G 2003 *Chem. Mater.* **15** 2854–60.

[7] Mi W, Tian J, Tian W, Dai J, Wang X and Liu X 2012 *Ceram. Int.* **38** 5575–83

[8] Sun Q, Fu S, Dong T, Liu S and Huang C 2012 *Molecules* **17** 8430–38

[9] Sharma P K, Dutta R K, Liu C H, Pandey R and Pandey A C 2010 *Mat. Lett.* **64** 1183–86

[10] Rong X L, Zhao Q and Tao G H 2012 *Chin. Chem. Lett.* **23** 961–964

Journal of Physics: Conference Series **572** (2014) 012029 doi:10.1088/1742-6596/572/1/012029

The study of CdSe colloidal quantum dots synthesized in aqueous and organic media

I I Mikhailov[1], S A Tarasov[1], A V Solomonov[1], O A Aleksandrova[1], L B Matyushkin[1], D S Mazing[1]

[1]Saint Petersburg Electrotechnical University «LETI»

(ETU) E-mail: iimihalov@gmail.com

Abstract. The samples of CdSe colloidal quantum dots (CQDs) synthesized in aqueous and organic media are studied. The possibility of luminescence peak position control depending on nanoparticle growth process is demonstrated. The samples synthesized in organic medium revealed the luminescence color variation effect with nanoparticle growth. The relation of this effect with processes of nucleation and defect formation in nanoparticles is considered. The CQDs of CdSe coated with CdS shell are fabricated. The use of inorganic shell can provide a double increase of the luminescence quantum yield.

1. Introduction

In the past decade, the studies on colloidal quantum dots (CQDs) have been developed widely. CQDs are semiconductor particles of nanometre-scale synthesized in liquid media and having unique properties. Because of the small geometric dimensions the CQDs reveal quantum size effects. These objects can be considered as "artificial atoms", their properties can be relatively easily controlled by varying the synthesis process parameters. The low cost and easiness of the techniques for synthesis CQDs stipulate broad prospects for their use in new nano- and optoelectronic devices. Colloidal quantum dots can be used for a high-performance LED and laser structures creation, as well as photosensitizers in solar cells or luminophores in micro-devices for biological tissues optical tomography. This article presents a method for the creation of colloidal quantum dots based on cadmium selenide, which allows controlling the wavelength of the radiation due to temperature and synthesis duration regulation.

2. Experiment

In order to study quantum dots properties and characteristics the non-invasive optical spectroscopy techniques were employed. These techniques are based on the use of photoluminescence effects and allow obtaining information regarding particle size, crystal structure quality, quantum yield, monodispersity and other CQDs parameters [1].

A measuring device based on a quick scan spectrometer (QSS) Ocean Optics USB-4000 allowing photoluminescence full range registration in a short time period was adjusted and configured in order to study the photoluminescence spectra. The sample in a quartz cell was placed into special holder and was excited by a focused laser radiation at a wavelength of 445 nm. The photoluminescence signal was collected by the collimating optical system and was directed to the input QSS by means of a fiber. Data from the spectrometer was analyzed automatically by means of specialized software designed in the LabView environment.

Several series of samples were synthesized. Synthesis was carried out in aqueous and organic media with the use of various precursors. CQDs which constitute CdSe cores covered with broad-band CdS were obtain in the same manner. The photoluminescence spectra measuring of fabricated samples was carried out.

The synthesis of colloidal CdSe quantum dots in organic medium was carried out in the first part of the work. Cadmium precursor solution was prepared by mixing CdO, oleic acid and liquid paraffin followed by heating to 150°C. Selenium precursor was obtained through heating selenium powder and liquid paraffin mixture to 220°C. The introduction of cadmium containing solution into selenium precursor was carried out at the temperature of 240°C. Samples were obtained at intervals of 1, 2 and 5 minutes. Figure 1 shows the photoluminescence spectra of the synthesized CQDs.

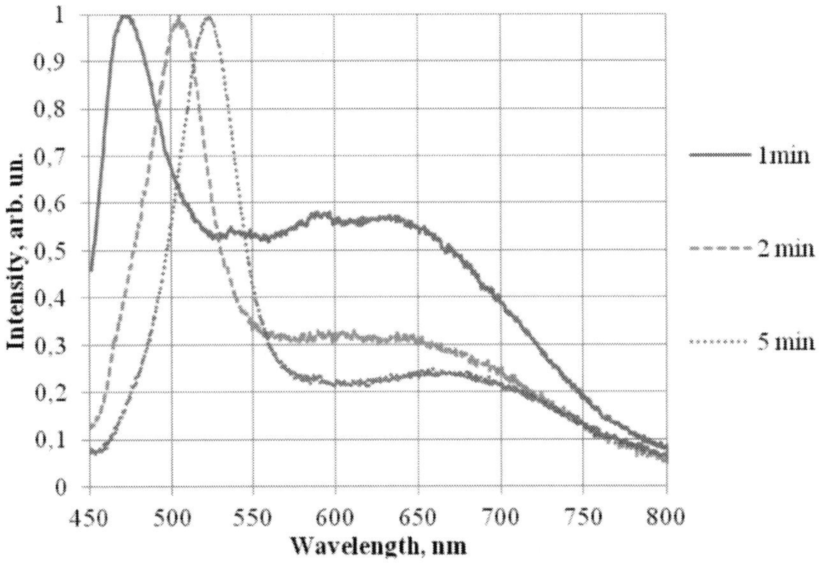

Figure 1. Normalized PL spectra of CdSe quantum dots as function of time: (a) 1 min, (b) 2 min, (c) 5 min after injection. Organic medium.

The spectra reveal two radiation regions: a high intensity peak on the left side of the characteristics and a broad plateau of a lower intensity on the right. We consider luminescence peak on the left that shifts from 470 nm to 520 nm wavelength with synthesis time to correspond to intraband transition. It is caused by the CQDs size increasing during the CQDs synthesis that leads to a decrease in the effective width of the band gap so that the emission shifts to a long wave region. The peak half-width was approximately 50 nm and decreased slightly in the course of the reaction which can be attributed to improved surface states passivation with oleic acid. The occurrence of plateau in the long wave region of the graph can be explained by existence of the QD surface defects that act as trap levels within band gap and provide the route for concurrent radiative recombination. With time oleic acid molecules passivate these surface states more efficiently and as a result the photoluminescence intensity caused by defects decreases comparatively to intraband emission.

The synthesis of CdSe colloidal quantum dots in aqueous medium was carried out in the second part of the work. Thioglycolic acid (TGA) was applied as a stabilizer. The synthesis was conducted during a long period with a gradual temperature increase. Samples were taken from the reaction medium in certain time intervals. CQDs were synthesized in almost boiling distilled water in the same manner. The molar ratio of the solution components was [TGA]:[Cd]:[Se] = 1,4:1:0,6. Temperature of the synthesis maintained at 97°C. Figure 2 shows the luminescence spectra of colloidal quantum dots built in relative units.

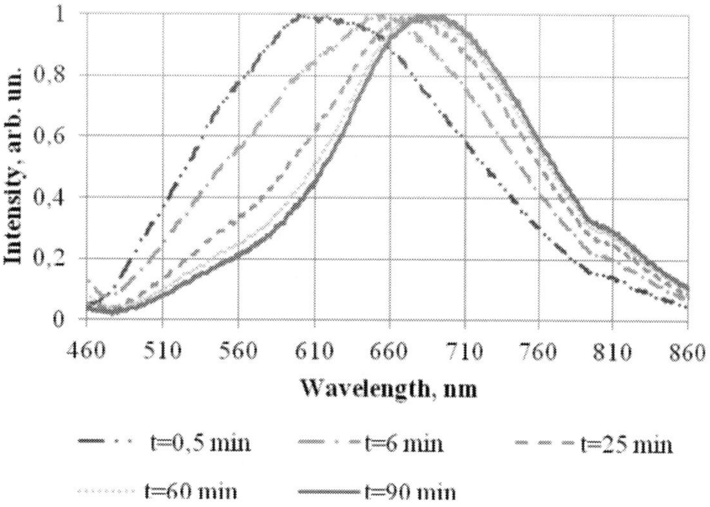

Figure 2. Normalized PL spectra of CdSe quantum dots as function of time: (a) 0,5 min, (b) 6 min, (c) 25 min, (d) 60 min, (e) 90 min after injection. Aqueous medium.

The luminescence peak shifted to long wave region with time. The spectrum half-width decreased from 200 nm for the first sample to 150 nm for the last one. The luminescence peak shift is caused by colloidal quantum dots growth in the course of the reaction. Rapidity of this process noticeably decreases with time. This is probably due to the mechanism of colloidal nanocrystal growth known as Ostwald ripening [2]. Nanoparticles grow not only due to the Cd and Se individual atoms, but also due to the fact that larger quantum dots absorb smaller ones. Over time the number of small-sized quantum dots is greatly reduced because most of them are absorbed by larger crystals. The complete depletion of Cd and Se occurs in the reaction medium. As a result further growth of quantum dots terminates or proceeds very slowly. The reduction of the spectra half-width can also be explained by the absorption of small nanoparticles by larger dots.

The main method to estimate the size of colloidal quantum dots into the liquid medium, is the study of optical absorption spectra [3-6]. Existence of exciton peak in this spectrum indicates that the sample contains CQD's [4]. Size of colloidal quantum dots can be calculated from exciton peak position [5]. Calculations show that for our sample particle size was 1.5 - 2.5 nm, which makes it difficult to study their structure directly. It is known [6], that the luminescence in the CQD's can occur in two ways – by interband transitions or by transitions through traps arising due to defects on the surface of the

Journal of Physics: Conference Series **572** (2014) 012029 doi:10.1088/1742-6596/572/1/012029

particles. In the first case, the Stokes shift is relatively small, in the second case, it may exceed 200 nm.

Figures 3 and 4 shows the absorption spectra of the CQD's synthesized in aqueous and organic media respectively. The spectra are very similar. The first exciton peak is near 450-500 nm. There is a shift of the peak to longer wavelengths, indicating an increase in particle size with increasing time of their synthesis. Comparison between absorption and luminescence spectra for CQD's synthesized in aqueous media shows that Stokes shift is quite large. The same mechanism of emission causes the appearance of a plateau on the luminescence spectra of the samples synthesized in organic media. Shortwave narrow peaks result from interband transitions. The occurrence of such peaks indicates a lower density of defects in the samples of this type.

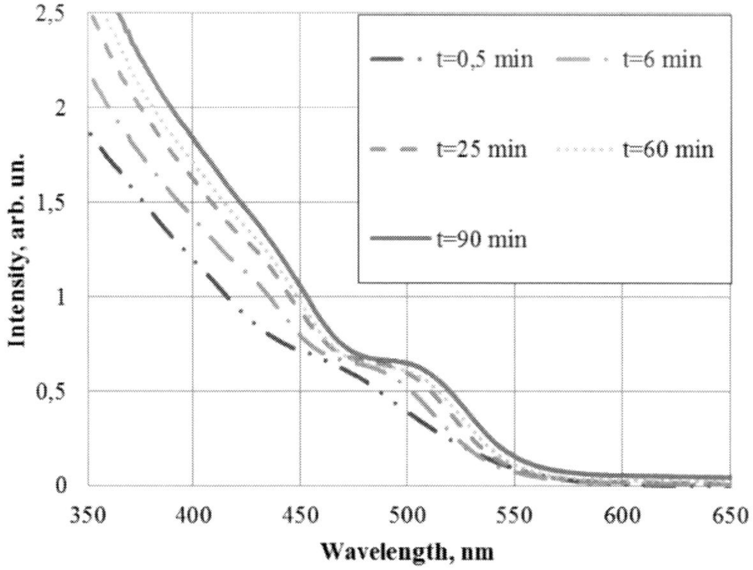

Figure 3. Normalized ABS spectra of CdSe quantum dots as function of time: (a) 0,5 min, (b) 6 min, (c) 25 min, (d) 60 min, (e) 90 min after injection. Aqueous medium.

The next experiment was carried out to coat CdSe CQDs with the broad-band CdS shell. Figure 5 shows the photoluminescence spectra of obtained samples.

Sample 0 represents the initial CdSe colloidal quantum dots obtained in aqueous medium by the same method as in the previous experiment. Sample 1 was prepared as follows. The initial CdSe solution and the Cd-TGA solution were mixed in a flask. Cd-TGA solution contains Cd 2.5 times as much as CdSe. Further, a solution of Na_2S ([Cd]/[S] = 1:1) was added dropwise during 25 min. A slight spectrum shift towards shorter wave regions was observable. The most noticeable result was the increase in photoluminescence intensity that is conditioned by the passivation of the surface states by broad-band CdS. Sample 2 was obtained by introducing the Na_2S ([Cd]: [S] = 10:1) in the initial solution during 2.5 min. Thus, we can conclude that intensity is significantly reduced by the surface states. The broad-band CdS with increased molar cadmium content passivated to the best advantage. A slight luminescence peak shift towards shorter wave region is probably associated with reduction in the crystal structure, and correspondingly influence thereof on the photoluminescence spectrum, and also due to the presence of broad-band CdS.

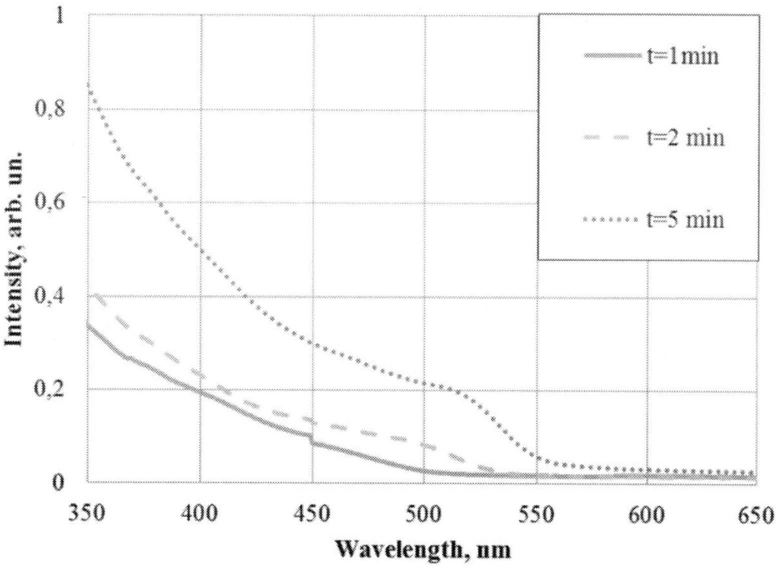

Figure 4. Normalized ABS spectra of CdSe quantum dots as function of time: (a) 1 min, (b) 2 min, (c) 5 min after injection. Organic medium.

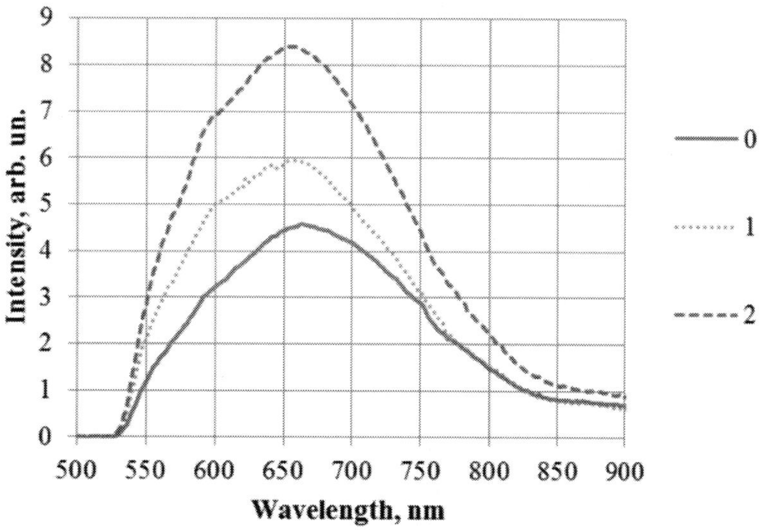

Figure 5. The photoluminescence spectra of CdSe nuclei without shell (0) and coated with shell (1, 2). Aqueous medium.

3. Conclusions.

The colloidal quantum dots of CdSe were created in organic and aqueous media, and further their parameters dependence on the synthesis was studied. The high temperatures of the synthesis were revealed to beneficially affect the quality of the CQD crystal structure. The photoluminescence emission intensity of the samples synthesized at high temperatures is significantly greater than the photoluminescence intensity of the samples synthesized at low temperatures. With increasing the synthesis time the particle concentration decreases. This should be taken into account when estimating the quantum yield. The broad-band CdS coating results in the surface states passivation and two times increase in intensity of photoluminescence.

The quality of the crystal structure may be estimated by measuring the intensity of photoluminescence. The defects in the crystal structure increase the possibility of non-emitting transitions. Thus, CQDs synthesized in organic media have less defective crystal structure as it allows applying higher temperatures.

We estimated the dispersion of the average size of the particles. Minimum particle size variance is observed in the samples synthesized in the organic medium. This phenomenon is caused by the high temperatures of the synthesis. In aqueous media, the luminescence peak width decreases with an increase in the synthesis time. This is due to the predominance of the growth mechanism called Ostwald ripening in the later stages of the synthesis.

The obtained array of CQDs are intended for use as luminophore [7] in order to create micro-devices for biological tissues optical tomography. Moreover, we carry out a work on the elaboration of white LEDs with high CRI based of nitride-structures and luminophores of colloidal quantum dots [8]. We actively develop tunable lasers in a wide range of wavelength. We design a new type solar cell with a selective absorption with using in their construction varied photodetectors including solar-blind detectors [9].

References

[1] I Mikhailov, S Tarasov, A Presnyakova and D Romanovsky, Izvestiya SPbGETU «LETI» **5**, 22 (2013).
[2] S Tarasov, I Gracheva, K Gareev, O Gordyushenkov, I Lamkin, E Men'kovich,V Moshnikov and A Presnyakova, Semiconductors **46**, 1584 (2012).
[3] A Rogach, A Kornowski, M Gao, A Eychmuller and H Weller, J. Phys. Chem. B **103**, 3065 (1999).
[4] C Burda, S Link, M Mohamed and M El-Sayed, 2001, J. Phys. Chem. B **105**, 12286 (2001).
[5] W Yu, L Qu, W Guo and X Peng, 2003, Chem. Mater. **15**, 2854 (2003).
[6] C Shen and W Tseng, 2009, Inorg. Chem. **48**, 8689 (2009).
[7] O Aleksandrova, A Maximov, E Maraeva, L Matyushkin, V Moshnikov, S Musikhin and S Tarasov, Nano- and Microsystems Technique **2**, 19 (2013).
[8] S Kurin, A Antipov, A Roenkov, I Barash, H Helava, E Menkovich, S Tarasov, I Lamkin, N Shmidt and Yu Makarov, Journal of Physics: Conference Series **461**, 012028 (2013).
[9] I Lamkin, S Tarasov, Journal of Physics: Conference Series **461**, 012025 (2013).

GaN nanowires growth with record growth rate using Ti thin film

M M Rozhavskaya, S I Troshkov, E Yu Lundina, A N Smirnov, V Yu Davydov and W V Lundin
Ioffe Physical Technical Institute, Politchnicheskaya 26, Saint-Petersburg, Russian Federation

E-mail: MRozhavskaya@gmail.com

Abstract. In this paper we present a new technological approach for GaN nanowires synthesis using metal-organic chemical vapor deposition. This aproach is to use thin titanium film, which promote nanowires seeds formation. Such an approach allows to synthesis nanowires with diametr in the range of 30 nm- 3 mkm, strongly oriented along saphire *c* axis and with regular hexagonal shape. A good quality of the nanowires material is confirmed by low temperature micro-photoluminescence spectrum of single nanowire, where spectral lines, corresponded to free and bound excitons, are observed.

Introduction. A great interest of researchers to the topic of III-N nanowires (NW) is caused by new possibilities of using these objects, which are not possible for classic thin films. A large aspect ratio of height to diameter and surface area to volume can significantly reduce the dislocation density in the upper part of NW [1]. Moreover, such structure can significantly reduce the stresses occurring due to the difference in lattice constants and coefficients of thermal expansion of substrate and epitaxial layers. Due to the three-dimensional geometry substantially increases the efficiency of light output from such structures.

In this paper we present a new method for producing GaN NWs using titanium nanofilm with a thickness of 30 nm. It is known from the literature about the use of titanium films in molecular-beam epitaxy process to form titanium nitride masks for selective epitaxy [2] , also TiN nanofilms were used for dislocations filtration in metal-organic vapor-phase epitaxy (MOCVD) [3] and in some papers titanium nitride layers were used to separate the thick epitaxial GaN layers from substrate in hydride vapor-phase epitaxy [4]. However, there is no mention of the possibility of the growth of GaN NWs through a solid titanium film. Thus, a fundamentally new approach is discovered to the synthesis of GaN NWs.

Experiment. All the samples were grown on sapphire substrates by metal- organic vapor phase epitaxy on the deeply modernized Epiquip VP50RP growth machine. Trimethylgallium and ammonia were used as precursors of gallium and nitrogen,respectively. The temperature of the substrate was 1040C , the growth time of NWs was 15 minutes. Titanium film was predeposited on a sapphire substrate or substrate with pregrown GaN layer using vacuum evaporation. The growth procedure began with the substrate heating in ammonia-nitrogen atmosphere, and then the atmosphere changed to ammonia- hydrogen and trimethylgallium was introduced in the reactor.

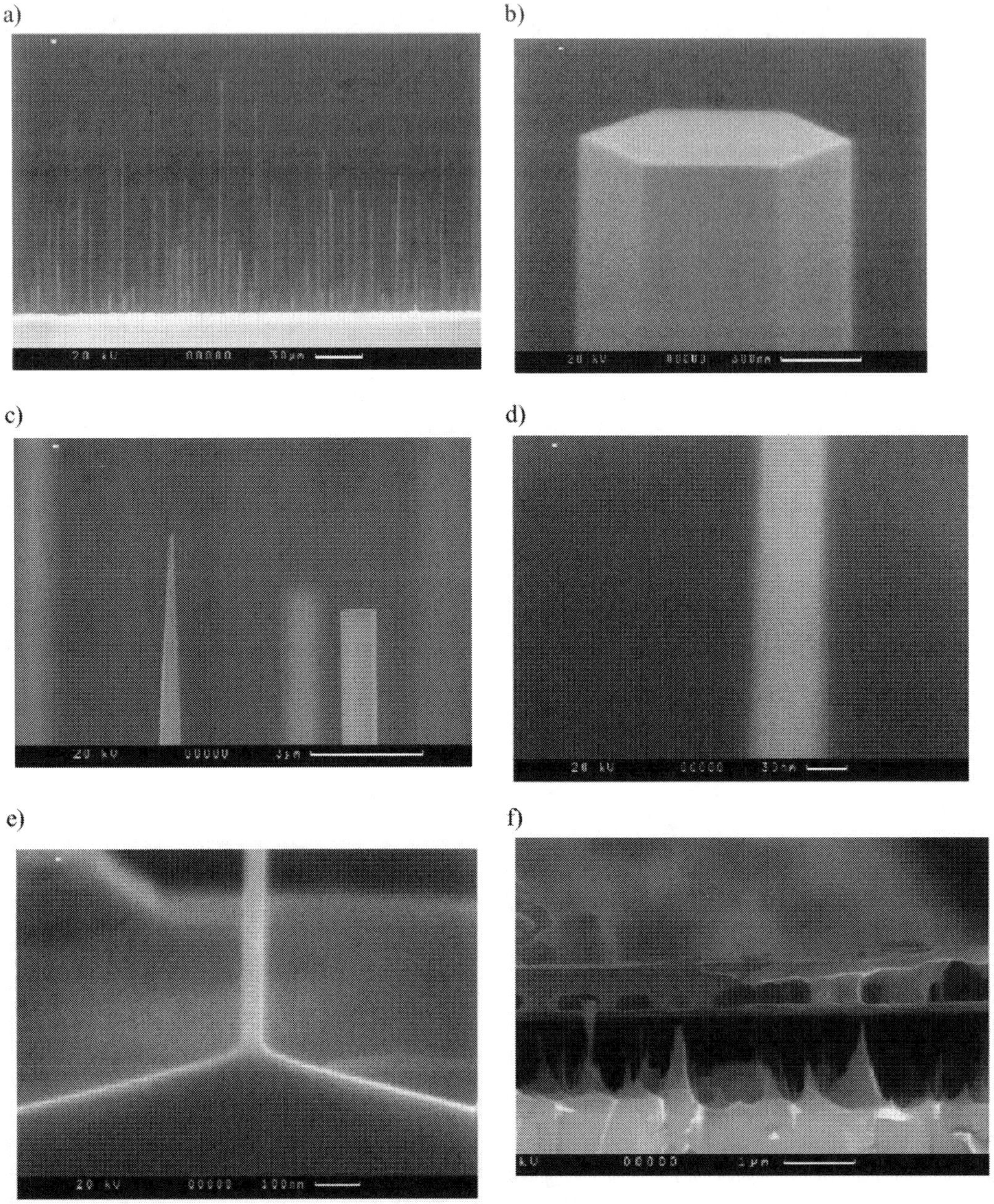

Figure1. SEM images of GaN NWs a) NW ensemble grown via Ti film b) all the NWs have regular hexagonal shape c) two types of NWs top d) the thinnest NW in the ensemble e) pyramid at the base of NW f) etching near the titanium film

Results and discussion. Figure 1 shows the scanning electron image of the NWs. All NWs have the regular hexagonal shape . NWs diameter in the ensemble is in the range of 30 nm - 3 mkm , NWs with a diameter greater than 100 nm were strongly oriented in the direction of the *c* axis of sapphire . The base of each NW was pyramidal island , the thickness of a continuous layer between the NWs and the substrate was 1-1.5 mm. It was also observed a results of etching process in the continious GaN layer upper the titanium film, and in the case of substrates with pre- deposited GaN layer etching took place also under the titanium film.

There was no metal droplets on the top of NW, which means that in this case growth take place not via well known VLS mechanism. [5] It is well known that at temperatures above 800C titanium reacts with nitrogen to form titanium nitride . We believe that in our method titanium film is partially converted into titanium nitride which is necessarily to from NW seeds. This mechanism is most similar to the self-organization of GaN NWs by molecular beam epitaxy using a AlN buffer layer, when the NW nucleation occurs on the edges of (1-102) hexagonal holes in a AlN layer[6,7]

Maximum height of NW in the ensemble was 200 mkm, which corresponds to the growth rate of 800 mkm / hour. This value is about one order of magnitude grater than the maximum published in the literature growth rate for GaN Nws [8,9] and 2 orders of magnitude greater than the standard rate of growth for the planar GaN MOCVD . We did some calculations to find out if it is possible to grow NW with such a high growth rate using material only from gas phase. In the MOCVD reactor there is a so called depletion layer which forms directly above the epitaxial layer. The concentration of precursors in this layer is lower than the concentration at the reactor inlet. The calculated growth rate of the planar layer on the non-activated substrate in the NW growth mode considering depletion layer is 4mkm/hour. This value was also confirmed by direct experiment.. Let us assume that all the substrate is covered with a dielectric mask , and there is only one window in mask with 2 microns in diameter , over which the depletion layer occurs. In this case the estimated growth rate of NW is equal to 24 microns / hour. Finally , we can estimate the maximum possible growth rate of NW from the gas phase using of the formula of the number of collisions between ideal gas atoms, if all the atoms have absolutely inelastic interaction with the upper bound of NW. The resulting growth rate of the NW in this case is equal to 90 mkm / hour. While experimental growth rate of 800 mkm / hour , which means that NW collects material not only from the gas phase but also from it walls . To estimate the surface roughness of the side walls, NW were investigated using atomic force microscopy . Single NW was mechanically separated from the sample and placed on a sapphire substrate. We didnt additionally fix NW on the sapphire substrate, however, due to van der waals interaction it was well held onto the substrate. The AFM investigation showed that NWs side walls are atomically flat. That means, the diffusion of gallium atoms along such a walls to the top of the NWs have to go very effectively.

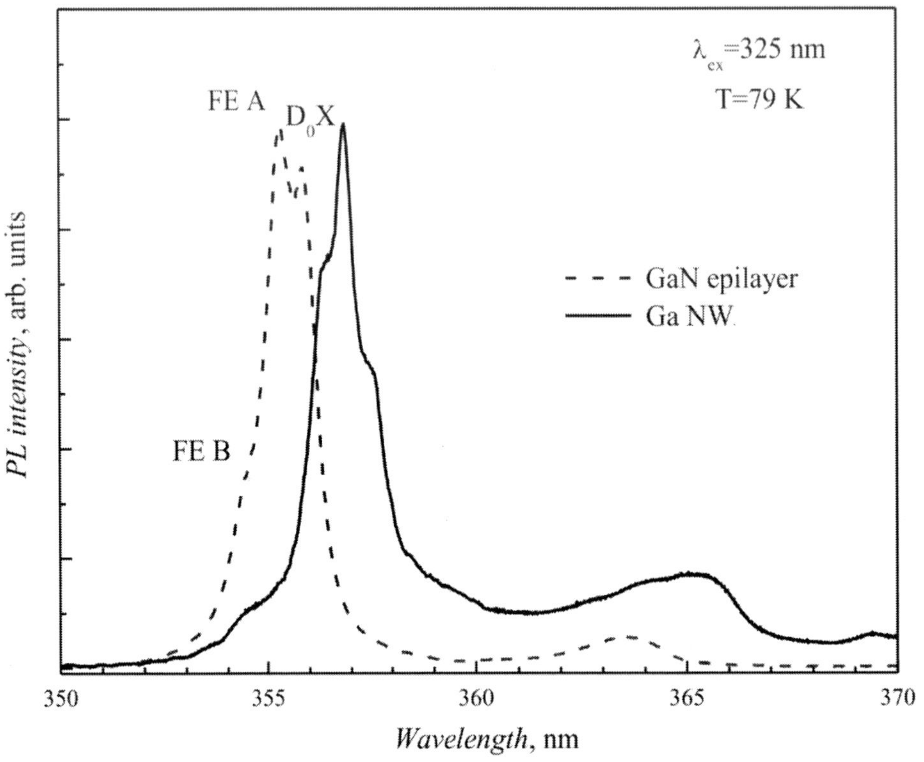

Figure 2 Micro-photoluminescence spectra of single GaN NWs (solid line) and the GaN epitaxial layer quality (dashed line), measured at a temperature of 79K

Figure 2 shows a photoluminescence spectrum of a single micro- NW, separated from the substrate . For comparison, the dotted line shows the spectrum of micro- PL solid GaN epitaxial layer of high quality. The spectra were measured at a temperature of 79K . The line widths of the edge luminescence NWs and a continuous layer of practically coincide. In both spectra allowed the free and bound excitons , which indicates the high quality of the material NOCs . The difference in the positions of the spectral lines associated with the mechanical stresses in the samples : a solid layer grown on a sapphire substrate is compressed in the plane of the substrate , while the position of the lines corresponds to the luminescence NOCs unstressed material.

Conclusion. Thus, a new approach to the synthesis of GaN NWs is desribed, which consists in the use of continuous thin (10 nm - 30 nm) titanium film. This method allows for the growth of NWs with high quality and record growth rate higher than 800 mkm / hour. This value is one order of magnitude higher than maximum growth rate of GaN NW published in the literature and two orders of magnitude higher than standard planar GaN growth rate for MOCVD.

This work was supported by RFBR (grants 13-02-12082 and 12-02-00453) and the program of the Presidium of Russian Academy of Sciences Physics of new materials and structures.

[1] Zubia D. and Hersee S. D. 1999 *J. Appl. Phys.* **85** 6492

[2] Kishino K, Sekiguchi S, and Kikuchi A 2009 *J. Cryst. Growth* **311** 2063

[3] Fu Y, Yun F, Moon Y T, Özgür Ü, Xie J Q, Ni X F, Biyikli N, Morkoç H, Zhou L., Smith David J, Inoki C. K. and Kuan T. S. 2006 *J. Appl. Phys.* **99** 033518

[4] Oshima Y., Eri Y. , Shibata M., Sunakawa H. and Kobayashi K. 2003 *Jpn. J. Appl. Phys.* **42** L1–L3

[5] Wagner R. S. and Ellis W. C. 1964 *Appl. Phys. Lett.* **4** 89

[6] Bertness K A, Roshko A., Mansfield L. M., Harvey T. E. Sanford N. A. 2007 *J. Cryst. Growth* **300** 94

[7] Sekiguchi H., Nakazato T. , Kikuchi A. and Kishino K. 2007 *J. Cryst. Growth* **300** 259

[8] Lekhal K, Avit G, André Y, Trassoudaine A, Gil E, Varenne C, Bougerol C, Monier G, Castelluci D. 2012 *Nanotechnology* **23** 405601.

[9] Avit G, Lekhal K, Andre Y, Bougerol C, Reveret F, Leymarie J, Gil E, Monier G, Castelluci D, Trassoudaine A 2014 *Nano Letters* **14** 559

Optical transmission spectra of porous aluminamembranes with different pore size

L B Matyushkin, E N Muratova, J M Spivak, V V Shimanova, S A Korlyakova and V A Moshnikov

Saint Petersburg Electrotechnical University "LETI", ul. Prof. Popova, 5, 197376 Russia

E-mail: leva.matyushkin@gmail.com, sokolovaeknik@yandex.ru, ymkanageeva@yandex.ru

Abstract. Membranes of nanoporous aluminum oxide (alumina) have been obtained using the electrochemical etching technique by varying technological regimes. The surface morphology and cleavages of obtained experimental samples are studied using scanning electron microscopy (SEM). The optical transmission measurements were performed on a spectrophotometer in the wavelength range of 190–1000 nm. It is possible to determine the average size and the dispersion of the pore diameter by UV-visible transmittance spectrum measuring.

1. Introduction

The electrochemically produced porous alumina (por-Al_2O_3) membranes have been widely used as catalyst carriers and templates for the fabrication of ordered nanostructures [1-5], owing to regular nanoscale structures and relatively easy and low-cost processing. Of considerable interest is the por-Al_2O_3 for the problems of biomedicine [6, 7], for various types of sensors [8, 9], in MEMS-devices [10, 11]. A special attention is paid to the study of the optical properties of both the porous structures [12, 13] and various composites based on them [14]. Due to their structure, these membranes have potential in various optical device applications, such as optical filters and photonic crystals [15]. It is known that aluminum anodized in acidic electrolytes may occur at different rates, efficiency, and stability, in general, depend on the nature and concentration of the electrolyte, temperature and anodic current density [16].

It is important to be able to give a quick characteristic of the porous structure parameters, to evaluate the pore diameter, size distribution and order. Therefore, the aim of this work is to study the dependence of the transmission spectrum of the porous alumina membranes of their structural parameters.

2. Experiment

In order to obtain a porous anodic alumina membrane electrochemical anodization of aluminum foil (40μm) was conducted in potentiostatic mode in electrolytes based on aqueous solutions of phosphoric (H_3PO_4) and sulfuric (H_2SO_4) acids. Geometric parameters of membranes obtained in different electrolytes are presented in Table 1.

Usually the investigation of the structure and basic parameters (diameter of pores, interpore distance, porous layer thickness) of the samples obtained are carried out with the help of scanning electron microscopy (SEM). As shown in figure 1, anodic alumina membranes contain many columnar pores with diameters are about of 20-90 nm.

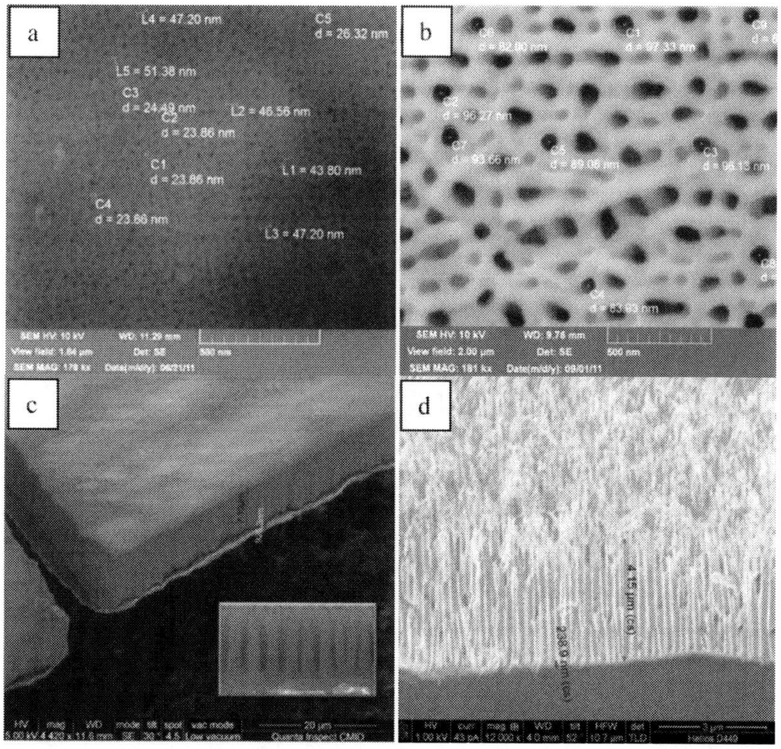

Figure 1. SEM images topology and cleavage of the membranes por-Al_2O_3 obtained in an electrolyte based on solutions of H_2SO_4 (a, c) and H_3PO_4 (b, d).

The optical transmission measurements were performed on a PE-5400UV spectrophotometer (LLC "Ekohim") in the wavelength range of 190–1000 nm.

3. Results and discussion

It was found that porous alumina layers produced in different electrolytes (based on H_2SO_4, H_3PO_4) have different colors. For example, samples produced in the sulfuric acid solution have yellow color, while the samples produced in the phosphoric are usually blue. The study of optical transmission spectra have been shown that for any membrane the transmission spectrum has the same form (figure 2): the membrane starts to pass radiation with a certain wavelength.

A comparison of the transmission spectra for the different membranes (figure 3) shows that the wavelength at which the membrane starts to pass radiation depends on the type of electrolyte in the presence of which it was produced. The band gap of alumina is too large ($\approx$ 9 eV) to observe the size effects in this optical range.

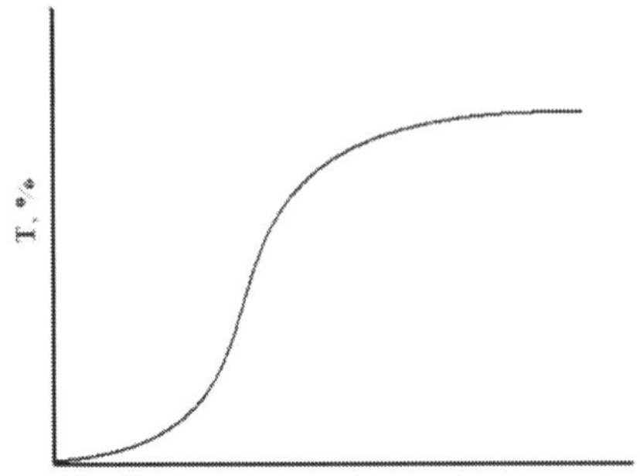

Figure 2. General view of the transmission spectra.

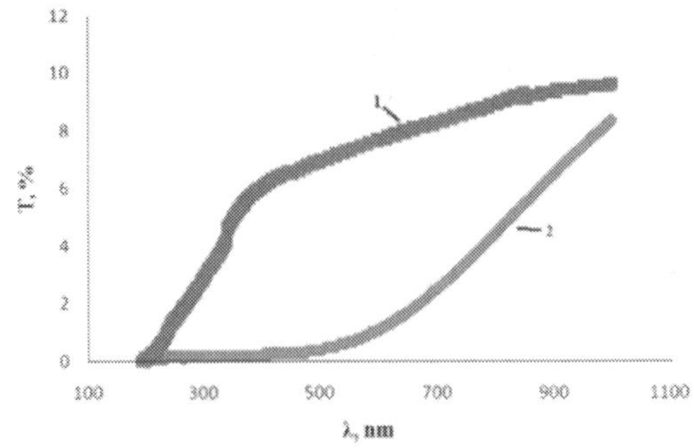

Figure 3. Transmission spectra of porous alumina membranes produced in presence of 1 – SO_4^{2-} and 2 – PO_4^{3-} anions.

So this phenomenon can be related to a) the pore size or b) the type of anion incorporated into the anodized alumina lattice (SO_4^{2-}, PO_4^{3-}). However, since it is acid or anion type, determine the pore size; we can assume that both of these explanations are interrelated. We found the following pattern: the smaller the pore diameter, the smaller the critical wavelength λ_c. It is important to note there is always a values range of the pore diameter, which could be described by a Gaussian function (figure 4).

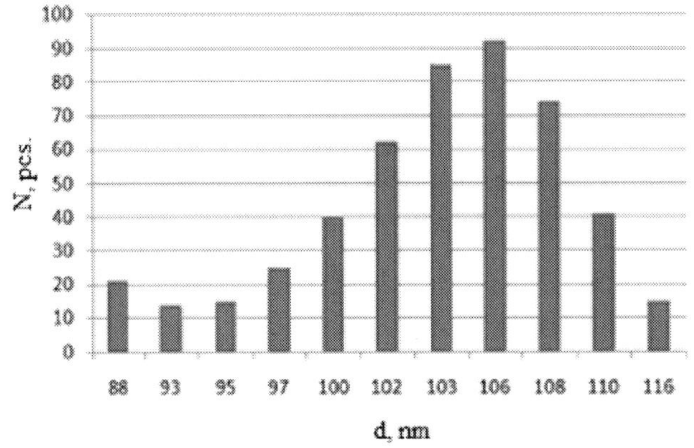

Figure 4. Figure size distribution of pore sizes in the samples produced in an electrolyte based on phosphoric acid.

Analysis of experimental data showed some correlation between λ_c and type of porous structure due to electrolyte. Summary of the experimental date presented in Table 1.

Table 1. Influence of the electrolyte type on the geometric parameters porous layer and λ_c.

Acid	Diameter of pore, nm	Thickness of membrane, nm	Thickness of the barrier layer, nm	Concentration of pores, pcs/μm^2	λ_c, nm
sulfuric	20-30	12-18	500	350	210…280
phosphoric	80-100	4-7	200	50	450…750

Thus, the presented of experimental data generalization allows to determine the average size and the dispersion of the pore diameter by UV-visible transmittance spectrum measuring.

Furthermore, it was noted that increasing of the anodizing time leads to improved transmission of the membranes (figure 5). This is because the longer the process is the more aluminum ions move from the bottom of alumina foil to alumina and the membrane is transparent.

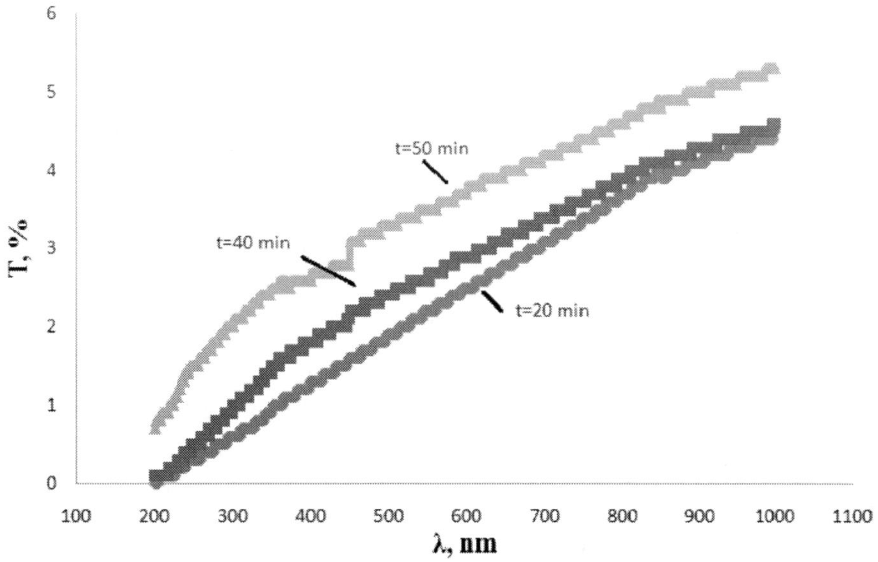

Figure 5. Transmission spectra porous alumina membranes obtained during different time anodization.

4. Conclusions

It is established that increasing of the anodizing time leads to improved transmission of the membranes. Investigations have shown that the membrane starts to pass radiation with a certain wavelength, which depends on the type of electrolyte in the presence of which membrane was produced. Thus, it is possible to determine the average size and the dispersion of the pore diameter by UV-visible transmittance spectrum measuring. It could be put in the basis of the non-destructive express control technique for the determination of pore size in por-Al_2O_3. The observed optical behavior of membranes may be used for cheap optical filters limiting the predetermined range radiation.

Acknowlegments

The reported study was supported by RFBR, research project No. 14-08-31609 mol_a.

The work was perfomed under the Russian Ministry of Education. Task (goszadanye) № 16.2112.2014/K (in the technological part, as well as the results of the influence of preparation conditions of the structure of porous layers).

References

[1]. Eftekhari A. 2008 *Nanostructured Materials in Electrochemistry* (WILEY-VCH Verlag GmbH & Co. KGaA, Weinheim)

[2]. Afanas'ev A V, Il'in, V A, Moshnikov V A, Sokolova E N and Spivak Yu M 2011 Synthesis of nanoporous and microporous structures by the electrochemical methods *Biotekhnosfera* **no 1–2** pp 20-26

[3]. S Y Jeong, J Y Kim, H D Yang, etc. 2003 Synthesis of Silicon Nanotubes on Porous Alumina Using Molecular Beam Epitaxy *Advanced Materials* **vol 15** is 14. pp1172–1176

[4]. W Zhan, Y Huangfu, G Ding, H Yea. 2013 Restrictions of Si-based Genanodots from porous alumina membranes *Superlattices and Microstructures* **vol 60** pp 73–82

[5]. Yu Piao, H Lima, Ji Yo Changb, et al. 2005 Nanostructured materials prepared by use of ordered porous alumina membranes *Electrochimica Acta* **50** pp 2997–3013

[6]. G Macias, L P Hernández-Eguía, J Ferré-Borrull 2013 Gold-Coated Ordered Nanoporous Anodic Alumina Bilayers for Future Label-Free Interferometric Biosensors *ACS Applied Materials & Interfaces* **5 (16)** pp 8093–8098

[7]. Y Song, Y Ju, Y Morita, G Song 2013 Effect of the nanostructure of porous alumina on growth behavior of MG63 osteoblast-like cells *Journal of Bioscience and Bioengineering* **vol 116** is 4 pp 509–515

[8]. Mozalev A, Calavia R, Vázquez R M 2013 MEMS-microhotplate-based hydrogen gas sensor utilizing the nanostructured porous-anodic-alumina-supported WO_3 active layer *International Journal of Hydrogen Energy* **vol 38** is 19 pp 8011–8021

[9]. Gracheva I E, Moshnikov V A, Karpova S S, et al. 2011 Net-like structured materials for gas sensors *Journal of Physics*: Conference Series **291** 012017

[10]. Zimina T M, Muratova E N, Spivak Yu M., et al. 2012 Technologies of formation and application of nano-layers and nano-porous Al_2O_3 compositions for micro-and nanotech *Nano and Microsystems* **no 12** pp 15-24

[11]. Akiyama T, Ishikawa Y, Hara K. 2013 Xylene sensor using double-layered thin film and Ni-deposited porous alumina *Sensors and Actuators B: Chemical*. **vol 181** pp 348–352

[12]. Tarasov S A, Gracheva I E, Gareev K G, et al. 2012 Atomic Force Microscopy and Photoluminescence Analysis of Porous Metal-Oxide Materials *Semiconductors* **vol. 46** no 13 pp 1588–

[13]. Lenshin A S, Kashkarov V M, Seredina P V et al. 2012 Study of the Morphological Growth Features and Optical Characteristics of Multilayer Porous Silicon Samples Grown on *n*_Type Substrates with an Epitaxially Deposited *p*+_Layer *Semiconductors* **vol 46** no 8 pp 1079–1084

[14]. Foss C A, Hornyak G L, Stocketr J A, et al. 1993 Optically Transparent Nanometal Composite Membranes *Advanced Materials* **5** no 2 pp 135-136

[15]. Choi J, Luo Y, Wehrspohn R B et al. 2003 Perfect two-dimensional porous alumina photonic crystals with duplex oxide layers // *J. Appl. Phys*. **94**, pp 4757

[16]. Muratova E N, Spivak Yu M, Moshnikov V A, Petrov D V, Shemukhin A A, Shimanova V V 2013 Influence of technological parameters of nanoporous Al2O3 layers' preparation on their structural characteristics *Glass Physics and Chemistry* **vol 39** no 3 pp 320-328

Electrical Conductivity Modeling and Research of Polypropylene Composites Filled with Carbon Black

A.S. Stepashkina[1], E.S. Tsobkallo[1], A.N. Alyoshin[2]
[1] Saint-Petersburg State University of Technology and Design, St. Petersburg, Russia
[2] Ioffe Physical-Technical Institute, St.Petersburg, Russia

E-mail: stepashkina.anna@yandex.ru

Abstract. Composites of polypropylene filled with carbon black (PP/CB composite) at different concentrations were prepared by melt mixing followed by compression molding. The dependence of electrical resistance on the filler mass fraction was experimentally received. It was shown that the received dependence had the threshold character. The composite kept dielectric properties at the filler concentration below the threshold and at the concentration above the threshold the electrical resistance decreased more than on 8-10 orders. The theoretical description of electrical conductivity of the composite was offered. Experimental data of the dependence between electrical resistance and the filler mass fraction agreed with the theoretical. The process of conductivity in the PP/CB composite was simulated by means of the Monte-Carlo method for threshold mass fraction estimation.

1. Introduction

Nowadays the great attention is paid to creation of new materials which have specially set properties included antistatic effect. One of the ways to achieve necessary properties is the preparation of polymeric composite materials [1-3]. Usually thermoplastic materials are used as a matrix because of their good physical and mechanical properties, availability and simply processing. One of the most widespread industrial thermoplastic polymers is polypropylene (PP) [4]. Conductive carbon fillers can considerably reduce the initial polymer resistance. So, carbon black (CB) is very appropriate as the electric conductive filler due to the number of characteristics: the high adsorption by the surface (the existence of foreign components worsens the structure), high void ratio. An average diameter of CB particles is from 10 to 300 nanometers, electrical conductivity fluctuates in the wide range, the most widespread trade names have conductivity from 1 to 100 $(Ohm \cdot m)^{-1}$ [5].

We received polypropylene/carbon black composite material. It was necessary to make a theoretical description of conductivity process in such materials because there are a lot of works devoted to different technologies of preparation, but it s not enough works which can illustrate the unusual dependence of electro conductivity between concentration of the filler [1-2].

2. Sample preparation, measuring technique of electrical resistance

Samples of PP/CB composite material were received in the form of films and blocks by compression molding. For this purpose micro-compounder 'DSM Xplore' was used. The polymer used in this study was isotactic polypropylene (PP) (trade name 01270 Balen), released in the form of granules. Carbon black (trade name P-805E) was used as a filler (table 1).

Table 1. Carbon black and polypropylene characteristics

	Carbon black	Polypropylene
Aspect ratio	~1	
Density, g/sm^3	1,8-2,1	0,9
Electrical resistivity, Ohm$\cdot m$	$1{,}3\cdot10^{-3}$	10^{15}
Modulus of elasticity, GPa	~15	1,2

For electrical resistance calculation volt-ampere characteristic was found under normal conditions and room temperature. The scheme of the automated device for resistance measurement based on double-probe system is shown on Figure 1. The sample (1) is placed in the screened chamber (2). Contacts on the sample were made with the silver paste. The power supply (3) gives constant voltage on one of electrodes (4). The electric current passes through a sample (1) on the second electrode (4) and further on the picoammeter (5), showing the value of current in the chain.

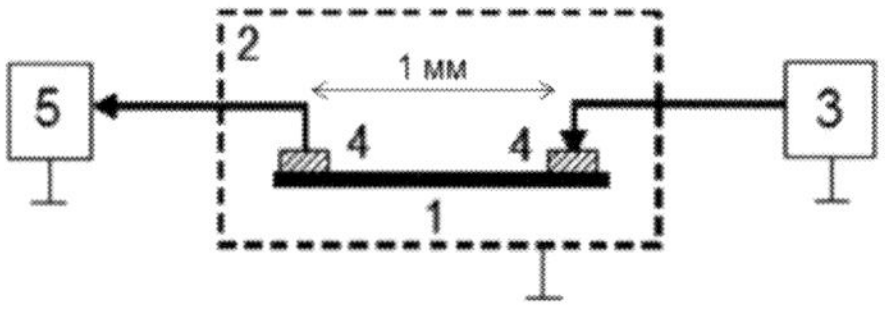

Figure 1. The scheme of the automated device for electrical resistance calculation volt-ampere characteristic: 1 – sample; 2 - screened chamber; 3 - power supply; 4 - electrode; 5 - picoammeter

3. Experimental Results

The dependence of electrical conductivity on the filler mass ratio was measured. Apparently from figure 2, change of electrical resistance from CB mass ratio has threshold character. At the small filler concentration the composite material was dielectric. If concentration was increased, at CB mass ratio 23-27% for film samples and 33-37% for blocks the threshold of electric resistance was observed. At high filler concentrations the composite behaved as carrying-out material.

4. The theoretical description of PP/CB composite conductivity. Task of spheres.

In the studied PP/CB material there are a few free electrons (charge carriers). So, the electrical conductivity in such structures is described within the theory of hopping conductivity where electron "jumps" from one donor to another [5]. In our case CB particles are donors. So, the electron can carry out transition from one donor to another passing the free condition and without occupying energy from thermal movement of atoms. Such process is called tunnelling. The event of tunnel transition is the improbable event therefore the dependence of electrical resistance from the concentration of conductive particles has to have exponential character [5-7].

Threshold nature of dependence between electrical resistance of composite from the mass fraction of the filler was experimentally observed, therefore the process of conductivity can be considered from the point of view of the percolation theory. According to this theory at the certain volume fraction of

CB, the CB particles are formed into the infinite carrying-out cluster at the expense of what considerable decrease in electrical resistance is observed.

As the studied PP/CB composite material is the polymorphic structure, i.e. there are areas of various degree of orderliness in material structure; the studied material has no ordered crystal lattice, that's why the process was considered by the task of spheres of the percolation [8].

It was made the averaging by the size of agglomerates of CB particles, also "spheres" was constructed for all agglomerates, all such spheres had identical radius. If the center of one sphere lied in other sphere, such spheres were considered as the connected. It allows to take into account the polymorphic structure and localization of wave functions in a studied composite. At the certain concentration of CB the spheres formed the infinite cluster at the expense of what considerable decrease in specific resistance of the PP/CB composite material was observed.

According to the solution of the task of spheres the dependence of electrical resistance on the filler concentration, described by expression [6], was found:

$$\rho = R_0 \cdot \sqrt[3]{\frac{3}{4\pi N}} \cdot \left(\sqrt[3]{\frac{3}{4\pi N}} \frac{1}{g a_b^*}\right)^{\gamma} e^{\frac{2 \cdot 1.39 \cdot \sqrt[3]{\frac{3}{4\pi N}}}{a_b^*}}$$

where R_0 - normalizing constant, R_0 = 1 Ohm, g=2, N — volume concentration of the filler, ν = 0.8 — for block samples; ν = 1.3 — for film samples, a_b^* — radius of the connected sphere.

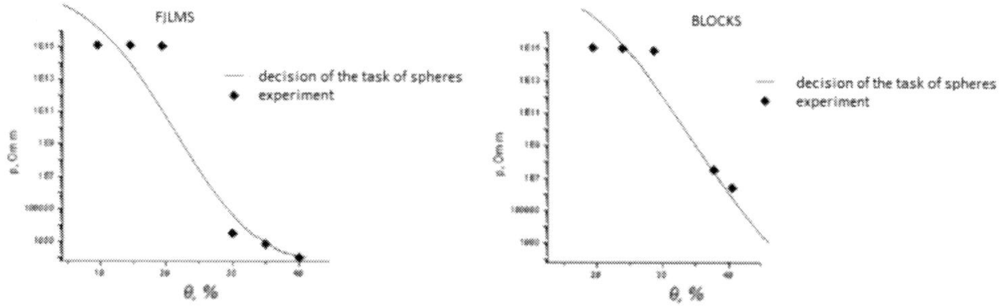

Figure 2. The dependence of electrical resistance on a filler mass fraction for films and blocks

The proximity of experimental and theoretical results was observed (fig. 2), however theoretical dependences had smoother character.

The model by means of the Monte-Carlo method was constructed for more accurate definition of the threshold in composite materials. This model allows to visualize filler distributions in the matrix and to define the threshold values of the concentration. The volume with randomly distributed CB particles was considered (see Figure 3: polypropylene – grey, carbon black –black and red, infinite cluster - red). It is necessary to consider the mechanism of hopping conductivity therefore the assumption was made: electrical contact between CB particles can be carried out not only at the direct contact between particles, but also through some dielectric layer, thickness of an equal one cell. Values of percolation thresholds were found. For film samples the threshold was observed at the CB mass fraction ~ 25%, for block samples ~ 36%. Experimental and theoretical values of the threshold coincided with rather good accuracy.

Figure 3. The dispersion of CB in PP matrix (polypropylene – grey, carbon black –black and red, infinite cluster – red)

5. Conclusion

The PP/CB composite material in the form of films and blocks was received. The dependence of the composite electrical resistance on the filler mass fraction was experimentally found. Such dependence had the threshold character. The proximity of experimental and theoretical results was observed (fig. 2), however theoretical dependences had smoother character. For more exact determination of threshold values in block and film samples the process of conductivity was simulated by means of the Monte-Carlo method. Set of the solution of the task of spheres and modeling by means of the Monte-Carlo method allows describing with high precision the conductivity process in the PP/CB composite materials.

References

[1] A.Mdahri, F. Carmona, C. Brossea, et al. 2008 *Journal of applied physics*
[2] Xiangcheng Luo, D.D.L. Chung 2001 *Composite Science and Technology*
[3] D.D.L. Chung, S. Wang 1999 *Smart mater struct*
[4] A. Levin 1961 *Polyethylene and polypropylene*.(Moscow: GOSINTI)
[5] N. F. Mott 1990 *Metal-Insulator Transitions* (London: Taylor & Francis)
[6] A.L. Efros 1987 *Physics and Geometry of Disorder. Percolation Theory* (Moscow: Mir)
[7] Shklovskii B.I., Efros A.L.1984 *Electronic properties of doped semiconductors* (Springer)
[8] I.K. Pilipovskij, I.K Jarceva 1967 *Polypropylene* (Leningrad: Himia)

Formation of A^3B^5 ternary solid solutions on GaSb plates by solid-state substitution reactions

V I Vasil'ev[1,2], G S Gagis[1], K K Soboleva[3], V I Kuchinskii[1,2]

[1] Ioffe Physical-Technical Institute of the Russian Academy of Sciences, St Petersburg, Russia
[2] St Petersburg Electrotechnical University "LETI", St Petersburg, Russia
[3] St Petersburg State Polytechnical University, St Petersburg, Russia

E-mail: giman@mail.ioffe.ru

Abstract. A method of ultra-thin GaAsSb and GaPSb solid solutions layers formation by supplying vapors of As or P to heated GaSb crystal plates is proposed.

1. Introduction

The paper [1] described that conversion efficiency of solar radiation into electricity in tandem solar cells composed of mechanical-joined upper photovoltaic converter (PC) based on GaAs and lower PC on the basis of GaSb is much higher than that in solar cells including only single GaAs-based PC. In this case, the presence of wide-bandgap optical window (WOW) on surface of each of PC element plays essential role [2]. WOW is the ultra-thin layer of solid solution isomorphous with the device structure and having wider bandgap comparing to the underlying layers. The nowadays layers of A^3B^5 solid solutions are obtained by epitaxial methods: liquid phase epitaxy, metal-organic chemical vapor deposition and molecular beam epitaxy. As this takes place, the crystal lattice of solid solution layer is completely made of components arriving at the crystal surface. Previously we have shown [3] that GaAsP ternary solid solution on the heated GaAs surface can be formed by solid-state substitution at which the solid solution forms in near-surface layer of GaAs plate at the expense of partial replacement of arsenic atoms by phosphorous atoms delivered to a crystal surface as vapors, at that this solid solution plays the role of WOW. The work describes the results of the processes of solid-state substitution in GaSb plates to obtain GaAsSb and GaPSb WOW on GaSb surfaces. In addition the possibility of simultaneous formation of WOW for GaSb and zinc diffusion implementing for p-n-junction creation in a single technological process has been considered.

2. Experiment

WOW epilayers and p-n-junctions were formed in GaSb:Te(100) plates. Thicknesses and compositions of obtained wide-bandgap layers are determined by three factors: a temperature, a duration of process and partial pressures of vapors of delivered elements. Processes were carried out at temperatures 520 - 580 °C. This temperature range is favorable for elements diffusion into crystal structure and subsequent solid-state substitution reactions. Phosphorous and arsenic vapor sources were unsaturated solutions-melts of $ZnSnAs_2$ and $ZnSnP_2$ in tin. The same solutions-melts were zinc vapors sources. Components concentration of the fifth group in the used solutions-melts were 0.01 - 0.05 atomic fractions. In doing so, concentrations of zinc were two times lower - 0.005 - 0.025 atomic fractions. The pressure values of the elements of the fifth group were estimated using data of saturated vapor pressures of arsenic and phosphorous at the temperatures of carrying out the process and taking into account the activity of these components in solutions-melts. Estimated values of pressures amounted up to 10^3 Pa for phosphorous and up to 10 Pa for arsenic. Solutions-melts and GaSb plates were in the closed graphite cassette. Processes were carried out in a bowed hydrogen quartz reactor to prevent material oxidation. Durations of processes were from 5 to 30 minutes.

Properties of the obtained samples were investigated using secondary ion mass-spectrometry (SIMS) method and photoluminescence (PL) measurements.

3. Results and discussion

Surfaces of obtained layers had mirror-smooth morphology, which indicated the absence of clusters and erosion processes.

Figure 1 shows the variation of arsenic concentration in crystal depth for the sample obtained by GaSb processing at 580 °C in arsenic vapors emitted from solution-melt containing 0.01 atomic fractions of arsenic. The shape of the curve indicates the absence of differed composition phases that is the evidence of formation of solid solution with continuous variation of composition from the surface into the depth of the sample.

Concentration of arsenic on this sample surface amounted 0.02 atomic fractions ($3.5 \cdot 10^{20}$ cm^{-3}), and its decrease deep into crystal had exponential shape, with reduction up to 10^{18} cm^{-3} occurring at 18 nm depth. As it can be seen in figure 1, zinc diffusion takes place much deeper than diffusion of arsenic, which provides p-n-junction formation outside the solid solution layer placed close to the surface.

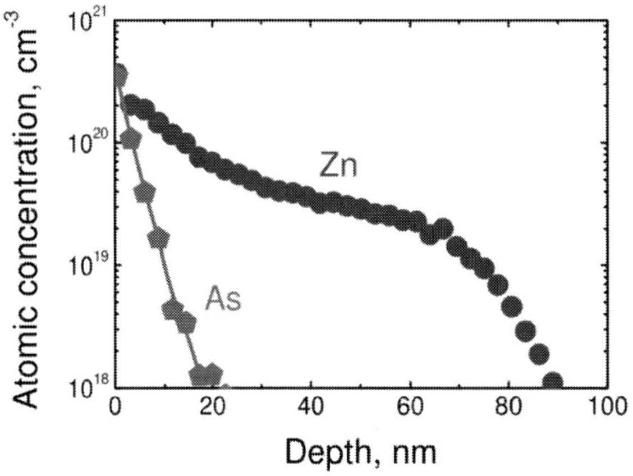

Figure 1. SIMS profiles of As and Zn vs. depth in $GaAs_xSb_{1-x}$/GaSb structure obtained by processing GaSb in arsenic vapors for 15 min at 580 °C.

According to calculations done using the paper [4] data, the bandgap value of $GaAs_xSb_{1-x}$ solid solution becomes greater than the bandgap value of GaSb at $x > 0.52$ and for bandgap value of GaP_xSb_{1-x} solid solution the same increase occurs at $x > 0.23$.

As evident from our investigations, the proposed method allows to obtain high concentrations of introduced components at increasing processing time and vapor pressures. Figure 2 gives the results of SIMS measurements for the sample obtained at processing GaSb at 580 °C in phosphorous vapors emitted from solution-melt containing about 0.04 atomic fractions of phosphorous, the process duration was 30 minutes. GaP_xSb_{1-x} solid solution formed at the depth up to 1200 nm, phosphorous concentration on the surface reaching $x = 0.8$ and smoothly decreasing deep into crystal from the surface.

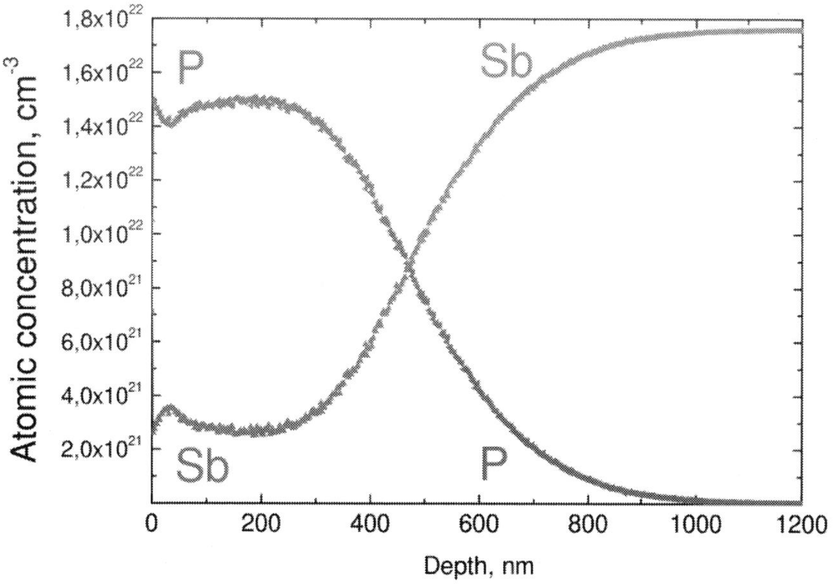

Figure 2. SIMS profiles of P and Sb vs. depth in GaP_xSb_{1-x}/GaSb structure obtained by processing GaSb in phosphorous vapors for 30 min at 580 °C.

These GaP_xSb_{1-x} solid solutions with compositions $x = 0.2 - 0.8$ seemingly cannot exist because of extensive miscibility gap. Nevertheless, in the articles [5, 6] the possibility of obtaining negative lattice-mismatched (about 10^{-3}) solid solution layers up to 50 nm thickness with compositions inside miscibility gap was demonstrated. The elastic deformation arising from the negative lattice mismatching stabilizes crystal lattice and prevents phase decomposition. It is $GaAs_xSb_{1-x}$ and GaP_xSb_{1-x} solid solutions that have negative value of lattice-mismatching to GaSb, and in the cases when their thickness does not exceed 50 nm the defectless solid solution layer formation can be suggested. Figure 2 demonstrates that there are the signs of phase decomposition in the region closed to the surface (below 40 nm). The phase decomposition on larger depths is hampered due to stabilizing influence of crystal lattice and presence of elastic deformation.

Crystal defects are known to decrease essentially PL intensity, whereas a layer formed without any defect can either demonstrate its own PL or play WOW role that results in significant increase of PL intensity of the underlying layer. WOW is transparent for underlying layer radiation and at the same time is a barrier which prevents currier migration on the surface where probability of their nonradiative recombination is great [2].

Previously we demonstrated the possibility of WOW obtaining on GaAs plates by supplying phosphorous vapors to their surfaces [3], therewithin the increase of PL intensity up to 25 times compared with PL intensity of unprocessed control GaAs plates was achieved. The given work demonstrates that PL intensity increase due to WOW effect was achieved as great as 5 times for GaSb processed in arsenic vapors. The same value was obtained for GaSb samples processed in phosphorous vapors. In this case one should take into account that PL-intensity decreases due to zinc doping process. The effect of increase of PL intensity without any change of spectrum shape in combination with SIMS data attest that the elements of the fifth group supplied to GaSb plate surface are embedded into crystal lattice instead of Sb atoms. In cases when the thicknesses of solid solutions layers and concentrations of embedded components are such that defectless conjunction with primary crystal

Journal of Physics: Conference Series **572** (2014) 012033 doi:10.1088/1742-6596/572/1/012033

becomes impossible, PL intensity increase due to WOW effect is either less intensive or entirely does not occur.

4. Conclusion

The experiments demonstrated that owing to solid-state substitution reactions it is possible to form GaAsSb or GaPSb solid solutions in the near-surface region of GaSb crystals which are continuous with primary crystal and play WOW role. Therewithin, in spite of the high zinc doping level of the structure, the effect of PL intensity increase due to WOW formation was observed.

References

[1] Fraas L M, Avery J E, Martin J, Sundaram V S, Girard G, Dinh V T, Davenport T M, Yerkes J W, O'neil M J 1990 Over 35-percent efficient GaAs/GaSb tandem solar cells *IEEE Trans. Electron. Dev.* **37**(2) 443

[2] Andreev V M, Khvostikov V P, Sorokina S V, Shvarts M Z, Vasil'ev V I 1997 GaSb Based PV Cells For Solar and Thermo-photovoltaic Applications *Proc. of the 14 th European Photovoltaic Solar Energy Confer-ence and Exhibition (Barcelona, 30 June-4 July)* 1763

[3] Vasil'ev V I, Gagis G S, Kuchinskii V I, Khvostikov V P, Marukhina E P 2013 Obtaining Nanodimensional Layers of GaAsP Solid Solutions on GaAs by Solid-State Substitution Reactions *Tech. Phys. Lett.* **39**(5) 472

[4] Vurgaftman I, Meyer J R, Ram-Mohan L R 2001 Band parameters for III–V compound semiconductors and their alloys *J. Appl. Phys.* **89**(11) 5815

[5] Vasil'ev V I, Akhmedov D, Deryagin A G, Kuchinskii V I, Nikitina I P, Smirnov V M, Tret'yakov D N 1999 GaInAsSb/GaSb heterostructures grown in the spinodal decay region by liquid-phase epitaxy from Sb-enriched solution-melts *Semiconductors* **33**(9) 1034

[6] Vasil'ev V I, Nikitina I P, Smirnov V M, Tret'yakov D N 1999 Isoperiodical heterostructures GaInAsSb/GaSb grown by LPE from Sb-rich melts in spinodal decomposition area *Materials Science and Engineering:B* **66**(1-3) 67

The Influence of Carrier Dynamics on Double-State Lasing in Quantum Dot Lasers at Variable Temperature

V V Korenev[1], **A V Savelyev**[1], **A E Zhukov**[1,2], **A V Omelchenko**[1], **M V Maximov**[1,2]
[1]Saint–Petersburg Academic University RAS, Saint–Petersburg 194021, Russia
[2]Ioffe Physico–Technical Institute RAS, Saint–Petersburg 194021, Russia

E-mail: korenev@spbau.ru

Abstract. It is shown in analytical form that the carrier capture from the matrix as well as carrier dynamics in quantum dots plays an important role in double-state lasing phenomenon. In particular, the de-synchronization of hole and electron captures allows one to describe recently observed quenching of ground-state lasing, which takes place in quantum dot lasers operating in double-state lasing regime at high injection. From the other side, the detailed analysis of charge carrier dynamics in the single quantum dot enables one to describe the observed light-current characteristics and key temperature dependences.

1. Introduction

Efficient semiconductor lasers having broadband emission spectra, which correspond to the transparency window of standard silicon optical fibre and SiGe-waveguide, are required for a vast range of practical applications, from optical coherence tomography to ultrafast data transmission. It turns out that the emission spectrum of long-wavelength emitting InAs/InGaAs quantum dot (QD) lasers can overlap this practically useful wavelength range (*O-band*) even at room temperature and low injection current [1 – 3]. Moreover, the usage of these multi-frequency laser diodes for the optical data transmission is a promising alternative to currently used DFB-lasers (*distributed feedback lasers*) due to the simplicity and cheapness of their production [3].

At low injection and sufficiently low optical loss, the emission spectrum of InAs/InGaAs QD laser is formed by the ground state (GS) optical transitions of QDs emitting near 1.3 μm. Such a broadband emission is obtained owing to the involvement of different quantum dots into the lasing, having the different energies of ground-state optical transition due to the inhomogeneous broadening [5 – 8]. As the injection increases, the population of the excited energy levels (ES) of electrons and holes and, consequently, the population inversion of the ES optical transition of QDs, also tend to increase with injection [6]. As a result, at sufficiently high pump current the population inversion of ES reaches its threshold value and simultaneous GS- and ES-lasing, i.e., *double-state lasing*, takes place. Such a behaviour is confirmed by the series of experiments, which at sufficiently high injection revealed in the lasing spectra an additional spectral line (with $\lambda < 1.2$ μm) associated with the first excited-state (ES) optical transition of QDs [5 – 12]. Moreover, the usage of this simultaneous lasing via QD GS and ES optical transitions is the simplest way to obtain extremely broad emission spectrum in QD laser [3, 10 – 12]. The lasers operating in such a regime are also promising candidates for multi-wavelength optical networks pump and new laser sources [3, 9]. Therefore, the theoretical investigation of the double-state lasing phenomenon is the subject of interest in the current Paper.

2. Double-state lasing phenomenon

For the first time the double-state lasing phenomenon in InAs/InGaAs QD lasers was theoretically predicted by *Grundmann et al.* in [4] and then experimentally observed in [5 – 6], while its generally recognized explanation was proposed by *Markus et al.* in [6] and then developed in [7]. This approach is based on the system of rate equations for the QDs, which are described in terms of excitonic energy levels. At low injection and sufficiently low optical loss, GS-lasing takes place. As soon as the injection current is above GS-lasing threshold, the occupancy of GS excitonic energy level becomes fixed. However, the occupancy and, hence, the population inversion of the ES excitonic energy level of QD tends to increase with injection. As a result, at sufficiently high injection, double-state lasing takes place. After the onset of the double-state lasing, carrier population of both GS and ES energy levels remains unchanged. The flux of excitons from ES to GS level becomes limited by the finite relaxation time due to the well-known phonon-bottleneck effect [15]. Therefore, output power, which is emitted via GS optical transitions, reaches its maximal value and then stays constant for all injection currents exceeding the double-state lasing threshold [5 – 8]. Thereby, it is commonly implied that the onset of simultaneous GS- and ES-lasing leads to the stabilization of the GS-lasing power.

However, there are a number of experiments for InAs/InGaAs QD lasers, where the quenching or even complete damping of GS-lasing was observed as the injection is above double-state lasing threshold [10 – 14]. Light-current characteristics of 100-µm-wide and 4-mm-long laser sample having a single plane of MBE-grown InAs/InGaAs QDs are presented in the figure 1. As it can be seen from the figure, there is a definite contradiction between the predicted and real behaviour of QD laser.

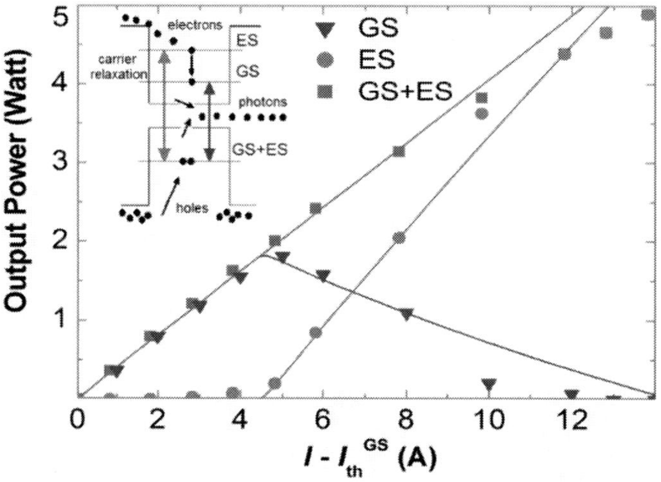

Figure 1. Experimental light-current characteristics (symbols) from [13] and their approximation (solid lines), where GS and ES correspond to the output power via GS and ES optical transitions of QDs, while GS+ES does to the total output power. Inset: QD energy levels diagram and optical transitions scheme.

To date, several attempts to explain the experimentally observed GS-lasing quenching were proposed. One of them is the self-heating of laser's active region operating in CW regime [10, 11]. If the self-heating effect takes place, with the increase of injection the red shift of lasing wavelength is expected to be observed in the experiments. However, as it was shown in [13] for the abovementioned samples, the increase of injection current from 5 to 12 A shifts the dominant laser emission peak only by 0.4 nm. This value corresponds to the increase of laser's active region temperature by only 1 °C and, in its turn, means that the self-heating effect cannot explain the observed GS-lasing quenching. Moreover, the experiments [13 – 14] show that the GS-lasing quenching occurs even in pulsed regime, when the self-heating is not pronounced.

At the same time, it was supposed in [12] that the increase of homogeneous broadening with the injection can lead to GS-lasing quenching due to the decrease of GS optical gain below the GS-lasing threshold. However, as it was shown in [13, 16], the dependence of homogeneous broadening on injection is not as strong as it was assumed in [12]. Thus, neither self-heating nor homogeneous broadening can completely describe the observed laser's behaviour in double-state lasing regime.

The more sophisticated models that do not imply both homogeneous broadening and self-heating effect are based on the assumption of the asymmetry in charge carrier distribution within a single QD [7, 14]. Indeed, the energy spacing between hole GS and ES energy levels is quite small (about 10 meV for typical InAs/InGaAs QDs [14]), hence an intensive exchange of carriers between these states takes place at room temperature so, that they can be considered as one effective hole reservoir – see the inset to figure 1. The "competition for the holes" between electrons occupying GS and ES levels takes place. Since the modal gain of the ES transition is approximately twice the gain of the GS transition, the ES-lasing line increases with pump at the expense of the GS one. To describe the observed GS-lasing quenching quantitatively authors of [14] had to introduce phenomenological linear emission rates of carriers from QD GS and ES into the matrix. This additional assumption seems to be artificial, since the authors did not verify it either by experiments or theoretical modelling.

It should be also noted that the complete quenching of GS-lasing in InAs/InGaAs QD lasers can be attributed to the growth of the internal loss with the increase of injection current as it was predicted in [17 – 18]. However, such a mechanism has to be accompanied by degradation of the laser's slope efficiency, which is not confirmed in the experiments – see figure 1. This, in its turn, means that the GS-lasing quenching cannot be described in the scopes of the assumption of the increase of internal loss and is determined by more subtle dynamic effects. Thus, the double-state lasing phenomenon and especially reasons for GS-lasing quenching still require an adequate theoretical description.

In order to explain the observed GS-lasing quenching we have recently proposed the new mechanism, which is based on the assumption of different hole and electron capture rates into QDs [19 – 20] and thereby takes into account the dynamics of charge carriers both in QDs and in matrix material. More precisely, we assume that the hole capture rate is lower than the electron one. The similar idea of de-synchronization between hole and electron captures was independently proposed in [16]. This assumption seems to be very natural, because the carrier capture rate is proportional to the velocity of their thermal motion in matrix material [21], while hole thermal velocity is lower than the electron one due to the higher effective mass of holes. The interplay between the difference in carrier capture rates and the aforementioned asymmetry [19 – 20] of their distribution in QD [14] intensifies the competition for the "common" holes between GS and ES electrons and, therefore, makes the process of depopulation of the effective hole level also more intensive. This, as a result, naturally accelerates the quenching of GS-lasing with the increase of injection.

To illustrate this qualitative understanding quantitatively, let us assume that each QD has only one hole and two electronic energy levels and there is no direct carrier exchange between the matrix and QD GS (see the inset to figure 1). Thus, neglecting the spontaneous recombination, one can obtain

$$\dot{f}_{e1}=\frac{D_2}{D_1}\cdot\left[g_{e21}\cdot f_{e2}\cdot\left(1-f_{e1}\right)-g_{e12}\cdot f_{e1}\cdot\left(1-f_{e2}\right)\right]-\frac{P_1}{D_1}, \tag{1}$$

$$\dot{f}_{e2}=g_{ec}\cdot\left(1-f_{e2}\right)-g_{ee}\cdot f_{e2}-g_{e21}\cdot f_{e2}\cdot\left(1-f_{e1}\right)+g_{e12}\cdot f_{e1}\cdot\left(1-f_{e2}\right)-\frac{P_2}{D_2}, \tag{2}$$

$$\dot{f}_{h}=g_{hc}\cdot\left(1-f_{h}\right)-g_{he}\cdot f_{h}-\frac{P_1+P_2}{D_1+D_2}. \tag{3}$$

Here, $P_{1(2)}$ is the rate of stimulated emission from QD GS(ES), $D_{1(2)}$ are the degrees of their degeneracy (for the common hole level the degeneracy is assumed to be D_1+D_2), $g_{e(h)c}$ and $g_{e(h)e}$ are the capture (into empty QD) and emission (from QD with one carrier) rates of an electron (hole), g_{e21} and g_{e12} are the rates of electron ES to GS and reverse transitions, $f_{e1(2)}$ and f_h are the occupancies of electronic GS (ES) and effective hole energy levels. System (1) – (3) should be completed with the following conditions, which correspond to different lasing regimes

$$(NL):\ P_1 = 0,\ P_2 = 0, \tag{4}$$

$$(GS):\ \gamma_1 = f_{e1} + f_h - 1,\ P_2 = 0, \tag{5}$$

$$(GS+ES):\ \gamma_1 = f_{e1} + f_h - 1,\ \gamma_2 = f_{e2} + f_h - 1, \tag{6}$$

$$(ES):\ P_1 = 0,\ \gamma_2 = f_{e2} + f_h - 1, \tag{7}$$

$$\gamma_{1(2)} = \frac{\alpha_{in} + \alpha_{out}}{G_{\max 1(2)}}. \tag{8}$$

Here, $\gamma_{1(2)}$ is the population inversion required for GS(ES)-lasing that is defined in accord with Eq. (8), α_{in} and α_{out} are the internal and output losses, $G_{max1(2)}$ is the maximal gain of the GS(ES) optical transition. Eq. (4) corresponds to the regime of no lasing (*NL*), (5) does to pure GS-lasing (*GS*), (6) does to simultaneous GS- and ES-lasing (*GS+ES*), and (7) does to pure ES-lasing (*ES*). Taking into account conditions (4) – (8) one can obtain the stimulated emission rates via GS and ES (P_1 and P_2) optical transitions of QDs at any given electron and hole capture rate values (g_{ec} and g_{hc}) by solving system (1) – (8) numerically. Output power of GS and ES emission (W_1 and W_2) and the injection current (I) can be then calculated using well-known equations [22]

$$W_{1,2} = \frac{\alpha_{out}}{\alpha_{in} + \alpha_{out}} \cdot N_{QD} \cdot E_{1,2} \cdot P_{1,2},\ I - I_{th}^{GS} = \frac{N_{QD} \cdot e}{\eta_{id}} \cdot (P_1 + P_2), \tag{9}$$

where $E_{1(2)}$ is the energy of the GS(ES) optical transition, N_{QD} is the total number of QDs in the laser's active region, I_{th}^{GS} is the threshold current of GS-lasing, η_{id} is the internal differential quantum efficiency, and e is the charge of the electron. The η_{id} can be directly obtained from the experimental data presented in figure 1 and is equal to 61%.

For the detailed analysis of double-state lasing, it is convenient to introduce the h-factor ($h \equiv g_{hc}/g_{ec}$). As it was previously stated, we expect that $h < 1$. Assuming that h does not depend on the injection, one can calculate the dependence of GS-power as a function of injection current for different values of h, in order to show the influence of hole-to-electron capture rate ratio on GS-lasing – see figure 2. The other parameters are in accord with [13]: g_{ee}= 219 ns^{-1}, g_{he}= 19 ns^{-1}, g_{e21}= 1250 ns^{-1}, α_{in}= 1.5 cm^{-1}, α_{out}= 3.1 cm^{-1}, D_1= 2, D_2= 4, E_1= 983 meV, E_2= 1050 meV, G_{max1}= 6 cm^{-1}, G_{max2}= 12 cm^{-1}.

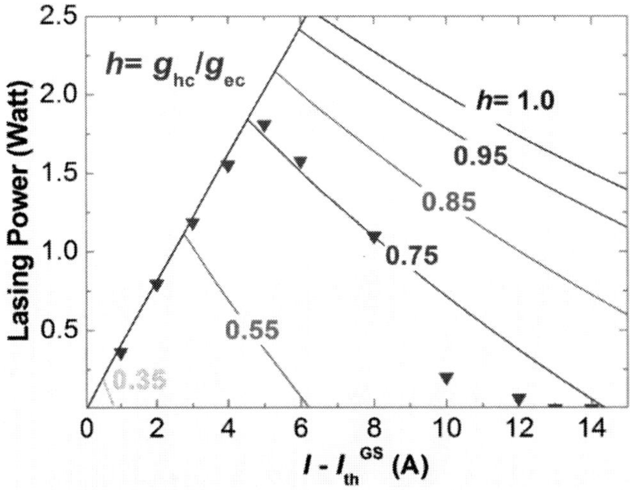

Figure 2. Light-current characteristics corresponding to GS-lasing component for different values of hole-to-electron capture rate ratio h (solid lines) and experimental data (symbols) from [13].

As it can be seen from the figure 2, even a small change of h (from unity to 0.75) leads to significant acceleration of GS-lasing damping and decrease of the maximum GS-lasing power by more than 40%. Therefore, we suppose that hole-to-electron capture rate ratio (h-factor) has a strong effect on the key lasing parameters. In particular, h can differ from the unity and should be taken into account to describe laser operation correctly. The best agreement between the proposed theory and the experimental data presented in the figure 1 is achieved for h= 0.75.

At the same time, the majority of works are based only on numerical analysis of the question that complicates the analysis and finding of key parameters, responsible for the double-state lasing. Therefore not only does numerical, but also analytical approach to this phenomenon is also important.

3. Analytical approach to the carrier dynamics in quantum dots

To solve the system of rate equations (1) – (9) analytically, it is convenient to introduce following dimensionless parameters

$$P_e = \frac{g_{ec}}{g_{ec}+g_{ee}},\ P_h = \frac{g_{hc}}{g_{hc}+g_{he}},\ P_{ve,vh} = 1 - P_{e,h},\ \tilde{P}_{1,2} = \frac{P_{1,2}}{D^2\cdot(g_{ec}+g_{ee})}, \tag{10}$$

$$D = \frac{D_2}{D_1},\ C = \left(\frac{1}{1+D^{-1}}\right)\cdot\frac{g_{ec}+g_{ee}}{g_{hc}+g_{he}},\ \eta = \frac{g_{ec}+g_{ee}}{g_{e21}},\ \xi = \frac{g_{e12}}{g_{e21}}. \tag{11}$$

In terms of introduced parameters the solution of the system (1) – (9) can be obtained in analytical form at any given electron and hole capture rates (g_{ec} and g_{hc}) in all possible lasing regimes

A) <u>NL-regime</u>:
$$\tilde{P}_1 = \tilde{P}_2 = 0,\ f_{e1} = \frac{P_e}{(1-\xi)\cdot P_e + \xi},\ f_{e2} = P_e,\ f_h = P_h, \tag{12}$$

where $\xi = \exp[(E_1 - E_2)/T]$, T is laser's temperature in meV; ξ is an order of 0.1 at room temperature.

B) <u>GS-regime</u>:
$$\begin{gathered}[C\cdot(1-\xi)]\cdot\tilde{P}_1^2 + [(1-\xi)\cdot\gamma_1 - \eta - C\cdot(P_e+\xi\cdot P_{ve}) - (P_h+\xi\cdot P_{vh})]\cdot\tilde{P}_1 + \ldots\\ \ldots + [(P_e\cdot P_h - \xi\cdot P_{ve}\cdot P_{vh}) - \gamma_1\cdot(P_e+\xi\cdot P_{ve})] = 0,\ \tilde{P}_2 = 0,\end{gathered} \tag{13}$$

$$f_{e1} = \frac{(1+\eta)\cdot\tilde{P}_1 - P_e}{(\xi-1)\cdot(P_e-\tilde{P}_1) - \xi},\ f_{e2} = P_e - \tilde{P}_1,\ f_h = P_h - C\cdot\tilde{P}_1, \tag{14}$$

where result for $\tilde{P}_1$ is presented in the form square equation for brevity (minor root should be used). In GS-regime the population inversion of QD GS is fixed.

C) <u>(GS+ES)-regime</u>:
$$\tilde{P}_1 = \frac{f_{e2}\cdot(1-f_{e1}) - \xi\cdot f_{e1}\cdot(1-f_{e2})}{\eta},\ \tilde{P}_2 = \frac{P_h - f_h}{C} - \tilde{P}_1, \tag{15}$$

$$f_{e1,2} = 1+\gamma_{1,2} - f_h,\ f_h = P_h - \frac{C\cdot[P_e + P_h - 1 - \gamma_2]}{C+1}. \tag{16}$$

In the case of (GS+ES)-regime, population inversion of both GS and ES optical transitions are fixed.

D) <u>ES-regime</u>:
$$\tilde{P}_1 = 0,\ \tilde{P}_2 = \frac{P_h - f_h}{C}, \tag{17}$$

$$f_{e1} = \frac{f_{e2}}{f_{e2}\cdot(1-\xi)+\xi},\ f_{e2} = \frac{C\cdot P_e - P_h + 1 + \gamma_2}{C+1},\ f_h = 1+\gamma_2 - f_{e2}. \tag{18}$$

Finally, when stimulated emission comes from the ES only, the population inversion of only ES persists equal to constant.

Moreover, using (10) – (18) it is possible to provide analytical conditions separating different regimes of laser emission and, therefore, to find out corresponding lasing regime for any given pair (g_{ec}, g_{hc}). This allows one to predict, when the switching between the regimes takes place. Thus, at sufficiently low injection corresponding to low values of g_{ec} and g_{hc}, both population and population inversion of GS and ES are sufficiently small, there is no stimulated emission can be seen

$$\begin{cases} \dfrac{P_e}{P_e \cdot (1-\xi)+\xi} + P_h - 1 < \gamma_1 \\ P_e + P_h - 1 < \gamma_2 \end{cases}. \tag{19}$$

Pure GS-lasing takes place at higher injection, when the population inversion between hole and electron GS of QD reaches its threshold value, while the population of electronic ES and, consequently, the population inversion of ES optical transition, stays still below corresponding lasing threshold so, that there is no ES-lasing. This corresponds to the following condition

$$(C+1)\cdot f_{e2} - C\cdot P_e + P_h - 1 < \gamma_2, \tag{20}$$

where f_{e2} is in accord with (14).

In order to reach the conditions of double-state lasing regime, the population inversion of ES ($f_{e2} + f_h - 1$) should become high enough for ES-lasing, while the population inversion of GS ($f_{e1} + f_h - 1$) should be still high enough in order to maintain GS-lasing. Therefore, simultaneous GS- and ES-lasing takes place, if the two following conditions are fulfilled

$$f_{e2} + f_h - 1 > \gamma_2, \tag{21}$$

where the values, used for f_{e2} and f_h are defined by (13) – (14), and the condition

$$f_{e1} + f_h - 1 > \gamma_1, \tag{22}$$

where the values, used for f_{e1} and f_h are in accord with (18).

After the onset of double-state lasing the population of electronic GS tends to decrease, and at sufficiently high injection the population inversion of GS optical transition drops below the threshold of GS-lasing. In this case pure ES-lasing regime takes place that corresponds to following condition

$$f_{e1} + f_h - 1 < \gamma_1, \tag{23}$$

where the values, which should be used for f_{e1} and f_h, are in accord with (18).

Within the framework of the discussion above (in the end of *Section 2*), it is possible to consider the question of separation between different lasing regimes in terms of hole-to-electron capture rate ratio (h). It should be emphasized that there is two critical values of h: h_{cr} and h^*. If $h > h_{cr}$ ($h_{cr} \sim 0.9$ for the parameters used), there is no GS-lasing quenching can be seen, however if $h < h_{cr}$, the complete quenching of GS-lasing early or later takes place depending on the value of h. Finally, if h is low enough ($h < h^* \sim 0.17$ for the parameters used), there is no GS-lasing seen at all. The expressions for h_{cr} and h^* can be obtained from Eqs. (1) – (8) in analytical form

$$h_{cr} = \frac{2}{3}\cdot\frac{t}{1-t-\gamma_2},\ h^* = \frac{g_{hc}^*}{g_{ec}^*},\ g_{ec}^* = g_{ee}\cdot\frac{1-t}{t},\ g_{hc}^* = g_{he}\cdot\frac{t+\gamma_2}{1-t-\gamma_2}, \tag{24}$$

where parameter t is given by the following equation

$$t^2 - (\gamma_1 - \gamma_2 + 1)\cdot t + (\gamma_1 - \gamma_2)\cdot g_{e21}/(g_{e21} - g_{e12}) = 0. \tag{25}$$

It can be also shown that the increase of output loss (decrease of the sample length) results in the increase of h_{cr} value. As a result, the range of the h-factor values, where GS-lasing can occur, becomes narrower up to complete shrinkage at sufficiently high loss.

4. Temperature dependence of threshold currents

It turns out that in terms of the proposed model it is also possible to describe the temperature dependences of threshold current densities of double-state lasing (j_{th}^{ES}) and complete quenching of the GS-lasing (j_{off}^{GS}) qualitatively and quantitatively, see figure 3.

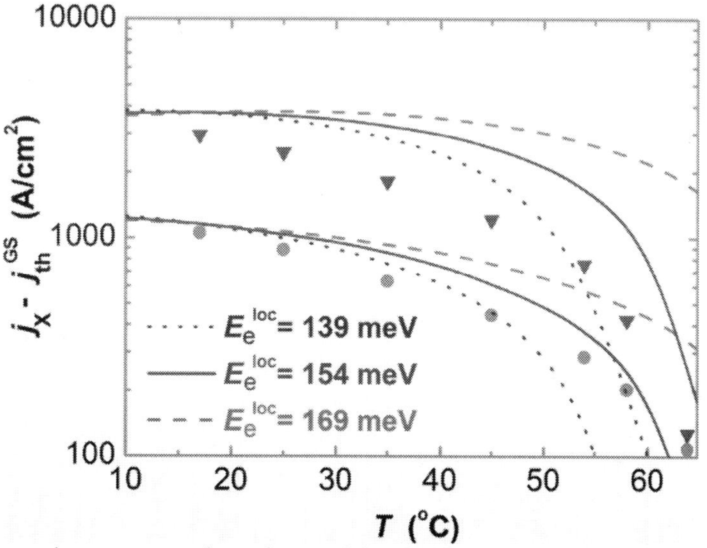

Figure 3. Experimental temperature dependence of threshold current density of double-state lasing (red circles) and complete quenching of GS-lasing (blue triangles), while the lines correspond to calculation.

We used in calculation the same set of parameters as for figure 1, in particular the value of h is assumed to be 0.75, whereas the localization energy of electrons (E_e^{loc}) is varied in the range from 139 to 169 meV. Taking into account that the ES-lasing line is 1.18 μm, the bandgap of GaAs is 0.870 μm and the asymmetry in hole and electron localization energies [23], one can calculate the temperature dependence of the threshold current densities. As it can be seen from the figure 3, the best agreement between the experimental data and its theoretical approximation is achieved for E_e^{loc}= 154 meV that is in accord with real experiment [23]. The fact that the parameters' values used for calculation of all figures 1 – 3 correspond to the real experiment also proves the proposed model of double-state lasing.

The decrease of both characteristic current densities with the temperature, which is seen from the figure 3, can be explained by the decrease of the ES-to-GS ($g_{e21}{\cdot}f_{e2}{\cdot}[1 - f_{e1}]$) and the increase of the GS-to-ES ($g_{e12}{\cdot}f_{e1}{\cdot}[1 - f_{e2}]$) electron transition rate at fixed injection. Thus, in double-state lasing regime the increase of the temperature leads to the decrease of the effective flux of carriers coming to the electron GS that, in its turn, is proportional to the output power via GS optical transition of QDs – see Eq. (1). Therefore, the GS-lasing power also tends to decrease with the increase of the temperature that ultimately leads to the complete damping of GS-lasing at sufficiently high temperature.

5. Conclusion

As a result, the influence of charge carrier dynamics on output power of QD laser operating in double-state lasing regime was studied analytically. Under the assumption of different hole and electron capture rates in QDs, the mechanism of GS-lasing quenching was proposed. The analytical conditions describing switching between different lasing regimes at variable injection were obtained. Temperature dependences of key threshold currents were also considered. It was figured out that only the simultaneous consideration of charge carrier dynamics in matrix and in quantum dots allows one to describe the observed temperature dependences, as well as ground-state lasing quenching, correctly.

Acknowledgements

This work was supported by the Program of Fundamental Research of Russian Academy of Sciences, Russian Foundation for Basic Research, by Federal Program "Scientific and scientific-pedagogical personnel of innovative Russia" of the Ministry of Education and Science of the Russian Federation and by the Dynasty Foundation.

References

[1] Zhukov A E and Kovsh A R 2008 *Quantum Electronics* **38** 409
[2] Zhukov A E, Kovsh A R, Nikitina E V, Ustinov V M, Alferov Zh I 2007 *Semicond.* **41** 606
[3] Djie H S, Ooi B S, Fang X, Wu Y, Fastenau J M, Liu W K, Hopkinson M 2007 *Opt. Lett.* **32** 44
[4] Grundmann M, Weber A, Goede K, Ustinov V M, Zhukov A E, Ledentsov N N, Kop'ev P S, Alferov Zh I 2000 *Appl. Phys. Lett.* **77** 4
[5] Zhukov A E, Kovsh A R, Livshits D A, Ustinov V M, Alferov Zh I 2003 *Semic. Sci. Tech.* **18** 774
[6] Markus A and Fiore A 2004 *Phys. Stat. Sol.* (a) **201** 338
[7] Shi L W, Chen Y H, Xu B, Wang Z C, Wang Z G 2007 *Physica E* **39** 203
[8] Jiang L and Asryan L V 2006 *IEEE Phot. Tech. Lett.* **18** 2611
[9] Naderi N A, Grillot F, Yang K, Wright J B, Gin A, Lester L F 2010 *Opt. Express* **18** 27028
[10] Kim Y J, Joshi Y K and Fedorov A G 2010 *J. Appl. Phys.* **107** 073104
[11] Ji H M, Yang T, Cao Y L, Xu P F, Gu Y X, Wang Z G 2010 *Jpn. J. Appl. Phys.* **49** 072103
[12] Sugawara M, Hatori N, Ebe H, Ishida M, Arakawa Y, Akiyama K, Otsubo T and Nakata Y 2005 *J. Appl. Phys.* **97** 043523
[13] Zhukov A E, Maximov M V, Shernyakov Yu M, Livshits D A, Savelyev A V, Zubov F I, Klimenko V V 2012 *Semicond.* **46** 231
[14] Viktorov E A, Mandel P, Tanguy Y, Houlihan J, Huyet G 2005 *Appl. Phys. Lett.* **87** 053113
[15] Sugawara M, Mukai K, Shoji H 1997 *Appl. Phys. Lett.* **71** 2791
[16] Gioannini M 2012 *J. Appl. Phys.* **111** 043108
[17] Asryan L V, Luryi S 2003 *Appl. Phys. Lett.* **83** 5368.
[18] Zhukov A E, Kovsh A R, Ustinov V M, Alferov Zh I 2003 *Laser Phys.***13** 39.
[19] Korenev V V, Savelyev A V, Zhukov A E, Omelchenko A V, Maximov M V 2013 *Appl. Phys. Lett.* **102** 112101
[20] Korenev V V, Savelyev A V, Zhukov A E, Omelchenko A V, Maximov M V, Shernyakov Yu M 2012 *Proc. of SPIE* **8432** 84321L
[21] Asryan L V and Suris R A 1996 *Semicond. Sci. Technol.* **11** 554
[22] Agrawal G P, Dutta N K 1993 *Semiconductor Lasers* New York (Van Nostrand Reinhold) 58
[23] Höglund L, Karlsson K F, Holtz P O, Pettersson H, Pistol M E, 2010 *Phys. Rev.* B **82** 035314

Journal of Physics: Conference Series **572** (2014) 012035 doi:10.1088/1742-6596/572/1/012035

Influence of the addition superlattices on the luminescence processes in nano-heterostructures based on nitrides

E A Menkovich[1] **S A Tarasov**[1], **A V Solomonov**[1], **S Suihnonen**[2], **O Svensk**[2], **L Riuttanen**[2], **H Nykanen**[2]
[1]Saint-Petersburg Electrotechnical University "LETI", Prof. Popova 5, St. Petersburg 197376, Russia
[2]Department of Micro and Nanosciences, Aalto University, Micronova, PL 13500, Aalto, 00076, Finland

E-mail: menkovichea@gmail.com

Abstract. The effect of addition of superlattice (SL) and the structure of the upper barrier layer on the luminescence processes occurring in light-emitting nanoheterostructures was studied. It was shown that the optimum is using of structures with two SL: InGaN / InGaN structure in the lower part and the AlGaN / GaN for the top p-layer. It is shown that the using of the SL InGaN / InGaN in the vicinity of the active region optimally compensate the elastic stresses and the piezoelectric field at the hetero boundaries. Such compensation of elastic stresses reduces the formation of dislocations in these structures, which increases the radiation intensity.

1. Introduction

Nano-heterostructures based on nitrides are widely used in modern nanoelectronics devices and photonics. Their applying in the creation of short-wavelength light emitting diodes is important. Difference of layers lattice constants in such structures forms significant elastic stresses in the heterointerfaces. Presence of such elastic stresses especially serious influences beside to the active region of the emitter. Also degradation accelerates for structures with significant elastic stresses. Degradation is caused by an increased defect formation occurring as a result of the relaxation of elastic stresses.

To reduce these effects in nitride based nanostructures for light-emitting diode typically use complex structure of the active region which contain several unrelated InGaN quantum wells. This allows to redistribute the elastic stresses between the layers and reduces the magnitude of their impact on the heterointerfaces. However, such methods do not fully allow to avoid the appearance of negative effects listed above.

One of the known methods which improve the parameters of emitting nanostructures is to use superlattices of different composition. Thus, the superlattice located between the substrate and the active region substantially reduce defects in the active region. This is primarily due to reduction of the dislocation density of threading through the epitaxial layers to the substrate surface. However, using the superlattice active region in that area, including above it might help offset the effect of elastic stresses and associated.

2. Sample and experimental technique

The authors studied the luminescence of the semiconductor nitride nano-heterostructures on the basis of solid solutions (Al, Ga, In)N having multiple quantum wells and superlattices, created by MOCVD method on the sapphire substrate (figure 1). The nano-heterostructures had the same structure except the last p-GaN epitaxial layer. The n-GaN buffer and contact layers were grown on the *c*-Al_2O_3 substrate having 4 μm thickness. Structures without superlattices were used as reference (type 0). The SL in the other structures consisting of 15 $In_{0,01}Ga_{0,99}N$ and $In_{0,001}Ga_{0,999}N$ layers each having 1 nm thickness were added before the active region. The active region comprised one 2.5 nm width $In_{0,15}Ga_{0,85}N$ quantum well. Further the low-alloyed 50 nm thickness i-GaN layer was grown. The final p-GaN layer was of four types: the first type represented SL comprising 9 pairs of $Al_{0,15}Ga_{0,85}N$ and GaN layers of 1.5 nm thickness (type A), the second type - p-AlGaN layer with 15% aluminum fraction (type B), the third type - p-AlGaN layer with 10% aluminum fraction (type C) and the forth type - p-GaN layer (type D). The final A, B, C, D layers thickness was 27 nm. Carrier concentration in these layers was $N = 3*10^{17}$ cm^{-3}.

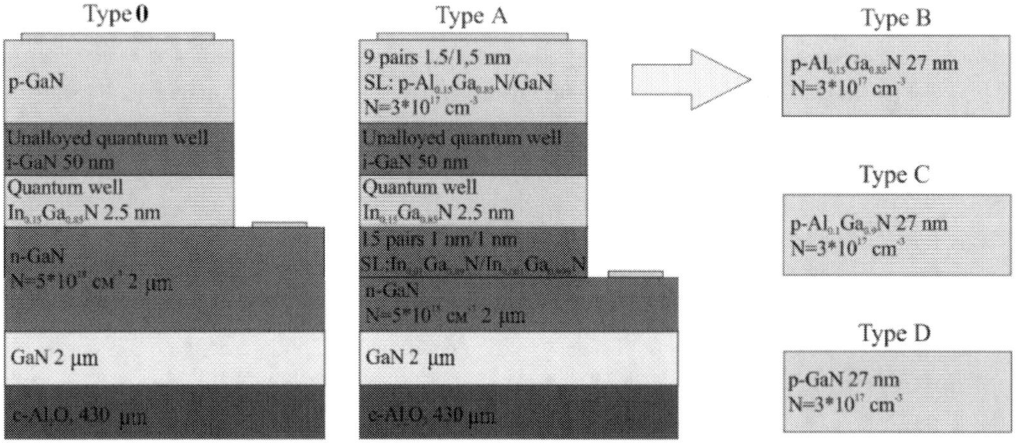

Figure 1. The structure of the studied samples.

The nano-heterostructures study were conducted by means of the author's elaborated diagnosis test system for the investigation nano-heterostructures operating parameters [3]. The system allows determining all general features of the solid-state emitters and products on its base: spectral, watt- and volt-ampere characteristics, luminescence intensity and efficiency. In the submitted research the samples represented undivided luminescent chips on the sapphire substrate. Therefore, for the diagnosis the probe station was applied. Several sets of substrates were investigated , each contained several tens of chips. Results for efficiency and power of radiation for the chips of the same type were almost close. The paper are presented experimental results summarizing the array of data obtained for a few hundred light-emitting nano-heterostructures.

3. Experimental results and discussions

According to [1, 2] the structures failing to comprise superlattices under low currents (up to 10 mA) demonstrated the significant shift up to 10 nm spectrum luminescence maximum forward to short-wave region. This effect can be explained by the influence of elastic stresses and piezo-fields arising at heteroboundaries in nanostructure as a result of great difference of the lattice parameters.

The active region and wells located there are influenced by three main components: p-n junction built-in electric field; built-in field induced by deformations; external field conditioned by an applied voltage. In spite of the essential p-n junction field influence on luminescent nanostructures properties in the whole, it cannot be considered as the main cause of the above mentioned effect arising. The nanostructure active region under low currents provides low external voltage drops. Against these drops built-in field contribution induced by deformations can be well recognized. Due to this the shape of well is strongly distorted and a quantum-confinement Stark shift arises. This leads to the significant distance decrease between working levels. With external voltage increasing the built-in field influence is compensated and the shift is decreased.

In structures containing no superlattices (type 0) a significant shift of the peak wavelength was observed (fig. 2). SL in the nanostructures allowed to partly compensate elastic stresses thereby the discussed effect was decreased to 2.3 nm for type A. For the type B structures the shift was 1.8 nm, 4.3 nm - for the type C structures and 3.4 nm for type D structures. This mean incomplete elastic stresses compensation at the single quantum well interface and higher layers in the type A structure. Probably, the residual elastic stresses within the structure relax without crystal lattice defects formation, therefore it had higher steadiness and operation efficiency. Furthermore, this assumption is confirmed by a lower luminescence efficiency of the type B structures where the influence of elastic stresses regarding the wavelength shift is decreased due to the dislocations formations and luminescent efficiency decreasing.

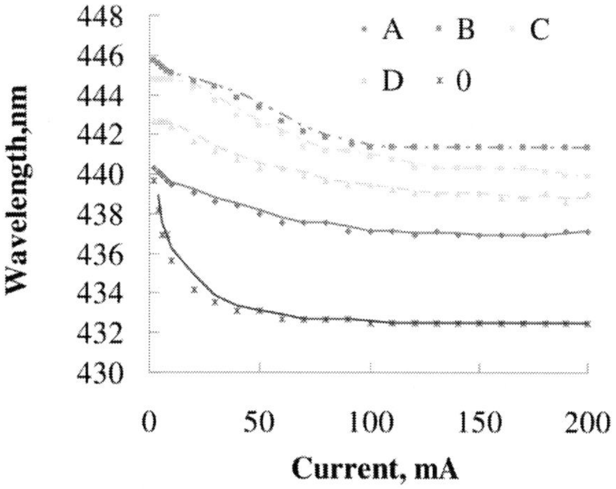

Figure 2. Wavelength peak dependence on the operating current.

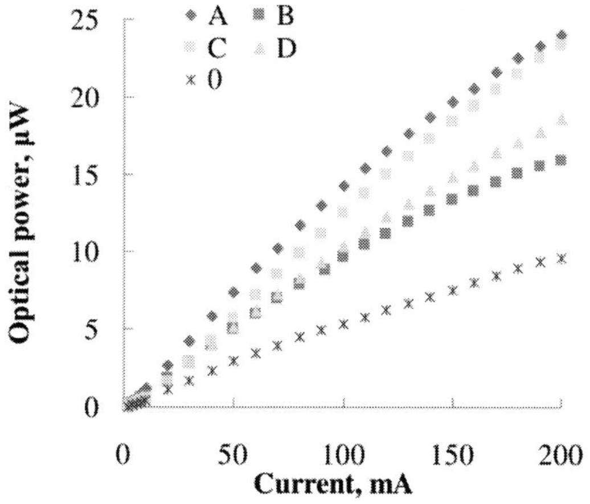

Figure 3. The optical capacity dependency on the current.

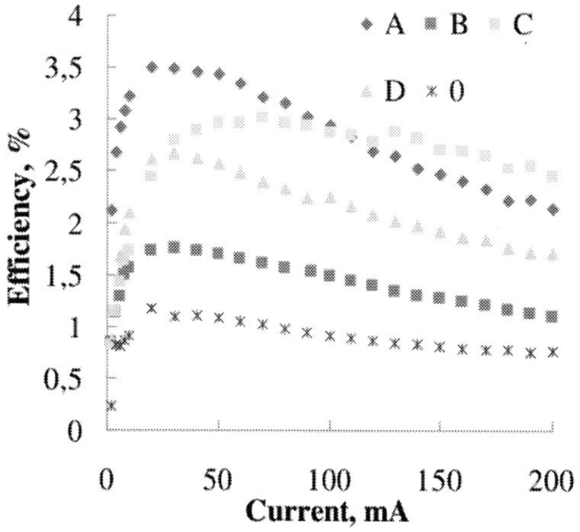

Figure 4. The dependency of the efficiency on the current.

Comparative analysis of A, B, C, D type structures also allowed studying of the final barrier electron intended layer structure influence of the regarding the radiators' luminescent and other operating features (figures 3 - 4). The luminescence efficiency of the samples was relatively low, since they were part of the unseparated wafer without additional lenses.

Considering the achieved optical power and efficiency values the addition of superlattices (type A) 3,5% and 2,7 mW correspondingly under the current 20 mA provides the best effect. A more weak effect is provided by the usage of p-AlGaN layer with aluminum portion 10% (type C) (2,4% and 1,6 mW correspondingly). Further increasing of Al concentration makes the parameters of obtained samples (type B) (1,7% and 1,8 mW correspondingly) significantly worse. In this case appears that negative influence of the lattice mismatch and increased defect formation associated both which exceed the advantages resulting from improved charge carrier confinement. Apparently, in this case, the negative influence of the lattice mismatch and the associated increased defect exceed benefits. This effect strongly proved itself so that optical power of reference type D structures with the barrier layer without aluminum be higher than the samples of type B (2,6% and 1,7 mW correspondingly). This effect could be reduced in the superlattice structures due to voltage rearrangement in thin layers consisting in superlattices.

Also it was demonstrated that all studied nanoheterostructures are operable up to the currents exceeding 150 mA. The optical power saturation was not observed in the type A, C, D structures except the type B structures which had the lowest efficiency proving their high imperfection. The optical power dependencies of the type A and type C structures coincide while the efficiency is distinguished. The addition of superlattice results in relatively small number of dislocations that doesn't affect the device operation efficiency.

4. Conclusion

Thus, it was established that the structure with two InGaN and AlGaN superlattices have high operation steadiness and the best operating characteristics. It is demonstrated that the InGaN/InGaN superlattices usage beside the nanostructure active region provides that the elastic stresses and piezofields at the heteroboundary are optimally compensated. Such stresses compensation reduces dislocations formation in the mentioned structures, what in turn increases the radiation intensity. It is shown the significant influence of the upper barrier layer on the radiators' parameters and is found that the AlGaN/GaN superlattices usage is optimal.

References

[1] Menkovich E A, Tarasov S A and Lamkin I A 2012 *Functional Materials* **19** (2) 233-237

[2] Menkovich E A, Tarasov S A, Lamkin I A, Kurin S Yu, Antipov A A, Roenkov A D, Barash I S, Helava H I and Roenkov Yu N 2013 *J. of Phys.: Conf. Ser.* **461** 012027

[3] Menkovich E A, Tarasov S A and Lamkin I A 2012 **10** *Proceedings of the ETU "LETI"* 17-23

Single-mode temperature and polarisation-stable high-speed 850nm vertical cavity surface emitting lasers

D E Nazaruk[1,2], S A Blokhin[1], N A Maleev[1], M A Bobrov[1,3], A G Kuzmenkov[4], A P Vasil'ev[4], A G Gladyshev[4], M M Pavlov[1], A A Blokhin[3], M M Kulagina[1], K A Vashanova[1], Yu M Zadiranov[1], A G Fefelov[5], V M Ustinov[1]

[1]Ioffe Physical-Technical Institute of the Russian Academy of Sciences, 26 Polytekhnicheskaya Str, St Petersburg 194021, Russia
[2]St Petersburg Academic University — Nanotechnology Research and Education Center of the Russian Academy of Sciences, 8/3 Khlopina Str, St Petersburg, 194021, Russia
[3] Saint Petersburg State Polytechnical University, 26 Polytekhnicheskaya Str, St Petersburg, 195251, Russia
[4]Submicron Heterostructures for Microelectronics, Research & Engineering Center of the Russian Academy of Sciences, 26 Polytekhnicheskaya Str, St Petersburg 194021, Russia
[5]JSV «Salut», 7 Larina Str, N Novgorod, 603950, Russia

Abstract. A new intracavity-contacted design to realize temperature and polarization-stable high-speed single-mode 850 nm vertical cavity surface emitting lasers (VCSELs) grown by molecular-beam epitaxy is proposed. Temperature dependences of static and dynamic characteristics of the 4.5 μm oxide aperture InGaAlAs VCSEL were investigated in detail. Due to optimal gain-cavity detuning and enhanced carrier localization in the active region the threshold current remains below 0.75 mA for the temperature range within 20-90°C, while the output power exceeds 1 mW up to 90°C. Single-mode operation with side-mode suppression ratio higher than 30 dB and orthogonal polarization suppression ratio more than 18 dB was obtained in the whole current and temperature operation range. Device demonstrates serial resistance less than 250 Ohm, which is rather low for any type of single-mode short-wavelength VCSELs. VCSEL demonstrates temperature robust high-speed operation with modulation bandwidth higher than 13 GHz in the entire temperature range of 20-90°C. Despite high resonance frequency the high-speed performance of developed VCSELs was limited by the cut-off frequency of the parasitic low pass filter created by device resistances and capacitances. The proposed design is promising for single-mode high-speed VCSEL applications in a wide spectral range.

1. Introduction

Vertical-cavity surface-emitting lasers (VCSELs) are widely used as low-cost high-performance light sources for data communication systems, optical interconnects and different sensors [1] for their unique characteristics such as small laser beam divergence, low threshold current, high speed properties, capability of fabricating dense 2-D laser arrays, and the testing on wafer level. New promising applications (such as atomic clock, gyroscope and etc.) have relatively demanding requirements on short-wavelength VCSELs: low power consumption, high frequency modulation bandwidth, narrow linewidth, stable linear polarization, precise output wavelength, low relative intensity noise, high temperature stability [2]. VCSELs with ultra-small apertures show stable fundamental transverse mode operation, however their modulation bandwidth is limited due to strong current-induced self-heating caused by the large resistance and gain suppression caused by non-uniform transverse optical field distribution [3]. Moreover the combination of cylindrical symmetry and isotropic gain leads to polarization uncertainty and instability in conventional GaAs-based VCSELs [4]. The above mentioned issues are quite complex, and still actual in spite of a number of efforts.

In this work we have focused on developing a new design and reliable technique to realize temperature and polarization-stable high-speed single-mode VCSEL grown by molecular-beam epitaxy (MBE). We discuss design optimization problems and demonstrate first approbation of our approach for 850-nm range VCSELs based on InAlGaAs material systems.

2. VCSEL design

Both MBE and metal-organic chemical vapor deposition (MOCVD) are widely used for VCSEL production. MBE-growth of the traditional short-wavelength oxide-confined VCSEL design based on fully-doped AlGaAs distributed Bragg reflectors (DBRs) is rather complicated due to large number digitally-graded heterointerfaces and precise doping control. Serial resistance of MBE-grown doped DBRs is usually higher than for MOCVD-grown doped DBRs. The possible approach to overcome the mentioned problems is intracavity-contacted VCSEL design with undoped DBRs [5]. However conventional intracavity-contacted VCSELs with contact layers close to active region suffer from non-uniform current spreading across the aperture and high optical losses in doped contact layers caused by free-carrier absorption. To improve intracavity design we propose to insert a few pairs of doped DBR between active region and contact layers. In addition a top dielectric DBR can be used to increase the optical confinement.

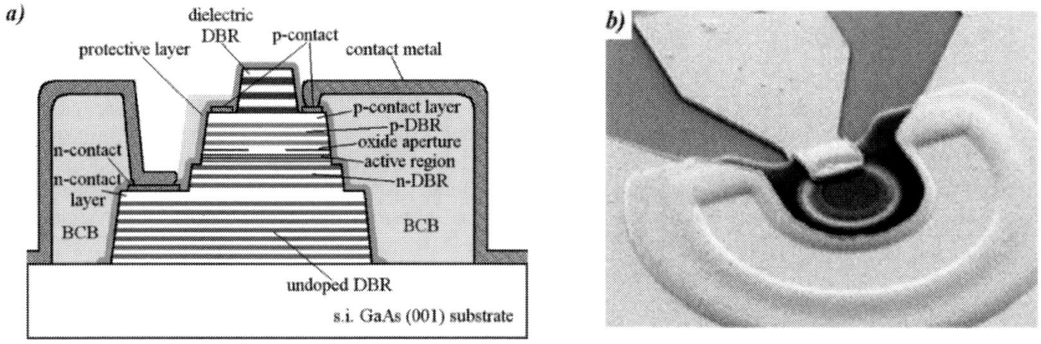

Figure 1. (a) Schematic cross-section of the oxide-confined InGaAlAs VCSEL with a dielectric DBR and intra-cavity contacts; (b) The scanning electron microscopy image of the fabricated VCSEL.

The VCSEL structures were grown on an undoped (001)-oriented GaAs substrate using MBE-systems. The epitaxial wafers consist of the top p^+-GaAs contact layer, the short p-doped $Al_{0.15}Ga_{0.85}As/Al_{0.9}Ga_{0.1}As$ DBR with a thin AlAs/AlGaAs aperture layer, the InGaAlAs -based active region, the short n-doped $Al_{0.15}Ga_{0.85}As/Al_{0.9}Ga_{0.1}As$ DBR, the buried n-$Al_{0.15}Ga_{0.85}As$ layer and the 30-pair undoped $Al_{0.15}Ga_{0.85}As/Al_{0.9}Ga_{0.1}As$ bottom DBR. The active region consists of multiple narrow InGaAs quantum wells (QWs) with an average In-composition of ~10% sandwiched between high energy bandgap AlGaAs cladding layers to suppress the thermal escape of nonequilibrium carriers. The number and the thickness of the QWs as well as the barrier thicknesses were chosen to be narrow enough to reach the maximum possible optical confinement factor. To improve the temperature stability of lasers the resonance wavelength of cavity mode was red shifted from the QW photoluminescence peak at 20°C by nominally 15 nm. The average doping concentration in the doped regions is about $2 \cdot 10^{18}cm^{-2}$ for p-type and $1 \cdot 10^{18}cm^{-2}$ for n-type, respectively. To maintain reasonable internal optical losses and provide low series resistance the intracavity contact layers were doped nonuniformly with higher level at nodes of optical field. To decrease the potential barrier height at DBR heterointerfaces, linear compositional digital-grading layers with modulated doping profile were used. Lasers with different mesa diameters were fabricated in the high-speed design using standard fabrication techniques including: optical lithography, metal and dielectric evaporation, dry and wet etching, etc. Selective wet oxidation of AlGaAs materials was applied to create lateral confinement both for the injected current and the optical field. Top dielectric SiO_2/TiO_2 DBR was formed by magnetron evaporation technology. The details of the VCSEL processing can be found in [6]. The schematic cross-section of the processed 850 nm oxide-confined 850 nm VCSEL with a ground-source-ground (GSG) pad configuration is shown in figure 1.

3. Results and discussion

3.1. Static device characteristics

Figure 3(a) shows the temperature evolution of the typical light output power-current-voltage (L-I-V) characteristics for the 4.5 µm oxide aperture InGaAlAs VCSEL. The maximum output power decreases from 2.4 mW at 20°C to 1.2 mW at 90°C, which is ~50% of the value at 20°C. The maximum slope efficiency is hardly temperature-dependent and exceeds 0.56 W/A. According to figure 3(b) the threshold current exhibits a clear minimum around 60°C, but still remains below 0.75 mA in the whole temperature range, helping to maintain the large differential gain at elevated temperatures. We believe that the obtained high-thermal stability of our VCSELs can be associated to the optimized gain-cavity detuning and the improved carrier localization in the active region. Note that the investigated lasers demonstrate relatively low serial resistance (<250 Ohm) as comparing with MBE-grown 980 nm VCSELs with fully doped DBRs and similar aperture size [7].

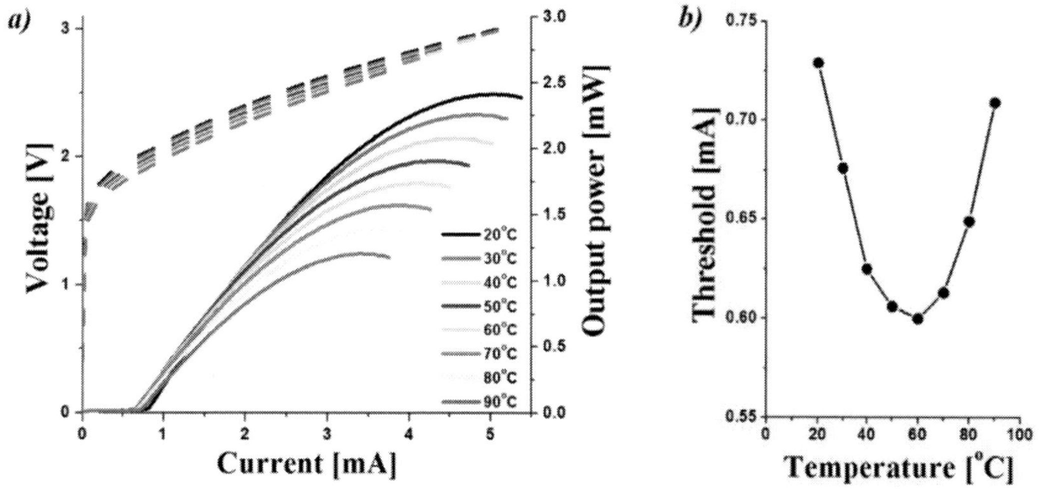

Figure 3. The InGaAlAs VCSEL with a 4.5 µm oxide aperture: (a) Measured light-current-voltage characteristics at different temperatures; (b) Threshold current as a function of temperature.

The lasing spectra at different currents and temperatures for the 4.5 µm oxide aperture InGaAlAs VCSEL are presented in figure 4. Despite the efficient longitudinal mode selection, the strong index guiding effect in oxide-confined VCSELs usually requires much smaller aperture sizes for single-mode operation, especially at short-wavelength range. However the laser emits via the fundamental mode at ~850 nm and demonstrates side-mode suppression ratio (SMSR) more than 30 dB throughout the whole current and temperature range. Such effective single-mode operation at relatively large current aperture can be explained by the enhanced transverse-mode discrimination in case of thin tapered oxide aperture [8].

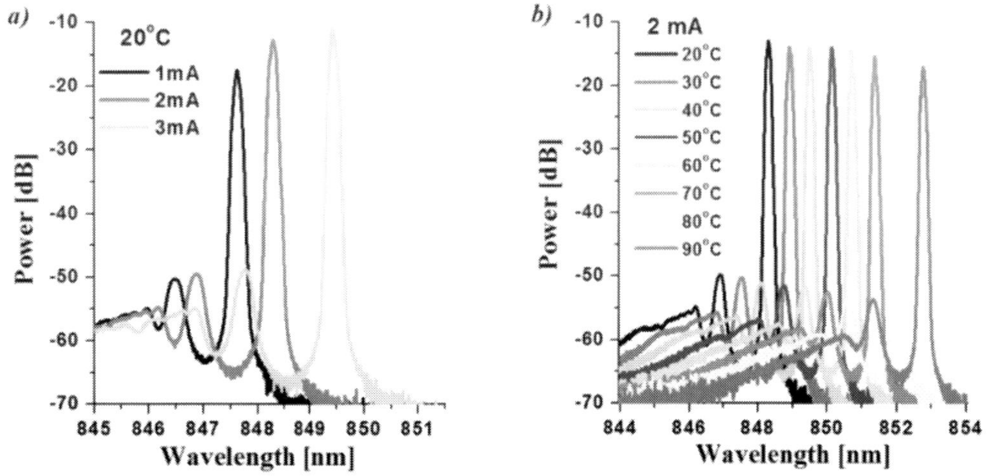

Figure 4. The InGaAlAs VCSEL with a 4.5 μm oxide aperture: (a) Lasing spectra at 20°C and different current; (b) Lasing spectra at 2 mA and different temperatures.

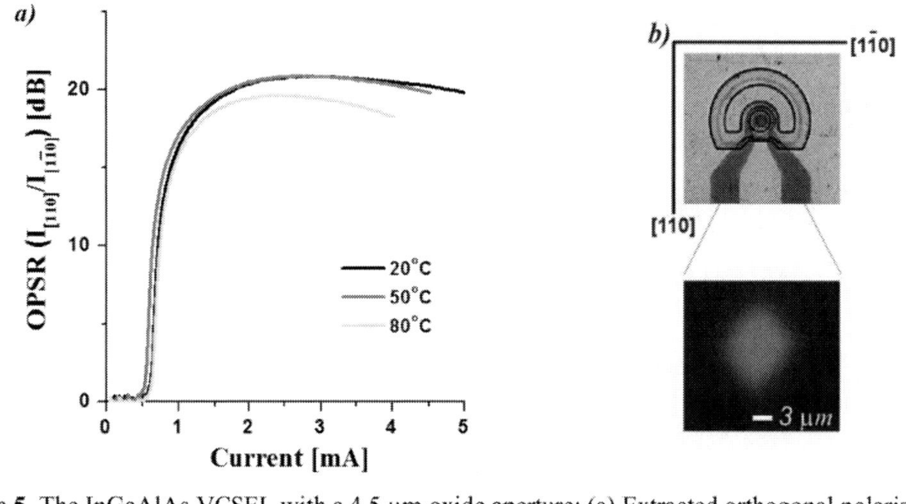

Figure 5. The InGaAlAs VCSEL with a 4.5 μm oxide aperture: (a) Extracted orthogonal polarization suppression ratio as a function of bias current at different temperatures; (b) Microscope image of fully processed VCSEL (top) and near-field pattern below threshold (bottom).

Figure 5.a shows the results of the polarization resolved light-current measurements for the 4.5 μm oxide aperture InGaAlAs VCSEL. As threshold is reached, polarization is immediately fixed along the [110] crystal axis with the orthogonal polarization suppression ratio (OPSR) over 20 dB. On the one hand, it can be attributed to the electro-optic effect in standard GaAs-based VCSELs grown on (001)-oriented substrate [9]. However this mechanism is inherently unstable and the polarization can change its orientation relative to the preferred crystal axis with varying the bias current or the temperature [4]. On the other hand, no polarization switching in our VCSELs was observed through the entire current and temperature range. As can be seen in Figure 5.b, the oxide current aperture has a rhombus shape despite the cylindrical symmetry of the mesa-structure, since the oxidation rate in our technology process is not the same in the different crystal directions. Hence one supposed that this strong polarization stability is due to in-plane non-uniform current injection and anisotropic transverse cavity geometries.

3.2. Dynamic device characteristics

To get a deeper understanding of high-speed capabilities of the developed VCSELs, especially at elevated temperatures, the small signal modulation response (S_{21}) and microwave reflection (S_{11}) were measured in the range from 50 MHz to 30 GHz. Figure 6(a) demonstrates the measured modulation bandwidth (or 3dB-frequncy) as a function of bias current at different temperatures. The maximum modulation bandwidth exceeds 13 GHz in whole temperature range. The modulation current efficiency factor (MCEF) is very temperature stable and beyond 10 GHz/mA$^{1/2}$, which is a result of the optimized cavity—gain detuning and advanced VCSEL design.

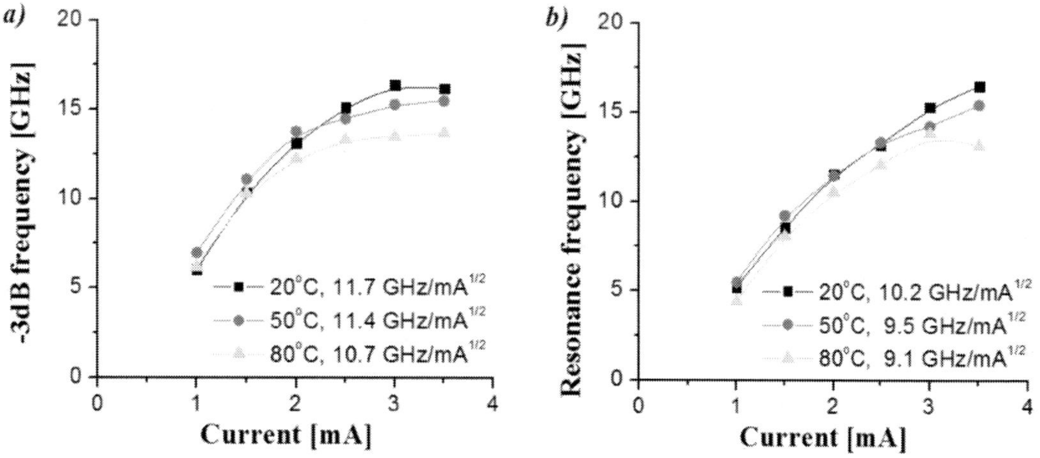

Figure 6. Modulation bandwidth (a) and resonance frequency (b) versus bias current at different temperatures for an InGaAlAs VCSEL with a 4.5 μm oxide aperture.

To clarify the limiting mechanism of high-speed performance, the dynamic physical properties of investigated lasers were extracted by fitting procedure similar to that described in [10]. Figure 6(b) shows the temperature evolution of the extracted resonance frequency at different bias current. The large gain—cavity detuning prevents the decrease in the differential gain at higher temperatures and result in the high temperature stability of the D-factor. Note that the relaxation resonance frequency reaches values comparable to the maximum bandwidth at the high temperature, which can be associated with a noticeable damping. However the extracted K-factor for our VCSELs is less than 0.25 ns corresponding to a maximum intrinsic bandwidth more than 35 GHz in the absence of thermal and parasitic limitations. Together with the small values of damping factor it clearly indicates the minor limiting effect of damping even at high temperature and currents.

According to the S_{11} data and the equivalent circuit model electrical, the parasitic cut-off frequency of the investigated VCSEL was found to be about 9 GHz over the entire measured current range. These electrical parasitic have a major impact on the VCSEL's high speed performance.

4. Conclusion

New intracavity-contacted design with composite DBRs was proposed and applied for MBE-grown 850-nm VCSELs based on InAlGaAs material systems. Temperature and polarization-stable high-speed single-mode VCSELs with submilliamp threshold current, slope efficiency more than 0.56 W/A, SMSR > 30 dB and modulation bandwidth higher than 13 GHz in the whole temperature range of 20-

80°C were realized. The proposed design can be easily adjusted to the desired spectral range in the promising VCSEL applications.

5. Aknowledgments

This work was partially supported by the Russian Academy of Sciences (RAS) Presidium Program (No. 24), the Russian Foundation for Basic Research (grant 13-02-12142-ofi_m) and the Ministry of Science and Education of the Russian Federation in the frame of the Federal contract 14.427.11.0004.

References

[1] Michalzik R 2013 *VCSELs: Fundamentals, Technology and Applications of Vertical-Cavity Surface-Emitting Lasers* (Berlin: Springer-Verlag)
[2] Serkland D et al. 2007 VCSELs for atomic sensors *Proceeding of SPIE* **6484** 648406
[3] Liu Y, Ng W – C, Oyafuso F, Klein B, and Hess K 2002 *IEE Proc.-Optoelectron* **149** 182
[4] Verschaffelt G, Panajotov K, Albert J, Nagler B, Peelers M, Danckaert J, Veretennicoff I and Thienpont H 2001 Polarisation switching in vertical-cavity surface-emitting lasers: From experimental observations to applications *Opto-Electron. Rev.* **9** 257–268
[5] Huffaker D L, Deppe D G 1999 Intracavity contacts for low-threshold oxide-confined vertical-cavity surface-emitting lasers *IEEE Photon. Technol. Lett.* **11** 934–936
[6] Maleev N A et al. 2013 Single-spatial-mode semiconductor VCSELs with a nonplanar upper dielectric DBR *Semiconductors* **47** 7 993-996
[7] Mutig A et al. 2009 Temperature-Dependent Small-Signal Analysis of High-Speed High-Temperature Stable 980-nm VCSELs *J. Sel. Topics Quantum Electron.* **15** 679–686
[8] Blokhin S A et al. 2006 Vertical-Cavity Surface-Emitting Lasers Based on Sub-Monolayer InGaAs Quantum Dots *Journal Quantum Electronics* **42** 851–858
[9] Van Exter M P, Jansen van Doorn A K and Woerdman J P 1997 Electrooptic effect and birefringence in semiconductor vertical-cavity lasers *Phys. Rev. A, Gen. Phys.* **56** 845–853
[10] Mutig A, Blokhin S A, Nadtochiy A M, Fiol G, Lott J A, Shchukin V A, Ledentsov N N, Bimberg D 2009 Frequency response of large aperture oxide-confined 850 nm vertical cavity surface emitting lasers *Applied Physical Letters* **95** 131101

Resonances in random reactance networks with fluctuating entries

N A Olekhno[1], Y M Beltukov[2] and D A Parshin[1]

[1]Saint Petersburg State Polytechnical University, 195251 Saint Petersburg, Russia
[2]A. F. Ioffe Physical-Technical Institute, 194021 Saint Petersburg, Russia

E-mail: [1]olekhnon@gmail.com

Abstract. It is well known that disordered LC networks are an appropriate model for describing giant fluctuations of electric fields in a random metal-dielectric composite. The fluctuations reflect an inherent multifractal internal structure of the local fields which arise due to dipole-dipole interactions between the clusters with "dielectric resonances". The resonances are poles of the conductance of disordered dissipationless LC networks. Though this topic is widely presented in the literature, described methods allow to study only the cases with fixed L and C values randomly distributed on a lattice. In this paper we generalize this approach and study random LC networks with fluctuating values of L and C entries. We demonstrate anticrossing of resonances due to their dipole-dipole interaction and splitting of the typical resonances for a discrete set of fluctuating L or C (or both) entries.

1. Introduction

Optical properties of metal-dielectric composites consisting of metallic nanoparticles incorporated into dielectric matrix are attracting considerable attention last years both from theoretical and experimental point of view [1, 2, 3, 4, 5, 6].

A simplest theoretical approach to such systems is to replace a continuous medium to a large discrete random network made of capacitors C and inductances L, the latter being in series with a weak resistor R. This network description is a discrete version of continuous Maxwell equations for scalar potential of electric field. The capacitors here model dielectric bridges between metallic nanoparticles, whereas isolated metallic granules are characterized by almost purely inductive response. This approximation is valid for frequencies ω of electromagnetic radiation satisfying to condition $\omega_\tau \ll \omega < \omega_p$. Here ω_p is the plasma frequency and ω_τ is the plasmon relaxation rate [3].

Such large random reactance networks (that is networks made of random mixture of capacitances C and inductances L) have a lot of resonance clusters with any size and shape and different resonance frequencies. The clusters have a finite real conductance at resonance frequencies and thus can disperse an electric power. These resonances show up as narrow peaks with Lorentzian line shapes, e.g. in the weak-dissipation regime of the $RL-C$ model [4]. The dielectric resonances provide a natural explanation for the anomalous fluctuations of the local electric field [1], which are responsible for giant surface-enhanced Raman scattering observed in semicontinuous metal films [7].

In the papers [4, 5] considering as an example a $2d$ square lattice it was shown that the problem of resonances of a random LC network (with random positions of capacitors C and inductances L) can be reduced to some generalized eigenvalue problem. However this reduction was done only for the very particular case when all values of the inductances L (and capacitors C) are equal to each other. Then in the system there is a typical frequency $\omega_0 = 1/\sqrt{LC}$. In the present paper we have extended these results to the general case when values of inductances L and capacitors C are also random and can fluctuate from bond to bond. First we consider the case of two different interacting resonances and show the typical anticrossing effect when two frequencies approach each other. Then it is shown that for discrete set of different L or C (or both) values the known resonances observed for equal values of L and C are split and spectrum becomes more reach than for non fluctuating cases investigated before. Finally, we have also studied the spatial structure of these resonances for 2d and 3d random LC lattices with fluctuating entries and have shown that they also experience giant fluctuations of local electric fields.

2. Kirchhoff equations

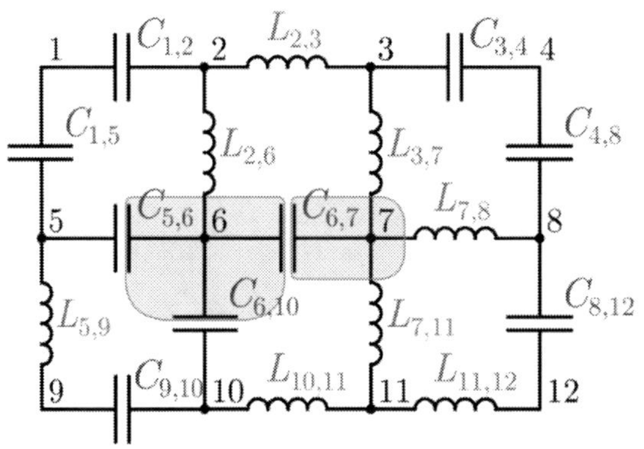

Figure 1. A piece of random LC network. Shaded are regions with charge accumulated on sites 6 and 7 respectively.

To derive Kirchhoff equations in the general case, let us numerate sites of our LC network using integer index i taking values from 1 to N, where N is the total number of cites (see Fig. 1). Let φ_i being the potential of site i. Then let us attribute to each site i a charge q_i. It is equal to the sum of charges of capacitor plates adjoined site i (see shaded regions shown on Fig. 1 around sites 6 and 7). If no capacitor adjoins site i, then the corresponding charge $q_i = 0$. The charges q_i and the potentials φ_i are satisfying to the system of equations

$$q_i = \sum_{j \neq i} C_{ij}(\varphi_i - \varphi_j), \tag{1}$$

where C_{ij} is a capacity of the capacitor connecting sites i and j. The matrix C is by definition symmetric $C_{ij} = C_{ji}$. If there is no capacitor between sites i and j then $C_{ij} = 0$. If we define the diagonal matrix element C_{ii} as a minus sum of the non-diagonal elements

$$C_{ii} \equiv -\sum_{j \neq i} C_{ij}, \tag{2}$$

then as follows from Eq. (1) we get

$$q_i = -\sum_{j} C_{ij}\varphi_j. \tag{3}$$

The time derivative $\dot{q}_i$ by definition is equal to the sum of incoming electric currents to the site i from all neighboring sites j

$$\dot{q}_i = \sum_{j \neq i} I_{ji}. \tag{4}$$

Here I_{ji} is the electric current from site j to site i. The matrix I_{ji} by definition is antisymmetric $I_{ij} = -I_{ji}$. In its turn the time derivative of electric current $\dot{I}_{ij}$ is related to voltage $\varphi_i - \varphi_j$ in the corresponding inductance coil

$$L_{ij}\dot{I}_{ij} = \varphi_i - \varphi_j. \tag{5}$$

Here L_{ij} is an inductance of the coil between sites i and j. The matrix L is a symmetric matrix, $L_{ij} = L_{ji}$. If there is no coil between sites i and j, then the corresponding value of $L_{ij} = \infty$.

Now using equations (4) and (5) we can relate the second time derivative of charge $\ddot{q}_i$ with voltages $\varphi_j - \varphi_i$

$$\ddot{q}_i = \sum_{j \neq i} \frac{1}{L_{ij}}(\varphi_j - \varphi_i). \tag{6}$$

To proceed further let us introduce matrix K of inverse inductances (between sites i and j)

$$K_{ij} = \frac{1}{L_{ij}}. \tag{7}$$

The matrix K is symmetric, i.e. $K_{ij} = K_{ji}$. If sites i and j are not connected by inductance coil, then by definition $K_{ij} = 0$. Similar with the case of matrix C let us define the diagonal matrix elements of the matrix of inverse inductances K as minus sum of the non-diagonal elements (compare with Eq. (2))

$$K_{ii} \equiv -\sum_{j \neq i} K_{ij}. \tag{8}$$

Then from (6) we get

$$\ddot{q}_i = \sum_j K_{ij}\varphi_j. \tag{9}$$

Comparing Eq. (3) and Eq. (9), we obtain finally a linear system of ordinary differential equations for site potentials

$$\sum_j C_{ij}\ddot{\varphi}_j = -\sum_j K_{ij}\varphi_j. \tag{10}$$

Looking for solution of these equations in the exponential form $\varphi_i \propto \exp(i\omega t)$ we get

$$\omega^2 \sum_j C_{ij}\varphi_j = \sum_j K_{ij}\varphi_j. \tag{11}$$

Non-zero solution of this linear system of equations exist when

$$\det\left(K - \omega^2 C\right) = 0. \tag{12}$$

As a result we have reduced calculation of eigenfrequencies of our random impedance network to a generalized eigenvalue problem.

Let us show finally that Eqs. (11) are equivalent to the Kirchhoff equations investigated in [3, 4, 5] and similar papers. For this purpose we can present them in the equivalent form

$$\sum_j i\omega C_{ij}\varphi_j = -\sum_j \frac{1}{i\omega} K_{ij}\varphi_j \tag{13}$$

Making use of properties (2) and (8) we transform them as follows

$$\sum_j i\omega C_{ij}(\varphi_i - \varphi_j) = -\sum_j \frac{1}{i\omega} K_{ij}(\varphi_i - \varphi_j) \tag{14}$$

or

$$\sum_j \sigma_{ij}(\varphi_i - \varphi_j) = 0, \quad \text{where} \quad \sigma_{ij} = i\omega C_{ij} + \frac{1}{i\omega L_{ij}}. \tag{15}$$

In the papers [3, 4, 5] it was investigated a particular case when all different from zero values $C_{ij} = C$ and all different from zero values $1/L_{ij} = 1/L$. Thus we have generalized these results to the general case of different values of capacitances and inductances.

3. Anticrossing effect between two resonances

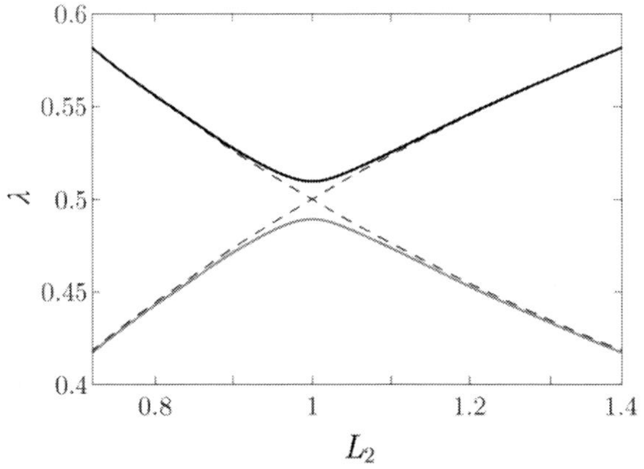

Figure 2. Anticrossing of frequencies for two interacting inductances L_1 and L_2 in a $2d$ lattice (50×50) of capacitors with $C = 1$ separated by 3 lattice constants.

Let us now apply the developed method to some particular cases. First we consider two inductance coils with inductances L_1 and L_2 (so that $L_1 L_2 = 1$) embedded into $2d$ lattice of capacitors with equal capacities C. To avoid interaction with boundaries we will use doubly periodic boundary conditions. As it follows from paper [5] in the case of one inductance coil L in an infinite $2d$ lattice of capacitors with equal capacities C we have one resonance frequency $\omega_0 = 1/\sqrt{LC}$. In the following instead of frequency ω it is convenient to use the dimensionless value of λ [4], so that $\lambda = 0.5$ for $\omega = \omega_0$

$$\lambda = \frac{(\omega/\omega_0)^2}{1 + (\omega/\omega_0)^2}. \tag{16}$$

As it was shown in [4, 8] the electric field from one inductance coil embedded into an infinite lattice of equal capacitors is a field of electric dipole with dipole moment parallel to the coil direction. In the case of two identical coils (oriented parallel to each other) we have two identical dipole oscillators which interact with each other. The interaction strength in $2d$ case is $I \propto 1/R^2$, where R is a distance between the dipoles. As a result the resonance frequency ω_0 splits on two values separated by interaction strength I. In the case of two different coils with inductances L_1 and L_2 we have two different harmonic oscillators interacting with each other. As a result changing bare frequencies of these oscillators we observe a typical anticrossing effect seen clearly on Fig. 2. The two asymptotes on the figure show two bare frequencies (or λ's) for non interacting oscillators $\lambda_1 = 1/(1 + L_1) = L_2/(1 + L_2)$ and $\lambda_2 = 1 \quad \lambda_1 = 1/(1 + L_2)$.

4. Splitting of the resonances

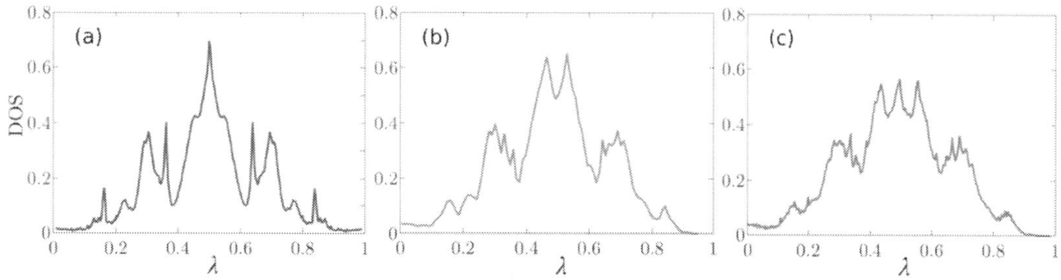

Figure 3. Splitting of the resonances in a random $2d$ network with fluctuating entries. (a) Random lattice with equal inductances $L = 1$, and concentration $p = 0.1$ embedded in a $2d$ square lattice with equal capacitances [5]; (b) The same lattice with two sorts of coils with inductances $L_1 = 1.13$ and $L_2 = 0.89$, and with concentrations of each type $p = 0.05$; (c) The same lattice with three sorts of coils with inductances $L_1 = 1.25$, $L_2 = 1$ and $L_3 = 0.8$, and with concentration of each type $p = 0.033$. In all three cases all capacitances C in the lattice are equal to each other ($C = 1$).

As it is well known random reactance networks have a complicated structure of "dielectric resonances" depending on the concentration of L and C entries [3]. In the case when all L and C values are equal the resonance structure for a $2d$ case was analyzed in [5]. But if L and C values can fluctuate from bond to bond the typical resonances should split and their structure becomes more reach. To demonstrate the splitting let us consider resonances in a $2d$ lattice of capacitors with a small concentration of inductance coils. On Fig. 3(a) is shown the case investigated in [5] when inductance coils with equal inductances are embedded in a $2d$ lattice with equal capacitances. On Fig. 3(b) is shown the same lattice but with fluctuating values of inductances L_1 and L_2 (so that $L_1 L_2 = 1$). We see how the central peak for $\lambda = 0.5$ splits into two peaks. The Fig. 3(c) shows splitting of the central peak into three peaks for the case when we use three different values of coil inductances L_1, L_2 and L_3 embedded randomly into the capacitors matrix. The density of states (DOS) shown on this figure was obtained on a lattice (10×10) with double periodic boundary conditions after averaging over 10^5 realizations.

5. Fluctuations of local fields

The last question we are going to touch in the paper is about the geometrical structure of resonances. Due to long-range dipole-dipole interaction the resonances are not localized [9] and have multifractal structure of eigenmodes [5, 10]. In the case with equal values of L and C entries the geometry of local fields was qualitatively described in [1] and investigated quantitatively in [5]. We have investigated the local structure of resonances in the case when values of L and C can fluctuate. For example in a $2d$ square lattice with equal capacitances $C = 1$ and concentration of inductance coils $p = 0.1$ for $L_1 = 0.8$, $p = 0.1$ for $L_2 = 1$ and $p = 0.1$ for $L_3 = 1.25$ the structure of resonances is shown on Fig. 4. Similar with the case of non fluctuating entries, we can see strong fluctuations of the local electric fields in the system demonstrating a multifractal structure of the eigenstates.

6. Conclusion

We generalized the problem of random LC networks to the case when values of capacitances C and inductances L can fluctuate in a lattice from bond to bond. We demonstrated the

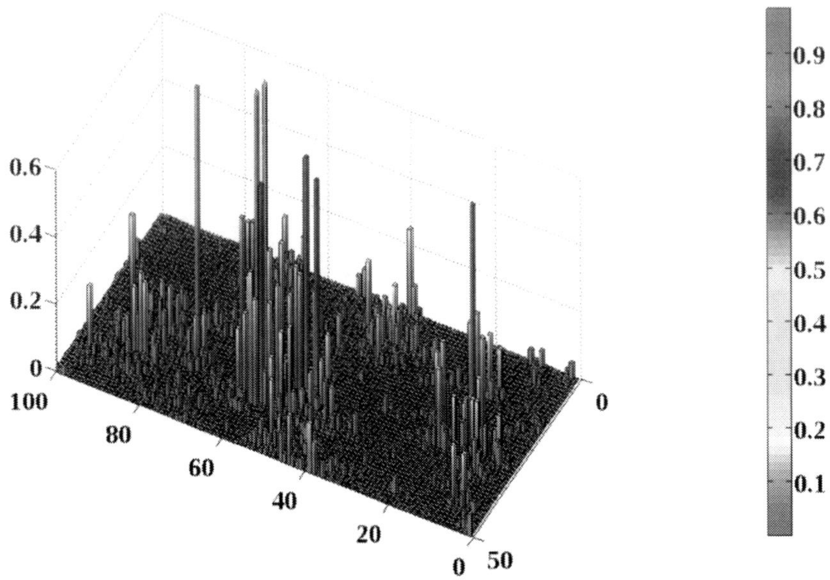

Figure 4. The geometry of local electric fields $|E|^2$ for the random lattice with 5000 bonds and total inductance concentration $p = 0.3$ of three different types (see text) for $\lambda = 0.48$.

anticrossing of two resonances and showed how typical resonances investigated before split in several peaks corresponding to different values of inductance L (or capacitance C) entries. The fluctuations of L and C values do not change qualitatively the multifractal local structure of resonances.

We are very grateful to Prof. V. I. Kozub for many fruitful discussions. One of the authors (YMB) thanks Dynasty Foundation for financial support.

References

[1] Brouers F, Sarychev A K, Blacher S and Lothaire O 1997 *J. Phys. A* **241** 146
[2] Sarychev A K and Shalaev V M 1999 *Physica A* **266** 115
[3] Fyodorov V V 2001 *Physica E* **9** 609
[4] Clerc J P, Giraud G, Luck J M and Robin Th 1996 *J. Phys. A* **29** 4781
[5] Jonckheere Th and Luck J M 1998 *J. Phys. A* **31** 3687
[6] Clerc J P, Giraud G, Laugier J M and Luck J M 1990 *Adv. Phys.* **39** 191
[7] Brouers F, Blacher S, Lagarkov A N, Sarychev A K, Gadenne P and Shalaev V M 1997 *Phys. Rev. B* **55** 13234
[8] Schafer S, Raymond L and Albinet G, 2005 *Eur. Phys. J. B* **43** 81
[9] Levitov L S, 1989 *Europhys. Lett.* **9** 83
[10] Parshin D A and Schober H R 1998 *Phys. Rev. B* **57** 10232

Journal of Physics: Conference Series **572** (2014) 012038 doi:10.1088/1742-6596/572/1/012038

Spatial and spectral mode filtration in phase-locked array of injection lasers with volume Bragg grating

A Payusov[1,2], N Gordeev[2,1], O Rumyantsev[3], Y Zadiranov[2] and M Maximov[1,2]

[1]Academic University – Nanotechnology Research and Education Center of the Russian Academy of Sciences, 8/3 Khlopina, St.Petersburg, 194021, Russia

[2]Ioffe Physical Technical Institute of the Russian Academy of Sciences, 26 Polytekhnicheskaya, St.Petersburg, 194021, Russia

[3]Stanford University - Department of Applied Physics, 348 Via Pueblo Mall, Stanford, CA, 94305, USA

plusov@mail.ioffe.ru

Abstract. High-power edge-emitting lasers with low divergence of the output beam are highly desired for a number of applications. In order to obtain high optical power one needs to expand the emitting area of the laser which usually leads to multi-mode lasing with broad spectrum and poor beam quality. Here we present the study of using a volume Bragg grating (VBG) as a spatial and spectral mode filter for phase-locked arrays of edge-emitting lasers based on 1D photonic bandgap crystals. Emitters in the arrays are synchronized due to the expanded vertical waveguide, which results in the lasing of high order oscillating modes. One lobe of high order mode was used to create an external cavity with VBG. We studied far-fields, spectra and L-I curves of the external cavity lasers and demonstrated that *500 mW* of the output optical power could be concentrated in one lobe with the divergence as low as 1° and spectral width of *0.6 nm*.

Introduction

High power semiconductor laser with high spatial and spectral brightness are desired for a number of applications such as pump systems for solid state and fiber lasers, material processing, medicine, etc. [1-3]. To achieve high power from edge-emitting lasers and reduce output beam divergence one needs to expand waveguide both in lateral and vertical directions. This usually results in poor spatial and spectral quality of laser beams. Several approaches have been proposed to keep single mode lasing in broadened waveguides [4,5]. However, many of them are rather sensitive to minor changes of refractive indexes of the waveguide layers [6]. On the other hand, such techniques as photonic band gap crystal (PBC) and tilted wave lasers (TWL) demonstrated ultra-narrow beam and robustness with respect to growth deviations and temperatures [7].

Broadened vertical waveguide allows obtaining low vertical divergence. However, for high power lasers it is critical to have broad emitting area and good beam quality in the lateral direction as well. Phase-locked arrays of edge-emitting lasers, which may produce narrower beam in lateral direction, have been investigated since early 1980s [8]. It has been shown that arrays of phase-locked semiconductor lasers tend to lase in oscillating (out-of-phase) modes rather than in fundamental (in-phase) modes. Recently PBC lasers were used to create phase-locked arrays emitting in single vertical mode with the optical power as high as *15 W* [9]. In these arrays modes of the adjacent emitters effectively overlap due to the broadened PBC waveguide. It has been shown that in PBC-based arrays, besides the loss selective mechanism [8], lateral modes are discriminated by their optical confinement factor [10]. Optical confinement factor is almost two times larger for the oscillating mode than for the fundamental one.

Oscillating modes of phase-locked arrays have two narrow lobes in the far field pattern. The lateral divergence of the output beam from the phase-locked arrays can be as low as 5° (FWHM). Two lobes from oscillating modes can be collimated in one with additional optical elements [11]. On the other hand, progress in recording volume Bragg gratings (VBG) in a glass makes it possible to create external cavities with minimum additional optical elements [12]. With modern VBG it is possible to achieve angular selectivity about 0.05^o and spectral selectivity about *0.1 nm* in wavelength range of *1 μm* , which should be enough to pick out single lateral modes from phase-locked arrays. Hence, one lobe of the oscillating mode can be used for the external feedback with VBG and another as an output beam, resulting in only one narrow lobe in far field of the external cavity laser.

In this paper we present the study of using a VBG as a spatial and spectral mode filter for arrays of phase-locked edge-emitting PBC-based lasers.

Laser design

The laser wafer was grown by MOCVD on GaAs substrate. Nine pairs of $Al_{0.3}Ga_{0.7}As/Al_{0.15}Ga_{0.85}As$ where deposited to form PBC structure. Three QW $In_{0.19}Ga_{0.81}As/GaAs_{0.86}P_{0.14}$ in the top $Al_{0.15}Ga_{0.85}As$ layer of PBC serve as an active medium and simultaneously as an "optical defect" to localize the mode. Bottom layers of PBC where doped with Si and serve as n-emitter. Thick p-doped layer of $Al_{0.5}Ga_{0.5}As$ was grown on the top of PBC to form p-emitter. The total thickness of the PBC waveguide is *~10 μm*. More details on the structure design can be found elsewhere [9].

The phase-locked arrays consisted of *23* emitters with the width of *4 μm* formed by standard photo-lithography (figure 1). Distance between emitters was *5 μm*. Therefore, the total width of the lateral waveguide was as high as *0.2 mm*. The cavity length was *1 mm*. Facets were coated by high-reflection (*95%*) and anti-reflection (*3%*) coatings. The lasers where mounted p-side down on copper heat-sinks. The lasing wavelength lies in the range of *970-980* nm at *20°C*.

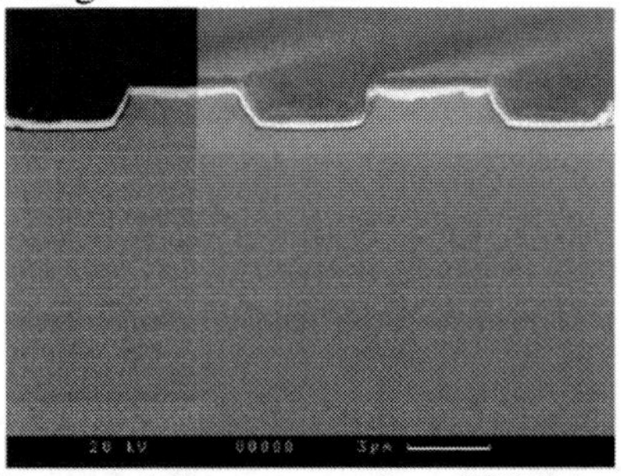

Figure 1. SEM image of the cross-section of PBC-based laser array.

Volume Bragg Grating was recorded in photo-thermo-refractive glass. The size of the grating is *10x10x1.5 mm*. It has maximum reflectivity of *94%* at *982.1* nm at room temperature. Spectral width of reflection on half maximum is *0.6 nm*, angular selectivity is less than 1°.

Figure 2 shows the scheme which we used to create the external cavity. First cylindrical lens with *3 mm* diameter collimated the beam in the vertical direction. Second plano-convex cylindrical lens was used to focus laser beam on the VBG. Lenses were uncoated and made of NBK-7 glass which is suitable for near IR range. Focal length of the second lens is *~24 mm* at *980 nm*. This scheme is very convenient for experiments and allows scanning lateral far fields of external cavity lasers. To adjust the cavity we rotated VBG. When VBG planes become normal to a certain mode lobe, emission of this lobe comes back into the array creating external feedback. Laser temperature was stabilized at *~50C* in order to match the lasing wavelength to the VBG reflection. As the laser temperature was quite high, all measurements have been done in pulsed regime in order to avoid additional overheating.

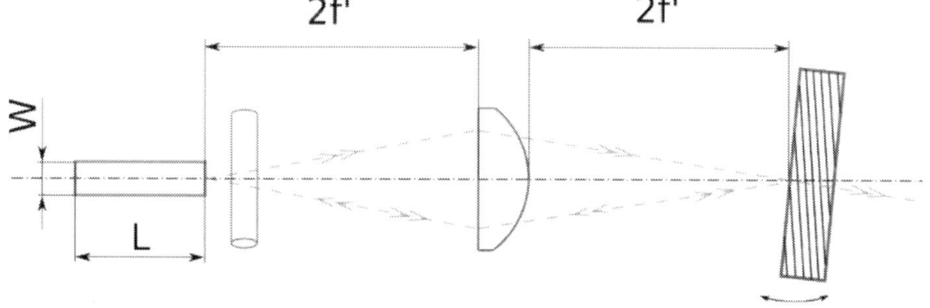

Figure 2. Scheme of external cavity with VBG. *W=0.2mm, L=1 mm, f' = 24 mm.* Layers of heterostructure are parallel to draw plane. Lateral far fields where measured by the rotating rod with a detector coaxial with VBG.

External cavity laser characteristics

We obtained similar results for all studied array samples. External cavity optical scheme has two symmetric VBG orientations. We have used an orientation with the best far field (*2.55°*, figure 3). Figure 3 shows lateral far fields (a) of free running array and external cavity laser. Pump current was slightly above the threshold. It is clearly seen that the free running array has two sharp lobes in the far field which indicates that it lases in oscillating modes. External cavity laser has one sharp (*1°*) lobe and two small satellites in the far field. Linewidth of *0.6 nm* (figure 3b) of the external cavity laser corresponds to the of VBG reflection linewidth.

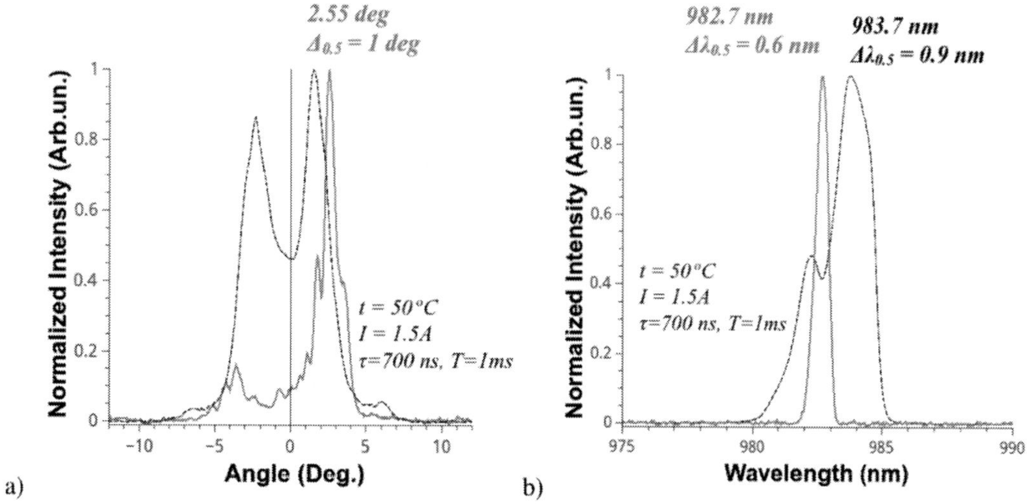

Figure 3. Lateral far field (a) and spectrum (b) of free running (dashed) and external cavity (solid) laser.

We measured L-I curve from free running array and external cavity laser (figure 4). Differential efficiency dropped by *30%*, threshold current almost unchanged. Differential efficiency can be improved by using anti-reflective coated lenses. Lateral far field and spectrum remained the same with rising pump current up to *8A*. At the current of *6A*, *500mW* of the optical power was concentrated in one lobe with the divergence of 1° and linewidth of *0.6 nm*.

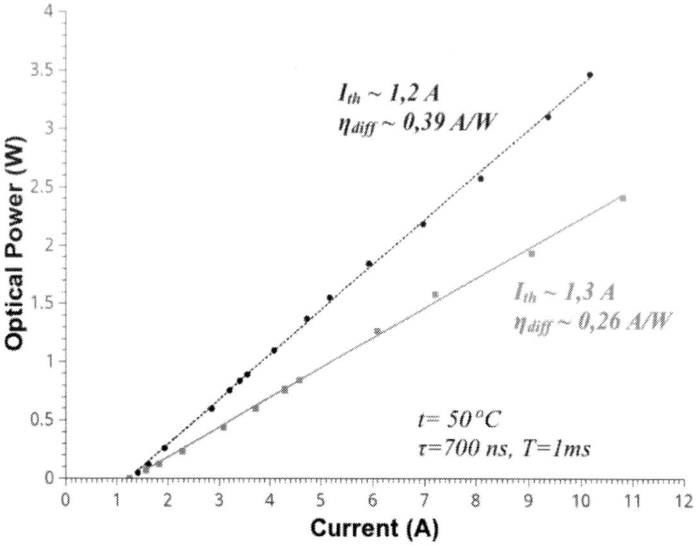

Figure 4. L-I characteristics of free running (dashed) and external cavity (solid) laser.

Discussion

The threshold current of the free running laser is almost the same as one of the external cavity laser (figure 4). We believe that the external cavity laser operates in a combined regime: it has feedbacks both from the laser chip facets and from the VBG. In this case of coupled cavities, the lasing wavelength is not entirely defined by the VBG spectral parameters, which may result in wavelength pulling from the VBG reflection maximum [13]. This effect is clearly seen in figure 3b. By suppressing a feedback from the laser chip facet a "standard" VBG laser operation reported elsewhere [12] can be obtained.

Satellite lobes in the far-field seem to indicate multi-mode operation of the external cavity laser. To avoid multi-mode lasing one can try to use a thicker VBG with better angular selectivity.

We believe that with optimized relation of feedback between two cavities one can improve characteristic of external cavity phase-locked laser array.

Conclusion

We have demonstrated for the first time that volume Bragg grating allows filtering oscillating modes in phase-locked arrays of PBC-based semiconductor lasers. VBG-based external cavity lasers have one narrow lobe in the far field with significant amount of optical power concentrated in this lobe. Spectral linewidth of external cavity lasers is defined by VBG and does not depend on the pump current. One can expect that the use of coated lenses in external cavity and optimization of anti-reflection coating on the laser facets may further improve characteristics of the external cavity lasers.

Acknowledgements

The authors are grateful to Prof. Leon Glebov, Photoinduced Processing laboratory, CREOL / The College of Optics & Photonics, University of Central Florida, for providing the volume Bragg gratings.

References

[1] Al-Muhanna A, Mawst L J, Botez D, Garbuzov D Z, Martinelli R U and Connolly J C 1998 Appl. Phys. Lett. **73**, 1182

[2] Pikhtin N A, Slipchenko S O, Sokolova Z N, Stankevich A L, Vinokurov D A, Tarasov I S and Alferov Zh I 2004 Electron. Lett. **40** 1413-1414

[3] Wang X, Crump P, Pietrzak A, Schultz C, Klehr A, Hoffmann T, Liero A, Ginolas A, Einfeldt S, Bugge F, Erbert G and Tränkle G 2009 *Proc. of SPIE* **7198** 719814

[4] Lin G, Yen S and Lee C 1996 IEEE Phot. Techn. Lett. **8**, 1588-1590

[5] Wenzel H, Bugge F, Erbert G, Hulsewede R, Staske R and Trankle G 2001 Electron. Lett. **37** 1024-1026

[6] Corbett B, Lambkin P, O'Callaghan J, Deubert S, Kaiser W, Reithmaier J P and Forchel A IEEE Phot. Techn. Lett. 2007 **19** 916-918

[7] Bimberg D, Posilovic K, Kalosha V, Kettler T, Seidlitz D, Shchukin V A, Ledentsov N N, Gordeev N Y, Karachinsky L Y, Novikov I I, Maximov M V, Shernyakov Yu M, Chunareva A V, Bugge F and Weyers M 2010 Proc. of SPIE Vol. **7616** 76161I-1

[8] Botez D and Connolly J C 1983 Appl. Phys. Lett. **43** 1096

[9] Gordeev N Y, Maximov M V, Shernyakov Yu M, Novikov I I, Karachinsky L Ya, Shchukin V A, Kettler T, Posilovic K, Ledentsov N N, Bimberg D, Duboc R, Sharon A, Arbiv D B and Ben-Ami U 2008 Proc. of SPIE **6889** 68890W.

[10] Rumyantsev O and Gordeev N 2010 Physics and Astronomy (PhysicA.SPb) Conference of the Young Scientists of St Petersburg 27-28th October Saint Petersburg Russia

[11] Machavariani G, Davidson N, Ishaaya A A, Friesem A A and Hasman E 2002 Optics letters **27**(17) 1501-1503

[12] Divliansky I, Jain A, Drachenberg D, Podvyaznyy A, Smirnov V, Venus G and Glebov L 2010 Proc. of SPIE **7751** 77510Z-1

[13] Siegman A E 1986 Oxford University Press Oxford ch. 12 sec 12.2 p. 462

Optical trapping with Bessel beams generated from semiconductor lasers

G S Sokolovskii,[1] V V Dudelev,[1] S N Losev,[1] K K Soboleva,[1,2] A G Deryagin,[1] V I Kuchinskii,[1,3] W Sibbett,[4] E U Rafailov[5]

[1]Ioffe Physical-Technical Institute, St.Petersburg, Russia
[2]Saint-Petersburg State Polytechnical University, St.Petersburg, Russia
[3]Saint-Petersburg State Electrotechnical University "LETI", St.Petersburg, Russia
[4]School of Physics and Astronomy, University of St.Andrews, St.Andrews, UK
[5]Aston Institute of Photonic Technologies, Aston University, Birmingham, UK

gs@mail.ioffe.ru

Abstract. In this paper, we study generation of Bessel beams from semiconductor lasers with high beam propagation parameter M^2 and their utilization for optical trapping and manipulation of microscopic particles including living cells. The demonstrated optical tweezing with diode-generated Bessel beams paves the way to replace their vibronic-generated counterparts for a range of applications towards novel lab-on-a-chip configurations.

1. Introduction

Optical trapping and tweezing technique is based on the forces that arise as a consequence of the law of conservation of momentum in the absorption, reflection and refraction of the laser beam at the particle [1]. A prerequisite for successful optical manipulation, i.e. reliable trapping of microparticles, is the high gradient of the optical power density. In other words, in order to trap the microparticle, one must sharply focus the laser light. This significantly reduces the workspace due to diffraction: tighter focusing results in faster divergence of radiation, and the problem can not be resolved in terms of the Gaussian optics. However, as was shown in the seminal work by Durnin [2] (and earlier by Zeldovich et al. [3]) diffractive divergence can be almost completely eliminated with a special class of non-diffracting light fields called Bessel beams. This class of propagation-invariant beams is generated through the interference of the conically converging rays taking place when a collimated Gaussian beam transits a cone-shaped lens (axicon [4]). The result of such interference is a bright spot surrounded by concentric fringes. The radial intensity profile of this distribution can be described by a zero-order Bessel function of the first kind as illustrated in Fig.1. The central lobe diameter of the Bessel beam is determined by the axicon apex angle and can be of the order of the optical wavelength. Practically, Bessel beams exhibit a finite propagation distance, which depends on the cross-sectional diameter of the initial collimated beam. Since Bessel beams maintain their focus along the line, the position of manipulated object can be significantly varied, thus making micromanipulation systems flexible and more attractive for practical implementations. The utilization of Bessel beams also opens new horizons in microporation [5], manipulation of micro-machines [6] micro-fabrication [7,8] and various 'lab-on-a-chip' applications [9,10].

Until recently, it was believed that generation of Bessel beams is only possible with highly coherent light sources, such as gas and solid-state lasers, but last years a number of publications reported on formation of Bessel beams from semiconductor light sources [11-13], including various types of edge-emitting laser diodes [11,12], vertical-cavity surface-emitting lasers (VCSELs) [11] and vertical external-cavity surface-emitting lasers (VECSELs) [13], and even light-emitting diodes (LEDs) [11].

2. Generation of Bessel beams from semiconductor lasers with high beam propagation parameter

Typically, the 'quality' of the laser beam can be described by the beam propagation parameter M^2 [14, 15], which is defined as the ratio of the beam divergence to that of the 'ideal' Gaussian beam (i.e. the beam with $M^2 = 1$) corresponding to the diffraction limit. Parameter M^2 allows the description of the quasi-Gaussian beams with mathematics developed for the 'ideal' Gaussian beams with simple replacement $\lambda \rightarrow M^2\lambda$, i.e. with M^2-fold numerical increase of the wavelength.

$$NA = \frac{M^2\lambda}{\pi\omega_0} \tag{1}$$

where NA is the numerical aperture, λ is the wavelength, and ω_0 is the beam waist size. With high parameter M^2 of the initial quasi-Gaussian beam typical for the laser diodes, the diameter of the central lobe of the Bessel beam gradually increases with propagation distance due to the high divergence of the forming beam. This leads to limitation of the propagation distance z_B of the generated Bessel beam (see Fig. 1). Here, one should bear in mind that without divergence of the forming beam taken into account, the propagation distance z_{B0} of the Bessel beam is defined by the geometrical parameters of the optical scheme [16].

In this paper, we assume that the initial quasi-Gaussian beam is coaxial with the axicon and collimated in the plane of its apex. Therefore in the paraxial approximation, the diameter of the central lobe of the Bessel beam, which is normally given by $d(z)=2.4\lambda/(\pi(n-1)\cos(\alpha/2))$, in view of (1) transforms to [17]:

$$d(z) \approx \frac{2.4\lambda}{\pi(n-1)\cos\alpha/2}\left[1+\left(\frac{M^2\lambda z}{\pi\omega_0^2}\right)^2\right] \tag{2}$$

where α is the apex angle of the axicon, n is the refractive index, and z is the axial distance. Equation (2) greatly simplifies the calculation of the propagation distance of the Bessel beam limited by the divergence of the forming beam [18]. Let's define the propagation distance z_B of the Bessel beam as the distance at which the transverse size of the central lobe increases $\sqrt{2}$-fold (similar to the definition of the Rayleigh range for the Gaussian beam).

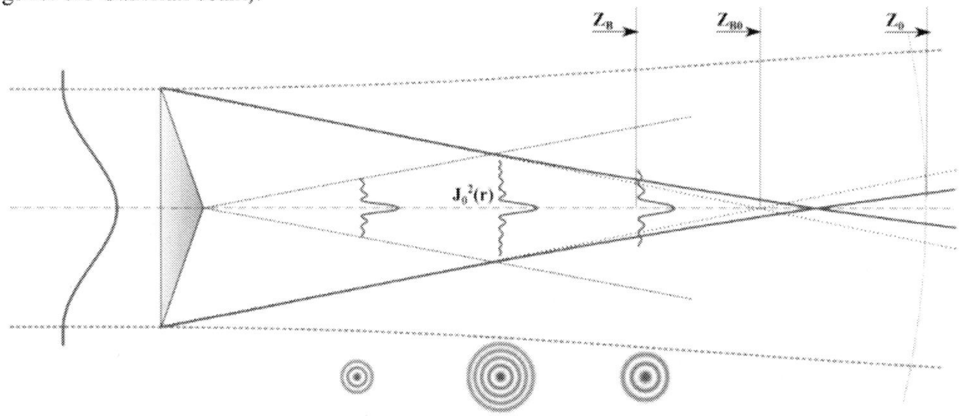

Figure. 1. Propagation of the Bessel beam formed from a quasi-Gaussian beam with a high parameter M^2: z_B is the Bessel beam propagation distance due to the beam divergence; z_{B0} is the geometrical propagation distance; z_0 is the Rayleigh range.

After simple transformations [16,17] equation (2) yields the propagation length of the Bessel beam generated from the quasi-Gaussian beam with high M^2:

$$z_B \approx \frac{2\omega_0^2}{M^2\lambda} \tag{3}$$

From (3) one can easily see that, despite the negative influence of the high beam propagation parameter, almost any required propagation length of the Bessel beam is achievable by increasing the radius of the initial quasi-Gaussian beam. Moreover, even with tight collimation of the quasi-Gaussian beam required for the power-demanding applications, the satisfying propagation length of the Bessel beam can be achieved as we confirm in the experiments below.

3. Optical trapping with Bessel beams generated from semiconductor lasers

In this paper, we demonstrate the use of the Bessel beams generated from the laser diodes for optical trapping and manipulation of microscopic particles including living cells which build on our recent results reported elsewhere [19]. A scheme of the experimental setup is shown in Fig. 2.

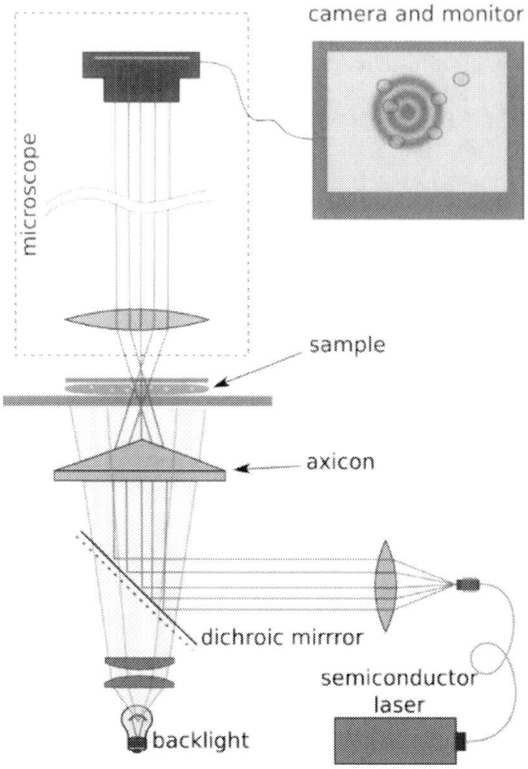

Figure 2. Scheme of the experimental setup for optical trapping and manipulation of microscopic particles with Bessel beams generated the laser diode.

For generation of Bessel beams in our experiments we used the fiber-coupled edge-emitting semiconductor laser with the output power up to 600 mW and the emission spectrum centered at 1065 nm. The output laser beam was collimated and refocused by the optical system including a set of interchangeable microscope lenses with magnification ranging from x8 to x60 and a dichroic mirror transmitting backlight (see Fig.2). The Bessel beam was generated with the axicon of 160^0 apex angle, which determined the transverse size of the central lobe of ~7 μm. The size of the Bessel beam features was measured with a graded transparent grid (so-called 'graticule') with scale divisions of 10 μm and the total length of 1 mm graded scale. The graticule was placed in the focal plane of the microscope lens in order to enable detection of the Bessel beam passing through the scaled region. This ensured precise control of magnification of the microscope detection system. The optical power in the central lobe of the Bessel beam with losses in the optical scheme taken into account was up to 20 mW.

The mechanical control for the optical trapping experiments was enabled by the high-precision micropositioners introduced into the optical scheme and a commercial microscope with a video-recording CCD camera. The micro-objects for the study were polystyrene beads (10 μm mean diameter) in a water solution as the high refractive index objects and water bubbles in oil as the low-index particles. The solution containing the micro-objects was placed on a glass slide covered with a cover glass and controlled with a micropositioner. Dichroic mirror enabled backlight illumination (see Fig. 2) .

We also studied the optical trapping of living red cells of rat blood. In the assessment, blood was dissolved in water with heparin added to prevent clotting. The mean size of the cells was approximately 5 μm (measured with graticule, similar to the measurement of the size of the features of the Bessel beam). The video in Fig. 3 demonstrates the two-dimensional optical trapping and manipulation of a living red cell of rat blood in the central lobe of the Bessel beam generated from the semiconductor laser. No damage of the living cells was detected in the available optical power range.

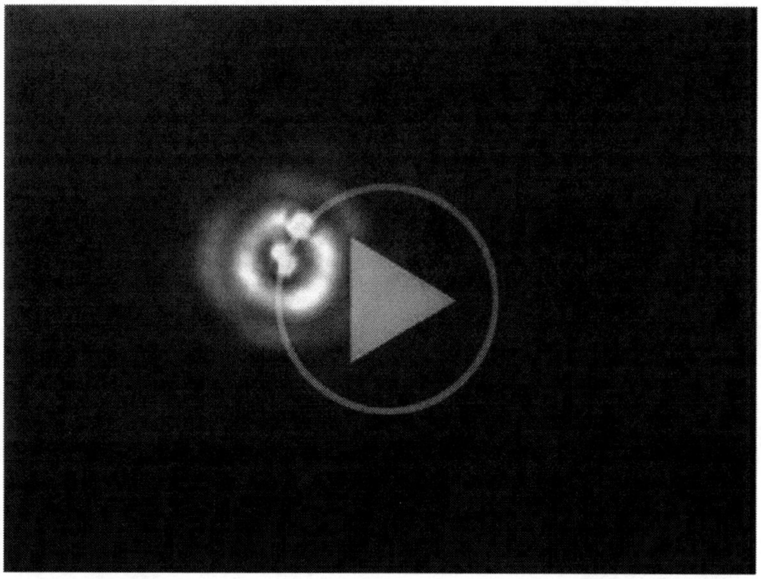

Figure 3. Optical trapping of a living red blood cell in the central lobe of the Bessel beam generated from the semiconductor laser.

Our assessments to date demonstrated very good potential for the Bessel beams generated from semiconductor lasers to replace their gas and solid-state laser-generated counterparts in applications relating to optical trapping and manipulation of microparticles including living biological micro-objects. The most attractive of these are so-called 'lab-on-a-chip' configurations. The important advantage in this sense is the recently demonstrated possibility of superfocusing of the laser diode beams with high parameter M^2 by generation of Bessel beams [18,20]. Combination of compactness and efficiency of semiconductor lasers with superfocusing and suppressed divergence of the beam opens up tremendous prospects for their 'lab-on-a-chip' applications.

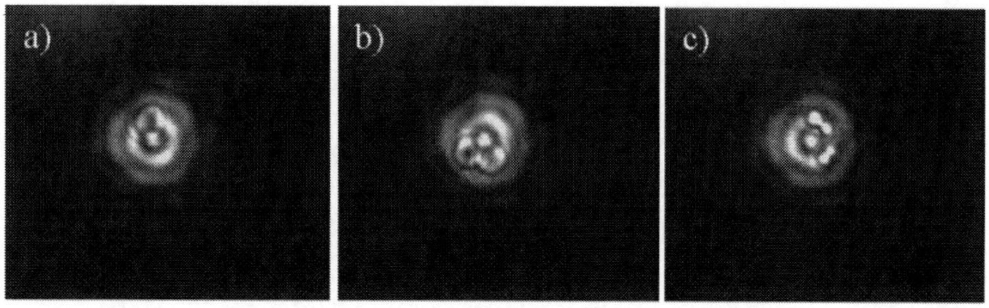

Figure 4. Trapping of multiple rat blood erythrocytes simultaneously using a Bessel beam: 2 captured particles (a), 3 captured particles (b), and 5 captured particles (c).

In our experiments, was also demonstrated the ability to capture and manipulate several microscopic objects with a single Bessel beam of sufficient optical power simultaneously. In Fig. 4 one can see one, two and four beads trapped in the first fringe of the Bessel beam in addition to the one trapped in its central lobe.

Conclusion

The properties of Bessel beams generated from laser diodes with high beam propagation parameter M^2 and their applications to optical tweezers are discussed. We have shown that the propagation length of such Bessel beams is defined by the divergence of the initial quasi-Gaussian beam rather than the parameters of the optical system. We have also demonstrated the two-dimensional optical trapping and manipulation of microparticles and small biological objects with diode-generated Bessel beams.

References

[1] Ashkin A 1970 Phys. Rev. Lett. vol 24 p 156.
[2] Durnin J 1987 J. Opt. Soc. Am. vol A4 p 651.
[3] Zel'dovich B Ya, Pilipetskii N F 1966 Izvestia VUZov Radiophysics vol 9 p 95.
[4] McLeod J H 1954 J. Opt. Soc. Am. vol 44 p 592.
[5] Tsampoula X, Garces-Chavez V, Comrie M, Stevenson D J, Agate B, Brown C T A, Gunn-Moore F, Dholakia K 2007 Appl. Phys. Lett. vol 91 053902.
[6] Larsen A E, Grier D G 1997 Nature vol 385, p 230.
[7] Friese M J, Rubinsztein-Dunlop H, Gold J, Hagberg P, Hanstorp D 2001 Appl. Phys. Lett. vol 78 p 547
[8] Amako J, Sawaki D, Fujii E 2003 J. Opt. Soc. Am. vol 20 p 2562
[9] Garcés-Chávez V, McGloin D, Melville H, Sibbett W, Dholakia K 2002 Nature vol 419, p 145.
[10] Rendall H A, Marchington R F, Praveen B B, Bergmann G, Arita Y, Heisterkamp A, Gunn-Moorec F J, Dholakia K 2012 Lab on a Chip vol 12, p 4816.
[11] Sokolovskii G S, Dudelev V V, Losev S N, Zolotovskaya S A, Deryagin A G, Kuchinskii V I, Rafailov E U, Sibbett W 2008 Tech. Phys. Lett. vol 34(12) p 1075.
[12] Sokolovskii G S, Dudelev V V, Losev S N, Deryagin A G, Vinokurov D A, Lyutetskiy A V, Pikhtin N A, Slipchenko S O, Tarasov I S, Zolotovskaya S A, Rafailov E U, Kuchinskii V I, Sibbett W 2010 Tech. Phys. Lett., vol 36(1) p 9.
[13] Sokolovskii G S, Butkus M, Losev S N, Dudelev V V, Deryagin A G, Kuchinskii V I, Sibbett W, Rafailov E U 2012 Proc. SPIE Int. Soc. Opt. Eng., vol 8242, 82420T.
[14] Siegman A E 1998 OSA Annual Meeting.
[15] ISO 11146
[16] Sokolovskii G S, Zolotovskaya S A, Losev S N, Dudelev V V, Deryagin A G, Kuchinskii V I, Sibbett W, Rafailov E U 2011 Proc. SPIE Int. Soc. Opt. Eng., 7919, 79190J
[17] Sokolovskii G S, Dudelev V V, Losev S N, Butkus M, Soboleva K K, Sobolev A I, Deryagin A G, Kuchinskii V I , Sibbett W, Rafailov E U 2013 Quant. El. vol 43 (5) p 423
[18] Sokolovskii G S, Dudelev V V, Losev S N, Deryagin A G, Kuchinskii V I, Sibbett W, Rafailov E U 2012 Tech. Phys. Lett. vol 38(5), p 402.
[19] Sokolovskii G S, Losev S N, Soboleva K K, Dudelev V V, Deryagin A G, Sibbett W, Kuchinskii V I, Rafailov E U 2014 Tech. Phys. Lett. vol 40(6), p 495.
[20] Sokolovskii G S, Melissinaki V, Dudelev V V, Losev S N, Soboleva K K, Kolykhalova E D, Deryagin A G, Kuchinskii V I, Viktorov E A, Farsari M, Sibbett W, Rafailov E U 2014 Proc. SPIE Int. Soc. Opt. Eng. vol 9134, 913457.

Journal of Physics: Conference Series **572** (2014) 012040 doi:10.1088/1742-6596/572/1/012040

Dynamical Diffraction of Light from 1D Photonic Crystals with Sinusoidal Profile of Permittivity

K O Romanenko[1,2] **and A V Sel'kin**[1,2]

[1] Solid State Physics Department, Saint Petersburg State University, Ulyanovskaya 1, Peterhof 198504, Russian Federation
[2] Spectroscopy of Solids Laboratory, Ioffe Physical Technical Institute, 26 Polytekhnicheskaya, St Petersburg 194021, Russian Federation

E-mail: st019276@student.spbu.ru, alexander.selkin@mail.ioffe.ru

Abstract. Bragg reflection and transmission spectra of the 1D photonic crystals characterized by a spatially sinusoidal profile of permittivity are studied as a function of the crystal-plate thickness. Applicability of the dynamical theory of diffraction in describing such spectra is considered. In the framework of the dynamical theory, we (i) calculated and analysed the reflection and transmission spectra for oblique incidence of polarized light, (ii) computed the spectra making use of the transfer matrix technique, and (iii) compared quantitatively the results of the two approaches. As a result, the analytical dynamical theory of diffraction is found to be correct in calculating the Bragg spectra in the vicinity of single photonic band-gap when the plate thickness is equal to the integer number of the spatial periods, or the permittivity is symmetric about the middle plane of the structure.

1. Introduction

Optical research of new artificial structures, studying their properties and applying them to modern technological devices is a trend in science nowadays. Photonic crystals belong to a large class of such structures [1, 2]. The photonic crystal (PhC) is a spatially periodic structure which permittivity is a spatially periodic function with the period equal-order to electromagnetic wavelength. In semiconductors, electronic properties are governed by the presence of allowed and forbidden energy bands for electrons. In the case of PhC one can control properties and propagation of electromagnetic waves. This unique feature of PhC can be utilized in various applications: photonics, lasers, optoelectronics, etc.

Propagation of light in PhC is very similar to propagation of X-rays in ordinary crystals for which the dynamical theory of diffraction is widely used to study optical properties [3]. Therefore it is of interest to apply the dynamical theory approach to PhCs taking into account high spatial modulation of the PhC dielectric function [4].

In this work, we discuss the model of the opal-like PhC characterized by the one-dimensional (1D) periodicity of permittivity $\varepsilon_s(z)$. As an example, we consider the permittivity of an opal-like PhC averaged along all the crystallographic directions except for [111]. In this case [5, 6]

$$\varepsilon_s(z)=\varepsilon_a f_s(z)+\varepsilon_b(1-f_s(z)), \tag{1}$$

where ε_a and ε_b are the permittivities of spheroidal particles that constitute the PhC and interparticle space, respectively, and $f_s(z)$ is the effective filling function (Figure 1).

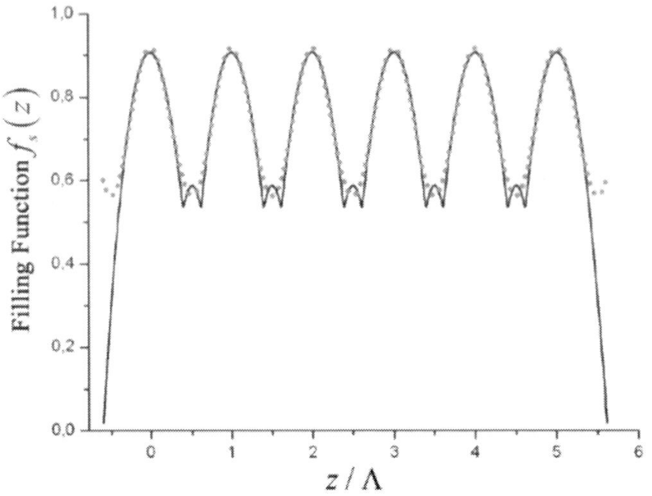

Figure 1. Effective filling function $f_s(z)$ of opal-like PhC along [111] direction. Solid line is given by exact calculation; dashed line corresponds to harmonic approximation.

It should be noted that the effective filling function can be approximated well by the harmonic one. It allows us to suppose that the dynamical theory can be applicable when describing Bragg reflection and transmission spectra of opal-like PhCs. The model proposed is of principal interest because it is closely associated with the previously performed theoretical and experimental studies [4-7] of the opal-like PhCs and allows one to answer the question why the dynamical theory of diffraction is applicable to the PhC with relatively high dielectric contrast.

The main purpose of our study is to elucidate the validity limits of the dynamical theory of diffraction, in particular, to find a minimum slab thickness that provides rather good numerical agreement between the analytical and full-electrodynamic computations of Bragg reflection and transmission spectra.

2. Dynamical theory of diffraction applied to photonic crystals

Let consider an infinite spatially periodic structure. The corresponding permittivity $\varepsilon(\vec{r})$, the electric, $\vec{E}(\vec{r})$, and magnetic, $\vec{H}(\vec{r})$, fields of an electromagnetic wave can be expanded into Fourier series over the reciprocal-lattice vectors $\vec{G}$. In the vicinity of a Bragg resonance determined by the Laue condition $\vec{k}^2 = (\vec{k}-\vec{G})^2$ for the wave vector $\vec{k}$ and a specified vector $\vec{G}$, such expansions take the form

$$\varepsilon(\vec{r}) = \varepsilon_0 + \varepsilon_{\vec{G}}\, e^{i\vec{G}\vec{r}} + \varepsilon_{\vec{G}}^{*}\, e^{-i\vec{G}\vec{r}}, \tag{2}$$

$$\vec{E}(\vec{r}) = \vec{A}_0\, e^{i\vec{k}\vec{r}} + \vec{A}_{\vec{G}}\, e^{i(\vec{k}-\vec{G})\vec{r}}, \tag{3}$$

$$\vec{H}(\vec{r}) = k_0^{-1}\left[\vec{k}\times\vec{A}_0\, e^{i\vec{k}\vec{r}} + (\vec{k}-\vec{G})\times\vec{A}_{\vec{G}}\, e^{i(\vec{k}-\vec{G})\vec{r}}\right], \tag{4}$$

where $k_0 = \omega/c = 2\pi/\lambda$ is the wavenumber of light in vacuum with the circular frequency ω (λ is the wavelength), ε_0 is the average dielectric constant. The amplitudes $\vec{A}_0$ and $\vec{A}_{\vec{G}}$ satisfy a set of equations

$$\begin{cases} \left(\vec{k}^2 - k_0^2\varepsilon_0\right)\vec{A}_0 - \vec{k}\left(\vec{k}\cdot\vec{A}_0\right) = k_0^2\varepsilon_{\vec{G}}\vec{A}_{\vec{G}} \\ \left((\vec{k}-\vec{G})^2 - k_0^2\varepsilon_0\right)\vec{A}_{\vec{G}} - (\vec{k}-\vec{G})\left((\vec{k}-\vec{G})\cdot\vec{A}_{\vec{G}}\right) = k_0^2\varepsilon_{\vec{G}}^*\vec{A}_0 \end{cases}, \tag{5}$$

and are related to an external fields through Maxwell's boundary conditions. On the other hand, the equality to zero of the determinant of equations (5) gives us the dispersion relations $\omega = \omega(\vec{k})$ for eigenmodes. So, the problem of finding the reflection and transmission coefficients for a PhC plate becomes, in principle, quite clear.

3. Numerical calculation and analytical approach

Now examine interaction between a monochromatic plane wave and 1D periodic structure of a thickness L, the permittivity being harmonically varied with the spatial period Λ

$$\varepsilon(z) = \varepsilon_0 + \varepsilon_G e^{iGz} + \varepsilon_G^* e^{-iGz}, \tag{6}$$

where $G = 2\pi/\Lambda$ and $\varepsilon_G^* = \varepsilon_{-G}$. Let the thickness of the slab be

$$L = l_1 + N\Lambda + l_2, \tag{7}$$

with l_1 and l_2 being fractions of Λ at the front and the back of the structure, respectively, and N be the integer number of spatial periods (Figure 2).

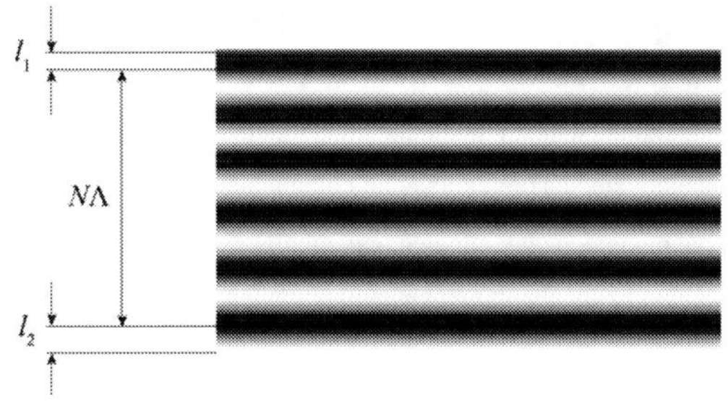

Figure 2. The structure under study: a sinusoidal profile of permittivity includes N periods, Λ, where N is integer and $l_1, l_2 \le \Lambda$.

We calculated reflectance and transmittance of such structure with different parameters l_1, l_2, N solving equations obtained from Maxwell's boundary conditions. Two approaches were considered based (i) on the dynamical theory of diffraction and (ii) on the numerical transfer matrix technique, in order to compare them to each other and draw conclusion about validity of the analytical model when describing the reflectance and transmittance spectra.

As a result, it was found that the dynamical theory of diffraction describes correctly the spectra in the vicinity of a single photonic band-gap as compared with the computations based on the transfer matrix technique, if

$$l_1 + l_2 = \Lambda \text{ or } l_1 = l_2. \tag{8}$$

In all other cases, conspicuous contradiction between analytical and numerical approaches takes place. It was shown analytically that

$$R + T = 1 + F(l_1, l_2), \tag{9}$$

where R is reflectance, T is transmittance and $F(l_1,l_2)$ □ $[\cos(Gl_1) - \cos(Gl_2)]$.

Figures 3 and 4 show, as an example, some results of computations of the reflectance and transmittance, respectively, when normal incidence of light on the plane surface of the structure is considered. The parameter values are taken close to that for an opal-like polystyrene PhC

($\varepsilon_0 = 2.127$, $|\varepsilon_G| = 0.135$) [6]. It can be noticed that at the conditions (8) both approaches give practically the same results (Figure 3) independent on N, among them the value $N = 0$. If permittivity is symmetric with respect to the middle plane of the structure ($l_1 = l_2$), this conclusion is valid at any thickness L of the slab including the limiting case $L \to 0$.

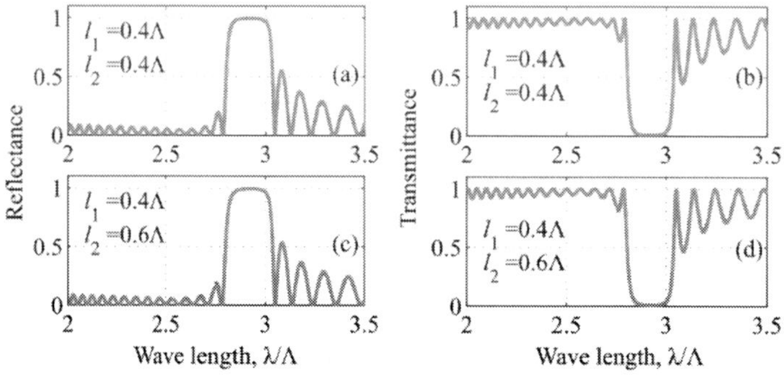

Figure 3. Reflectance and transmittance spectra at normal incidence of light on the PhC-plate (the number of the spatial periods, Λ, is *N*=30): (a,b) $l_1 = l_2$, (c,d) $l_1 + l_2 = \Lambda$. Red solid curves correspond to the dynamical model; blue dotted curves are computed with the transfer matrix technique.

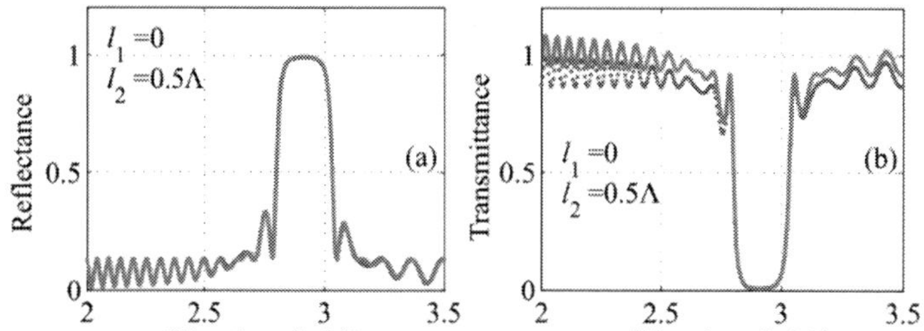

Figure 4. Reflectance (a) and transmittance (b) spectra at normal incidence of light on the PhC-plate (the number of the spatial periods, Λ, is *N*=30): here the conditions (8) are not satisfied. Red solid curves correspond to the dynamical model; blue dotted curves are computed with the transfer matrix technique.

When the conditions (8) are not fulfilled, the analytical approach doesn't agree well with the numerical one (figure 4), which is most pronounced in the case of transmission spectra. Moreover, the transmittance exceeds unity on some frequencies, which is in contradiction with the energy balance principle.

4. Conclusion

The reflectance and transmittance spectra have been calculated within two approaches based on the dynamical theory of diffraction as applied to 1D photonic crystals and on the numerical modeling using the transfer matrix technique. The dynamical theory is shown to be correct if a photonic crystal plate is symmetric in its dielectric properties about plate boundaries or the thickness of the plate is a multiple of the spatial period of the structure. The conditions obtained are consistent with the energy balance and time-reversal symmetry considerations.

References

[1] Joannopoulos J D, Johnson S G, Winn J N and Meade R D 2008 *Photonic Crystals. Molding the Flow of Light* (Princeton University Press, Princeton, Oxford)
[2] Sibilia C, Benson T M, Marciniak M and Szopik T 2008 *Photonic Crystals: Physics and Technology* (Springer-Verlag, Italia)
[3] Cowly J M *Diffraction Physics* 1995 (Elsevier Science B. V., Amsterdam)
[4] Sel'kin A V 2004 *Proc. of 12th Int. Symp. "Nanostructures: Physics and Technology"* (St. Petersburg) p 111
[5] Gajiev G M, Golubev V G, Kurdyukov D A, Medvedev A V, Pevtsov A B, Sel'kin A V, Travnikov V V 2005 *Phys. Rev.* B **72** 205115
[6] Bazhenova A G, Sel'kin A V, Men'shikova A Yu and Shevchenko N N 2007 *Solid State Physics* **49** 2010
[7] Fedotov V G, Sel'kin A V, Ukleev T A, Men'shikova A Yu and Shevchenko N N 2011*Phys. Status Solidi* B **248** 2175

Journal of Physics: Conference Series **572** (2014) 012041 doi:10.1088/1742-6596/572/1/012041

Dielectric properties and structure particularity of silver stearate

M A Goryaev, R A Castro, A P Smirnov[1]
Herzen Russian State Pedagogical University, St Petersburg, Russia

E-mail: veter8808@mail.ru

Abstract. Dielectric properties, parameters and structure particularity of silver stearate are studied. The silver stearate sample in monolithic formation was researched by method of dielectric spectroscopy. Measurements were realized by spectrometer Novocontrol Technologies "Concept 41" in wide range of frequencies and temperatures. Dielectric parameters of silver stearate were measured and calculated. γ-relaxation process in temperature band 80 – 130^0 C and first phase transition at 126^0 C. are found.

1.Introduction

The photothermographic materials wide used in operate registration of image. Those materials are consist of silver halides and silver salts of fatty acid, in particularly silver stearate. During the manufacture process of photosensitive composition silver halides are synthesized on surface of silver stearate particle. By different dyes-sensitizators the heat-developable composition may be sensitized for visible-light and for nearest infra-red spectrum [1, 2]. The silver stearate takes parts as in the process of sensitization according the lightguide mechanism, and in the heat-developing process of latent images at temperature range 120 – 140^0 C [3, 4, 5]. Therefore investigations of the silver stearate structure peculiarities, optical and electric properties are actual problems.
The structure of the silver stearate molecule is showed in figure 1.

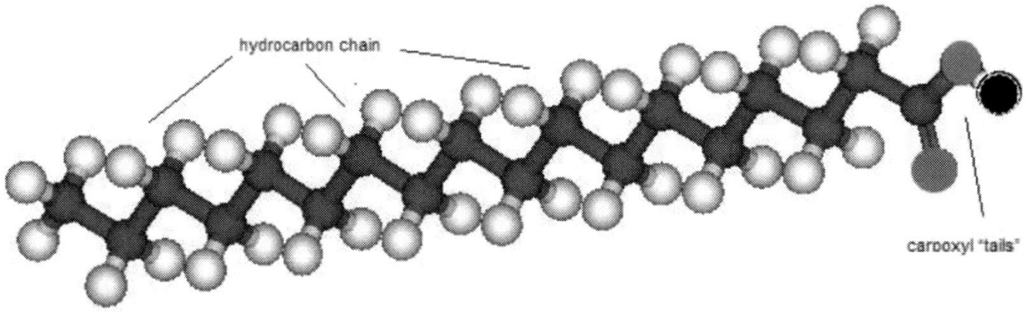

Figure 1. The molecule of silver stearate

[1] To whom any correspondence should be addressed.

2. Experiment

The dielectric measurements of silver stearate sample were realized with help dielectric spectroscope Novocontrol Technologies "Concept 41". Dielectric spectroscopy method allows to measure and to calculate dielectric and other parameters of complex materials [6]. More importantly, dielectric spectroscopy method allows studying structure particularities in such materials. The silver stearate ($CH_3-(CH_2)_{16}-COOAg$) are synthesized by exchange reaction of replacement sodium at superfluous concentration nitrate silver in microparticle formation with size around one micrometer, so it was necessary to prepare monolithic sample. The monolithic sample of silver stearate was prepared by pressing method. Then received sample was inserted in sample cell between gold-plate electrodes. Sample diameter – 10 mm, sample thickness – 0.5 mm. Work voltage – 1 V. Dielectric permittivity, loss factor, tanδ and other dielectric parameters were measured in frequency range $10^{-2}-10^{2}$ Hz in isothermic regulation. Temperature range-0^0C-130^0C. Temperature step – 10^0C in temperature range 0^0C-120^0C, and temperature points 125^0C and 130^0C in region of supposed phase transition. Temperature stability of the instrument was controlled within $\pm 0.2^0C$.

The measurement results were approximated by Havriliak - Negami function [7, 8], using software Novocontrol WinFit:

$$\varepsilon^*(\omega) = \varepsilon_{00} + \frac{\Delta\varepsilon}{\left[1+(i\omega\tau)^{\alpha_{PT}}\right]^{\beta_{PT}}} \tag{1}$$

where ε_{00} – nonrelaxing dielectric permittivity, $\Delta\varepsilon$ – dielectric increment, τ – relaxation time, ω – circular frequency, α_{HN} and β_{HN} – HN parameters, describe symmetrical and asymmetrical expansion of relaxation function corresponding [9].

3. Results and discussion

The measurements of dielectric spectrums in wide range frequency and temperature allowed studying γ-relaxation dielectric process in the silver stearate [10]. The values of different parameters such as, relaxation strength, relaxation time, nonrelaxing dielectric permittivity, parameters of HN-function are showed in table 1.

Table 1. Different dielectric parameters of silver stearate

$T, ^0C$	τ_{max}, s	ΔE	τ, s	ε_∞	α	β
80	0.57	0.15	0.80	0.56	0.94	0.68
90	0.27	0.02	0.34	0.24	0.92	0.77
100	0.16	0.03	0.19	0.28	0.91	0.83
110	0.11	0.02	0.13	0.22	0.91	0.85
120	0.08	0.06	0.10	0.37	0.92	0.86
125	0.09	0.78	0.10	1.02	0.95	0.82
130	0.09	0.84	0.11	1.13	0.98	0.80

In many occurrences dielectric modules are used for analyze. In figure 2 frequency dependence dielectric module losses are showed for different temperature. In set figure visible displacement maximums dielectric module losses in size of increase frequency with increase temperature is observed. This displacement allows studying γ-relaxation dielectric process silver stearate in considered temperature range [10].

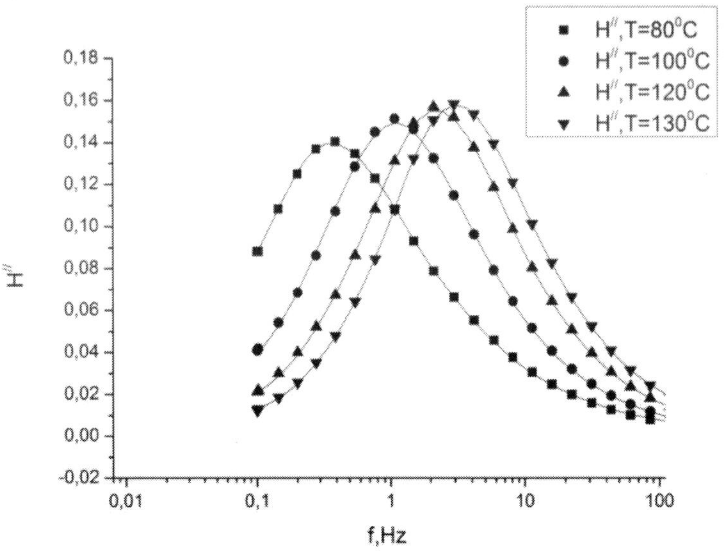

Figure 2. Dependence of dielectric module losses from frequency for different temperature

According to the available literature the dielectric relaxation process stipulated, apparently, by progressive absorting in carboxyl "tails" [11].

Also dependence $-\ln(\tau)$ from reverse temperature was plotted (figure 3), where $\tau = \tau_0 \exp\left(-\frac{E_a}{kT}\right)$.

This graph shows existence γ-relaxation too. Also activation energy was calculated. For silver stearate the activation energy is $(0.46 \pm 0.01)eV$.

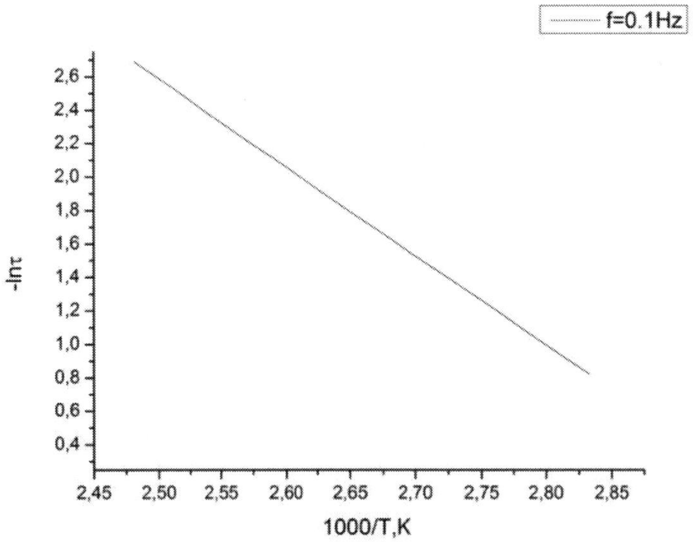

Figure 3. Dependence $-\ln(\tau)$ from reverse temperature for $f = 0.1Hz$

The measurements of dielectric spectrums and approximation experimental results allowed detecting phase transition in silver steatrate. In figure 4 the dependence of relaxation strength from reverse temperature is showed. On set figure visible surge leap of relaxation strength at $126^0 C$ is observed.

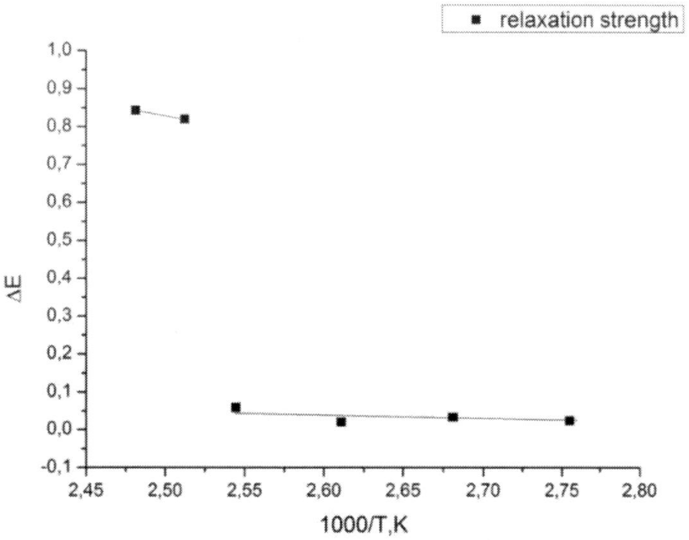

Figure 4. Reverse temperature dependence of relaxation strength

According previous literature [11, 12] this surge leap shows phase transition, defined "crystal-curd" in [11], or first phase transition in [12]. Also according previous articles this phase transition is stipulated by melting of hydrocarbon chains. However in [12] authors said, that destruction of ion silver layers was not arise at this phase transition. This fact confirms, that usage of silver stearate in photothermographic materials is optimum solution.

4. Conclusion

The dielectric parameters of silver stearate, fabricated by pressing method, were measured by dielectric spectroscope method. Dielectric measurements were realized by dielectric spectroscope Novocontrol Technologies "Concept 41" in wide frequency and temperature range. Due to results of measurements and calculations γ – relaxation process in range $80^0 C - 130^0 C$ at low frequency range was detected. Also activation energy of silver stearate was calculated. The different dielectric parameters of silver stearate, such as relaxation strength, relaxation time, nonrelaxing dielectric permittivity, HN parameters, were obtained. The first phase transition was detected at $126^0 C$, that accords with available reference very well. The silver stearate has sufficient complex structure and pronounced dielectric properties. There are shown the silver ion structure is not distract at temperature of phase transition, so usage of silver stearate in photothermographic compositions is reasonable.

5. Reference

[1] Goryaev M A 1993 *Sci Appl Photo,* **33 №5** 759.
[2] Goryaev M A, Shapiro B I 1997 *Sci Appl Photo,* **39 №2** 201.
[3] Morgan D A 1993 *J. Phot. Sci.* **41 №1** 108.
[4] Goryaev M A 1994 *Rus. J.Appl. Chem* **67 №6** 868
[5] Goryaev M A, Smirnov A P 2012 *J. Izv. Herzen* **№144** 29.

[6] Castro R A, Bordovsky G A, Bordovsky V A, Anisimova N I 2006 *J. Non-Crystalline Solids*. **32 № 9-20** 1560.
[7] Havriliak S, Negami S 1966 *J. Polym. Sci. Part C* **№ 14** 99.
[8] Havriliak S, Negami S 1967 *J. Polym.* **№8** 161.
[9] Kremer K, Shonhals A (Eds). 2003 Broadband dielectric spectroscopy, Springer, Berlin Heidelberg.
[10] Karabanova L V, Sergeeva L M, Sviatyna A V, Seytre G, Boiteux G, Stevenson I 2007 *Polymer J.* **29 № 4** 286.
[11] Sharafutdinov M R, Bohonov B B 2007 Symposium "Phase transitions in solid solution and fusion". Abstracts **2** 223.
[12] Haifutdinova D N, Sergeeva E A, Kovalenko V I 2002 Materials of reports VIII All-Russian conference "Structure and dynamic of molecular system" Chemistry and computer simulation. Butler messages **№6**.

Journal of Physics: Conference Series **572** (2014) 012042 doi:10.1088/1742-6596/572/1/012042

Variants of the light pulses controlling under conditions of double radio-optical resonance

A S Timofeev

Department of Theoretical Physics and Astronomy, Herzen University, Moika emb., 48, St. Petersburg, 191186 Russia

E-mail: Alextimofeev272@gmail.com

Abstract. On the bases of semi-classical theory of the laser light interaction with a cold atomic cloud, specifically, of the electromagnetically induced transparency (EIT), repeated reproduction and modulation of the light pulses are investigated. A scheme of the light pulses control by the use of radio-frequency (RF) pulses is shown to be very effective. Short RF pulses of appropriate areas stimulate repeated playbacks of the output light pulses similar to the input one.

1. Introduction

The purpose of this work is to develop an idea of the combined use of optical – radio-frequency double resonance and EIT to control the characteristics of optical pulses. We analyze the repeated reproduction and modulation of the light pulses. Storage, reproduction and control of light pulses are contemporary challenges of quantum optical informatics. One of the possible methods to control the light pulses is controllable change of properties of a medium (cold alkali atoms) under condition of EIT [1, 2]. Recall the main physical conditions related to EIT. Two laser pulses (control and signal) are injected to a gas cloud of alkali metal atoms. The atomic transitions resonant to these pulses form a Λ – scheme. The most pronounced effect under such conditions is extremely strong reduction of the group velocity of the signal pulses [3], with their corresponding localization inside the medium. In most of the models using the EIT, the dark pauses are proposed to control the light pulses. This possibility was demonstrated experimentally [4, 5]. In this article, the results of detailed calculations are summarized concerning the light pulses controlling by the use of RF pulses, as it was firstly proposed in [6]. We use the semiclassical approach, i.e., the self – consistent system of Maxwell – Bloch equations. Some of our results were reported in [7, 8, 9].

2. The model and basic equations

In the used model, the system is an atomic cloud that is cooled to low temperatures and placed in a constant magnetic field. Atoms are distributed in the cloud homogeneously. We consider transitions between levels of the hyperfine structure of the terms $3^2S_{1/2}$ and $3^2P_{3/2}$ of the D_2 line of sodium. The scheme of levels and transitions is presented in figure 1.

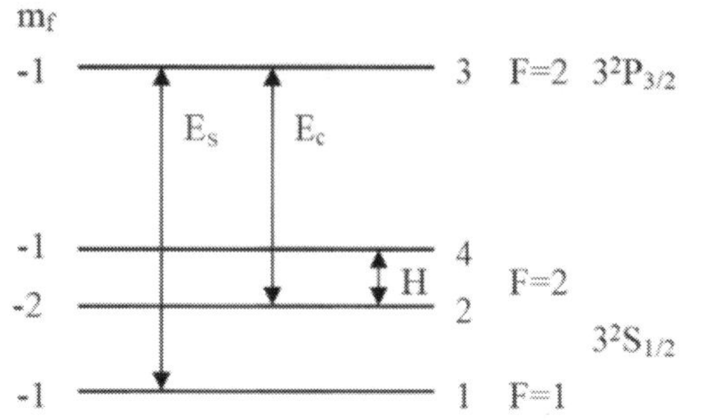

Figure 1. Scheme of levels and transitions. E_s – signal pulse, E_c – control pulse, H – RF pulse.

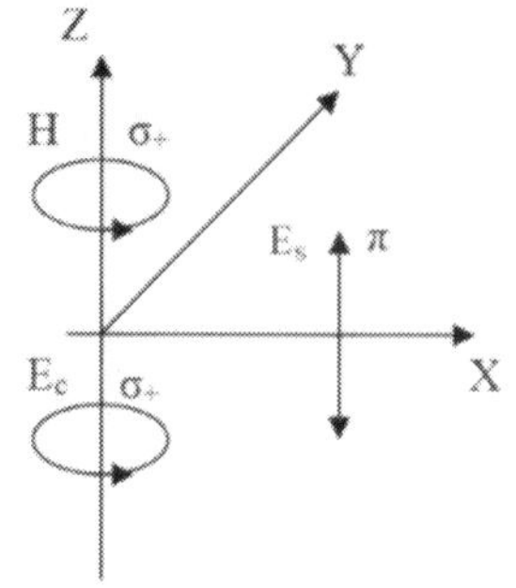

Figure 2. Scheme of he pulses propagation directions and their polarization.

Two laser pulses (signal and control) are injected into the atomic cloud. The quasi – monochromatic controlling pulse is at the sharp resonance to the 2 – 3 transition. The probe pulse is much shorter, and its carrier frequency is tuned to the transition 1 – 3. The linearly polarized signal pulse propagates along the OX axis; the controlling and radio-frequency fields are both circularly polarized and propagate along the OZ axis (figure 2).

The fields of the controlling light wave as well as those of the RF pulses are considered to be given, and their fronts, parallel to the XY plane, are assumed to be sufficiently broad to cover the atomic cloud (or at least its main area along the OX axis). The latter (geometrical) conditions are usual in EIT experiments, including one of the first and most convincing of them [3].

The model is described by the following system of Maxwell – Bloch equations, (1) – (10) for the one-atomic density matrix elements and (11) – (15) for the electromagnetic field inside the medium:

$$\dot{\rho}_{11} = \frac{1}{3}\rho_{33} + i[V_p^*\rho_{31} - c.c], \tag{1}$$

$$\dot{\rho}_{22} = \frac{1}{3}\rho_{33} + i[V_c^*\rho_{32} - c.c] + i[H^*\rho_{42} - c.c], \tag{2}$$

$$\dot{\rho}_{33} = -\rho_{33} - i[V_p^*\rho_{31} - c.c] - i[V_c^*\rho_{32} - c.c], \tag{3}$$

$$\dot{\rho}_{44} = \frac{1}{3}\rho_{33} - i[H^*\rho_{42} - c.c], \tag{4}$$

$$\dot{\rho}_{31} = -\frac{1}{2}\rho_{31} - iV_p[\rho_{33} - \rho_{11}] + iV_c\rho_{21}, \tag{5}$$

$$\dot{\rho}_{32} = -\frac{1}{2}\rho_{32} - iV_c[\rho_{33} - \rho_{22}] + iV_p\rho_{12} - iH\rho_{34}, \tag{6}$$

$$\dot{\rho}_{21} = -iV_p\rho_{23} + iV_c^*\rho_{31} + iH^*\rho_{41}, \tag{7}$$

$$\dot{\rho}_{41} = -iV_p\rho_{43} + iH\rho_{21}, \tag{8}$$

$$\dot{\rho}_{42} = -iH[\rho_{44} - \rho_{22}] - iV_c\rho_{43}, \tag{9}$$

$$\dot{\rho}_{43} = -\frac{1}{2}\rho_{43} - iV_p^*\rho_{41} - iV_c^*\rho_{42} + iH\rho_{23}, \tag{10}$$

$$V_p(x,t)=\frac{d_{32}E_p(t)}{\hbar\gamma_3}+iB_p\int_0^x\rho_{31}(x',t)dx',\ B_p=\frac{4\pi n_0 d_{31}^2}{\hbar\gamma_3}; \tag{11}$$

$$E_p(t)=E_{p0}\exp(-\frac{t^2}{2(Tp)^2}), \tag{12}$$

$$V_c(x,t)=\frac{d_{31}E_{c0}}{\hbar\gamma_3}, \tag{13}$$

$$H(t)=\frac{\mu_{24}H'_{1,2}(t)}{\hbar\gamma_3}, \tag{14}$$

$$H'_{1,2}(t)=H_{01,2}\exp(-\frac{t^2}{2(Tp')^2}). \tag{15}$$

Here, d_{31},d_{32} are the dipole moments of transitions $3\leftrightarrow 1$ and $3\leftrightarrow 2$, respectively; μ_{24} is the magnetic dipole moment of transition $2\leftrightarrow 4$; and n is the concentration of atoms. The unit of time was chosen as γ_3^{-1}, where γ_3 is the inverse lifetime of the state 3. For sodium, $\gamma_3=6\cdot 10^7 c^{-1}$ [12]. The relaxation of the lowest state was neglected, since the lifetimes of states $|2\rangle$ and $|4\rangle$ are rather long [13]. Since the level 3 population stays very low under EIT conditions and under the action of the radio-frequency field, the relaxation coefficients for spontaneous transitions are conventionally assumed to be equal, and other open relaxation channels can be not taken into account.

In concrete calculations we have set

$$\frac{d_{31}E_{p0}}{\hbar\gamma_3}=0.1;\ \frac{d_{32}E_{c0}}{\hbar\gamma_3}=0.1;\ B_p=0.02. \tag{16}$$

For two RF pulses, the following their areas and amplitudes were taken as most suitable:

$$\int_0^t H_1'(t')dt'\approx\frac{\pi}{4};\ \int_0^t H_2'(t')dt'\approx\frac{3\pi}{2}, \tag{17}$$

$$\frac{\mu_{24}H_{01}}{\hbar\gamma_3}=0{,}045;\ \frac{\mu_{24}H_{02}}{\hbar\gamma_3}=0{,}27;\ t_1\approx t_2=40\gamma_3^{-1}. \tag{18}$$

3. The field evolution

The joint action of two pulses provides the EIT effect, so that the medium becomes almost transparent for both of them. When signal pulse has already entered the medium, but has not yet begun to leave it, an RF pulse is turned on. The second and some subsequent short RF pulses of the appropriate areas stimulate repeated playbacks of the output light pulses similar to the input one. As concerned to the RF pulses operation, we initially consider the simplified and somewhat artificial situation similar to the well-known Rabi problem for a two–level system [10, 11]: we do not take into account that the radio-frequency field redistributes the population between all Zeeman sublevels.

Consider the evolution of the signal pulse (figure 3). Let the controlling and signal pulses enter the atomic cloud and EIT conditions are realized. At the moment $200\gamma_3^{-1}$ the whole signal pulse has already entered the medium, but did not yet begin to leave it. In this moment, by a short RF pulses we transfer half of the population of the meta-stable level 2 to another hyperfine level 4, also meta-stable one (figure 4). After some delay time, which is long enough to ensure that the signal pulse has already left the medium, we send again the RF pulse to bring back the atomic population from level 4 to level

2. The second RF pulse is chosen so that population transfer would be complete. A new two light pulses are produced inside the sample. The second pulse was absorbed by the medium. That second, anti-phase pulse is weak, and the shorter the second RF pulse, the weaker the second light pulse at the output. The third light pulse is identical to the first (figure 3). It is necessary to preserve the phase; so the second RF pulse effective area is to be $\frac{3\pi}{2}$ instead of $\frac{\pi}{2}$. The information about the form of the input pulse is stored in the coherency of the collective atomic state.

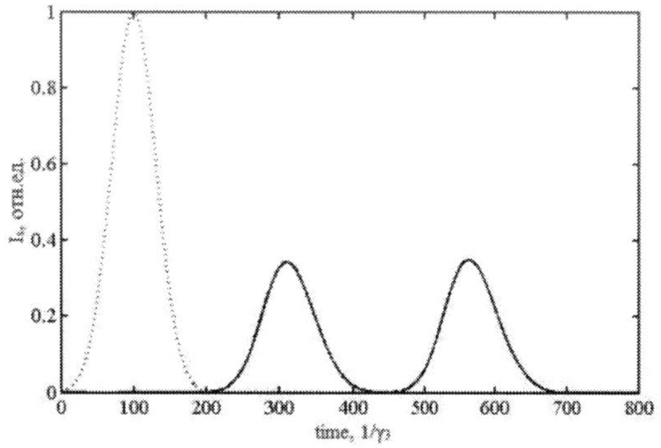

Figure 3. Intensity of signal pulse at the entrance into a medium (dashed-and-dotted curve) and at the exit of the medium (solid curve).

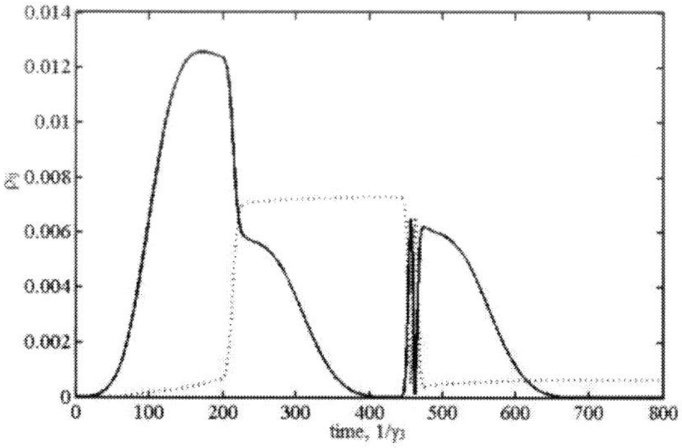

Figure 4. Diagonal elements of the density matrix averaged over the atoms of a medium: ρ_{22} (solid curve) and ρ_{44} (dashed curve)

3. Variants of control and a more complicated structure of levels

The amplitude and phase of signal pulses can be controlled by changing the areas of RF pulses, i.e. changing the Bloch vector rotation angle in the subspace spanned upon states $|2\rangle$ and $|4\rangle$. In the model considered, repeated reproduction of the light pulses may be achieved. For example, N RF pulses stimulate N playbacks of the output light pulses similar to the input one (figure 5).

Another variant of the light pulses control is the use of the long RF pulse leading to modulation of the output light pulse (figure 6); this can be considered as a transfer of Rabi oscillations from the RF to the optical frequency range. This effect was described in [9].

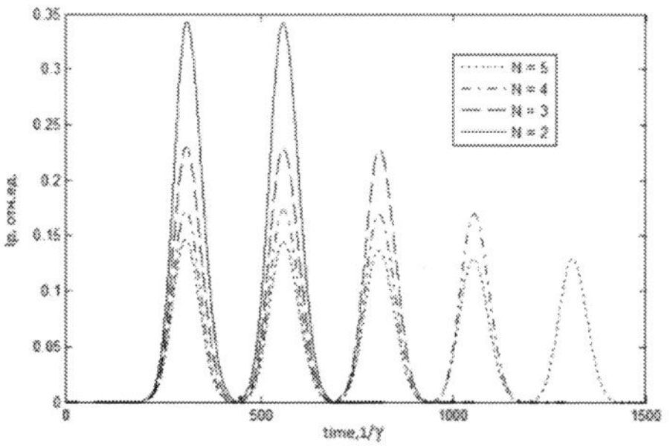

Figure 5. Intensity of signal pulses at the exit of the medium after the action of N RF pulses.

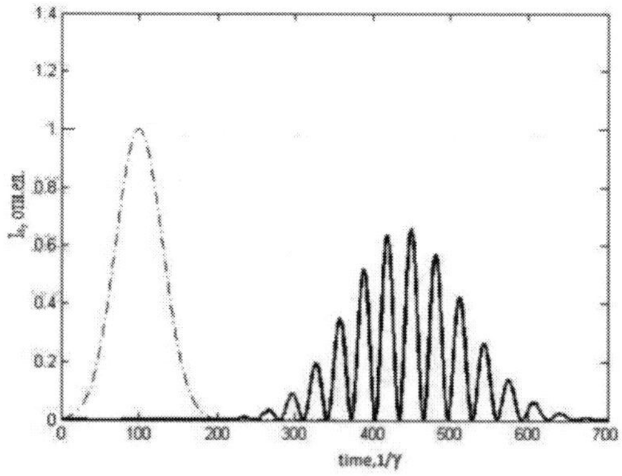

Figure 6. Intensity of signal pulse (dashed-and-dotted curve) at the entrance into a medium and (solid curve) at the exit of the medium.

As it was already noted above, it is necessary to take into account that the RF field acts between all Zeeman sublevels. Without any intension of an exhaustive analysis of real particular schemes, we will take into account full Zeeman splitting for the sub-levels F = 2 and F = 1 of the sodium term $3^2S_{1/2}$ hyperfine structure (figure 7).

In this case, the system of equations for the atomic density matrix elements becomes too cumbersome to be completely presented here. The equations for the field are similar to eqns. (11) – (15). The evolution of the pulse assumes a more complicated character. The population of level 2 is now redistributed among all the Zeeman sublevels under the RF field influence. When the second RF pulse is turned on, not only the direct the states population transfer 4 → 2 occurs, but to another Zeeman sub-levels as well. Thus it becomes impossible to restore completely the population of the level 2, which plays a decisive role in the EIT. Therefore, the output signal pulse intensity is reduced.

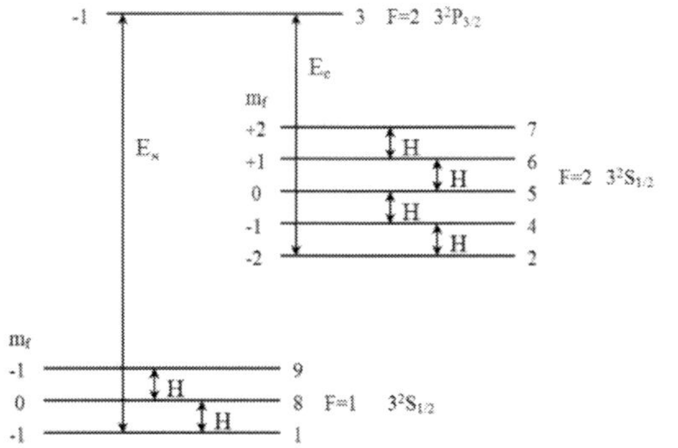

Figure 7. Scheme of levels and transitions. E_s – signal pulse, E_c – control pulse, H – RF pulse.

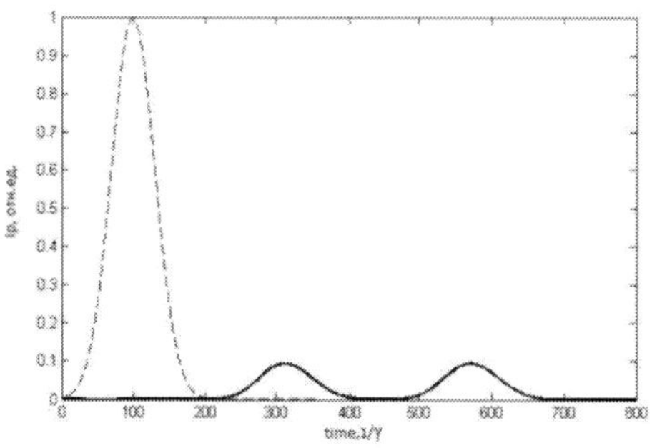

Figure 8. Intensity of signal pulse (dashed-and-dotted curve) at the entrance into a medium and (solid curve) at the exit of the medium.

The population redistribution of levels 1, 8, 9 practically does not occur because their initial populations are equal and the durations of RF pulses are small. Due to the influence of the lowest triplet is a slight increase in the duration of the second output signal pulse. The shorter are RF pulses, the less significant is the role of the low triplet.

When choosing two RF pulses such that

$$\int_0^t H_1'(t')dt' \approx 0.4\pi \, , \, \int_0^t H_2'(t')dt' \approx 1.3\pi \, , \tag{19}$$

we obtain two output pulses, but with a significant decrease in the intensity (figure 8).

Acknowledgments
The author is grateful to N.A. Vasil'ev and I.V. Sokolov for discussion of results, to A.S. Troshin for useful advices.

References

[1] E. Arimondo 1996 *Progress in Optics*, ed. by E. Wolf (Elsevier, Amsterdam)
[2] Scully M.O. and Zubairy M.S. 1997 *Quantum Optics* (Cambridge Univ. Press, Cambridge)
[3] Hau L.V., Harris S.E., Dutton Z., Behroozi C.H. 1999 Lett. Nature **397** (3) 594
[4] Phillips D.F., Fleischhauer A., Mair A., Walsworth R.L., Lukin M.D. 2001 *Phys. Rev. Lett.* **86** (5) 783

[5] Liu C., Dutton Z., Behroozi C.H., Hau L.V. 2001 Nature **409** (8) 490
[6] Shakhmuratov R. N., Kalachev A. A., Odeurs J. 2007 *Phys. Rev.* A **76** 031802.
[7] Timofeev A.S., Troshin A.S., in Materials of the 8th International Conference "Basic Problems of Optics 2012" (St. Petersburg, 2012), p. 12.
[8] Timofeev A.S., Troshin A.S. 2012 *Fiz. Vestnik RGPU,* **6**, 13
[9] TimofeevA.S., Troshin A.S. *Optics and Spectroscopy* 2013 **115** (3) 368
[10] Allen L., Eberly J. 1975 *Optical Resonance and Two – Level Atoms* (Wiley, New York).
[11] L.D. Landay, E.M. Lifshits 1977 *Quantum Mecanics (Non-relativistic Theory)* (Pergamon Press, Oxford)
[12] Steck D.A. Sodium D Line Data. //http://steck.us/alkalidata/Revision 2.1.4. 2010.
[13] Matsko B., Novikova I., Scully M. O., Welch G. R. 2001 *Phys. Rev. Lett.* **87** 133601.

Journal of Physics: Conference Series **572** (2014) 012043 doi:10.1088/1742-6596/572/1/012043

Femtosecond spectroscopy of optical Kerr effect with multipulse excitation in chloroform

D K Zharkov[1], A G Shmelev[1], V G Nikiforov[1], V S Lobkov[1]

Kazan Physical-Technical Institute, Russian Academy of Sciences, 10, Sibirsky trakt str., Kazan 420029, Russia

E-mail: dzharkov@list.ru

Abstract. Optical Kerr effect in the spectrum of chloroform is investigated by the method of two-pulse pumping. Here it's shown that during one-pulse pumping hyperpolarizability response, libration, the orientation response and the response of the intramolecular vibrations take place. Also we have achieved the suppression of intramolecular vibrations by two-pulse pumping with corresponding intensity and time delay.

1. Introduction

The femtosecond laser control nowadays is the object of intensive theoretical and practical research. The main aim of laser control is to manipulate the molecular dynamics, which can create the necessary molecular state. This state is determined by the conditions of the problem which is being solved: electronic, oscillatory or rotatory state, the necessary way of dissociation or ionization, the beginning of the chemical reaction [1] etc. At present the femtosecond technique is actively used for realization of laser control. The reason for the use of ultrashort pulses is their duration, which is comparable to molecular oscillation period.

2. Experimental setup

The experimental setup is showed at Fig. 1.

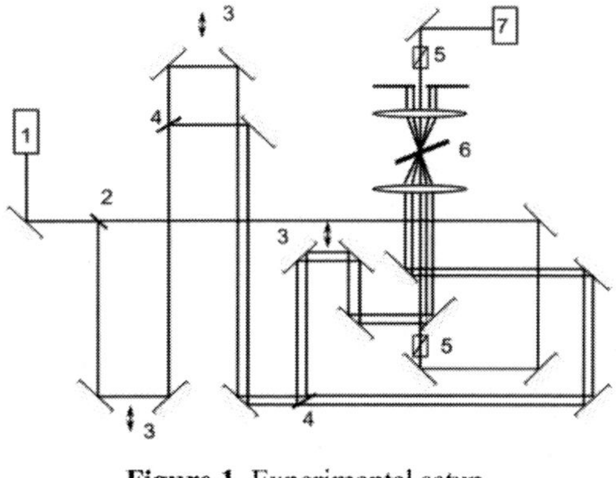

Figure 1. Experimental setup.

Mode-locked Ti:Sapphire femtosecond generator of laser pulses (1) has got the following characteristics: pulse duration 35 fs, wavelength 790 nm, output power 400 mW, pulse rate 80 MHz. Mirror 2 divides laser beam in ratio 98:2, mirror 4 – in ratio 50:50. Lines of delay (3) allow to control delay time between the pulses. Polarizers (5) allow to detect the changes in isotropism of analyzed liquid with detector (7). Optical Kerr effect (OKE) concludes in registration of the light transmitted through crossed polarizers and investigated liquid between them [2,3,4].

3. Investigated liquid and sample preparation

Chloroform was selected for demonstrating the control of the vibrational dynamics of the molecules. The laser pulse duration was maximally reduced to 35 fs since this fluid has a Raman-active modes near 400 cm^{-1}. The direction of the linear polarization is in accord with pump pulses. Vibrational-rotational wave packets are created in the medium by the pump pulse. Thus the effect of non-stationary polarizability anisotropy is induced into the liquid. The wave packets evolution is fixed by the low probe pulse, acting on medium after a specified delay.

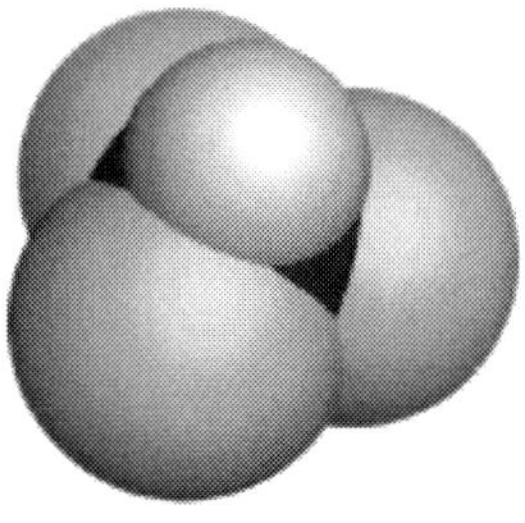

Figure 2. Chloroform molecule

Chloroform is widely used in the practice of chemical laboratories as a dissolver. In normal conditions it's a transparent easily evaporating liquid, not soluble in water. Chloroform molecule has comparatively small dipole momentum (1,08 Debay). As it's shown in Fig. 2, the molecule has a structure of a tetrahedron with a shortened top, resulting to small contribution of rotational diffusion and librational movements in the OKE spectrum. It's possible to excite two Raman-active modes (273 cm^{-1} and 375 cm^{-1}) by laser pulses with a duration of 35 femtoseconds. Thereby this liquid allows to learn and control the vibration dynamics by the experiments with OKE registration.

Only high purity liquids were used in the experiment. The quartz cuvette was abundantly washed with acetone. The cuvette was washed first with distilled water (in the case if the substances poorly soluble in acetone were in the cuvette before). After the drying the cuvette was washed with filtered acetone and dried again. Afterwards the filtered liquid was poured into the cell and the cuvette was sealed. The surface of the cuvette was washed with acetone and wiped with the special soft paper for optical elements soaked in acetone. After that the cell was placed in a holder. The holder was put onto the table, so that it became possible to move the cell in two directions (the cuvette was moved so that the focus lens was in the thick liquid which helped to minimize dispersion made by cuvette surfaces), and to rotate the cuvette around its vertical axis (to avoid spurious reflections).

4. OKE spectrum analysis

OKE spectrum analysis is based on the kinetic theory of fluids. OKE signal can be decomposed into four components (different in dynamics) [5]. Firstly, it's non-inertion response, which can be described by an easy scaling of the autocorrelation function. Secondly, it's the response (main contribution in OKE signal) which is associated with the rapid libration motion of the molecule. Thirdly, after a long time (1 picosecond and more) the main contribution comes from the rotational diffusion response. If the molecule has Raman-active modes, and stimulating pulse duration allows to excite it, the following effect takes place: well-visible OKE signal oscillations with a frequency corresponding to excited Raman-active modes superimposed on all previous responses. So, it should be obvious that to demonstrate the control of vibrational-rotational dynamics of the molecules we have to find molecules and experimental parameters for which the corresponding contribution to the OKE signal (e.g. response of rotational diffusion or coherent oscillations) would be maximized.

5. OKE spectroscopy and vibrations suppression of chloroform

At Fig. 3 the OKE signal with one-pulse excitation of the medium grade is shown. The analysis of experimental data shows that in chloroform all the molecular responses discussed above (response of hyperpolarisability, libration, the orientation response and the response of the intramolecular vibrations) are presented. The subject of major interest is the response of coherent intramolecular oscillations, because the molecules of chloroform in our experiment (two Raman-active modes with frequencies of 273 cm^{-1} and 375 cm^{-1}) are excited. The main contribution is made by low-frequency mode, as the excitement of the intramolecular oscillations in the region of 400 cm^{-1} by pulses about 35 fs is less effective. We should note that there occurred the integration of the molecular response and the probe pulse at the coherent oscillations registration. Also it reduces the amplitude of higher frequency response in the aggregate OKE signal.

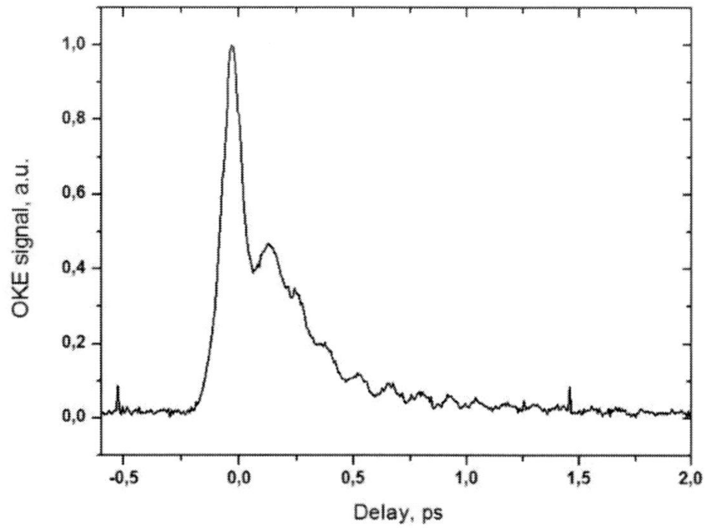

Figure 3. OKE signal spectrum with one-pulse pumping (duration 35 fs).

Fig. 4 shows examples of the suppression scenarios of coherent intramolecular oscillations in chloroform. Appropriate settings of the mode is the delay between the first and second exciting pulses and the relative intensity of the pump pulse. In any case the contributions related to the molecular rotations (libration and the orientation anisotropy of the molecule) are amplified as a result of the second pump pulse. For long delays it's necessary to reduce the intensity of the second pump pulse because of the damped oscillatory response for reduction mode. Thus, the method of two-pulse pumping enables the manipulation of vibrational dynamics of molecules with different delays between the exciting pulses up to a time of full oscillatory response damping.

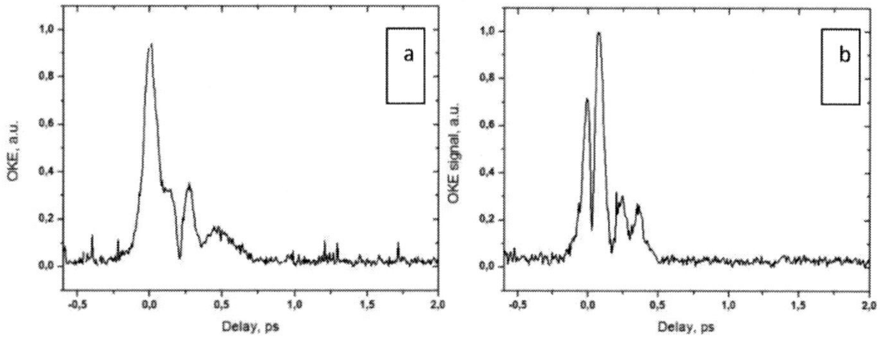

Figure 4. Two-pulse pumping (35 fs), a – delay between pulses 60 fs, intensity ratio 0,7; b – delay between pulses 260 fs, intensity ratio 2,5.

6. Conclusion

The control of coherent intramolecular vibrations was carried out in pure chloroform liquid at room temperature. For this purpose, non-resonant excitation by sequences of the two linearly polarized laser pulses (duration of 35 fs) was used. Probing the medium state was achieved by the weak third pulse with the use of recording ultrafast optical Kerr effect. It's shown that by using two parameters (delay between exciting pulses and their relative intensity) the control of the molecular vibrational dynamics in the subpicosecond range is achieved. The control of the vibrational dynamics in the subpicosecond range and even with long delays (compared to the decay time of the intramolecular vibrations) between the pump pulses is possible. The intensity of the second pump pulse must be reduced. When we use a double-pulse pumping, quantity of controlled parameters increases (relative to single-pulse excitation), and the number of constants for medium behavior modeling isn't changed. It results in the simplification of the OKE spectrum analysis.

References

[1] Nuernberger P Vogt G Brixner T Gerber G 2007 Femtosecond quantum control of molecular dynamics in the condensed phase *Physical Chemistry Chemical Physics* **9** 2470-97
[2] Duguay M Hansen J 1969An ultrafast light gate *Applied Physics Letters* **15** 192-194
[3] Ho P Yu W Alfano R 1976 Relaxation of the optical kerr effect of anisotropic molecules in mixed liquids *Chemical Physics Letters* **37** 91–96
[4] Ho P Alfano R 1979 Optical kerr effect in liquids *Physical Review A.* **20** 2170-87
[5] McMorrow D Lotshaw W Kenney-Wallace G 1988 Femtosecond optical kerr studies on the origin of the nonlinear responses in simple liquids *IEEE Journal of Quantum Electronics* **24** 443–454

Journal of Physics: Conference Series **572** (2014) 012044 doi:10.1088/1742-6596/572/1/012044

Study of a novel type of the optical modes in VCSELs

G S Sokolovskii[1], V V Dudelev[1], A M Monakhov[1], A Yu Savenko[2], S A Blokhin[1], A G Deryagin[1], K K Soboleva[3], A G Kuzmenkov[1], S N Losev[1], V V Luchinin[2], N A Maleev[1], E U Rafailov[4], W Sibbett[5], V I Kuchinskii[1,2]
[1] Ioffe Physical-Technical Institute, St Petersburg, Russia
[2] St Petersburg State Electrotechnical University "LETI", St Petersburg,, Russia
[3] St Petersburg State Polytechnical University, St Petersburg, Russia
[4] Aston University, Bermingham, United Kingdom
[5] University of St Andrews, St Andrews, United Kingdom

E-mail: gs@mail.ioffe.ru

Abstract. We study novel side-emitting modes in VCSEL microcavities. These modes correspond to π-shaped propagation along the mesa diameter, reflection from angled mesa walls and bottom Bragg reflector. We believe this study of π-modes is important for optimization of VCSEL design for improvement of efficiency.

1. Introduction

Recent progress of the vertical-cavity surface-emitting lasers (VCSELs) has made possible their wide expansion into fiber-optical communication systems, optical input/output devices, transducers and sensors [1,2]. However, further optimization of VCSEL parameters including increase of lasing efficiency and improvement of dynamic characteristics is of high demand for the most of applications mentioned above.

Unfortunately, many VCSEL microcavity configurations support not only the modes of the vertical resonator, but also the disk-propagating modes associated with the circular mesa structure. Such side-emitting light is typically associated with the whispering-gallery modes (WGM) that obviously compromise VCSEL efficiency. However, laser diodes with WGM-operating disk microcavities are very attractive due to the typically high Q-factor and robustness of fabrication [3,4]. Therefore a lot of attention is paid to investigation of the mode structure of the side-emitted light from the VCSEL mesa and study of possibility of WGM generation in VCSEL structures.

Quantum dot (QD) VCSELs with selectively oxidized aperture are the most attractive due to relatively small threshold current (~ 1 mA) and a big expansion of spectral range for device based on GaAs substrate [5]. Threshold current reduction is achieved due to using QD in active region and extremely narrow current aperture, which can be used in consequence to small lateral diffusion of current carriers in QD structures relative to quantum well (QW) structures. But it must be noted that thoughtful optimization of QD VCSELs design is crucial for effective exploitation of their advantages.

2. Experimental samples and study of the side emission from VCSEL microcavity

Our experiments were performed on VCSEL structures grown by MBE on GaAs substrate [1]. Doped top and bottom Bragg mirrors had 20 and 33 pairs $GaAs/Al_{0.9}Ga_{0.1}As$ with gradient composition on boundaries. Active region consisted of 3 QD layers (InGaAs), which were placed in maximum of optical field in undoped optical cavity with composition $Al_{0.15}Ga_{0.85}As$. Aperture layer p-($AlAs/Al_{0.9}Ga_{0.1}As$) was placed in the minimum of the optical field. Thickness and composition of layers were designed for generation at the wavelength of 980 nm. Mesa structures of 48 μm diameter were manufactured by etching with Ar+ ion beam. P-doped Bragg mirror was dry-etched with 5% accuracy for investigation of the possibility of WGM generation. Laser generation in VCSEL microcavity could be suppressed by removal of the odd number of quarter-wavelength layers due to decrease of the reflection coefficient.

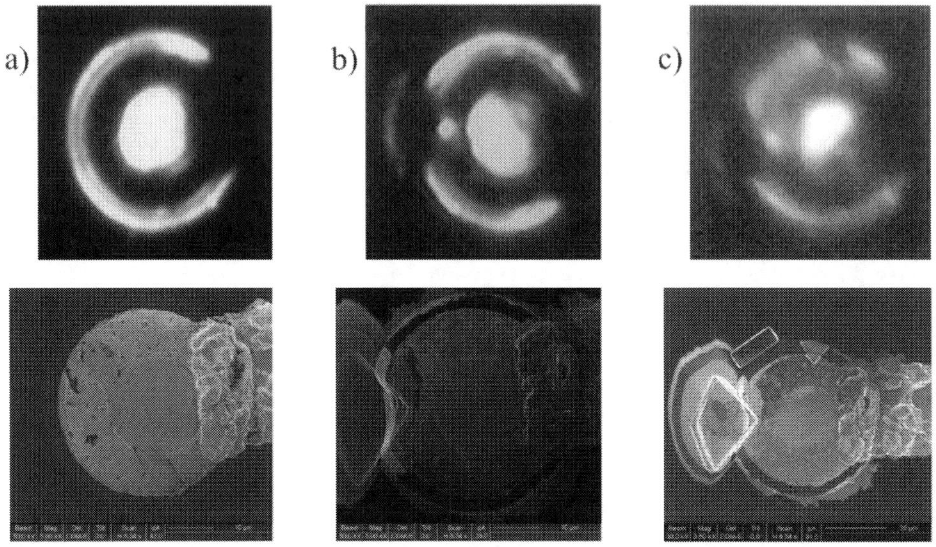

Figure 1 Images of VCSEL near-field near threshold and mesa without etching (a) after focused ion-beam etching of a section (b) and after etching a section and Bragg reflector (c).

Figure 1a shows the typical near-threshold distribution of the near-field of VCSEL with very pronounced emission from the mesa side surface. The spectra of the vertical cavity and of the side-emission are shown in figure 2.

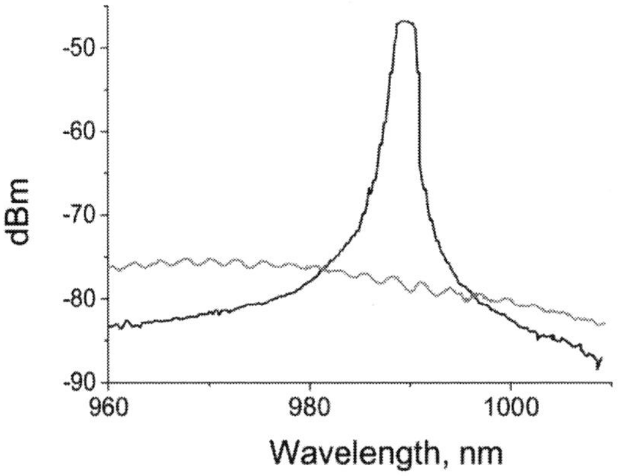

Figure 2 Spatially-resolved spectra of the vertical-cavity and side-emission of VCSEL structure at pump current I=0.53 mA.

Spatially resolved spectral measurements were made by 1:1 projection of VCSEL emission with two 30x microscope lenses to the input of the single-mode optical fiber. High-precision positioning of the optical fiber input in the focal plane of the second lens allowed for sampling of the VCSEL near-field with spatial resolution of ~ 5 μm corresponding to the effective diameter of the fiber core. Fiber output was plugged to the optical spectrum analyzer with 90 dB dynamical range at spectral resolution of 0.1 nm. In fig 2, one can clearly see that the spectrum from the mesa center is attributed to the VCSEL cavity with good side-mode suppression of almost 40 dB. In contrast to the vertical-cavity radiation, the emission from the mesa side walls features pronounced ripples corresponding to the mode spacing of 2.7 nm.

3. Theoretical analysis of the spectral mode spacing of the side-emission

First, the cone-shaped microcavity must be considered in order to find the possibility for generation of WGM modes and dependence of the spectral mode spacing of the cone angle and diameter of the mesa. Details of this analysis are published elsewhere [6] and here we only briefly note that no considerable dependence of the WGM spectra on the mesa cone angle can be reported and the spectrum of the cone-shaped mesa can be well approximated with that of the 'classical' disk one of the same diameter. For the high-order WGM mode propagating very close to the outer surface of the circular mesa, the spectral mode spacing can be written as:

$$\Delta\lambda_O = \frac{\lambda^2}{\pi n D} \qquad (1)$$

where λ is the emission wavelength, n is the effective refractive index and D is the diameter of the mesa. With 48 μm mesa and refractive index of 3.5 one can see the WGM mode spacing is approximately 1.9 nm and does not correspond to the experimental data. However, for the lowest-order WGM modes close to the center of the circular mesa (that can be approximated by triangular propagation) the spectral mode spacing:

$$\Delta\lambda_\Delta = \frac{2\lambda^2}{3\sqrt{3} n D} \qquad (2)$$

for the same mesa parameters can be as high as 2.3 nm, much closer to the experimental value.

One should also keep in mind the possibility of generation of the Fabri-Perot modes corresponding to propagation of the light via mesa diameter. However, the spectral mode spacing in such a case:

$$\Delta\lambda_{FP} = \frac{\lambda^2}{2nD} \qquad (3)$$

values to approximately 3 nm, higher than in the experiment.

In order to explain the experimental spectral mode spacing, the π-shaped propagation of the microcavity mode was suggested recently [7]. These π-modes propagate along the mesa diameter with reflection from angled mesa walls and the bottom Bragg reflector. Taking the effective thickness of the Bragg reflector L_B into account, one can write the spectral mode spacing of π-modes in the form:

$$\Delta\lambda_\pi = \frac{\lambda^2}{2n(D + 2L_B)} \qquad (4)$$

With practical value of $L_B \approx 2$μm, the π-mode spacing according to (4) yields 2.7 nm, which is in perfect agreement with the experiment.

4. Study of the side emission from the modified VCSEL microcavity

In this section we describe the measurements that were carried out in order to suppress emission of the side-mode of the microcavity and to reveal its origin experimentally. In order to verify the possibility of WGM generation, we prepared a cavity with a section removed from the circular mesa structure (Figure 1b). This was done by focused ion beam etching with the Strata FIB 205 setup [8]. With WGM nature of VCSEL side-emission, one should expect suppression of fringes seen in its spectrum in Figure 2. However, neither the measured near field distribution shown significant changes (figure 1b), nor the fringes disappeared in the spectrum of the side-emission as can be seen in the spectra at 12 and 1 o'clock in Figure 3.

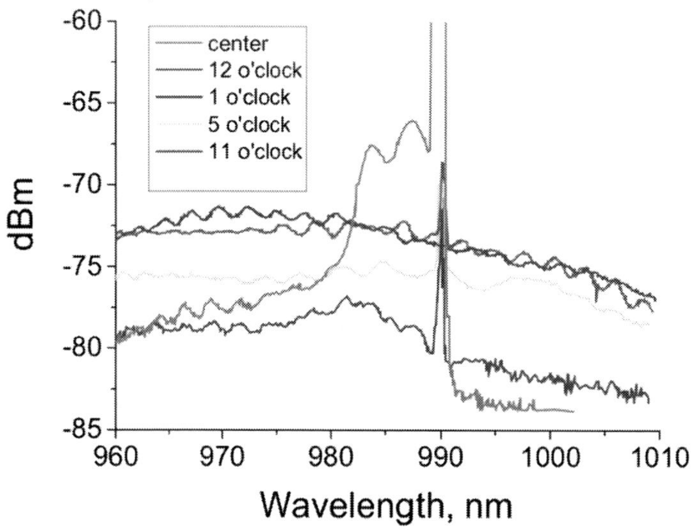

Figure 3 Spatially-resolved spectra of emission from different parts of etched VCSEL mesa. Red line corresponds to the vertical-cavity mode (emission from the mesa center). All other lines correspond to the side-emission from different parts of the etched mesa (Figure 1c): blue line – 12 o'clock; brown line – 1 o'clock; green line – 5 o'clock; purple line – 11 o'clock.

In order to experiment with the π-shaped propagation of the mode over the VCSEL cavity, we performed additional processing of the experimental samples. We made angled etching of the bottom Bragg reflector under the mesa as shown in Figure 4 in order to suppress the generation of π-modes. The detailed scheme of the bottom Bragg reflector etching is shown in figure 4. This angled etching was located in such a way to enable measurement of emission from the opposite side of the mesa circle. The near field of VECEL structure with such sophisticated etching is shown in figure 1c.

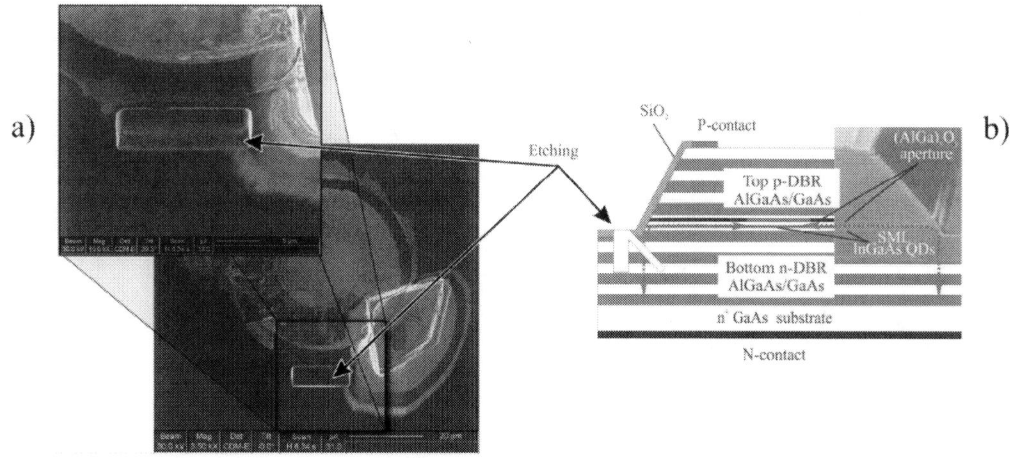

Figure 4 VCSEL with etched mesa and bottom Bragg reflector (a) and π–shaped light propagation in VCSEL cavity with angled mesa walls (b).

From the spectra at 11 and 5 o'clock in figure 3 one can see that not only did the etching of the Bragg reflector suppress the fringes in the spectrum measured at the point of etching (11 o'clock corresponding to figure 1c), but also the opposite side of the mesa circle (5 o'clock corresponding to figure 1c) featured absence of the π-mode signature in the side-emission spectrum around 970 nm.

5. Conclusion

In summary, we have demonstrated and studied the novel side-emitting modes in VCSEL microcavities. These modes correspond to π-shaped propagation of radiation along the mesa diameter with subsequent reflections from angled mesa walls and bottom Bragg reflector. Excitation of these modes should be taken into account for optimization of VCSEL microcavities for improvement of efficiency.

References

[1] Blokhin S A, Maleev N A, Kuz`menkov A G, Shernyakov Y M, Novikov I I, Gordeev N Y, Dyudelev V V, Sokolovskii G S, Kuchinskii V I, Kulagina M M, Maximov M V, Ustinov V.M., Kovsh A R, Mikhrin S S, Ledentsov N N, "VCSELs based on arrays of sub-monolayer InGaAs quantum ddots", *Semiconductors*, **40**(5), 615.

[2] Wilmsen C W, Temkin H, Coldren L A "Vertical cavity surface emitting lasers", New York: *Cambridge Univ. Press*, 1999.

[3] Sherstnev V V, Monakhov A M, Astakhova A P, Kislyakova A Y, Yakovlev Y P, Averkiev N S, Krier A, Hill G, "Semiconductor WGM lasers for the mid-IR spectral range", *Semiconductors*, **39**(9), 1087.

[4] Averkiev N S, Astakhova A P, Grebenshchikov E A, Il`inskaya N D, Kalinina K V, Kizhaev S S, Kislyakova A Y, Monakhov A M, Sherstnev V V, Yakovlev Y P, "Continuous-wave disk WGM lasers (lambda=3.0 um) based on InAs/InAsSbP heterostructures", *Semiconductors*, **43**(1), 117.

[5] Ustinov V M, Maleev N A, Kovsh A R, and Zhukov A E, "Quantum dot VCSELs", *Phys. Stat. Sol. (a)*, **202**(3), 396.

[6] Alekseenko V, Monakhov A M, Rozhanskii I V, "Whispering Gallery Modes in a Conical Resonator" *Tech. Phys.*, **54**(11), 1633.
[7] Sokolovskii G S, Dudelev V V, Monakhov A M, Savenko A Yu, Blokhin S A, Deryagin A G, Zolotovskaya S A, Kuzmenkov A G, Losev S N, Luchinin V V, Maleev N A, Rafailov E U, Sibbett W, Kuchinskii V I 2009 Generation of π modes in semiconductor vertical-cavity surface-emitting lasers, *Tech. Phys. Lett.* **35**(12), 1133
[8] Luchinin V V, Savenko A Yu, "Technology of local precision etching by focused ion beam"*Vac. Tech. and Technology*, **18**(1), 191.

The influence of the adhesive bonding on the magnetoelectric effect in bilayer magnetostrictive-piezoelectric structure

T A Galichyan and D A Filippov

Novgorod State University, 173003 Velikiy Novgorod, Russia

E-mail: tigrangalichyan@yahoo.com

Abstract. The influence of the interlayer adhesive bonding is considered in bilayer magnetostrictive-piezoelectric structure. The expression for the frequency dependence of the magnetoelectric voltage coefficient in the electromechanical resonance region is obtained using the simultaneous solution of the motion equations for the magnetostrictive, adhesive, piezoelectric phases and material equations. It is shown that in the passage to the limits this expression for the coefficient transforms to the expression for ideal connection between the layers.

1. Introduction

Ferrite-piezoelectric structures excite great interest, because in these structures there are effects which can be absent in ferrite and piezoelectric components separately. These effects appear due to mechanical interaction between the magnetostrictive and piezoelectric subsystems. One of such effects is the magnetoelectric (ME) effect, which is the change of the electric polarization under applied magnetic field: so called linear ME effect, or the occurrence of magnetization under applied electric field (inverse ME effect). The advantage of bilayer ferrite-piezoelectric structures is that the value of ME effect in such samples is greater than in bulk composites, thereby they are of great interest for the creation of various devices based on solid-state electronics. The ME effect is caused by the elastic interaction of magnetostrictive and piezoelectric subsystems in bilayer structures. The applied magnetic field induces mechanical deformations in the magnetostrictive phase which pass via mechanical coupling into the piezoelectric phase and lead to the change of the polarization of the sample. The frequency dependence of the effect is defined by the dispersion relation for such type of waves, because the ME effect in composites is caused by the elastic interaction of magnetostrictive and piezoelectric phases. Magnetostrictive-piezoelectric bilayer structures offer certain advantages over bulk composites [1], because they can be polarized much easier and exhibit practically no leakage currents as the magnetostrictive phase is insulated by the high-resistivity piezoelectric layer. The theory of the ME effect in bulk and multilayer composites based on the effective parameters method at the low frequency region was presented in works [2-6] and in the region of electromechanical resonance was first developed in works [7-9]. But the method of effective parameters can be used when the characteristic size of the structure units of the composite is much smaller than the acoustic wavelength and in this case the composite can be regarded as a homogeneous medium. Consideration of the magnetoelectric effect in layered structures based on the simultaneous solution of the equations of motion and constitutive relations for the magnetostrictive and piezoelectric phases was previously presented in [10-16]. Therein, an expression for ME voltage coefficient was derived based on the simultaneous solution of motion and material equations for the magnetostrictive and piezoelectric phases taking into account the boundary conditions on the interface. Thus, the consideration of the interlayer adhesive bonding was taken into account formally by an interface coupling coefficient in [10-12] or by an assumption of ideal coupling and equal oscillation amplitude over the thickness of both magnetostrictive and piezoelectric phases [13]. The ME effect in bilayer magnetostrictive-piezoelectric structure was presented in case of a perfect bonding taking into account the changes of the amplitude in oscillations over the thickness of the sample [14-16]. Recently the ME effect was investigated in laminated composite structure coupled by a bonding material considering the interlayer bonding material in work [17]. But in this work the amplitude change of the oscillations over the thickness of the sample (in direction perpendicular to the interface) was not assumed. In this work, the parameters of the adhesive bonding material and the amplitude change of the oscillations over the thickness are taken into account. Using motion equation, elastodynamics and electrostatic equations for the magnetostrictive piezoelectric phases and taking into account the boundary conditions on the interface, the equation for frequency dependence of the ME effect was obtained. It is shown that the value of the effect depends not only on parameters of magnetostrictive and piezoelectric phases, but also the parameters of interlayer adhesive material. For the case of ideal coupling this equation transforms to the equation previously obtained.

2. Model

As a model, we consider a bilayer structure in the form of a rectangular plate of length *L* and width *W*, consisting of mechanically interacting magnetostrictive, adhesive and piezoelectric layers of thickness ${}^{m}t$, ${}^{G}t$ and ${}^{p}t$ respectively, the values which are not assumed to be small. Thin metal contacts are applied on the top and the bottom of the plate. The direction of the wave propagation *X* coincides with the interface of the piezoelectric and adhesive layers. The direction *Z* is perpendicular to that interface as it is shown in figure 1. Preliminarily, the piezoelectric layer is polarized perpendicularly to the contacts (*Z* axis). Let us consider the longitudinal ME effect. In this case, the magnetic field coincides with the direction of polarization.

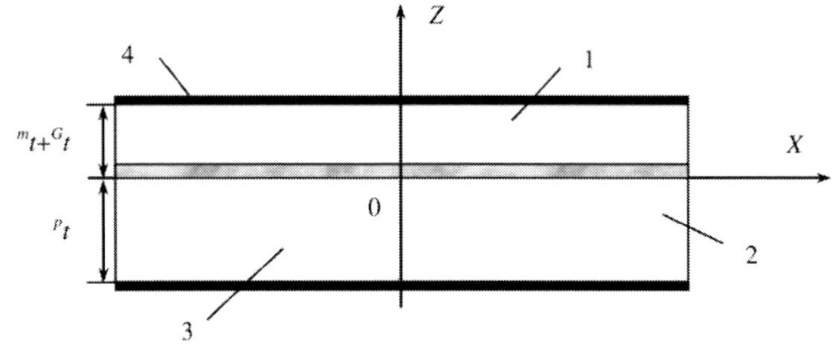

Figure 1 Schematic view of the sample
1 - Magnetostrictive phase, 2 - Piezoelectric phase, 3 - Adhesive phase, 4 - Electrodes.

The alternating magnetic field with angular frequency ω causes elastic oscillations in the magnetostrictive components, transferring through the interface of the partition to the piezoelectric component through the shear stresses, which brings to interconnected oscillations of magnetostrictive and piezoelectric subsystems. As there is a nonuniformity along *Z* axis and the width of the plate *W* (*Y* axis) is small, under the first approximation we can assume that the displacements along plane *Y* are homogeneous and the nonzero components of stress are only ${}^{\alpha}T_{xx}$ and ${}^{\alpha}T_{xz}$. The value of the stresses will depend on the thickness of the sample, perpendicularly to the interface between magnetostrictive and piezoelectric phases. Therefore the equations of motion for the *x* projection of the ${}^{\alpha}u_x$ displacement vector for the magnetostrictive, adhesive and piezoelectric phases are of the form

$${}^{\alpha}\rho\frac{\partial^2\,{}^{\alpha}u_x}{\partial t^2}=\frac{\partial\,{}^{\alpha}T_{xx}}{\partial x}+\frac{\partial\,{}^{\alpha}T_{xz}}{\partial z}, \tag{1}$$

where upper-case index "α" is correspondingly "m" for ferrite, "G" for glue and "p" for piezoelectric; ${}^{\alpha}\rho$ is the density of ferrite, adhesive and piezoelectric layers; ${}^{\alpha}T_{ij}$ are the components of the stress tensor. The constitutive equations for the magnetostrictive, adhesive and piezoelectric phases are

$$ {}^{m}S_{xx} = \frac{1}{{}^{m}Y}{}^{m}T_{xx} + {}^{m}q_{xx,z}{}^{m}H_{z}, \tag{2} $$

$$ {}^{m}S_{xz} = \frac{1}{{}^{m}G}{}^{m}T_{xz}, \tag{3} $$

$$ {}^{G}S_{xx} = \frac{1}{{}^{G}Y}{}^{G}T_{xx}, \tag{4} $$

$$ {}^{G}S_{xz} = \frac{1}{{}^{G}G}{}^{G}T_{xz}, \tag{5} $$

$$ {}^{p}S_{xx} = \frac{1}{{}^{p}Y}{}^{p}T_{xx} + {}^{p}d_{xx,z}{}^{p}E_{z}, \tag{6} $$

$$ {}^{p}S_{xz} = \frac{1}{{}^{p}G}{}^{p}T_{xz}, \tag{7} $$

$$ {}^{p}D_{z} = {}^{p}\varepsilon_{zz}{}^{p}E_{z} + {}^{p}d_{xx,z}{}^{p}T_{xx}, \tag{8} $$

where ${}^{\alpha}S_{ij}$ are the strain tensor of the magnetostrictive, adhesive and piezoelectric phases; ${}^{p}E_{z}$ and ${}^{m}H_{z}$ are the electric and magnetic fields; ${}^{p}D_{z}$ is the electric displacement; ${}^{\alpha}Y$ and ${}^{\alpha}G$ are the Young's modulus and shear modulus; ${}^{p}d_{xx,z}$ and ${}^{m}q_{xx,z}$ are the piezoelectric and piezomagnetic coefficients; ${}^{p}\varepsilon_{zz}$ is the tensor component of permittivity.

Taking into account the nonuniformity along Z axis, the solution of equation for the displacement vector will be presented as planar waves, whose amplitudes are changed on the thickness of the sample and have the following form:

$$ {}^{\alpha}u(x,z) = {}^{\alpha}g(z)({}^{\alpha}A\cos(\omega t - kx) + {}^{\alpha}B\sin(\omega t - kx)), \tag{9} $$

where ${}^{\alpha}A$ and ${}^{\alpha}B$ are the constants of integration and ${}^{\alpha}g(z)$ are some functions.

The substitution of (9) into the expression for the motion of the medium gives the equation for function ${}^{\alpha}g(z)$. Solution of this equation depends on the elastic wave velocity in ferrite, piezoelectric and adhesive phases. For the sake of definiteness, we choose a structure of ferrite, piezoelectric and adhesive, in which the velocity of elastic waves in the ferrite will be higher than that in the piezoelectric and adhesive phases. The boundary conditions, namely, the normal components of stress tensor on top and bottom surfaces are zero. The displacement and the shear stresses are equal on the ferrite-adhesive and adhesive-piezoelectric interfaces. These boundary conditions determine a system of equations. Solution of these equations gives the dispersion relation in the following form:

$$ {}^{m}Y{}^{m}\chi\,\mathrm{th}({}^{m}\kappa)\left(1 - \mathrm{tg}({}^{G}\kappa)\mathrm{tg}({}^{p}\kappa)\frac{{}^{p}Y{}^{p}\chi}{{}^{G}Y{}^{G}\chi}\right) = {}^{p}Y{}^{p}\chi\,\mathrm{tg}({}^{p}\kappa)\left(1 + \frac{\mathrm{tg}({}^{G}\kappa)}{\mathrm{tg}({}^{p}\kappa)}\frac{{}^{G}Y{}^{G}\chi}{{}^{p}Y{}^{p}\chi}\right), \tag{10} $$

where ${}^{\alpha}\kappa={}^{\alpha}\chi\,{}^{\alpha}t$, are non-dimensional parameters,

$${}^{m}\chi^2=-2(1+\nu)\left[\frac{\omega^2}{{}^{m}V_L^2}-k^2\right],\ {}^{p}\chi^2=2(1+\nu)\left[\frac{\omega^2}{{}^{p}V_L^2}-k^2\right],\ {}^{G}\chi^2=2(1+\nu)\left[\frac{\omega^2}{{}^{G}V_L^2}-k^2\right],\ \frac{1}{{}^{\alpha}V_L^2}=\frac{{}^{\alpha}\rho}{{}^{\alpha}Y},$$

${}^{\alpha}V_L$ are the velocities of longitudinal waves respectively in the magnetostrictive, piezoelectric and adhesive medium, and ν is the Poisson's ratio.

Equation (10) defines the dependence of angular frequency ω from the wave number k in an implicit form for the propagation of elastic waves in bilayer magnetostrictive-piezoelectric structures taking into account the adhesive phase. As we can see this equation transforms to the equation previously obtained for the case of ideal contact [14-16].

3. The ME voltage coefficient

Magnetoelectric voltage coefficient is defined as the ratio of the average value of the electric field to the average value of the external magnetic field, which induced it, i.e.:

$$<\alpha_E>=<E>/<H>, \tag{11}$$

where $<E>=U/({}^{m}t+{}^{G}t+{}^{p}t)$ is the average value of the electric field of the structure. With this method of determination, the ME voltage coefficient characterizes the efficiency of ME transformation of the entire structure. In order to define a theoretical expression for the ME voltage coefficient, we use the method developed earlier [15,16]. The condition of mechanical equilibrium at the free surfaces of the plate, i.e. points $x=\mp L/2$, gives boundary conditions in the form

$$\int_{-{}^{p}t}^{0}{}^{p}T_{xx}(\pm L/2,z)dz+\int_{0}^{{}^{G}t}{}^{G}T_{xx}(\pm L/2,z)dz+\int_{{}^{G}t}^{{}^{m}t+{}^{G}t}{}^{m}T_{xx}(\pm L/2,z)dz=0. \tag{12}$$

Using (12) we can calculate the constants of integration A and B in (9). We will find the ME voltage coefficient using the open circuit condition I=0, where the electric current is determined by the expression

$$I=\frac{\partial}{\partial t}\int_{0}^{W}dy\int_{-L/2}^{L/2}{}^{p}D_z dx. \tag{13}$$

Expressing the stress tensor component ${}^{p}T_{xx}$ by the strain tensor component ${}^{p}S_{xx}$ from (6) and substituting it into (8), we obtain expression for the electric displacement ${}^{p}D_z$. Then, substituting the expression obtained for ${}^{p}D_z$ into (13) and using the condition of open circuit I=0, we obtain the equation for the electric field $<E>$ taking into account (9) and (12). Thus, we find the ME voltage coefficient using (11) and the fact that the potential difference between the electrodes is defined as $U=<{}^{p}E_z>{}^{p}t$. Finally, in the case of longitudinal orientation of the fields, we obtain the expression for the ME voltage coefficient in the form:

$$\alpha_E=\frac{{}^{p}Y\,{}^{p}d_{xx,z}\,{}^{m}q_{xx,z}}{{}^{p}\varepsilon_{zz}\Delta_a}\frac{{}^{m}Y\,{}^{m}t}{{}^{m}Y\,{}^{m}t\frac{\text{th}({}^{m}\kappa)}{{}^{m}\kappa}C_m+{}^{G}Y\,{}^{G}t\frac{\sin({}^{G}\kappa)}{{}^{G}\kappa}C_G+{}^{p}Y\,{}^{p}t\frac{\text{tg}({}^{p}\kappa)}{{}^{p}\kappa}}\frac{\text{tg}(\kappa)}{\kappa}\frac{\text{tg}({}^{p}\kappa)}{{}^{p}\kappa}\frac{{}^{p}t}{{}^{m}t+{}^{G}t+{}^{p}t}, \tag{14}$$

where $C_m=\cos({}^{G}\kappa)\left(1-\text{tg}({}^{p}\kappa)\text{tg}({}^{G}\kappa)\frac{{}^{p}Y\,{}^{p}\chi}{{}^{G}Y\,{}^{G}\chi}\right)$, $C_G=1-\text{tg}({}^{p}\kappa)\text{tg}\left({}^{G}\kappa/2\right)\frac{{}^{p}Y\,{}^{p}\chi}{{}^{G}Y\,{}^{G}\chi}$

and $$\Delta_a = 1 - K_p^2 \left(1 - \frac{{}^pY\,{}^pt}{{}^mY\,{}^mt \frac{\mathrm{th}({}^m\kappa)}{{}^m\kappa} C_m + {}^GY\,{}^Gt \frac{\sin({}^G\kappa)}{{}^G\kappa} C_G + {}^pY\,{}^pt \frac{\mathrm{tg}({}^p\kappa)}{{}^p\kappa}} \frac{\mathrm{tg}(\kappa)}{\kappa} \frac{\mathrm{tg}({}^p\kappa)}{{}^p\kappa} \right),$$ with

$K_p^2 = \frac{{}^pY({}^pd_{xx,z})^2}{{}^p\varepsilon_{zz}}$ the squared coefficient of electromechanical coupling and $\kappa = kL/2$ the dimensionless parameter.

As follows from (14), the frequency dependence of the ME voltage coefficient has resonant character in bilayer magnetostrictive-piezoelectric structures considering the adhesive phase. At the so-called antiresonance frequencies when $\Delta_a = 0$, there is a peak increase in the ME voltage coefficient. It can also be seen that the value of ME voltage coefficient depends on each, parameters of magnetostrictive, adhesive and piezoelectric phases and the length of the structure.

In the low frequency region, the magnetoelectric voltage coefficient is nearly independent of frequency. Expanding Eq. (14) in powers of the small parameters ${}^m\kappa$, ${}^G\kappa$, ${}^p\kappa$ and retaining only the first-order terms of the expansion, we find the low-frequency value of the coefficient

$$\alpha_{low} = \frac{{}^pY\,{}^pd_{xx,z}\,{}^mq_{xx,z}}{{}^p\varepsilon_{zz}} \frac{1}{1 - K_p^2\left(1 - \frac{{}^pY\,{}^pt}{{}^mY\,{}^mt + {}^GY\,{}^Gt + {}^pY\,{}^pt}\right)} \frac{{}^mY\,{}^mt}{{}^mY\,{}^mt + {}^GY\,{}^Gt + {}^pY\,{}^pt} \frac{{}^pt}{{}^mt + {}^Gt + {}^pt}. \quad (15)$$

As we can see from (15), the value of the ME voltage coefficient decreases by the increase of the thickness of the adhesive layer.

4. Conclusion

The consideration of the adhesive bonding leads to a change of the results for the value of the ME voltage coefficient in a bilayer magnetostrictive-piezoelectric structure. It also leads to the nonlinear dispersion relation between the angular frequency and the wave number. The derived relations in the limiting case, when the thickness of the adhesive layer tends to zero, transform to the relations derived earlier for the case of ideal contact.The value of the ME voltage coefficient decreases with the increase of the thickness of the adhesive layer.

References

[1] Filippov D A, Laletin V M and Srinivasan G 2012 "Low-frequency and resonance magnetoelectric effects in nickel ferrite-PZT bulk composites" *Tech. Phys.* **57** 44-47

[2] Harshe G, Dougherty J P and Newnham R E 1993 "Theoretical modelling of multilayer magnetoelectric composites" *Int. J. Appl. Electromagn. Mater.* **4** 145-59

[3] Harshe G, Dougherty J P and Newnham R E 1993 "Theoretical modelling of 3-0/0-3 magnetoelectric composites" *Int. J. Appl. Electromagn. Mater.* **4** 161-71

[4] Bichurin M I, Petrov V M and Srinivasan G 2002 "Theory of low-frequency magnetoelectric effects in ferromagnetic-ferroelectric layered composites" *J. Appl. Phys.* **92** 7681-83

[5] Osaretin I A and Rojas R G 2010 "Theoretical model for the magnetoelectric effect in magnetostrictive/piezoelectric composites" *Phys. Rev.* B **82** 174415

[6] Wang Y, Hasanyan D, Li M, Gao J, Li J, Viehland D and Luo H 2012 "Theoretical model for geometry-dependent magnetoelectric effect in magnetostrictive/piezoelectric composites" *J. Appl. Phys.* **111** 124513

[7] Filippov D A, Bichurin M I, Petrov V M, Laletin V M and Srinivasan G 2004 "Resonant amplification of the magnetoelectric effect in ferrite-piezoelectric composites" *Phys. Solid State* **46** 1674-80

[8] Filippov D A, Galkina T A, Laletin V M, Srinivasan G 2011 "Inverse magnetoelectric effect in disk-shaped samples of ferrite piezoelectric composites" *Phys. Solid State* **53** 1832-38

[9] Bichurin M I, Filippov D A, Petrov V M, Laletsin V M, Paddubnaya N N and Srinivasan G 2003 "Resonance magnetoelectric effects in layered magnetostrictive-piezoelectric composites" *Phys. Rev.* B **68** 132408

[10] Filippov D A 2005 "Theory of the magnetoelectric effect in ferromagnetic-piezoelectric heterostructures" *Phys. Solid State* **47** 1118-21

[11] Filippov D A, Laletsin U and Srinivasan G 2007 "Resonance magnetoelectric effects in magnetostrictive-piezoelectric three-layer structures" *J. Appl. Phys.* **102** 093901

[12] Filippov D A, Srinivasan G and Gupta A 2008 "Magnetoelectric effects in ferromagnetic films on ferroelectric substrates" *J. Phys.: Condens. Matter* **20** 425206

[13] Bichurin M I, Petrov V M, Averkin S V and Filippov A V 2010 "Electromechanical resonance in magnetoelectric layered structures" *Phys. Solid State* **52** 2116-22

[14] Galichyan T A and Firsova T O 2013 "Propagation of elastic waves in bilayer ferrite-piezoelectric structure" *J. Phys.: Conf. Ser.* **461** 012016

[15] Filippov D A, Laletin V M and Galichyan T A 2013 "Magnetoelectric effect in a magnetostrictive-piezoelectric bilayer structure" *Phys. Solid State* **55** 1840-45

[16] Filippov D A, Galichyan T A and Laletin V M 2013 "Magnetoelectric effect in bilayer magnetostrictive-piezoelectric structure. Theory and experiment" *Appl. Phys.* A doi:10.1007/s00339-013-7957-z

[17] Hasanyan D, Wang Y, Gao J, Li M, Shen Y, Li J and Viehland D 2012 "Modeling of resonant magneto-electric effect in a magnetostrictive and piezoelectric laminate composite structure coupled by a bonding material" *J. Appl. Phys.* **112** 064109

16th Russian Youth Conference on Physics and Astronomy (PhysicA.SPb/2013) IOP Publishing
Journal of Physics: Conference Series **572** (2014) 012046 doi:10.1088/1742-6596/572/1/012046

Giant polarization and magnetic dynamics in $GdMn_2O_5$

B.Kh. Khannanov[1], E.I. Golovenchits[1], and V.A. Sanina[1]

[1]Ioffe Physical Technical Institute of RAS, Polytechnicheskaya str. 26, Saint Petersburg, RF, 194021

E-mail: boris.khannanov@gmail.com

Abstract. Polarization, dielectric permittivity, and low-frequency magnetic dynamics (30 - 50 GHz) of $GdMn_2O_5$ are presented. The emergence of an abnormally high polarization along the b axis reported recently by N. Lee et al. is verified. It is found that this polarization is rigid and is observed in the internal field. Superlattices formed due to phase separation and charge carrier self-organization in the bulk of $GdMn_2O_5$ the single crystals are revealed. The superlattices manifest themselves in the form of a set of ferromagnetic resonances which are responses of individual superlattice layers. A weak polarization along the c axis at T≤ 130 K caused by the superlattices is observed. The superlattices that occupy a small crystal volume are likely to be domain walls of the rigid bulk multiferroic domains. The correlation between magnetic structure, magnetic dynamics, and polarization in $GdMn_2O_5$ is analyzed.

1. Introduction

Orthorhombic manganites RMn_2O_5 (R – rare earth ion) [1, 2] are examples of multiferroics in which ferroelectricity occurs at the same temperature as a special type of magnetic ordering and is driven by it. These multiferroics are characterized by a strong magnetoelectric effect. The possibility to control electric (magnetic) properties by applying a magnetic (electric) field attracts considerable attention to these materials.

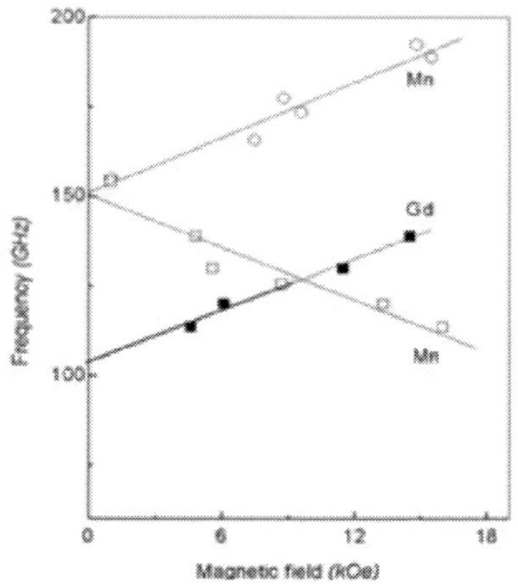

Figure 1. AFMR of the Mn-subsystem and ferromagnetic resonance of the Gd-subsystem (T = 5K) [6].

The polar order along the b axis in RMn_2O_5 occurs due to charge and magnetic orderings. Non-centrosymmetric ion positions in the lattice and polarization in RMn_2O_5 result from the exchange striction mechanism [3]. The polarization is typically of the order of 0.03 - 0.05 $\mu C/cm^2$ [2]. The polar and magnetic ordering temperatures in RMn_2O_5 are $T_C \approx$ 35 K and $T_N \approx 40$ K, respectively.

Among RMn_2O_5, $GdMn_2O_5$ is of special interest. Recently, an observation of the abnormally high polarization along the b axis in $GdMn_2O_5$ (P_b = 0.36 $\mu C/cm^2$) was reported [4]. This is very unusual because a complicated magnetic structure (incommensurate and spiral structure) typically

Journal of Physics: Conference Series **572** (2014) 012046 doi:10.1088/1742-6596/572/1/012046

induces a much lower polarization in multiferroics of this family. The Gd-Mn symmetric exchange striction plays a major role in the polarization of $GdMn_2O_5$ [4].

As the x-ray resonance magnetic scattering study has shown [4], $GdMn_2O_5$ is characterized by a uniform antiferromagnetic structure with the wave vector q = [1/2, 0, 0]. In this case, the magnetic order parameter contributes to both the Mn and Gd ions which are interrelated by a strong exchange interaction. Thus, the homogeneous magnetic structure of $GdMn_2O_5$ leads to an abnormally strong polarization. This paper is concerned with further studies of the magnetic structure and polarization features in $GdMn_2O_5$.

Note that earlier magnetic dynamics (antiferromagnetic resonances (AFMR) and mixed magneto-electric modes) studies also revealed unusual excitation spectra in $GdMn_2O_5$ [5, 6]. Two branches of resonance excitations were observed. One of them, with a gap near 150 GHz, was typical of AFMR for the uniaxial collinear antiferromagnetic without a weak ferromagnetic moment. The second branch with the gap near 110 GHz was caused by the paramagnetic Gd subsystem magnetized by the Mn subsystem (Fig.1). Thus, the magnetic dynamics spectra observed in [5, 6] were consistent with the magnetic structure observed in x-ray resonance magnetic scattering in [4].

In this paper we consider polarization, dielectric properties, and low-frequency magnetic dynamics (30 – 50 GHz) of $GdMn_2O_5$. We confirm the existence of a high polarization in $GdMn_2O_5$ oriented along the b axis and study its properties. Specific features of $GdMn_2O_5$ magnetic state resulting from the ground state of the Gd^{3+} ion ($^8S_{7/2}$) and the Gd-Mn exchange interaction are analyzed. A set of ferromagnetic resonances (FMRs) in the frequency range (30 – 50 GHz) similar to that observed earlier for other RMn_2O_5 is revealed in $GdMn_2O_5$ [7, 8]. Like in [7, 8, 9], we attribute the set of FMRs to individual superlattice layers formed in the $GdMn_2O_5$ bulk due to phase separation and charge carrier self-organization.

2. Specific features of magnetic structure of $GdMn_2O_5$

All ions in RMn_2O_5 are in the layers perpendicular to the c axis [10]. The unit cell of RMn_2O_5 includes two manganese ions. They are Mn^{3+} ion and Mn^{4+} ion with ($3t_{2g}$ $1e_g$) and ($3t_{2g}$ $0e_g$) 3d electron states, respectively. Mn^{4+} ions occupy positions z = 0.25 c and (1-z) = 0.75 c in the octahedral oxygen environment (MnO_6); Mn^{3+} ions are in positions z = 0.5~c and have the environment in the form of a pentagonal pyramid (MnO_5); R^{3+} ions occupy positions z=0 c.

Since the polarization in RMn_2O_5 is typically induced by a special type of magnetic ordering, the unusual properties of the polar order in $GdMn_2O_5$ should be sought for in specific features of its magnetic structure.

Because of a strong absorption of neutrons by Gd ions an information on the magnetic structures of the Mn and Gd ion subsystems of $GdMn_2O_5$ can be obtained by the x-ray magnetic scattering technique [4] and from magnetic dynamics spectra [5, 6]. X-ray magnetic scattering in off-resonance (NRMS) and at the Gd L_3-edge (resonance – RMS) conditions in $GdMn_2O_5$ [4] have shown that both magnetic subsystems (Mn and Gd ions) are highly interrelated and form a common order parameter. An incommensurate magnetic structure with wave vector q = (0.49, 0, 0.18) arises in the temperature range T_{N1}= 33 K – T_{N2} = 30 K. Near T_{N2} a lock-in transition into a commensurate magnetic phase with wave vector q = (1/2, 0, 0) which exists to the lowest temperatures occurs. Comparison of NRMS and RMS has shown that, along with this ordered state, some Gd ions which are in the paramagnetic state exists [4].

Note that the AFMR branch typical of the uniaxial antiferromagnetic collinear structure along the a axis observed for $GdMn_2O_5$ [5, 6] most likely originated from the antiferromagnetic Gd-Mn order parameter revealed in [4]. In [5, 6], the spectrum attributed to the paramagnetic Gd subsystem was also observed.

The reason for the emergence of this magnetic structure of $GdMn_2O_5$ should be sought for in properties of the ground state $^8S_{7/2}$ of the Gd^{3+} ions with a large spin S=7/2 which weakly interact with the lattice. The magnetic state of the Gd subsystem is mainly determined by the Gd – Mn exchange interaction. This differs $GdMn_2O_5$ from other RMn_2O_5 with R ions characterized by strong spin-orbital

Journal of Physics: Conference Series **572** (2014) 012046 doi:10.1088/1742-6596/572/1/012046

interactions in which complex magnetic structures are typically formed [2]. The state of the R subsystem in these RMn_2O_5 is mainly determined by the crystal field and, to a lesser extent, the R-Mn exchange.

3. Polarization of $GdMn_2O_5$

To explain the magnetic origin of ferroelectricity in RMn_2O_5, the model of exchange striction caused by the characteristic charge and spin orderings of Mn^{3+} and Mn^{4+} ions along the b axis is typically used. These orderings result from a paired arrangement of the Mn^{3+}-Mn^{4+} ions along the b axis. The mutual orientation of the spins of these ion pairs alternates, i.e., ferromagnetic pairs alternate with antiferromagnetic ones. A strong double exchange arises between the ferromagnetic spins of Mn^{3+} - Mn^{4+} ion pairs. A weaker indirect exchange is characteristic of the antiferromagnetic spins of such pairs. As a result of the exchange striction, distances between ferromagnetic and antiferromagnetic pairs change. This breaks the lattice centrosymmetricity and gives rise to the polarization along the b axis [3]. This mechanism can explain only a small contribution of the Mn-subsystem into the polarization (typically of the order of 0.01 – 0.05 $\mu C/cm^2$) usually observed in RMn_2O_5 [2]. As noted above, an order of magnitude stronger polarization induced by a symmetric Gd-Mn exchange is observed in $GdMn_2O_5$. Since the Gd spin configuration is similar to the manganese one, the difference in the Gd-Mn exchange interactions for the Gd- Mn^{3+} and Gd- Mn^{4+} ion pairs along the b axis leads to additional exchange striction, non-centrosymmetricity, and growth in the total polarization.

The polarization in single crystals of $GdMn_2O_5$ along the b and c axes was studied by measuring the pyrocurrent after application of a polarizing electric field and without it (at E = 0). Polarization was calculated by integrating the pyrocurrent measured during sample heating with a constant temperature variation rate.

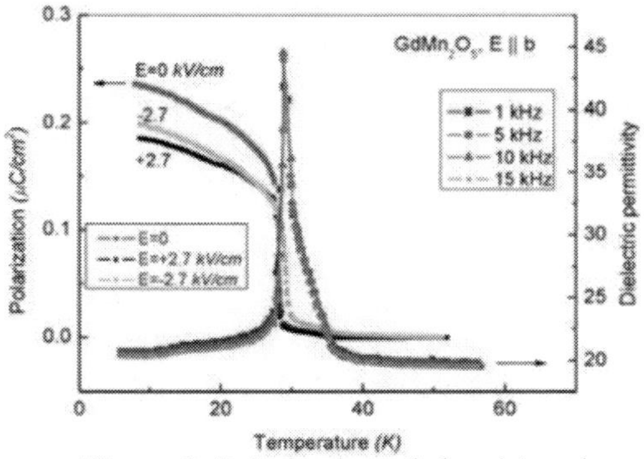

Figure 2. Polarizations (left axis) and dielectric permittivities (right axis) in some electric fields, at set of frequencies. E||b.

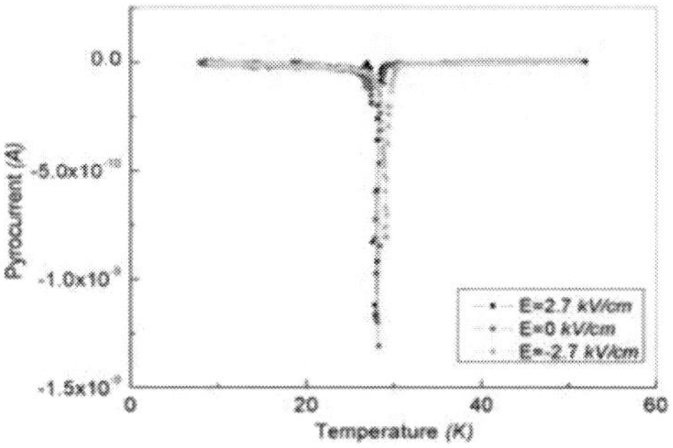

Figure 3. Pyrocurrent in some electric fields. E||b.

An abnormally high polarization in $GdMn_2O_5$ along the b axis was confirmed (Fig. 2). Along with this, it was found that the maxima in pyrocurrent near 30 K in the case of E= 0 and after application of a polarizing electric field E = ± 2.7 kV/cm (Fig. 3) were similar. This means that rigid polarization is formed in the high internal field and the usual 180-degree domain structure is not formed in $GdMn_2O_5$. Probably, only sufficiently large rigid bulk domains with different polarization orientations along the b axis that arise due to elastic strains and defects in real crystals can exist in $GdMn_2O_5$. Such a picture is characteristic of antiferromagnetic domains, for example, in Cr_2O_3 [11].

4. Low-frequency (30 – 50 GHz) magnetic dynamics

Let us consider the low-frequency magnetic dynamics of $GdMn_2O_5$. Studies of the dynamics of a set of RMn_2O_5 crystals with different R ions revealed similar sets of FMRs for these crystals [7, 8]. We attributed these resonances to responses of individual superlattice layers formed due to phase separation and charge carrier selforganization in the crystal bulk. Superlattices occupied a small volume of the initial crystal and at T < 30-40 K did not interact with each other. Superlattice layers were ferromagnetic and contained different amounts of manganese ions with different valences and charge carriers. An additional small polarization was formed in the superlattice layers along the crystal c axis. We supposed that superlattices were domain walls between bulk domains [8, 9].

Since manganese ions with different valences are located in the layers perpendicular to the c axis at distance ¼ c, there is a finite probability of tunneling of the e_g electron from the Mn^{3+} ion to the Mn^{4+} ion lying in the neighboring layers along the c axis (double exchange). In this situation ferromagnetic Mn^{3+}-Mn^{4+} ion pairs and electrons that recharge these pairs emerge in all the layers containing Mn ions. Because of an energetically favorable phase separation typical of all the manganites containing Mn^{3+} and Mn^{4+} ions [12, 13], these ferromagnetic pairs and electrons are accumulated in isolated regions inside the antiferromagnetic dielectric matrix. The double exchange and the Jahn-Teller effect (for the Mn^{3+} ions in octahedra) leads to attraction of charge carriers and ferromagnetic pairs, while the Coulomb repulsion forms dynamically equilibrated isolated superlattices (selforganization process).

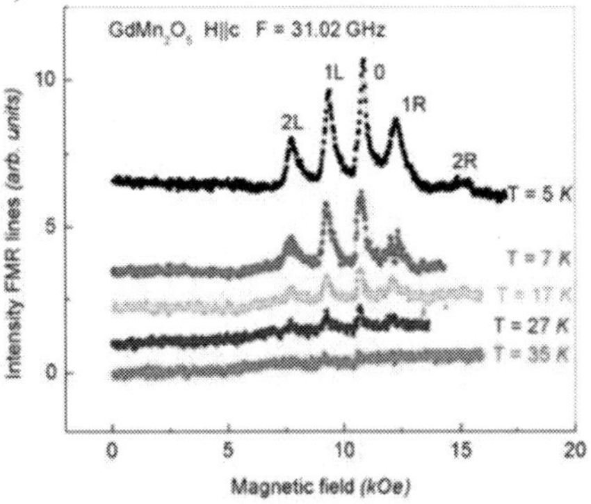

Figure 4. A set of the FMR as responses of the superlattice layers vs magnetic field (H ||c) at several temperatures. F = 31 GHz.

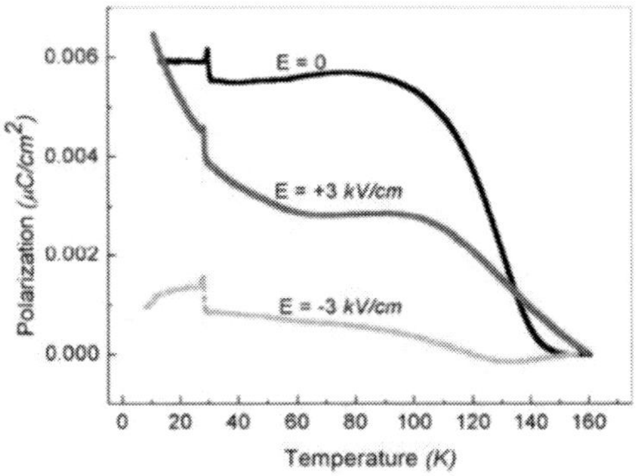

Figure 5. Polarizations along the *c* axis vs temperature at different polarizing fields induced by superlattices.

Fig. 4 shows a set of FMR lines pointing to formation of superlattices in $GdMn_2O_5$. We suppose that the Gd and Mn ions that are present in the crystal regions occupied by the superlattices form a subsystem of paramagnetic Gd ions magnetized by ferromagnetic Mn ions. This system is responsible for the additional paramagnetic branch of the spectrum of the Gd subsystem reported in [6] and a difference between the RMS and NRMS spectra observed in [4].

Fig. 5 presents a small polarization along the c axis induced by the superlattices. Like in the case of a giant polarization along the b axis of the basic crystal bulk, the polarization of the superlattices is induced by the internal field. Small anomalies in the pyrocurrent and polarization near T_{N2} are seen. In addition, there are pyrocurrent peaks near 130 K. We attribute the peaks to the disappearance of the polar order in the superlattices at this temperature.

To summarize, the presence of an abnormally high polarization along the b axis in $GdMn_2O_5$ as compared with other RMn_2O_5 has been confirmed. It has been found that this unidirectional rigid polarization is induced by the internal field. Like in all multiferroics with manganese ions of variable valences, uniaxial superlattices (normal to the c axis) are formed in $GdMn_2O_5$ due to phase separation and charge carrier selforganization. We suppose that the superlattices are domain walls between rigid bulk domains. A weak polarization along the c axis is formed inside the superlattice layers. The interrelated Gd – Mn antiferromagnetic order forming the basic rigid domains with a giant polarization along the b axis gives rise to the AFMR branch [6]. The superlattices are responsible for the paramagnetic branch of excitations [6] and a difference in the NRMS and RMS spectra [4].

Acknowledgments
This work is supported by Russian Foundation for Basic Research (grant No. 11-02-00218), Presidium of RAS, and the Government of RF (project N0.14.B25.31.0025).

References

[1] Hur N, Park S, Sharma P A, Ahn J S, Guba S and Cheong S-W 2004 *Nature (London)* **429** 392
[2] Noda Y, Kimura H, Fukunaga M, Kobayashi S, Kagomiya I and Kohn K 2008 *J. Phys.: Condens. Matter* **20** 434206
[3] Van den Brink J and Khomskii D I 2008 *J. Phys.: Cond. Matter* **20** 434217
[4] Lee N, Vecchini C, Choi Y J, Chapon L C, Bombardi A, Radaelli P G, ans S-W. Cheong 2013, *Phys. Rev. Lett.* **110,** 137203
[5] Golovenchits E I and Sanina V A 2003 *JETP Lett.* **78** 88
[6] Golovenchits E I and Sanina V A 2004 *J. Phys.: Cond. Matter* **16** 4325
[7] Golovenchits E I, Sanina V A and Zalesskii V G 2012 *JETPLett.* **95** 386
[8] Sanina V A, Golovenchits E I and Zalesskii V G 2012 *J. Phys.: Cond. Matter* **24** 346002
[9] Sanina V A, Golovenchits E I, Zalesskii V G and B Kh Khannanov 2013 *J. Phys.: Cond. Matter* **25** 336001
[10] Radaelli P G and Chapon L C 2008 *J. Phys.: Condens. Matter* **20** 434213
[11] Fiebig M, Frohlich D, Leute S, and Pisarev R.V. 1998 *Appl. Phys. B. Lasers and Optics 1998* **66** 265
[12] Gor'kov L P 1998 *Phys.-Usp.* **41** 589
[13] Kagan M Yu and Kugel K I 2001 *Phys.-Usp.* **44** 553

Electrical properties of silver iodide nanoparticles system embedded into opal porous matrix

A E Lukin, E N Ivanova, S V Pan'kova, V G Solovyev, V L Veisman

Pskov State University, Lenin square, 2, 180000 Pskov, Russia

E-mail: richardstone@yandex.ru

Abstract. Opal-based composite with silver iodide nanoparticles (AgI/opal) has been prepared by host-guest technology. Temperature and frequency dependences of electrical conductivity and those of dielectric permittivity of AgI/opal samples were measured. Size effects in this opal-based nanocomposite have been discussed.

1. Introduction

Silver iodide is a well-known model superionic compound in solid state physics [1-3]. At room temperature AgI is a mixture of low-conducting β (wurtzite) and γ (zinc blende) phases. At transition temperature $T_t \approx 420\ K$ the phase transition to the high-conducting (superionic) α-AgI occurs. In this high-temperature α phase large iodine anions form a body-centred cubic (bcc) lattice, whereas small silver cations move over numerous sites, forming "liquid-like" system.

In last decades, superionic conductor AgI has been thoroughly studied under restricted geometry conditions. A giant dielectric permittivity values as well as significant broadening of the melting-freezing phase transitions were found, in particular, in opal-based composites containing nanoparticles of AgI and those of other ionic compounds [4-6]. To discover new size effects as well as to design novel switching and memory devices operating at lowered temperatures, silver iodide nanoparticles [7-11], nanowires [12, 13], nanoplates [14, 15] and thin films [16-18] have been investigated.

In this work, we use regular porous opal matrix for preparing nanosize objects within the regular systems of voids. This idea proposed by Bogomolov and co-workers [19-21] was successfully used in [7-9] where silver iodide particles were prepared in voids and channels of molecular sieves MCM-41 and SBA-15, porous glasses and opals by chemical synthesis [8, 9] or high-pressure technique [7]. Unlike these technologies, we filled opal matrix directly with silver iodide melt under the ambient conditions.

A three-dimensional ordered face centered cubic (fcc) structure of self-assembled artificial opal matrix [22] is formed by closely packed silica globes with their volume filling factor $\rho = \pi\sqrt{2}/6 \approx 0.74$ and the first-order porosity $f_1 = 1 - \rho \approx 0.26$. One can also find smaller voids and channels in some types of "open" (fractal) opal structures with high order ($m \geq 1$) porosity:

$$f_m = (1-\rho)(1+\rho+...+\rho^{m-1}) = 1-\rho^m. \qquad (1)$$

2. Experimental procedure

To prepare AgI/opal nanocomposite (i.e., host opal matrix filled with silver iodide guest substance) we inserted initial opal matrix into the melt of silver iodide ($T = (836 \pm 3)\,\mathrm{K}$) for three hours. Then the sample ($5 \times 5 \times 0.7\ mm^3$ plate) was cleaned and polished mechanically and its opposite plane surfaces were covered with graphite (used as electrode material).

In our experiments the gravimetric density of the polished composite sample was found to be $\rho_c = 3.3\ g/cm^3$, so one can calculate the volume fraction of the guest substance in the composite: $\frac{\rho_c - \rho_o}{\rho_{AgI}} \approx 0.35$, where $\rho_{AgI} = 5.7\ g/cm^3$ and $\rho_o = 1.3\ g/cm^3$ are the gravimetric densities of bulk AgI and bare opal, correspondingly. Thus, the sample under study possessed not only the first-order porosity, but partially the second-order one (see formula (1)). According to this estimation, the concentration of nanostructured silver iodide in our composite based on "open" opal structure was about 40% more than that in the sample obtained by high-pressure technique in [7] and one order of magnitude higher than that in the sample synthesized chemically in [8].

Electrical measurements of the samples under study (initial opal host matrix, bulk AgI guest substance and AgI/opal nanocomposite) were carried out at the frequency of 1 kHz by an E7-13 impedance meter as well as by impedance bridge MPP300 and generator Philips GM6012 within the frequency range 100 Hz – 300 kHz. Direct current was measured by electrometer Keithley 6517B and micro ammeter M95. The temperature dependences of all the electric sample characteristics were measured upon continuous heating or cooling with a rate of 1 to 2 degrees per minute.

3. Results

Temperature dependences of alternating-current (AC) ionic conductivity of AgI/opal nanocomposite material (AC frequency $\nu = 1$ kHz) are shown in figures 1 and 2. It is clearly seen that the temperature of the phase transition to the superionic state upon heating of the nanocomposite sample (figure 1, curve 1) is close to the corresponding temperature in the bulk guest material ($T_t \approx 420\ K$). On the other hand, the T_t value shifts to approximately 400 K upon cooling (figure 1, curve 2). This rather wide hysteresis loop, as well as the values of activation energy in the Arrhenius plot (figure 2) agrees with recently published data [8]. On the other hand, the AC ionic conductivity values measured in [8] at $\nu = 1.5$ kHz are much smaller ($\sigma_{AC} \approx (10^{-4} - 10^{-3})\ S/m$ above the temperature of the phase transition) than corresponding σ_{AC} values plotted in figure 1. This fact as well as the nonmonotonic temperature dependence $\sigma_{AC}(T)$ observed in [8] at middle and low frequencies is probably due to the low concentration and nonhomogeneous distribution of silver iodide nanoparticles chemically synthesized within opal pores.

In figure 3 direct-current (DC) ionic conductivity values for bulk silver iodide (figure 3, *a*) and those for AgI/opal nanocomposite with much more pronounced hysteresis (figure 3, *b*) are displayed. It should be noted that contrary to AC conductivity demonstrating distinct step-like conductivity enhancement at elevated temperature (figure 1), estimated DC conductivity values are much lower, decreasing monotonically during measurement process and showing only peaks at corresponding temperatures (figure 3). In our opinion, these effects are due to the polarisation processes.

Nevertheless, DC conductivity of AgI/opal nanocomposite at $T = 450\ K$ is at least five orders of magnitude higher than that of initial opal host matrix [6].

Figure 4 illustrates, in addition, the low-temperature shift of the phase transition point for AgI/opal nanocomposite upon cooling when compared with that for the bulk guest substance. The shift value $\Delta T_t \approx 20\ K$ also agrees with the result published in [8] for AgI/opal nanocomposite. It should be mentioned that for AgI nanowires grown with the usage of porous alumina template this shift is much larger ($\Delta T_t \approx 70\ K$) and it is thought that plenty of stacking defects and adsorption/segregation of Ag^+ ions at the alumina surface could be the main factors for abnormal phase transition in this composite structure [12].

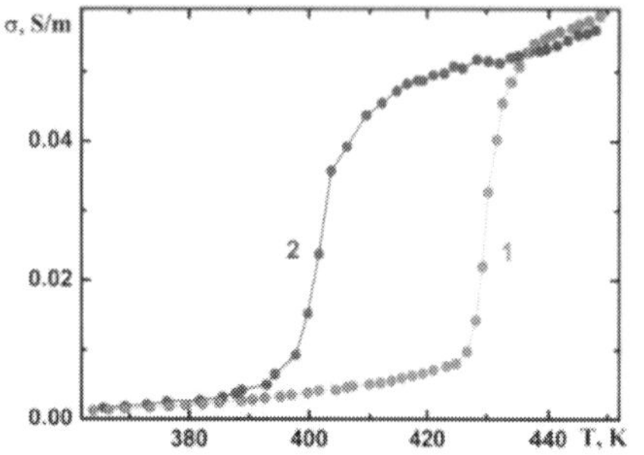

Figure 1. Temperature dependence of AC ionic conductivity (ν = 1 kHz) of AgI/opal nanocomposite (1 - heating, 2 - cooling).

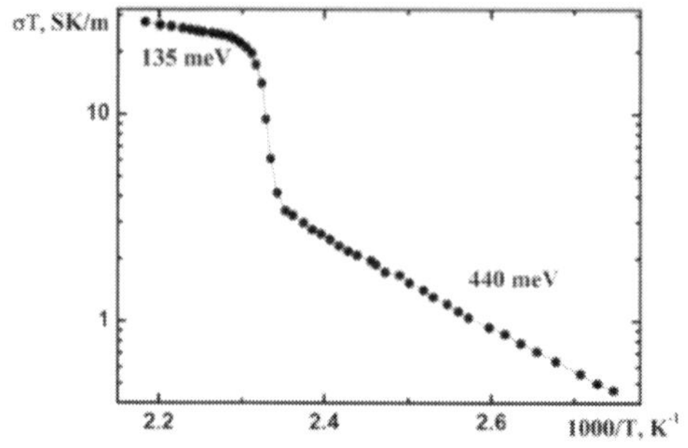

Figure 2. Arrhenius plot of AC ionic conductivity (ν = 1 kHz) of AgI/opal (heating). Activation energies are shown for two linear segments.

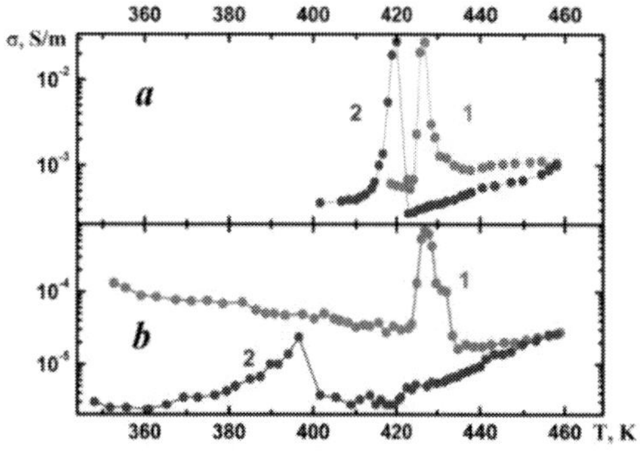

Figure 3. Temperature dependence of DC ionic conductivity of *a*) bulk AgI and *b*) AgI/opal (1 - heating, 2 - cooling)

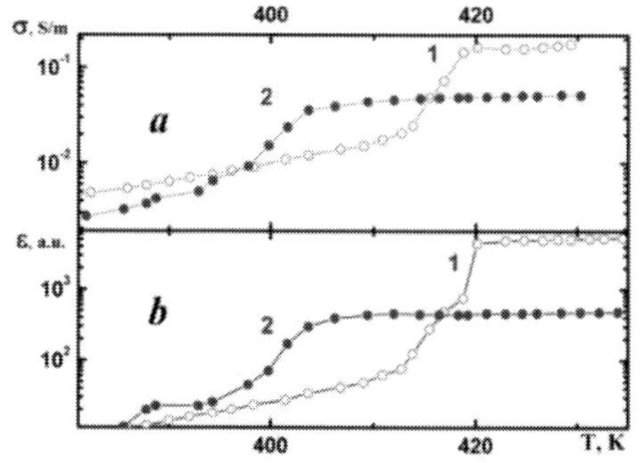

Figure 4. Temperature dependences of AC ionic conductivity (*a*) and dielectric permittivity (*b*) of bulk AgI (1) and those of AgI/opal nanocomposite (2) on cooling (ν = 1 kHz).

The frequency dependence $\varepsilon(\nu)$ of dielectric permittivity within the frequency range from 10^2 to 10^5 Hz is in agreement with the power law approximation ($\varepsilon(\nu) = C\nu^{-n}$), which is typical for ionic compounds dispersed within the voids of opal matrices [6].

4. Acknowledgments

The authors are grateful to Dr. S.G. Romanov for providing bulk opal matrices.

This work was supported by the Ministry of Education and Science of Russian Federation according to the program "Development of Scientific Potential of Higher Educational Institutions".

References

[1] Mott N F and Gurney R F 1948 *Electronic processes in ionic crystals* (Oxford)
[2] Lidiard A 1957 *Ionic conductivity* (Berlin: Springer)
[3] Salamon M B, Beyeler H U, Boyce J B *et al* 1979 *Physics of superionic conductors*

ed M B Salamon (Berlin: Springer)
[4] Pan'kova S V, Poborchii V V, Solov'ev V G 1996 *Journal of Physics: Condensed Matter* **8** L203
[5] Fokin A V, Kumzerov Yu A, Okuneva N M *et al* 2002 *Physical Review Letters* **89** 175503
[6] Solovyev V, Kumzerov Y, Khanin S 2011 *Physics of regular matrix composites (Electrical and optical phenomena in nanocomposite materials based on porous dielectric matrices)* (Saarbrücken, Germany: LAP, in Russian)
[7] Il'inskii A V, Aliev R A, Kurdyukov D A *et al* 2006 *Physica Status Solidi A* **203** 2073
[8] Vergent'ev T Yu, Koroleva E Yu, Kurdyukov D A *et al* 2013 *Physics of the Solid State* **55** 175
[9] Baryshnikov S V, Tien C, Charnaya E V *et al* 2008 *Physics of the Solid State* **50** 1342
[10] Wang Y, Huang L, He H *et al* 2003 *Physica B* **325** 357
[11] Yamasaki S, Yamada T, Kobayashi H *et al* 2013 *Chemistry. An Asian Journal* **8** 73
[12] Liang C, Terabe K, Hasegawa T *et al* 2007 *Journal of Applied Physics* **102** 124308
[13] Liang C, Terabe K, Tsuruoka T *et al* 2007 *Advanced Functional Materials* **17** 1466
[14] Guo Yu G, Lee J S and Maier J 2005 *Advanced Materials* **17** 2815
[15] Guo Yu G, Lee J S, Hu Y S *et al* 2007 *Journal of the Electrochemical Society* **154** K51
[16] Tappertzhofen S, Valov I and Waser R 2012 *Nanotechnology* **23** 145703
[17] Liang X F, Chen Y, Shi L *et al* 2007 *Journal of Physics D: Applied Physics* **40** 4767
[18] Furusawa S and Sakai Y 1999 *Journal of the Physical Society of Japan* **68** 976
[19] Bogomolov V N 1978 *Soviet Physics Uspekhi* **21** 77
[20] Astratov V N, Bogomolov V N, Kaplyanskii A A *et al* 1995 *Il Nuovo Cimento* **17D** 1349
[21] Kumzerov Y and Vakhrushev S 2004 *Encyclopedia of Nanoscience and Nanotechnology* (vol 7) ed H S Nalwa (American Scientific Publishers) pp 811–849
[22] Balakirev V G, Bogomolov V N, Zhuravlev V V *et al* 1993 *Crystallography Reports* **38** 348

Journal of Physics: Conference Series **572** (2014) 012048 doi:10.1088/1742-6596/572/1/012048

Effects of nickel ions implantation and subsequent thermal annealing on structural and magnetic properties of titanium dioxide

I R Vakhitov[1*], N M Lyadov[2], V F Valeev[2], V I Nuzhdin[2], L R Tagirov[1], R I Khaibullin[1,2]

[1] Institute of Physics, Kazan Federal University, Kazan, Russian Federation
[2] Laboratory of Radiation Physics, Zavoisky Physical-Technical Institute of Russian Academy of Science, Kazan, Russian Federation

*E-mail: ujay@mail.ru

Abstract. Wide bandgap semiconducting rutile (TiO_2) doped with 3d-elements is a promising material for spintronic applications. In our work a composite material of TiO_2:Ni has been formed by using implantation of Ni^+ ions into single-crystalline (100)- and (001)- plates of TiO_2. Sub-micron magnetic layers of TiO_2 containing nickel dopant have been obtained at high implantation fluence of 1×10^{17} ion/cm^2. A part of the implanted samples was then annealed in vacuum at different temperatures T_{ann} = 450-1200 K for 30 min. The influence of the implantation fluence, crystalline orientation, as well as subsequent annealing on the structural and magnetic properties of the nickel-implanted TiO_2 have been investigated by using X-ray photoelectron spectroscopy, scanning electron microscopy and coil magnetometry techniques.

1. Introduction

Synthesis and study of a new class of materials known in literature as oxide diluted magnetic semiconductors (DMS) have great scientific interest today. It is new class of semiconductors, considered as most promising material for applications in rapidly developing branches of science like spintronics and magnetic optoelectronics [1,2]. Various techniques have been developed for synthesis of DMS systems. Among them ion implantation is a promising method [3]. This refers both to technological capabilities of the method, as well as wide implementation of this technology in manufacturing of modern integrated circuits. Earlier, we studied semiconducting rutile TiO_2 single crystals implanted by cobalt ions in a wide range of implantation fluencies [4-6]. It was shown that depending on the implantation regimes, crystallographic orientation of the substrate, substrate temperature during the irradiation, either anisotropic ferromagnetic nanocomposite material TiO_2:Co, or isotropic DMS $Co_xTi_{1-x}O_{2-\delta}$ with a high Curie temperature (~700 K) were obtained [6-8]. In this paper we present results of the study of Ni implantation and subsequent thermal annealing on the structural and magnetic properties of rutile TiO_2.

2. Experimental details

Ni^+ ions with the energy of 40 keV were implanted into (100)- and (001)- oriented plates of single crystalline rutile TiO_2 to the fluence of 1×10^{17} ion/cm^2 at ion current density of 8 μA. The implantation was carried at room temperature of TiO_2 substrate by using ILU-3 ion accelerator (ZPhTI of RAS). Then, the implanted samples were annealed under high vacuum ~10^{-6} Torr at different temperatures ($T_{ann.}$) in the range from 450 to1200 K for the fixed time interval of t_{ann} = 30 min.

The elemental composition and surface morphology were studied by using scanning electron microscope (SEM) Carl Zeiss EVO'50 equipped with X-ray energy-dispersive spectrometer Oxford INCA Energy 330. The valences of Ni and Ti ions were determined by X-ray photoelectron spectroscopy (XPS). XPS measurements were performed in a UHV chamber (base pressure ~ 3×10^{-10} mbar) equipped with a monochromatic Mg K_α X-ray source operated at 12.5 kV and 250 W, and a Phoibos 150 hemispherical energy analyzer (all from SPECS). The energy analyzer was operating with the pass energy of 25 eV and the step size of 0.1 eV. All spectra were analyzed by using the CasaXPS software [9]. Peak shifts due to any apparent charging were calibrated with the carbon 1s peak set to 284.6 eV. After the calibration, the background from each spectrum was subtracted using a Shirley-type background. Magnetic properties of the samples were studied using the experimental home-made coercive magnetometer. The dependences of the induced and remanent magnetic moments on magnetic field were recorded at room temperature upon sweep of magnetic field up to 500 mT. The magnetic field was applied either in the plane of the sample (in-plane geometry) or perpendicular to the plane of the sample (out-of-plane geometry). Diamagnetic contribution from the substrate was subtracted during the processing of the results. The registered magnetic moment was normalized by the number of nickel atoms contained in the samples.

3. Results and discussions

SEM and XPS elemental analysis of the implanted rutile has shown the presence of only the implanted nickel and structure forming elements: titanium and oxygen with a reduced content of the last. This indicates that main radiation defects in titanium dioxide are oxygen vacancies. Areal concentration of the implanted impurity does not change after the thermal annealing. As a result of the SEM study it was found that the surface of the samples is generally smooth without any new growths both after ion implantation and thermal annealing.

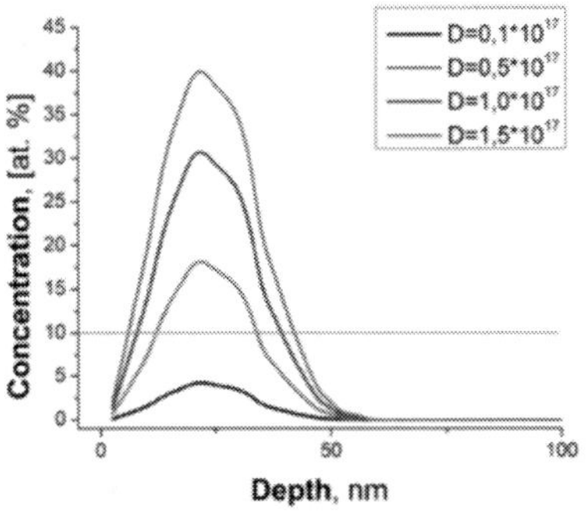

Figure 1. Depth profiles of nickel concentration at different values of the implantation fluence.

Computer package SRIM-2008 [10] was used to calculate depth distribution of nickel implanted into the rutile structure TiO_2 matrix that has density equal to 4.24 g/cm^3. Figure 1 shows that the concentration of Ni implants increases with increasing the implantation fluence. At the fluence more than 0.2×10^{17} ion/cm^2 the concentration of the implanted impurity exceeds the average value of maximum solubility of cobalt in the rutile matrix, which is equal to ~10 at.% (orange line in fig. 1). This means that at higher values of implantation fluence the implanted impurity may be found in the form of nanosize particles of nickel metal as the solubility limit is exceeded.

XPS measurements were taken in order to check valence states of titanium and nickel in the Ni-implanted TiO_2 samples. First of all, the analysis of XPS spectra have shown that there is no influence of orientation of the samples on the XPS spectra of Ti 2p and Ni 2p. Figure 2 shows high-resolution XPS spectrum of the Ti 2p region for (100) Ni implanted TiO_2 sample and the same sample subsequently annealed in vacuum at 1050 K, respectively. The Ti $2p_{3/2}$ and Ti $2p_{1/2}$ peaks are identified at 458.1 eV and 463.8 eV, respectively. The splitting of the binding energy between Ti $2p_{3/2}$ and Ti $2p_{1/2}$ core levels is about 5.7 eV, which matches the reported value of 5.7 eV [11]. This indicates a valence state of Ti^{4+} in TiO_2 [11].

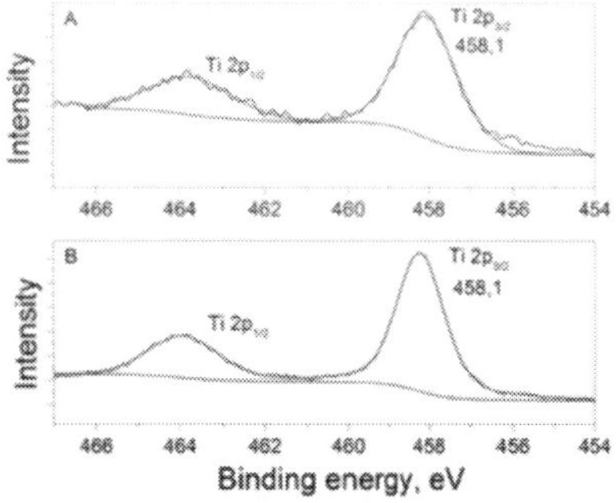

Figure 2. High-resolution XPS spectrum of the Ti 2p region for (A) (100)-oriented rutile implanted by Ni^+ ions and (B) subsequently annealed in vacuum at 1050 K.

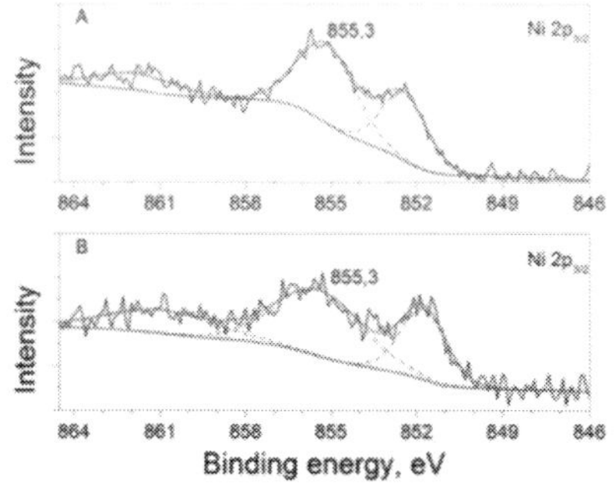

Figure 3. High-resolution XPS spectrum of the Ni $2p_{3/2}$ region for (A) (100)-oriented rutile implanted by Ni^+ ions and (B) subsequently annealed in vacuum at 1050 K.

Figure 3 shows Ni 2p core level X-ray photoelectron spectra of (100)-oriented TiO_2 samples: as-implanted and post-annealead (T_{ann}=1050 K) sample, respectively. From our analysis of the chemical states of nickel, the Ni $2p_{3/2}$ main peak located at 855.3 eV (corresponding to the Ni^{2+} in $NiCO_3$) and more weak peak was assigned to Ni^0 ("pure" Ni metal binding energy – 852.4 eV) [11]. This indicates that the chemical valence of Ni in as-implanted sample is 0 (metal) or 2+ (ion). The results for the post-annealed Ni-implanted TiO_2 are in agreement with the literature data for the Ni $2p_{3/2}$ main peak located at 855.3 eV that corresponds to Ni^{2+} ions.

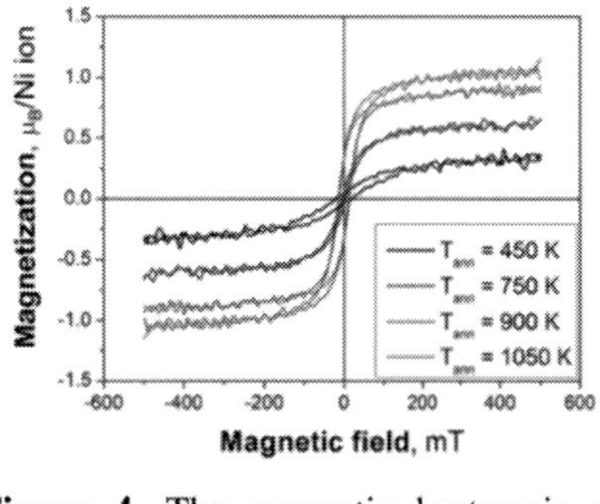

Figure 4. The magnetic hysteresis curve evolution for Ni-implanted TiO_2 after the annealing of the sample in high vacuum at different temperatures ($T_{ann.}$).

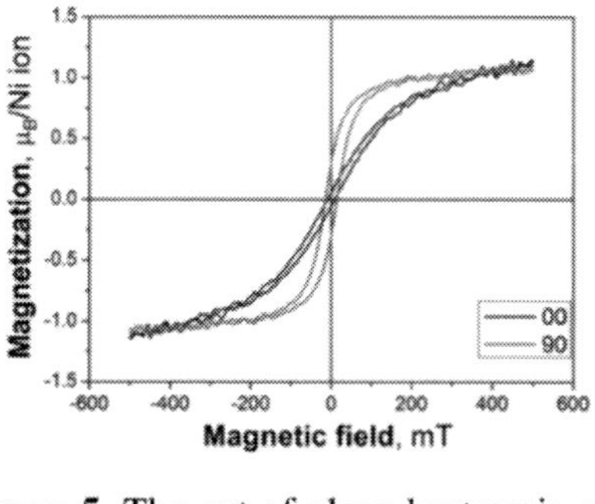

Figure 5. The out-of-plane hysteresis curves of the (001)- TiO_2 sample implanted with nickel ions and then annealed in vacuum at 900 K. The curves for two polar angles of θ=0° and θ=90° are shown.

Magnetic measurements have shown that the virgin (non-implanted) TiO_2 plates reveal diamagnetic response. The measurements of the as-prepared samples implanted with nickel ions have shown superparamagnetic behavior for all samples in any orientation. However, subsequent thermal annealing in vacuum resulted in the appearance of room temperature ferromagnetism of the samples. Figure 4 shows magnetic evolution of Ni-implanted (100) TiO_2 samples after the annealing in vacuum at different temperatures ($T_{ann.}$). The magnetization increases with increasing the temperature of annealing. It reaches maximum of about 1.1 μ_B at the annealing temperature of $T_{ann.}$ = 900 K. After that, the saturation magnetization starts to the decrease with following increase of the annealing temperature.

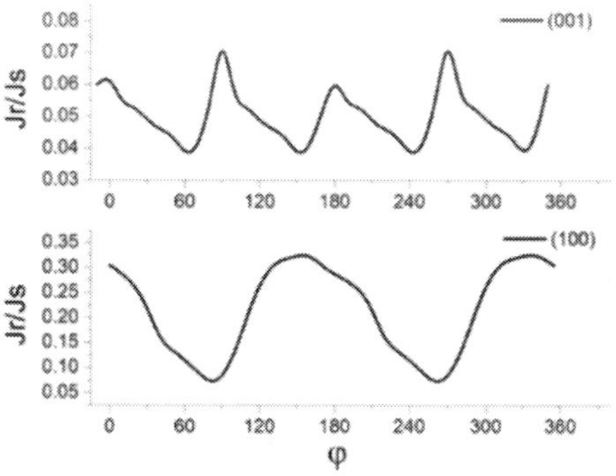

Figure 6. The in-plane azimuthal angular dependences of the ratio of remanent magnetic moment (J_r) to the spontaneous saturation moment (J_s) for different orientations of the post-annealed Ni-implanted TiO_2 plates.

Figure 5 shows out-of-plane magnetization curves of the (001)-oriented TiO_2 plate implanted with nickel ions and subsequently annealed in vacuum at 900 K (the curve for (100) orientation is the same). These dependences show that the samples demonstrate shape anisotropy that is typical for ferromagnetic thin films. Namely, samples quickly and easily magnetize to saturation by an applied magnetic field in the plane (θ =90°) of the plate. In contrast, application of magnetic field along the normal to the plate (θ=0°) does not saturate the sample even at maximum magnetic field of 500 mT.

In addition, the samples show magneto-crystalline anisotropy in the plane of the Ni-implanted plate of TiO_2 after high-temperature annealing at $T_{ann.}$ =750 K. Figure 6 displays in-plane angular dependences of the ratio of the remanent magnetic moment (J_r) to the spontaneous saturation moment (J_s) in the TiO_2 plates with different orientations, (100) and (001), respectively. It is clearly seen that the (100)-oriented samples reveal two-fold symmetry of the ferromagnetic response, and the (001)-oriented plates of TiO_2 show four-fold magnetic anisotropy. Taking into account the tetragonal symmetry of rutile, in which [100] and [001] are the axes of second and fourth orders of symmetry, respectively, we can conclude that the ion-synthesized DMS material TiO_2:Ni is coherently embedded in rutile matrix. Both the shape and magneto-crystallographic anisotropies become less pronounced with increasing the annealing temperature, and disappear completely at 900 $T_{ann.}$=1200 K for (100)-oriented TiO_2 plates and at $T_{ann.}$ = 900 K for (001) orientation of the samples.

It is important to note that the similar magnetic crystallographic anisotropies were early observed in single crystalline rutile (TiO_2) heavily implanted with Co [5, 12, 13], Fe [14, 15] or Ni [16]. Authors of these works have shown that in-plane magnetic anisotropy in 3d-ion implanted TiO_2 originates from secondary phases: metal nanoclusters of cobalt, iron or nickel grown coherently in the rutile matrix. Therefore, we may conclude that the magnetic crystallographic anisotropy observed in our post-annealed Ni-implanted samples could be related to the formation of magnetic Ni nanoparticles with orientations correlated with theTiO_2 rutile structure.

4. Conclusion

The effects of Ni ions implantation and subsequent vacuum annealing on structural and magnetic properties of TiO_2 single-crystalline plates were studied. XPS measurements show the mixed, 2+ and 0, valence states of Ni which corresponds to Ni^{2+} ions and Ni metal nanoparticles implanted in rutile matrix. All as-implanted plates of TiO_2 reveal superparamagnetic response at room temperature. Subsequent thermal annealing of the samples in vacuum leads to room temperature ferromagnetism with magnetic shape anisotropy characteristic for thin magnetic films. Strong influence of the crystallographic orientation of the TiO_2 plates on the in-plane magnetic anisotropy of post-annealed samples, was observed: (100) TiO_2 plate showing two-fold, and (001) TiO_2 – four-fold symmetry of the ferromagnetic response, respectively.

Acknowledgments

This work was funded by the subsidy allocated to Kazan Federal University for the state assignment in the sphere of scientific activities, and support in part by Russian Youth-Innovation Program U.M.N.I.K. is acknowledged.

References

[1] Janisch R, Gopal P, Spaldin N 2005 *J. Phys.: Cond. Matter.* **17** R657
[2] Prellier W, Fouchet A, Mercey B 2003 *J. Phys.: Cond. Matter.* **15** R1583
[3] Theodoropoulou N, Hebard A, Chu S, Overberg M, Abernathy, Pearton S, Wilson R, Zavada J 2002 *J. Appl. Phys.* **91** 7499
[4] Khaibullin R, Tagirov L, Rameev B, Ibragimov Sh, Yildiz F, Aktas B 2004 *J. Phys.: Cond. Matter.* **16** L443
[5] Khaibullin R, Ibragimov Sh, Tagirov L, Popok V, Khaibullin I 2007 *Nucl. Instr. and Meth. in Phys. Res. Sect. B* **257** 369
[6] Akdogan N, Nefedov A, Zabel H, Westerholt K, Becker H, Somsen C, Gok S, Bashir A, Khaibullin R, Tagirov L 2009 J. Phys. D: Appl. Phys. **42** 115005
[7] Khaibullin R, Tagirov L, Ibragimov Sh, Valeev V, Nuzhdin V, Osin Yu, Achkeev A, Faizrakhmanov I, Cherkashin N 2007 *Kazan. Gos. Univ. Uchen. Zap. Ser. Fiz.-Mat. Nauki* **149** 31 (English abstract is accessible at http://mi.mathnet.ru/eng/uzku/v149/i3/p31)
[8] Achkeev A, Khaibullin R, Tagirov L, Mackova A, Hnatowicz V, Cherkashin N 2011 *Phys. Solid State* **53** 543
[9] CasaXPS: Processing Software for XPS, AES, SIMS and More (http://www.casaxps.com)
[10] Ziegler J, Biersack J, Littmark U 1985 (http://www.srim.org/#SRIM)
[11] Wagner C, Riggs W, Davis L, Moulder J, Muilenberg 1979 *Handbook of X-ray Photoelectron Spectroscopy* (Minnesota, Perkin-Elmer Corporation)
[12] Akdogan N, Rameev B, Khaibullin R, Westphalen A, Tagirov L, Aktas B, Zabel H 2006 J. *Magn. Magn. Mater.* **300** e4
[13] Pinto J, Cruz M, da Silva R, Alves E, Gonzalez R, Godinho M 2007 Eur. Phys. J. B **55**, 253
[14] Vakhitov I, Nujdin V, Osin Yu, Khaibullin R 2010 *Kazan. Gos. Univ. Uchen. Zap. Ser. Fiz.-Mat. Nauki* **152** 7 (English abstract is accessible at http://mi.mathnet.ru/eng/uzku/v152/i3/p7)
[15] Zhou S, Talut G, Potzger K, Shalimov A, Grenzer J, Skorupa W, Helm M, Fassbender J, Cizmar E, Zvyagin S, Wosnitza J 2008 *J. Appl. Phys.* **103** 083907
[16] Cruz M, da Silva R, Pinto J, Borges R, Franco N, Casaca A, Alves E, Godinho M 2013 *J. Magn. Magn. Mater.* **340** 102

Semitransparent solar modules based on amorphous and microcrystalline silicon.

G M Ablayev[1,3], A I Kosarev[1], A V Kukin[2,4], M Y Semerukhin[1,3], M Z Shvarts[1,2], E I Terukov[1,2], D V Zhilina[1,4]

[1] Ioffe Physical Technical Institute of the Russian Academy of Sciences, Saint Petersburg, Russian Federation
[2] Research and development center for thin-film technologies in energetics under Ioffe Institute LLC, Saint Petersburg, Russian Federation
[3] Saint Petersburg Academic University - Nanotechnology Research and Education Centre of the Russian Academy of Sciences, Saint Petersburg, Russian Federation
[4] Saint Petersburg Electrotechnical University "LETI", Saint Petersburg, Russian Federation
E-mail: gani.ablay@gmail.com

Abstract. After modification of standard "Oerlikon" thin film solar cells producing technology we have managed to fabricate two experimental (10x10 cm^2) semitransparent solar modules based on amorphous and microcrystalline silicon with efficiency 6.5% and 6.7% and transparency about 20% and 9% in the visible range.

1. Introduction

According to Pike Research`s forecast [1] the global BIPV market will grow from just over 400MW in 2012 to 2,250MW in 2017 representing a five-fold increase. Thin-film solar cell technologies based on large areas are particularly well-suited to applications in the building industry. The use of a-Si modules in building-integrated photovoltaics (BIPV) is manifold and extends to all elements of a building shell, namely roofs, facades, windows and awnings. As a particularly attractive design option, thin-film technology offers the unique possibility of providing semitransparent modules by making very thin amorphous silicon layers on glass with transparent contacts. Obviously, the efficiency of semitransparent modules is reduced by the percentage of their optical transmission. However, besides electricity generation, they serve additional functions in terms of light and temperature management in buildings, i.e. by providing shading and thermal insulation [2].

Typical a-Si:H single junction solar cell is composed of five principal layers: Si p-i-n diode sandwiched between two conductive layers. Stacking two a-Si:H/μc-Si:H cells on top of each other forms the tandem junction structure, which is also sandwiched between the front and back contacts[3]. Figure 1 shows the schematic structure of single and tandem solar cells based on a-Si:H and μc-Si:H.

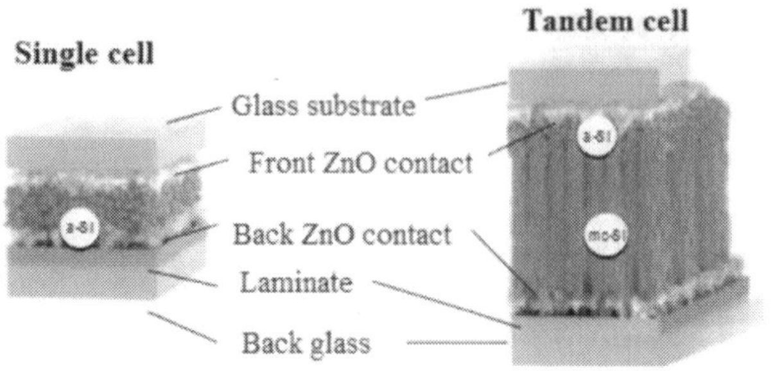

Figure 1. Structure of single and tandem solar cells based on a-Si:H and μc-Si:H.

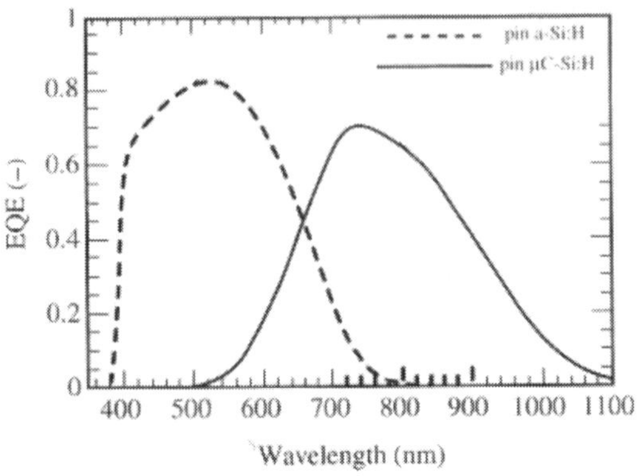

Figure 2. External quantum efficiency of a-Si:H and μc-Si:H pin structures (Source: IMT, Neuchatel) [4].

Figure 2 shows that a-Si:H layer absorbs short-wave solar radiation (380-700nm) and μc-Si:H layer makes better use of the long wavelengths (700-1100nm). The a-Si:H layer has a thickness of 0.3 μm, μc-Si layer - 2 μm.

«NovoVellum» laminate (figure 1) normally used at laminating stage has high reflectivity in spectral range 400-1100 nm. Thus it increases the optical path length and improves efficiency of solar cell.

2. Results and Discussion

We have replaced «NovoVellum» material to «Radeva» laminate, produced by Moscow company «Plastic». This material has high transparency within the visible and NIR range.

That allowed us to fabricate two experimental semitransparent solar modules $10\times10cm^2$ with single and tandem junction structure (figure 3). To compare the impact of reflecting laminate absence on electrical parameters two similar nontransparent solar modules were prepared as reference devices.

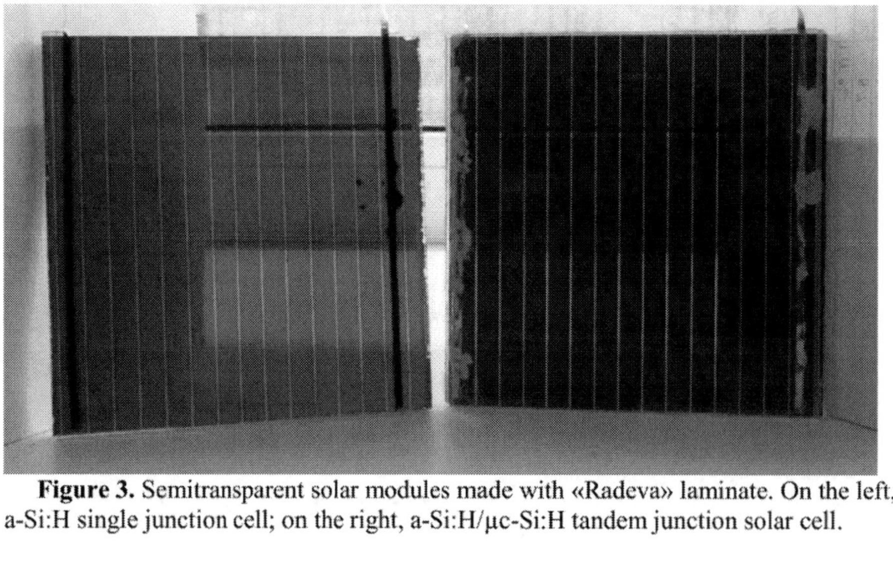

Figure 3. Semitransparent solar modules made with «Radeva» laminate. On the left, a-Si:H single junction cell; on the right, a-Si:H/μc-Si:H tandem junction solar cell.

Transmitted light has a red tint because the blue and green portion of the spectrum is absorbed in silicon layers. You can see it on figure 4, which shows transmission spectra of semitransparent solar modules.

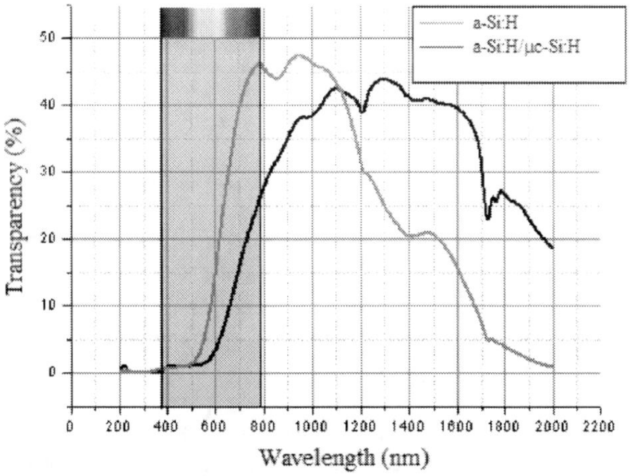

Figure 4. Transmission spectra of semitransparent solar modules.

a-Si:H semitransparent solar module has an average light transmission of 19.8% over the 400 to 800 nm range, and a-Si:H/μc-Si:H - 8.4%. It should be noted, that we can also increase color-neutral transmission up to 10% by laser scribing: the cell layers are removed along narrow lines perpendicular to the scribe lines of the integral series connection of the cells [2].

Figure 5 demonstrates the current-voltage (I-V) curves of semitransparent and nontransparent solar modules measured under simulated AM 1.5G illumination with intensity of 1000 $W \cdot m^{-2}$.

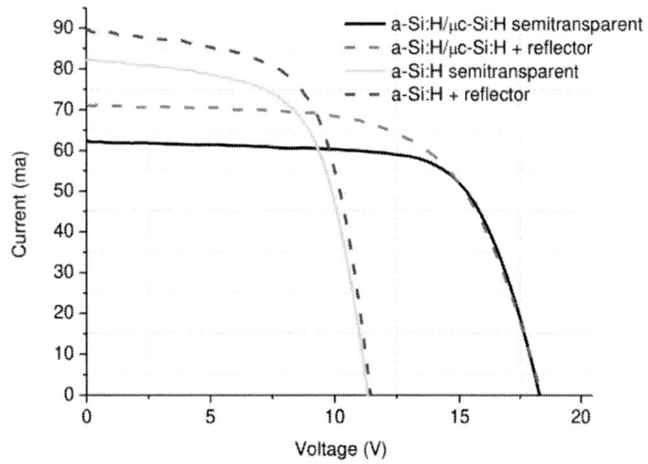

Figure 5. Current-Voltage characterization of semitransparent and nontransparent solar modules.

Electrical characteristics are demonstrated in table 1.

Table 1. Electrical characteristics of semitransparent and nontransparent solar modules.

Structure	Laminate	Isc (mA)	Uoc (V)	Pmax (mW)	Iopt (mA)	Uopt (V)	Seff (cm^2)
a-Si:H	«NovoVellum»	89.11	11.4	652	73.39	8.9	84,5
a-Si:H	«Radeva»	82.79	11.3	593	68.23	8.7	84,5
a-Si:H/μc-Si:H	«NovoVellum»	71.00	18.4	825	60.99	13.5	91
a-Si:H/μc-Si:H	«Radeva»	62.03	18.3	794	55.64	14.3	91

You can see that short-circuit current of nontransparent solar module is higher than I_{sc} of semitransparent analogue, because NovoVellum laminate acts like a reflector. Thus it increases optical path length and light absorption. However, if you take a look on table 2 you will see, that there is a slight efficiency difference between a-Si:H structures (1%). For a-Si:H/μc-Si:H structures this difference is greater (2%). This is due to the fact that most of the short-wave radiation is absorbed by a-Si:H layer before it reaches the laminate. Also after laminate replacement we have unbalanced photocurrents of the top and bottom cell in a-Si:H/μc-Si:H semitransparent solar module. Considering the fact that the bottom cell benefits from the light enhancement from the back reflector, while the top cell receives little benefit from the back reflector [5], the μc-Si:H cell limits the total current of semitransparent tandem junction solar module.

Table 2. Efficiencies of semitransparent and nontransparent solar modules.

Structure	Laminate	Efficiency (%)
a-Si:H	«NovoVellum»	7.5
a-Si:H	«Radeva»	6.5
a-Si:H/μc-Si:H	«NovoVellum»	8.7
a-Si:H/μc-Si:H	«Radeva»	6.7

In table 3 we compare characteristics of our a-Si:H module with characteristics of available BIPV products on the market [6].

Table 3. Comparison of the efficiency and transparency of semitransparent solar modules.

Solar module	Transparency (%) 400-800 nm	Efficiency (%)
a-Si:H (our)	20	6.5
Schott	10	5.2
SunTech	10	4.5
Smile Solar	10	7.1
Sun Well [7]	20	6.3
Tianwei [8]	20	4.9

In [9] research group from Korea demonstrated a-Si:H cell with efficiency of 5.5% and average transmittance of 21.7%.

Thus, our solar modules have quite competitive characteristics for international market.

3. Methods

Solar modules were obtained by PECVD in a KAI R&D reactor from the company Oerlikon Solar.

Optical and Electrical Characterization. The transmission spectra were recorded using a Cary 5000 UV-Vis-NIR spectrophotometer. Current-voltage characteristics of photovoltaic cells were measured using a Keithley 2400 source unit under a simulated AM1.5G spectrum with an Wacom WXS-90S-L2. The light intensity was ~100 mW·cm^{-2}, as calibrated using a Si photodiode.

4. Conclusions

We have achieved 6.5% power-conversion efficiency and transparency about 20% in visible range for a-Si:H semitransparent solar modules. The results in efficiency and transparency make these modules an extremely competitive product for BIPV technologies.

References

[1] Building Integrated Photovoltaics Report 2012, http://www.navigantresearch.com
[2] Schade H 2010 Examples of solar module applications *Thin-film Silicon Solar Cells* ed A V Shah (Lausanne: EPFL Press) chapter 7 pp 369 - 371
[3] Yang F 2011 Large Area a-Si/μc-Si Thin Film Solar Cells *Solar Cells - Thin-Film Technologies* ed L A Kosyachenko (Rijeka: Intech) chapter 16 pp 338 - 339
[4] Baliff C 2010 Introduction *Thin-film Silicon Solar Cells* ed A V Shah (Lausanne: EPFL Press) chapter 1 p 10
[5] Schiff E A, Hegedus S, Deng X 2011 Amorphous Silicon-based Solar Cells *Handbook of Photovoltaic Science and Engineering* ed A. Luque and S. Hegedus (Chippenham: Wiley) chapter 12 pp 524 – 525
[6] K Sinapis, M van den Donker SEAC BIPV Report 2013 pp 97 - 102
[7] http://www.sunwellsolar.com
[8] http://www.btw-solarfilms.com
[9] J W Lim, S H Lee, D J Lee, Y L Lee, S J Yun 2013 *Thin Solid Films* **547** 212

Conversion efficiency in a solar splitting system

S Yu Kurin[1], V D Doronin[2], S A Ivanov[3], H I Helava[4], B P Papchenko[3], A A Antipov[1], A S Usikov[3, 4], and Yu N Makarov[1, 4]

[1]Nitride Crystals Ltd., 27 Engels Ave., St. Petersburg 194156, Russia,
[2]National Research Nuclear University «MEPhi», 31 Kashirskoye shosse, Moscow 115409, Russia,
[3]National Research University «ITMO», 49 Kronverkskiy Ave., St. Petersburg 197101, Russia,
[4]Nitride Crystals Inc., 181 E Industry Court, Suite B, Deer Park, NY 11729, USA,

E-mail: sergey.kurin@nitride-crystals.com

Abstract. In this paper we report on concentrator photovoltaic system made by splitting the solar system based on separate Si, GaAs, and InGaN solar cells. The SSCPV module was fabricated and conversion efficiency up to 24.8% was achieved for the concentration factor of 12.8 that is in correlation with theoretical predictions.

1. Introduction

Concentration of solar radiation is a well-recognized and attractive method to increase the efficiency of photovoltaic systems. It reduces the area of expensive solar cells of a factor roughly equal to the concentration factor and makes feasible to use high-efficiency semiconductor materials of the solar cell cost-effectively. In addition, the concentration permits the collection of solar energy with low cost focusing systems such as concave mirrors or plastic Fresnel lenses. The greater the concentration, the higher the efficiency becomes. Higher concentration allows using smaller multi-junction (MJ) cells, thus making higher efficiency solar cells more cost attractive which further increases the photovoltaic system efficiency. The most efficient concentrator photovoltaic (CPV) cells today are 4-junction III-V tandem cells [1]. However, these cells have a number of deficiencies and difficulties in a fabrication:

1. The bandgaps of the solar cells (SCs) junctions that can be used are limited by the materials system;
2. Since the subcells are monolithically connected in tandem, the current of the entire cell is determined by the subcell with the lowest current;
3. Variations in the solar spectrum due to environmental changes such as atmospheric vapor change the current distribution in the cells and the overall efficiency;
4. The cells have to be electrically connected by tunnel junctions;
5. The III-V materials used are sensitive to the thermal environment; therefore efficient cooling is needed.

A higher efficiency and more flexible CPV system can be made by splitting the solar spectrum using dichroic mirrors between several single-junction (SJ) SCs [2-5]. Instead of the use of various materials connected in tandem, they can be optimized for the spectral components independently and the outputs combined remotely. Thus, spectral splitting concentrator photovoltaic system (SSCPV) uses the best available solar cell technology over a narrow spectral range where the cells can be more easily optimized. It avoids problems of current matching and makes it more feasible to optimize the arrangement of bandgaps.

2. Calculations of spectral splitting efficiency

In this paper we discuss conversion efficiency of the SSCPV based on separate SCs. Figure 1 shows the simplest concept of splitting the solar spectrum into three ranges. Main components of the SSCPV module (figure 2) are given below:

1. Fresnel lens to concentrate the incident light;
2. Collecting lens to focus the solar beam on the solar cell;
3. Cold mirror to separate infrared and visible portions of the solar spectrum;
4. Dichroic mirror to split the spectrum into ultraviolet + blue and the rest of the visible spectrum;
5. InGaN, GaAs, Si (Ge) SCs.

The optical elements (cold mirror, dichroic mirror) split the solar radiation into three light beams. The SCs for the short-wavelength spectral component could be based on the ternary compounds InGaN/GaN (or InGaP/GaP) and GaAs/AlGaAs. Possible semiconductor materials for long-wavelength spectral ranges were considered to be crystalline silicon (c-Si), germanium (Ge) or the binary compound GaSb.

The advantage of such spectral splitting assemblies is that the SC for each spectral range and the spectral range itself can be optimized independently.

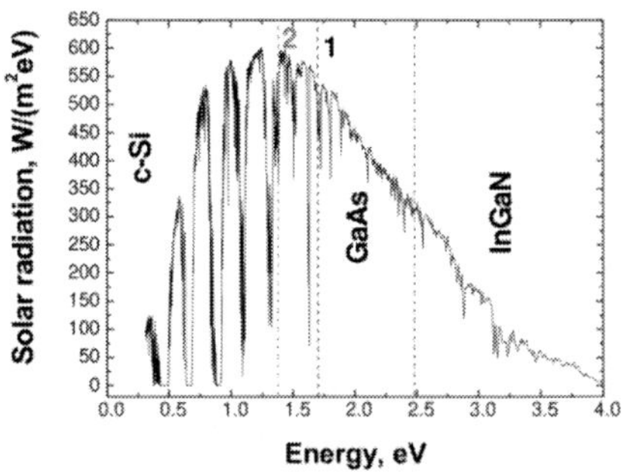

Figure 1. Standard solar emission spectrum AM1.5-Global. Dash lines show boundaries of spectral ranges. The numbers 1, 2 correspond to the different variants of the spectral splitting.

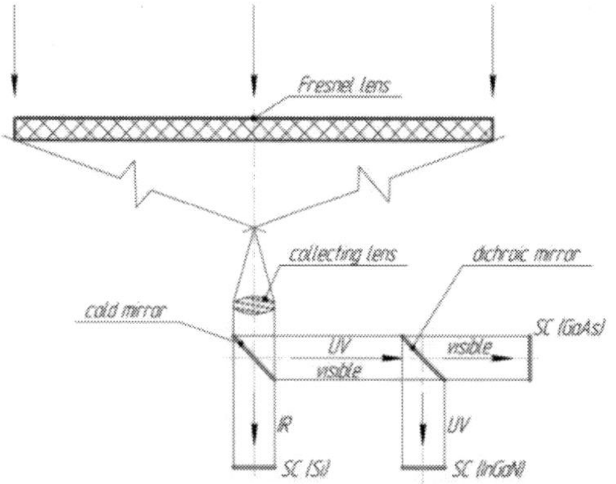

Figure 2. Schematic of the assembled SSCPV module.

The calculations were made by using software based on finite element method. The software was developed to simulate nitride-based and arsenide-based SCs. It allows two-dimensional (2D) simulation of the band diagram, generation rate of electron-hole pairs inside the structure, electron and hole transport inside the structure, radiative and non-radiative carrier recombination rate, electric field distribution, current-voltage (*I-V*) and power-current (*P-I*) characteristics. The software implements 2D drift-diffusion model with account for such features as elastic strain in the heterostructure layers and its effect on the valence band structure, as well as existence of spontaneous electric polarization and piezoeffect in III-nitride materials. According to the model, electrons and holes obey Fermi-Dirac statistics. Electron-hole generation rate at the point with z-coordinate within the *j*-th layer is calculated using the expression (1):

$$G_j(z) = M \int_0^{\infty} \frac{dE}{E} \cdot \alpha_j(z) n_j(z) I_j(z, E) P(E), \tag{1}$$

where M is the concentration degree of solar irradiation, $\alpha_j(z)$ is the absorption coefficient of the j-th layer, $n_j(z)$ is the refraction coefficient of the j-th layer, $I_j(z, E)$ is the light intensity at the point with z-coordinate, $P(E)$ is the spectral power density corresponding to the AM1.5-Global solar spectrum.

For simulation, the solar spectrum was divided into three spectral ranges: larger than 2.48 eV (less than 500 nm), in the range from 1.7 eV to 2.48 eV (500-725 nm), and smaller than 1.7 eV (more than 725 nm) (variant 1) and larger than 2.48 eV (less than 500 nm), in the range from 1.38 eV to 2.48 eV (500-900 nm), and smaller than 1.38 eV (more than 900 nm) (variant 2). A commercially available cold mirror having cut-off wavelength of 725 nm is used in variant 1. The cold mirror having cut-off wavelenrth of 900 nm in the variant 2 corresponds to the bandgap of the GaAs SC. The InGaN SC that intends for operation in the spectral rage less than 500 nm is a SJ structure containing two layers with a gradient change of In content having n-type conductivity and p-type conductivity, respectively, and an active layer sandwiched between them. The GaAs SC that is for spectral range from 500 nm to 725 nm (or 900 nm) consists of the following layers: p-AlGaAs, p-GaAs, n-GaAs, n-AlGaAs. The c-Si SC that is for the spectral range more than 725 nm (or 900 nm) consists of two layers of different conductivity types. The properties of the SCs based on the data presented in [6-8] are summarized in Table 1. It is necessary to note that hole mobility in the InGaN material reflects rather optimistic scenario. However lower values do not seriously influence on the results obtained.

Table 1. Properties of InGaN, GaAs, c-Si used in the simulation.

Property	InGaN	GaAs	c-Si
Electron mobility, $cm^2/(V \cdot s)$	400	6000	1350
Hole mobility, $cm^2/(V \cdot s)$	50	400	400
Dislocation density, cm^{-2}	10^8	10^5	10^8

The total conversion efficiency (η) for both variants was calculated in accordance with the expression (2):

$$\eta = \sum_i (\eta_i \cdot k_i) = \eta_{InGaN} \cdot k_{InGaN} + \eta_{GaAs} \cdot k_{GaAs} + \eta_{Si} \cdot k_{Si} \tag{2}$$

where η_i is the efficiency of i-th SC and k_i is the integrated power density in the operating range of i-th SC. Table 2 and Table 3 summarize the simulation results.

Table 2. Simulated SSCPV total conversion efficiency (variant 1).

SC (i)	Operating range $\Delta\lambda$, nm	In-band efficiency at concentration of 1000 (η_i), %	Part of the integrated power density (k_i), %	($\eta_i \cdot k_i$), %
InGaN (1)	< 500	62.5	18.3	11.4
GaAs (2)	500-725	53.9	32.8	17.7
c-Si (3)	> 725	2.6	48.9	1.3
η ($\sum_i(\eta_i \cdot k_i)$), %				30.4

Table 3. Simulated SSCPV total conversion efficiency (variant 2).

SC (i)	Operating range $\Delta\lambda$, nm	In-band efficiency at concentration of 1000 (η_i), %	Part of the integrated power density (k_i), %	$(\eta_i \cdot k_i)$, %
InGaN (1)	< 500	62.5	18.3	11.4
GaAs (2)	500-900	58.4	50.1	29.3
c-Si (3)	> 900	0.4	31.6	0.1
η $(\Sigma_i(\eta_i \cdot k_i))$, %				40.8

Low in-band efficiency of c-Si SC was due to the low absorption coefficient of c-Si in the considered spectral range. However, c-Si could be easily replaced with Ge to achieve higher total efficiency.

Figures 3, 4 show the following dependencies: in-band efficiency vs. voltage and *I-V* characteristics for the GaAs, InGaN SCs in the module.

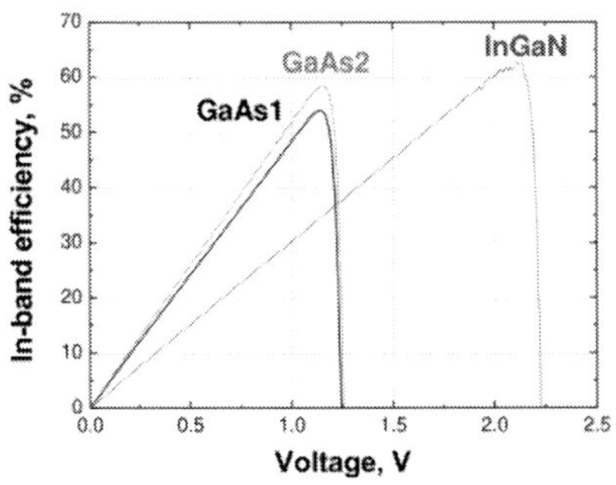

Figure 3. In-band efficiency simulated for the GaAs and InGaN SCs. GaAs1 and GaAs2 correspond to variant 1 and variant 2 of the spectral splitting, respectively.

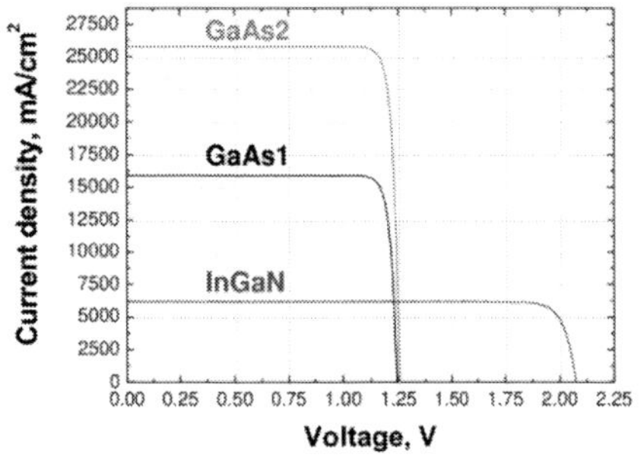

Figure 4. *I-V* characteristics simulated for the GaAs and InGaN SCs. GaAs1 and GaAs2 correspond to variant 1 and variant 2 of the spectral splitting, respectively.

The total conversion efficiency taking into account the optical losses dropped to 19.3-20.8% for variant 1 and 25.7-27.8% (variant 2) depending on the type of transmissive material for cold and dichroic mirrors.

In a more complex concept, although still simple to implement, the spectrum can be split into an arbitrary number of components for which optimized cells can be made. We guess the subcells can all be made using InGaN. Since the general acceptance that the bandgap of InN is ~0.65 eV, InGaN has been recognized as an attractive candidate for efficient covering practically full solar spectrum. The main limitation is the difficulty of growing high-quality InGaN layers with high In content due to the segregation of In during growth. InGaN is not yet sufficiently developed to address the longer wavelength portion of the solar spectrum. For the solar spectrum from near UV to blue, the efficiency of InGaN solar cells is very high. A number of groups have demonstrated very high internal and external quantum efficiencies [9, 10].

3. Experimental results

Figures 5, 6 show the SSCPV module that was fabricated to verify a concept of the spectral splitting system. The measurements were done under solar simulator (Newport) supplied with a Xe lamp, a collimator and a filter to fully simulate the solar spectrum (AM1.5-Global).

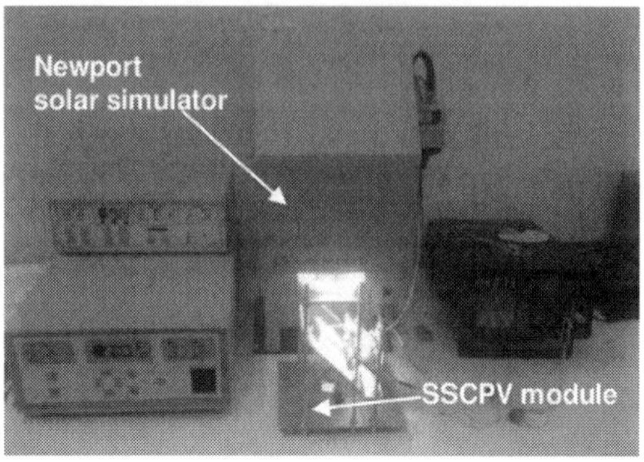

Figure 5. Newport solar simulator and SSCPV module.

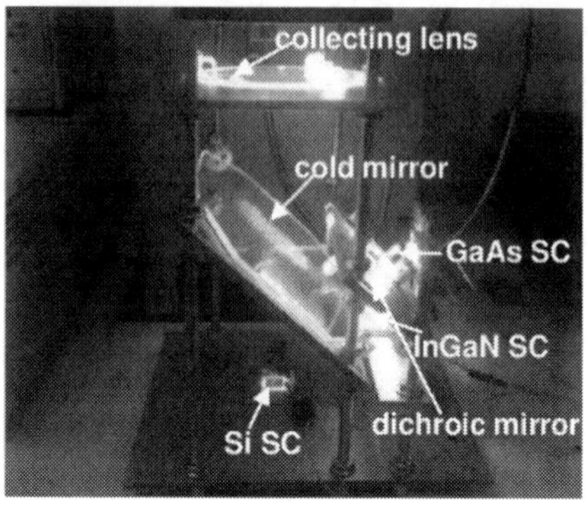

Figure 6. SSCPV module.

The total efficiency of the InGaN, GaAs, Si SCs for the module with the splitting of the spectrum optimized for the concentration ratio of 12.8 suns is given in Table 4. Thus, total efficiency of system with developed cells can be estimated as about 24.8%.

Table 4. Measured total conversion efficiency of the SSCPV module.

SC (i)	Operating range $\Delta\lambda$, nm	In-band efficiency at concentration factor of 12.8 (η_i), %	Part of the integrated power density (k_i), %	$(\eta_i \cdot k_i)$, %
InGaN (1)	< 500	33.3	18.3	6.1
GaAs (2)	500-725	24.7	32.8	8.1
Si (3)	> 725	21.7	48.9	10.6
η $(\Sigma_i(\eta_i \cdot k_i))$, %				24.8

4. Conclusions

The solar spectrum can be split into three spectral ranges for which optimized cells can be fabricated. This frees the MJ cell from the current-matching constraint, eliminates tunnel diodes and each subcell can be optimized independently. Experimental overall conversion efficiency for the SSCPV module consisting of InGaN, GaAs, Si SCs is equal to 24.8% at the concentration ratio of 12.8 suns. This result correlates with the theoretical predictions.

References

[1] http://www.ise.fraunhofer.de/en/press-and-media/press-releases/presseinformationen-2013/world-record-solar-cell-with-44.7-efficiency.

[2] Fraas I, Avery J, Huang H, Minkin L and Shifman E 2006 *Proc. IEEE 4th World Conf. on Photovoltaic Energy Conversion (Hawaii)*

[3] Barnett A, Kirkpatrick D, Honsberg C, Moore D, Wanlass M, Emery K *et al* 2009 *Progr. Photovolt.: Res. Appl.* **17** 75

[4] Groβ B, Peharz G, Siefer G, Peters M, Goldschmidt J S, Steiner M, Guter W, Klinger V, George B and Dimroth F 2009 *Proc. 24th Europ. Photovoltaic Solar Energy Conf. (Hamburg)* p 130

[5] Wang X, Wait N, Murcia P, Emery K, Steiner M, Kiamilev F, Goossen K, Honsberg C and Barnett A 2009 *Proc. 24th Europ. Photovoltaic Solar Energy Conf. (Hamburg)* p 811

[6] Brown G F, Ager J W, Walukiewicz W and Wu J 2010 *Solar Energy Materials & Solar Cells* **94** 478

Journal of Physics: Conference Series **572** (2014) 012050 doi:10.1088/1742-6596/572/1/012050

[7] Shen X, Lin S, Li F, Wei Y, Zhong S, Wan H and Li J 2008 *Proc. SPIE 7045 Photovoltaic Cell and Module Technologies II* 70450E
[8] http://www.ioffe.rssi.ru/SVA/NSM/Semicond.
[9] Jani O, Ferguson I, Honsberg C and Kurtz S 2007 *Appl. Phys. Lett.* **91** 132117
[10] Neufeld C J, Toledo N G, Cruz S C, Iza M, DenBaars S P and Mishra W K 2008 *Appl. Phys. Lett.* **93** 143502

Development of technology obtaining layers A^3B^5 compounds with variable bandgap for use in solar cells

A N Svistunov[1,2], R V Levin[1], B V Pushny[1]
[1]Ioffe Institute, Saint-Petersburg, Russia
[2]Saint Petersburg Electrotechnical University "LETI", Saint-Petersburg, Russia

e-mail: ANSwistunow@gmail.com

Abstract. Developed technology of obtaining layers with variable band gap by metalorganic vapor phase epitaxy (MOCVD). Made structure graded-gap layer with a thickness of 0.5 microns. Internal pulling electrical field the resultant structure was 5.8 kV/cm. In the resultant structure observed an increase circuit voltage due to the contribution of the gradient of the band gap.

1. Introduction

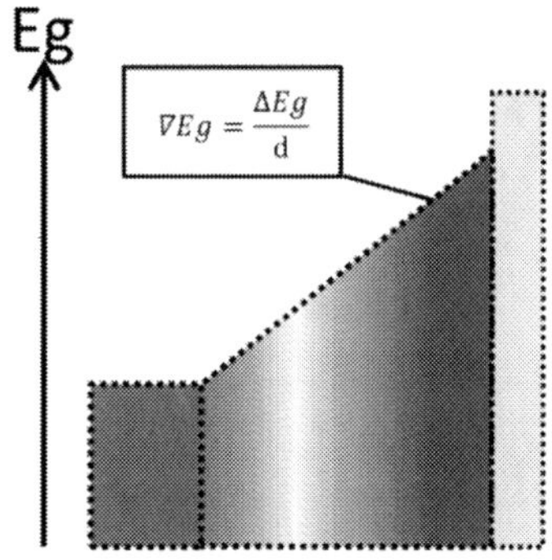

Fig. 1. Schematic view of graded-gap photoelectric converter

One of the main concerns was the issue of the energy crisis in the 21st century. One solution to the problem is to convert solar energy into electricity.
Currently, photovoltaic cells with a p-n junction thin-film silicon-based have a maximum coefficient of performance (COP) ~ 20%. Difficult in the manufacture multijunction photovoltaic cells ~ 44.7%. One possible option the construction of photovoltaic cells is the use of smooth heterostructures. In these heterostructures, the band gap increases to the illuminated surface (Fig. 1.)[1].
In 1957, in a theoretical paper [1] it was shown that smooth heterostructure at high light levels photovoltage ($U = \frac{b}{b+1}\frac{\Delta Eg}{q}$) appears that when $b = \frac{\mu n}{\mu p}$>>1 is similar to that bandgap difference.
In such a heterostructure is broadening of the spectral characteristics of photosensitivity and better utilization of solar energy[2]. However, from 1978 to the present, were not carried out experiments on the production of graded-gap structures, due to the imperfection technology for production. Purpose of our work obtaining structures with variable band gap, the modern method by metalorganic vapor phase epitaxy.

2. Obtaining of solid solutions based on gallium antimonide (GaSb)

Model material was selected on the basis of a solid solution of gallium antimonide (GaSb). The lattice parameter coincides with gallium antimonide solid solution parameters, the quantity gap which lies in the range from 0.3 eV to 1.7 eV (0.7-4 microns) (Fig. 2.)[3].
At the beginning of our work, experiments were conducted to obtain solid solutions lattice-matched to GaSb with a band gap of 0.3 eV to 0.9 eV. All experiments were carried out on the installation phase epitaxy AIXTRON AIX 200.

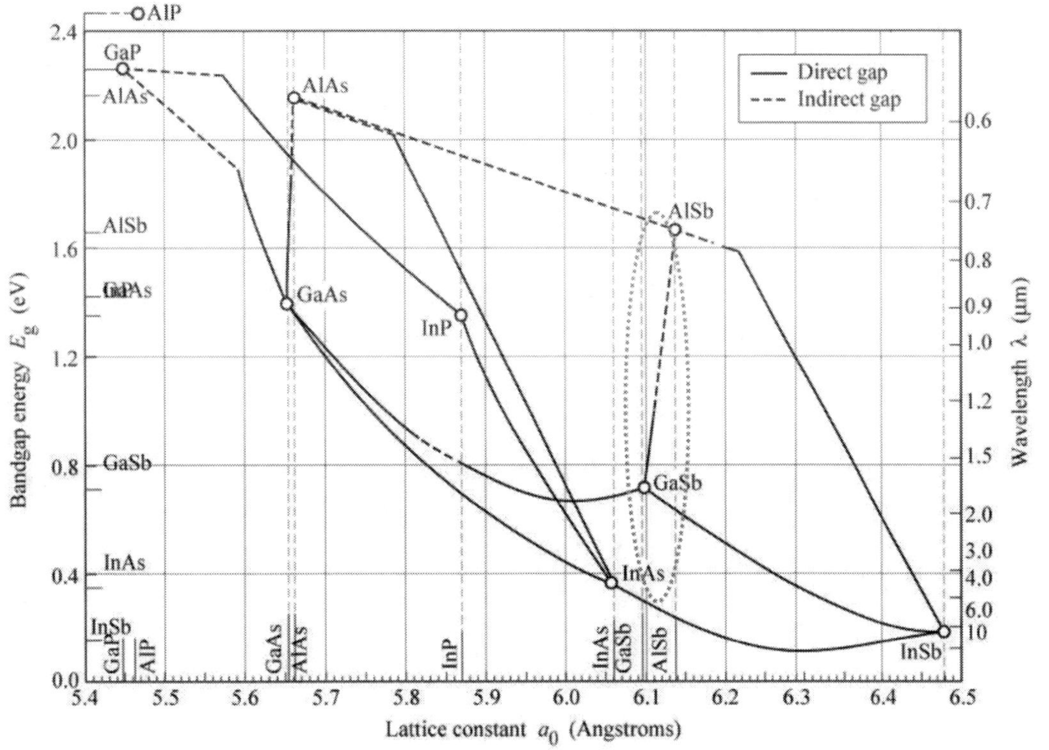

Fig. 2. Diagram of solid solutions

Table 1. The bandgap of the obtained solid solutions

№	Solid solution	Eg, эВ
1	$InAs_{0,917}Sb_{0,083}$	0,29
2	$Ga_{0,074}In_{0,92}As_{0,89}Sb_{0,11}$	0,32
3	$Ga_{0,58}In_{0,42}As_{0,406}Sb_{0,594}$	0,43
4	$Ga_{0,79}In_{0,21}As_{0,127}Sb_{0,873}$	0,48
5	$Ga_{0,848}In_{0,152}As_{0,096}Sb_{0,904}$	0,54
6	$Ga_{0,85}In_{0,15}As_{0,11}Sb_{0,89}$	0,57
7	$Ga_{0,923}In_{0,077}As_{0,072}Sb_{0,928}$	0,64
8	GaSb	0,72
9	$Al_{0,016}Ga_{0,983}As_{0,005}Sb_{0,995}$	0,74
10	$Al_{0,111}Ga_{0,889}As_{0,013}Sb_{0,987}$	0,84
11	$Al_{0,197}Ga_{0,803}As_{0,024}Sb_{0,976}$	0,93

Growth of solid solutions held at a temperature of T = 600 ^{0} C and a pressure of P = 100 mbar. The flow of carrier gas (H2) was - Fc = 5 l/ min. The source of arsenic (As) to a solid solution served - arsine (AsH3), the source of antimony (Sb) - trimetilsurma (TMSb). Ga source served - trietilgally (TEGa), served as a source of In - TMIn, served as a source of Al - trimethylaluminum (TMAl). The quality of the solid solutions has been examined by X-ray diffraction analysis (Fig. 3.) and photoluminescence (Fig. 4). Table 1. shows the compositions and the bandgap of the solid solutions.

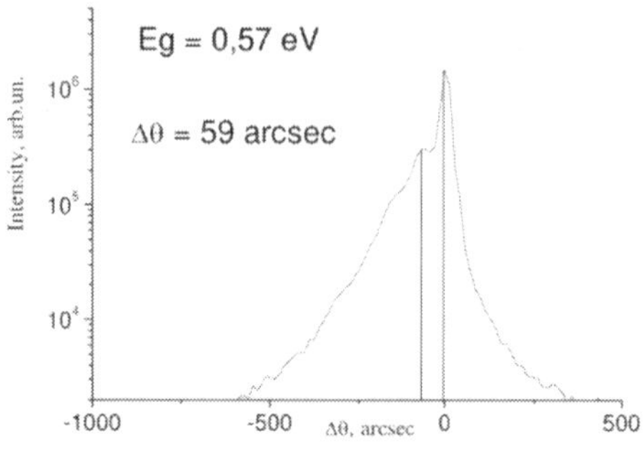

Fig. 3. XRD spectr of the sample (Eg = 0.57 eV)

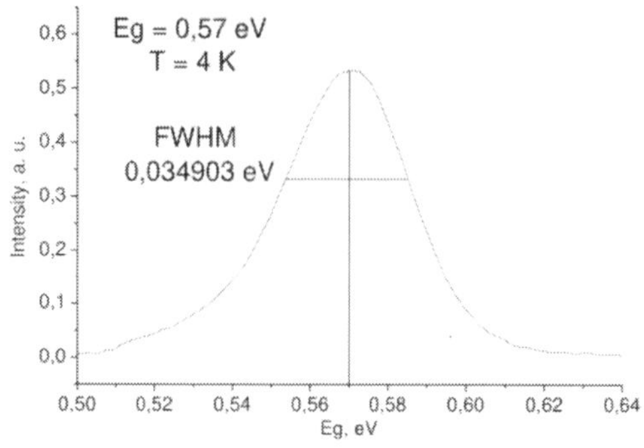

Fig. 4. PL spectr of the sample (Eg = 0.57 eV)

3. Obtaining of solid solutions with different gradient of the band gap and based on their structures.

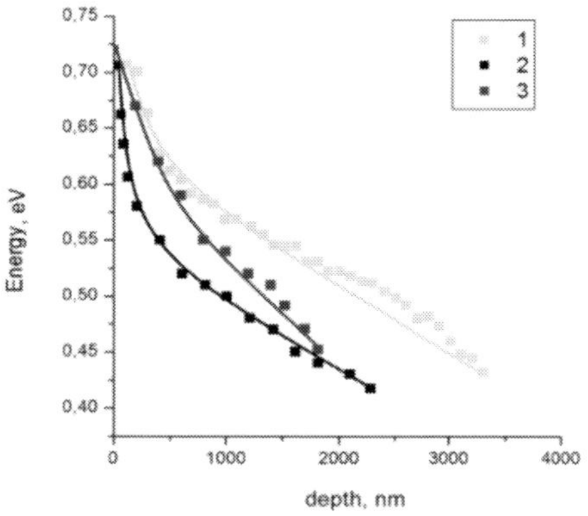

Fig. 5. Profile bandgap through the thickness (gradient Eg)

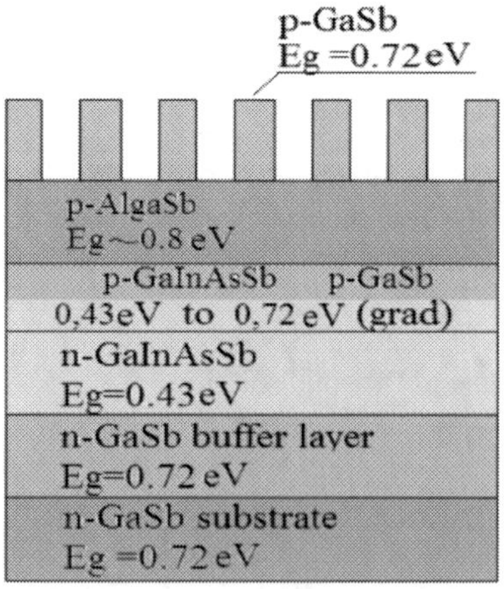

Fig. 6. Schematic representation structure under study.

Table 2. The value of the gradients for different thicknesses

d, μm	1.7	2.2	3.4
∇E, kV/cm	1.7	1.3	0.85

Based on these results were identified technological modes of manufacturing graded-gap structure with a smooth change of the band gap of 0.43 eV to 0.72 eV, and different changes in the gradient of the band gap (Fig. 5). Profile of the band gap was calculated according to a study obtained by mass spectrometry of secondary ions (SIMS).

Based on these studies, the technology was developed and manufactured structure with graded-gap layer thickness of 0.5 μm (Fig. 6). The internal electric field pulling the obtained structure was 5.8 kV/cm (Fig. 7.). Fig. 7. shows the band diagram the obtained structure, the real and the assumption of fixing the valence band. The open circuit voltage in such a structure, it V_{oc} = 0.33 V (300 K), V_{oc} = 0.5 V (77 K) (Fig. 8.). Indicating that the increase in the open circuit voltage due to the contribution of the gradient of the band gap.

Fig. 9. shows the spectral characteristics structure under study in comparison with the structure of gallium antimonide. The comparison shows that the graded-gap structure of the obtained absorption spectrum is wider due to the narrow-band and the contribution of the gradient of the band gap.

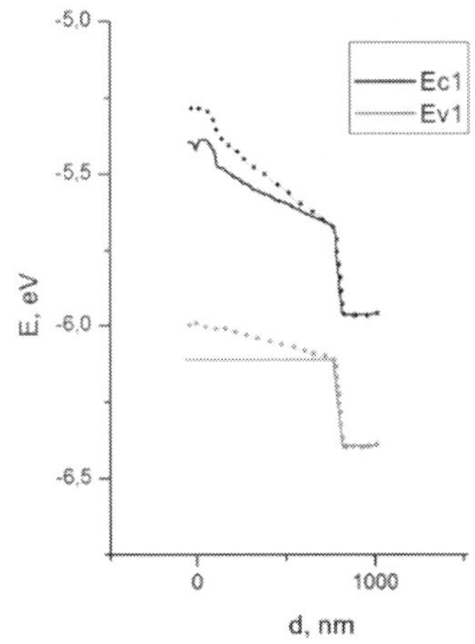

Fig. 7. Band diagram of the obtained structure

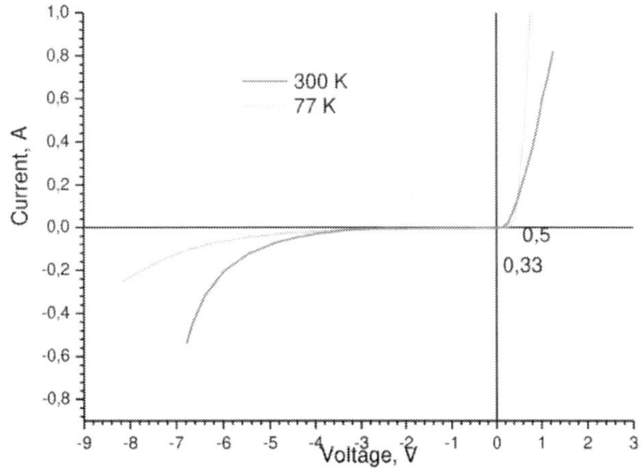

Fig. 8. The current-voltage characteristics of the obtained structure

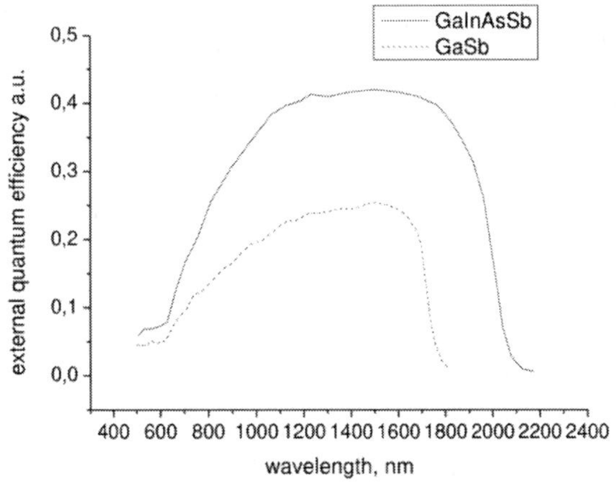

Fig. 9. The spectral characteristics of photovoltaic cells

4. Conclusion

Were identified technological modes of manufacturing graded-gap structure with a smooth change of the band gap of 0.43 eV to 0.72 eV, and different changes in the gradient of the band gap. Based on these studies, the technology was developed and manufactured structure with graded-gap layer thickness of 0.5 μm. The internal electric field pulling the obtained structure was 5.8 kV/ cm.

The open circuit voltage in such a structure, it Uhh = 0.33 V (T = 300 K) Uhh = 0.5 V (T = 77 A). Indicating that the increase in the open circuit voltage due to the contribution of the gradient of the band gap. Due to the narrow-band contribution to the gradient of the band gap, there is a broadening of the absorption spectrum. A subsequent study is planned to expand the range of variation of the band gap of 0.43 to 1.2 eV.

The work was supported in part by the Russian Foundation for the Basic Research (grant No 12-08-00219).

References

[1] J Tauc. Rev. 1957 Moden Phys., 29, chapter 3, p 308.

[2] Z I Alferov, V M Andreev and V I Korol'kov. 1978 Technical Physics Letters, 4, chapter 7, p 369.

[3] P A Houston. 1981 Journal of Materials Science, 16, 11, pp 2935-2961

Journal of Physics: Conference Series **572** (2014) 012052 doi:10.1088/1742-6596/572/1/012052

An 8-band Kane model for quantum-sized heterostructures based on cubic semiconductors A^3B^5

G F Glinskii and M S Mironova

Department of Micro- and Nanoelectronics, St. Petersburg State Electrotechnical University, 5 Professora Popova Street, St. Petersburg 197376, Russia

E-mail: mironova.m.s@gmail.com

Abstract. This paper presents an 8-band Kane model for arbitrary heterostructures with abrupt interfaces. The method of invariants for heterostructures is used to construct Γ_1-Γ_{15} effective kp-Hamiltonian, taking into account both conductance and valence bands kp-interaction and charge carriers scattering at an interface. The Γ_6-Γ_8-Γ_7 effective kp-Hamiltonian is obtained from the Γ_1-Γ_{15} effective kp-Hamiltonian within the assumption that spin-orbit interaction of the bands Γ_6, Γ_8 and Γ_7 with all the rest band can be neglected.

1. Introduction

At the present time, the effective mass approach (EMA) is a highly popular method for calculation of energy states in quantum-sized semiconductor heterostructures. However, bands parameters dependence on coordinate doesn't allow form of the kinetic energy operator to be uniquely determined [1]. Moreover, bands parameters of bulk materials are not enough to describe Hamiltonian of heterostructure [2] and it is necessary to introduce additional parameters, which don't have any analogous in bulk material and exists only in presence of an interface. There are two different approaches to take into account charge carriers scattering on a heterointerface: determining of boundary conditions [3-5] or determining of form of Hamiltonian [6-15]. In [12, 16] a general symmetry approach to determine corrections to EMA, which caused by short-range part of an interface potential, was introduced by one of authors of the present paper.

Besides charge carriers scattering effects, in heterostructures based on narrow-gap semiconductors a kp-interaction of conductance and valence bands should be taken into account. This can be done within an 8-band Kane model for heterostructures. Such models for superlattices were proposed in [14, 15].

In this paper, we introduce 8-band effective kp-Hamiltonian of an arbitrary heterostructure with abrupt interfaces, which includes both kp-interaction and charge carriers scattering on an interface and takes into account microscopic symmetry of an interface. First, we use the method of invariants[16] for heterostructures to construct an effective kp-Hamiltonian Γ_1-Γ_{15}. Then, we obtain Γ_6-Γ_8-Γ_7 Kane Hamiltonian from Γ_1-Γ_{15} via unitary transformation. The thus obtained Hamiltonian does not include spin-orbit interaction of bands in question with all other bands.

2. Hamiltonian of heterostructure

We considered heterostructure based on two semiconductors I and II as a bulk material I, some atoms of which are substituted by atoms of material II. As an approximation, we assume that the lattice parameters of bulk materials are the same. As shown in [16, 17], Hamiltonian of such structure can be expressed as

$$\hat{H} = \frac{\hat{\mathbf{p}}^2}{2m_0} + \sum_l \sum_{\mathbf{a}} V_l^I(\hat{\mathbf{x}} - \mathbf{a} - \chi_l) + \sum_l \sum_{\mathbf{a}} f_l(\mathbf{a}) \Delta V_l(\hat{\mathbf{x}} - \mathbf{a} - \chi_l), \tag{1}$$

where $\hat{\mathbf{p}}$ is the momentum operator; m_0 is the electron rest mass; $V_l^I(\mathbf{x})$ is a local potential produced by an atom l-th sublattice of material I and defined within a unit cell; $\mathbf{a}$ are direct lattice vectors, indexing unit cells, $l = 1,2$ for A^3B^5; χ_l are vectors indicating a position of an atom of l-th sublattice in a unit cell; $\Delta V_l(\mathbf{x}) = V_l^{II}(\mathbf{x}) - V_l^I(\mathbf{x})$ is the difference between the local potentials produced by atoms of l-th sublattice of material II and I; $f_l(\mathbf{a})$ is a characteristic functions indicating the replacement of atoms in the l-th sublattice:

$$f_l(\mathbf{a}) = \begin{cases} 1, & \textit{if the unit cell with number } \mathbf{a} \textit{ occupied by atom of material I} \\ 0, & \textit{if the unit cell with number } \mathbf{a} \textit{ occupied by atom of material II} \end{cases}.$$

Let us consider $f_1(\mathbf{a}) = f(\mathbf{a})$, $f_2(\mathbf{a}) = 0$ and $\chi_1 = 0$. Thus, Hamiltonian (1) in Luttinger-Kohn basis $e^{i\mathbf{k}\hat{\mathbf{x}}}|\alpha,n\rangle$ can be rewritten as follows[KH,CT]

$$\langle \alpha,n | e^{-i\mathbf{k}\hat{\mathbf{x}}} \hat{H} e^{i\mathbf{k}'\hat{\mathbf{x}}} | \alpha',n' \rangle = \left\{ \left[E_A^{(\alpha)} + \frac{\hbar^2 k^2}{2m_0} \right] \delta_{\alpha\alpha'} \delta_{nn'} + \frac{\hbar}{m_0} \mathbf{k}\mathbf{p}_{nn'}^{\alpha\alpha'} \right\} \delta_{\mathbf{k}\mathbf{k}'} + \Delta U_{nn'}^{\alpha\alpha'}(\mathbf{k} - \mathbf{k}'). \tag{2}$$

The first term in (2), which is diagonal in the indices $\mathbf{k}$ and $\mathbf{k}'$, is an ordinary kp-Hamiltonian of the material I. The second term $\Delta U_{nn'}^{\alpha\alpha'}(\mathbf{k}-\mathbf{k}')$ determines charge carriers scattering at the I/II interface:

$$\Delta U_{nn'}^{\alpha\alpha'}(\mathbf{k}-\mathbf{k}') = f(\mathbf{k}-\mathbf{k}') \frac{1}{\Omega} \int_{\Omega} u_n^{\alpha *}(\mathbf{x}) \Delta V(\mathbf{x}) u_{n'}^{\alpha'}(\mathbf{x}) e^{-i(\mathbf{k}-\mathbf{k}')\mathbf{x}} d\mathbf{x}, \tag{3}$$

$$f(\mathbf{k}-\mathbf{k}') = \frac{1}{N} \sum_{\mathbf{a}} f(\mathbf{a}) e^{-i(\mathbf{k}-\mathbf{k}')\mathbf{a}}.$$

If we assume that $\mathbf{k}-\mathbf{k}'$ is small value compared with Brillouin zone, we can expand the exponential in a Taylor series:

$$e^{-i(\mathbf{k}-\mathbf{k}')\mathbf{x}} = 1 - i(\mathbf{k}-\mathbf{k}')\mathbf{x} - \frac{1}{2}(\mathbf{k}-\mathbf{k}')_i(\mathbf{k}-\mathbf{k}')_j \mathbf{x}_i \mathbf{x}_j + \ldots$$

Thus, we can rewrite matrix element (3) as

$$\Delta U_{nn'}^{\alpha\alpha'}(\mathbf{k}-\mathbf{k}') = \left[\Delta V_{nn'}^{\alpha\alpha'} + i\left(\Delta L_{nn'}^{\alpha\alpha'}\right)_i (k_i - k_i') - \left(\Delta Q_{nn'}^{\alpha\alpha'}\right)_{ij} (k_i - k_i')(k_j - k_j') \right] f(\mathbf{k}-\mathbf{k}').$$

The band offset $\Delta E^{(\alpha)}$ of a band $E^{(\alpha)}$ in the first order of perturbation theory is determined by the diagonal matrix element $\Delta V_{nn}^{\alpha\alpha}$

$$\Delta V_{nn}^{\alpha\alpha} = \frac{1}{\Omega} \int_{\Omega} \left| u_n^{\alpha}(\mathbf{x}) \right|^2 \Delta V(\mathbf{x}) d\mathbf{x} = \Delta E^{(\alpha)}.$$

3. Effective Hamiltonian of heterostructure and the method of invariants.

The Hamiltonian (2) contains all of the bands of material I. If the main contribution to the formation of electronic states is provided by states of bands α and β, which are separated from the rest bands γ by large energy gaps, it is convenient to use an effective multi-band Hamiltonian $H_{mn}^{\alpha\beta}(\mathbf{k},\mathbf{k}')$ instead of complete Hamiltonian (2). In this Hamiltonian the interaction between states of interest $|\alpha,m\rangle$ and $|\beta,n\rangle$ should be taken into account accurately and the interaction of these states with any other states

$|\gamma, s\rangle$ can be considered within the desired order of perturbation theory. Such Hamiltonian can be derived using either the perturbation theory or the method of invariants.

At the heart of the method of invariants lie three conditions imposed on the effective Hamiltonian: Hamiltonian must be Hermitian, invariant under time reversal and invariant under point transformations belonging to the crystal point symmetry group of bulk materials I and II. The Hamiltonian of a heterostructure, obtained via the method of invariants, contains set of parameters ("constants" of the method of invariants), which should be determined either from experiment or from microscopic calculations.

The effective multi-band Hamiltonian of an arbitrary heterostructure might be written as

$$H^{\alpha\beta}(\mathbf{k},\mathbf{k}') = H_I^{\alpha\beta}(\mathbf{k})\delta_{\mathbf{kk}'} + \Delta H_1^{\alpha\beta}(\mathbf{k},\mathbf{k}') + \Delta H_2^{\alpha\beta}(\mathbf{k},\mathbf{k}'),$$

where $H_I^{\alpha\beta}(\mathbf{k})$ is a matrix of an effective kp-Hamiltonian of bulk material I containing band parameters γ_i^I; $\Delta H_{1,2}^{\alpha\beta}(\mathbf{k},\mathbf{k}')$ are corrections to EMA due to the presence of the interface. Term $\Delta H_1^{\alpha\beta}(\mathbf{k},\mathbf{k}')$ is proportional to $\Delta\gamma_i f(\mathbf{k}-\mathbf{k}')$, where $\Delta\gamma_i = \gamma_i^{II} - \gamma_i^I$ are differences between band parameters of bulk materials II and I. Term $\Delta H_2^{\alpha\beta}(\mathbf{k},\mathbf{k}')$ is proportional to $\Delta\beta_j f(\mathbf{k}-\mathbf{k}')$, where $\Delta\beta_j$ are parameters, which do not have any analogous in bulk materials.

Table 1. Hermitian forms made up of components of vectors of $\mathbf{k}$ and $\mathbf{k}'$ that transform according to the irreducible representations of the point group Td.

Representation	Even under time reversal	Odd under time reversal
Γ_1	$\mathbf{kk}'$, $(\mathbf{k}-\mathbf{k}')^2$	$i(\mathbf{k}^2-\mathbf{k}'^2)$
Γ_{12}	$\begin{cases} 2k_zk_z'-k_xk_x'-k_yk_y' \\ \sqrt{3}(k_xk_x'-k_yk_y') \end{cases}$, $\begin{cases} 2(k_z-k_z')^2-(k_x-k_x')^2-(k_y-k_y')^2 \\ \sqrt{3}\left[(k_x-k_x')^2-(k_y-k_y')^2\right] \end{cases}$	$\begin{cases} i\left[2(k_z^2-k_z'^2)-(k_x^2-k_x'^2)-(k_y^2-k_y'^2)\right] \\ \sqrt{3}i\left[(k_x^2-k_x'^2)-(k_y^2-k_y'^2)\right] \end{cases}$
Γ_{15}	$i(\mathbf{k}-\mathbf{k}')$, $\begin{cases} k_yk_z'+k_zk_y' \\ k_zk_x'+k_xk_z' \\ k_xk_y'+k_yk_x' \end{cases}$, $\begin{cases} (k_y-k_y')(k_z-k_z') \\ (k_z-k_z')(k_x-k_x') \\ (k_x-k_x')(k_y-k_y') \end{cases}$	$\frac{1}{2}(\mathbf{k}+\mathbf{k}')$, $\begin{cases} i(k_yk_z-k_y'k_z') \\ i(k_zk_x-k_z'k_x') \\ i(k_xk_y-k_x'k_y') \end{cases}$
Γ_{15}	-	$i[\mathbf{k}\times\mathbf{k}']$

To use the method of invariants, first, we need to determine all possible Hermitian forms made up of components of vectors $\mathbf{k}$ and $\mathbf{k}'$, which transform according to the irreducible representations Γ_α of crystal point group. Then, determine a complete set of matrices, which transform according to the irreducible representations Γ_α, using the Clebsch-Gordan coefficients. Finally, using these Hermitian forms and basis matrices, we need to construct all possible invariants, satisfying the conditions imposed on the effective Hamiltonian. Hermitian forms of $\mathbf{k}$ and $\mathbf{k}'$ that transform according to the irreducible representations of the point group Td are shown in table 1.

The Hermitian forms from table 1, which are not equal to zero if $\mathbf{k}=\mathbf{k}'$, are contained in the effective Hamiltonian of a heterostructure with parameters $\Delta\gamma_i$. And Hermitian forms, which vanish if $\mathbf{k}=\mathbf{k}'$, are contained in the effective Hamiltonian with parameters $\Delta\beta_j$.

4. Kane model for heterostructures based on cubic semiconductors A^3B^5.

An 8-band Kane model accurately takes into account an interaction between the conductance band Γ_6, the valence band Γ_8 and the spin-orbit split band Γ_7. The interaction of these bands with the rest is considered within the perturbation theory:

$$H=\begin{pmatrix} H^{(\Gamma_6)} & H^{\Gamma_6\Gamma_8} & H^{\Gamma_6\Gamma_7} \\ \left[H^{\Gamma_6\Gamma_8}\right]^\dagger & H^{(\Gamma_8)} & H^{\Gamma_8\Gamma_7} \\ \left[H^{\Gamma_6\Gamma_7}\right]^\dagger & \left[H^{\Gamma_8\Gamma_7}\right]^\dagger & H^{(\Gamma_7)} \end{pmatrix}.$$

If we neglect the contribution due of the spin-orbit interaction of the bands in question with all the others, we can obtain the Kane effective Hamiltonian of a heterostructure from Hamiltonian for bands Γ_1 and Γ_{15}, which should include both kp-interaction and interface scattering terms. This Hamiltonian might be obtained using the method of invariants.

We omit the intermediate results and present only corrections to EMA $\Delta H_2^{\alpha\beta}(\mathbf{k},\mathbf{k}')$ ($\alpha,\beta=\Gamma_6,\Gamma_7,\Gamma_8$), which vanish in bulk materials

$$\Delta H_2^{(\Gamma_6)}=\begin{pmatrix} \Delta\beta_c(\mathbf{k}-\mathbf{k}')^2 & 0 \\ 0 & \Delta\beta_c(\mathbf{k}-\mathbf{k}')^2 \end{pmatrix} f(\mathbf{k}-\mathbf{k}'),$$

$$\Delta H_2^{(\Gamma_7)}=\begin{pmatrix} \Delta P_2+\frac{2}{3}\Delta F & \frac{2}{3}\Delta G \\ \frac{2}{3}\Delta G^\dagger & \Delta P_2-\frac{2}{3}\Delta F \end{pmatrix} f(\mathbf{k}-\mathbf{k}'),$$

$$\Delta H_2^{(\Gamma_8)}=\begin{pmatrix} \Delta P_2+\Delta Q_2+\Delta F & \Delta W_2+\frac{1}{\sqrt{3}}\Delta G & \Delta V_2 & 0 \\ \Delta W_2^\dagger+\frac{1}{\sqrt{3}}\Delta G^\dagger & \Delta P_2-\Delta Q_2+\frac{1}{3}\Delta F & \frac{2}{3}\Delta G & \Delta V_2 \\ \Delta V_2^\dagger & \frac{2}{3}\Delta G^\dagger & \Delta P_2-\Delta Q_2-\frac{1}{3}\Delta F & -\Delta W_2+\frac{1}{\sqrt{3}}\Delta G \\ 0 & \Delta V_2^\dagger & -\Delta W_2^\dagger+\frac{1}{\sqrt{3}}\Delta G^\dagger & \Delta P_2+\Delta Q_2-\Delta F \end{pmatrix} f(\mathbf{k}-\mathbf{k}'),$$

$$\Delta H_2^{\Gamma_6\Gamma_8}=\begin{pmatrix} \frac{\sqrt{3}}{\sqrt{2}}\Delta R_2^* & \sqrt{2}\Delta U & \frac{1}{\sqrt{2}}\Delta R_2 & 0 \\ 0 & \frac{1}{\sqrt{2}}\Delta R_2^* & \sqrt{2}\Delta U & \frac{\sqrt{3}}{\sqrt{2}}\Delta R_2 \end{pmatrix} f(\mathbf{k}-\mathbf{k}'),$$

$$\Delta H_2^{\Gamma_6\Gamma_7}=\begin{pmatrix} \Delta U_2 & \Delta R_2 \\ -\Delta R_2^* & -\Delta U_2 \end{pmatrix} f(\mathbf{k}-\mathbf{k}'),$$

$$\Delta H_2^{\Gamma_8\Gamma_7} = \begin{pmatrix} \frac{1}{\sqrt{2}}\Delta W_2 + \frac{1}{\sqrt{6}}\Delta G & \sqrt{2}\Delta V_2 \\ -\sqrt{2}\Delta Q_2 - \frac{\sqrt{2}}{3}\Delta F & -\frac{\sqrt{3}}{\sqrt{2}}\Delta W_2 + \frac{\sqrt{2}}{6}\Delta G \\ -\frac{\sqrt{3}}{\sqrt{2}}\Delta W_2^* + \frac{\sqrt{2}}{6}\Delta G^* & \sqrt{2}\Delta Q_2 - \frac{\sqrt{2}}{3}\Delta F \\ -\sqrt{2}\Delta V_2^* & \frac{1}{\sqrt{2}}\Delta W_2 + \frac{1}{\sqrt{6}}\Delta G^* \end{pmatrix} f(\mathbf{k}-\mathbf{k}'),$$

where

$$\Delta P_2 = \Delta\beta_1(\mathbf{k}-\mathbf{k}')^2,$$

$$\Delta Q_2 = \Delta\beta_2\left((k_x-k'_x)^2+(k_y-k'_y)^2-2(k_z-k'_z)^2\right),$$

$$\Delta V_2 = -\Delta\beta_2\sqrt{3}\left((k_x-k'_x)^2-(k_y-k'_y)^2\right)+i2\sqrt{3}\Delta\beta_3(k_x-k'_x)(k_y-k'_y)-2\sqrt{3}\Delta\beta_4(k_z-k'_z),$$

$$\Delta W_2 = -2\sqrt{3}\Delta\beta_3(k_z-k'_z)\left((k_x-k'_x)-i(k_y-k'_y)\right)-2\sqrt{3}\Delta\beta_4\left(i(k_y-k'_y)+(k_x-k'_x)\right),$$

$$\Delta F = i\Delta\beta_4[\mathbf{k}\times\mathbf{k}']_z,$$

$$\Delta G = i\Delta\beta_4\left([\mathbf{k}\times\mathbf{k}']_x - i[\mathbf{k}\times\mathbf{k}']_y\right),$$

$$\Delta R_2 = \frac{i}{\sqrt{3}}\Delta\beta_{cv}\left((k_x-k'_x)-i(k_y-k'_y)\right),$$

$$\Delta U_2 = \frac{i}{\sqrt{3}}\Delta\beta_{cv}(k_z-k'_z).$$

Note that the thus obtained effective 8-band Hamiltonian includes all of possible effects in considered heterostructures that caused by charge carriers scattering on an interface, such as the heavy-light hole mixing in (001) quantum wells [18] (term $\sim\Delta\beta_4$ in $\Delta H_2^{(\Gamma_8)}$) and an interface Rashba effect for holes (terms $\sim\Delta F,\Delta G$ in $\Delta H_2^{(\Gamma_8)}$ and $\Delta H_2^{(\Gamma_7)}$). We also note that matrix element of light holes mixing is zero in bulk materials for any wave vector $\mathbf{k}$, but it exists for heterostructures (term $\sim\frac{2}{3}\Delta G$ in $\Delta H_2^{(\Gamma_8)}$). Although this term has little effect in structures with superlattices and quantum wells grown along the [001] directions, because of its proportionality to $k_{x,y}$, it can possibly make a significant contribution in case of quantum dots or (011) and (111) interfaces.

5. Conclusions

In this paper we demonstrated the possibilities of the method of invariants for heterostructures, introduced in [16], to construct effective Hamiltonians of arbitrary heterostructures, taking into account both kp-interaction and charge carriers scattering at an interface. We introduced a table of all possible Hermitian forms of $\mathbf{k}$ and $\mathbf{k}'$, which transform according to the irreducible representations of the point group Td. Using the method of invariants we obtained the effective 8-band Kane Hamiltonian for heterostructures with abrupt interfaces within, assuming interaction of the conductance band Γ_6, valence band Γ_8 and spin-orbit split band Γ_7 with all other bands is negligible. We determined corrections to 8-band Kane Hamiltonian, which caused only by charge carriers scattering at an interface and vanish in bulk materials.

References

[1] von Roos O 1983 *Phys. Rev. B* **27** 7547
[2] Foreman B A 1993 *Phys. Rev. B* **48** 4969
[3] Kisin M V Gelmont B L and Luryi S 1998 *Phys. Rev. B* **58** 4605.
[4] Rodina A V *et al* 2002 *Phys. Rev. B* **65** 125302
[5] Rodina A V Alekseev A Yu 2006 *Phys. Rev. B* **73** 115312
[6] Volkov V A and Takhtamirov E E 1991 *Semicond. Sci. Technol.* **12** 77
[7] Volkov V A and Takhtamirov E E 1997 *Phys. Usp.* **40** 1071
[8] Volkov V A and Takhtamirov E E 1999 *JETP* **89** 1000
[9] Volkov V A and Takhtamirov E E 2000 *JETP* **90** 1063
[10] Glinskii G F *et al* 2000 *Nanotechnology* **11** 233
[11] Glinskii G F and Kravchenko K O 1998 *arXiv preprint cond-mat/9808174* (unpublished)
[12] Glinskii G F and Kravchenko K O 1998 *Phys. Solid State* **40** 803
[13] Foreman B A 2005 *Phys. Rev. B* **72** 165345
[14] Foreman B A 2007 *Phys. Rev. B* **76** 045326
[15] Klipstein P C 2010 *Phys. Rev. B* **81** 235314
[16] Glinskii G F 2008 Semiconductors *and semiconductor nanoelectronics: symmetry and* electronic states (Saint Petersburg: Tehnolit)
[17] Glinskii G F and Mironova M S 2013 *J. Phys.: Conf. Series* **461** 012040
[18] Ivchenko E L Kaminski A Y and Rössler U 1996 *Phys Rev B* **54** 5852

Propagation of the chirped microwave pulse in the long-distance sliding-mode plasma waveguide in air.

A V Bogatskaya[1,2], I V Smetanin[1], V D Zvorykin[1], A O Levchenko[1] and N N Ustinovskii[1]

[1]P.N. Lebedev Physical Institute, Leninskii prosp. 53, 119991 Moscow, Russia
[2]M. V. Lomonosov Moscow State Univ., Department of Physics, 119991, Moscow, Russia

E-mail: annabogatskaya@gmail.com

Abstract. Propagation of the phase-modulated microwave pulse in the sliding-mode plasma waveguide in air is considered. It is shown that with the proper choice of the chirp parameter the peak intensity of the microwave pulse can be significantly enhanced with respect to the non-phase-modulated pulse thus compensating the attenuation by the pulse compression. Calculations show that the microwave pulse peak intensity at the end of plasma waveguide can be made larger than the initial peak intensity, even for the plasma waveguides of a few kilometres length.

1. Introduction

Impressive progress in creation of long ionized channels in atmosphere with the use of filamentation effect [1,2] has renewed interest in such promising applications as initiation of lightning and triggering of high-voltage discharges [3-6], microwave transport [7,8], remote sensing, biological and chemical detection [1,9]. At the present moment, a numerous schemes of microwave transport utilizing different plasma configurations have been studied theoretically, some of them have been tested experimentally. High-conductivity dense plasma hollow waveguide scheme has been studied theoretically in [10], the experimental test of which was performed with the 4.5 cm diameter waveguide formed in laboratory air by multiple filaments, and as a result, it has been demonstrated microwave (at λ=3 cm the wavelength, 10 GHz the frequency) transport on the distance of ~16 cm [7]. Using of periodically arranged filaments can provide somewhat longer propagation distances of microwaves [11, 12]. Prospects of microwave transmission lines based on Zernike-Zommerfeld surface wave excited at a single plasma filament have been studied in [13], the two-filament (i.e., analog of double-wire line) configuration has been considered in [14].

The most prospective for the long-distance (up to few kilometers) microwave transport is the concept of the hollow plasma sliding-mode waveguide which has been developed in [8, 15-17]. The concept is based on the effect of total reflection at the interface with an optically less dense medium (i.e., plasma wall of the waveguide): When the inner radius of tubular plasma waveguide is large enough with respect to the microwave wavelength $R>>\lambda$, the effective angle of incidence for the lowest-order modes exceeds the critical angle determined by the ratio of refractive indices of air and plasma. In contrast with [7, 10], high conductivity of the plasma is not necessary, and one can provide effective transport of sub-mm and mm-range radiation at low degree of air ionization, with the waveguide wall plasma density of $10^{11} \div 10^{14}$ cm^{-3} . This sliding-mode microwave transport scheme was demonstrated for the first time in the experiment [8, 15], in which the 35.5 GHz signal has been transported over 60 m distance with the use of plasma tubular waveguide of 10 cm diameter created in the laboratory air by KrF laser radiation. Theoretical calculations show [17] the characteristic propagation distance of the mm-range wave can be enlarged up to few kilometers with ~30 cm waveguide radius and plasma wall density of ~ 10^{12} cm^{-3}.

Theoretical analysis [16, 17] shows that sliding-mode plasma waveguides are characterized by rather strong dispersion. At the air plasma densities below 10^{14} cm^{-3} the dominant mechanism of the plasma decay is the three-body attachment of electrons, the characteristic time of which is ~10 ns [18], thus constraining the microwave pulse width. This fact means that such a rather short microwave pulses can exhibit dispersive growth in their pulse widths when propagates over large (kilometer-range) distances, which leads to an increase in pulse peak intensity with respect to the case of monochromatic signal.

In this paper we propose, in analogy with the conventional nonlinear optics (see, for example, [19]), the technique of phase-modulated microwave pulses to compensate this effect of dispersive spreading and even enlarge the peak intensity at the end of the sliding-mode waveguide.

2. Dispersion characteristics of the sliding mode plasma waveguide

To characterize the dispersion properties of the sliding-mode plasma waveguide, let us consider the dispersion equations for the lowest axial-symmetric modes E_{01} ($\chi = \varepsilon_p$) and H_{01} ($\chi = 1$),

$$\frac{1}{\kappa_1 R}\frac{J_1(\kappa_1 R)}{J_0(\kappa_1 R)} = \frac{\chi}{\kappa_2 R}\frac{H_1^{(1)}(\kappa_2 R)}{H_0^{(1)}(\kappa_2 R)}, \tag{1}$$

and for the lowest axial-asymmetric mode EH_{11},

$$\left[\left(\frac{1}{\kappa_1}\frac{J_0(\kappa_1 R)}{J_1(\kappa_1 R)} - \frac{\varepsilon_p}{\kappa_2 R}\frac{H_1^{(1)}(\kappa_2 R)}{H_0^{(1)}(\kappa_2 R)}\right) - \left(\frac{1}{\kappa_1^2 R} - \frac{\varepsilon_p}{\kappa_2^2 R}\right)\right] \times$$
$$\left[\left(\frac{1}{\kappa_1}\frac{J_0(\kappa_1 R)}{J_1(\kappa_1 R)} - \frac{1}{\kappa_2 R}\frac{H_1^{(1)}(\kappa_2 R)}{H_0^{(1)}(\kappa_2 R)}\right) - \left(\frac{1}{\kappa_1^2 R} - \frac{1}{\kappa_2^2 R}\right)\right] = \left(\frac{1}{\kappa_1^2 R} - \frac{\varepsilon_p}{\kappa_2^2 R}\right)\left(\frac{1}{\kappa_1^2 R} - \frac{1}{\kappa_2^2 R}\right) \tag{2}$$

Here, $\kappa_1^2 = \omega^2/c^2 - h^2$ and $\kappa_2^2 = \varepsilon_p \omega^2/c^2 - h^2$ are the transverse wavenumbers within the waveguide (the dielectric susceptibility of air is assumed to be equal 1) and the waveguide wall, respectively, h is the propagation constant of the microwave signal of the frequency ω, $J_n(x)$ and $H_n^{(1)}$ are the Bessel functions and the Hankel functions of the first kind. The wall plasma dielectric susceptibility is $\varepsilon_p = 1 - \Omega_p^2/(\omega(\omega + i\nu))$, $\Omega_p = (4\pi e^2 n_e/m)^{1/2}$ is the plasma frequency, ν is the electron transport collision rate, n_e is the plasma density. Dispersion equations (1) and (2) are written under assumption of the abrupt plasma density profile. A real plasma waveguide wall has a spread density profile which is caused by the UV laser intensity profile, diffraction, plasma diffusion etc. It has been shown in [17] that this density spreading doesn't effect on the sliding microwave mode dispersion characteristics if $\kappa_2 d \ll 1$, where d is the characteristic size of the plasma wall density spreading.

It has been shown in [8, 16, 17] that the sliding mode regime in the hollow plasma waveguide is possible when the parameter $\mu = \xi k_0 R > 1$, $\xi = \Omega_p/(\omega^2 + \nu^2)^{1/2}$. In the range of parameters $\mu \sim 0.5 \div 1$, the characteristic propagation length sharply decreases and vanishes with $\mu \to 0$.

At the limit $\mu >> 1$, $\nu/\omega >> 1$ the roots of dispersion equations (1), (2) can be approximated by the following analytic expressions [16, 17].

(a) At sufficiently low densities, when $|1 - \varepsilon_p| \ll k_0 R\sqrt{\xi\nu/\omega}$, we find for the microwave signal wavenumber:

Journal of Physics: Conference Series **572** (2014) 012053 doi:10.1088/1742-6596/572/1/012053

$$h-k_0 \approx -\frac{\alpha^2}{2k_0R^2}\left(1-i\frac{\chi}{\mu}\sqrt{\frac{2\omega_0}{\nu}}\right) \qquad (3)$$

Here $\alpha=\alpha_1$ for the axially symmetric modes H_{01} ($\chi=1$) and E_{01} ($\chi=\varepsilon_p$), $\alpha_1\approx 3.83$ is the first root of the Bessel function $J_1(x)=0$ and $\alpha=\alpha_0$ for the lowest asymmetric mode $EH_{11}, \alpha_0\approx 2.405$ is the first root of $J_0(x)=0$. The characteristic propagation length for the EH_{11} sliding mode is thus $(\alpha_1/\alpha_0)^2\approx 2.54$ times larger than that for the axially-symmetric modes.

(b) At moderate densities, $k_0R\gg\xi\sqrt{\nu/\omega}\gg 1$, we find the solution for the H_{01} mode is given by the formula (3) whereas for E_{01} and EH_{11} modes wavenumbers are

$$h-k_0 \approx -\frac{\alpha^2}{2k_0R^2}\left(1-i\frac{\xi}{k_0R}\sqrt{\frac{\nu}{2\omega_0}}\right) \qquad (4)$$

with $\alpha=\alpha_1$ for the E_{01} mode and $\alpha=\alpha_0$ for the EH_{11} mode, respectively.

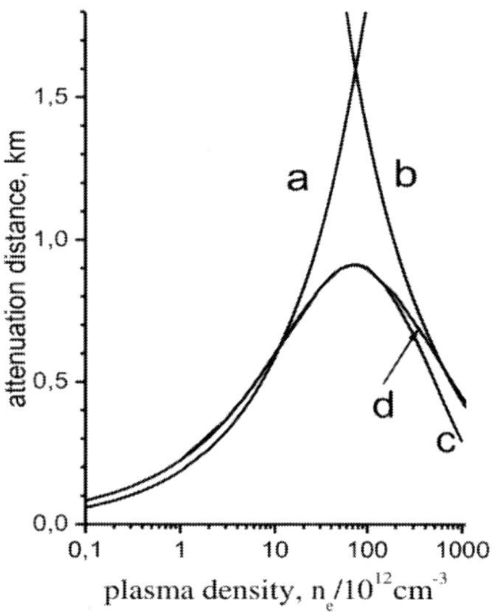

Figure 1a. Characteristic attenuation length of E_{01} mode vs plasma density calculated according analytical formulas (3) and (4) (curves a and b, respectively) and numerical solution of dispersion equation (1) with (c) and without (d) Coulomb collisions taking into account.

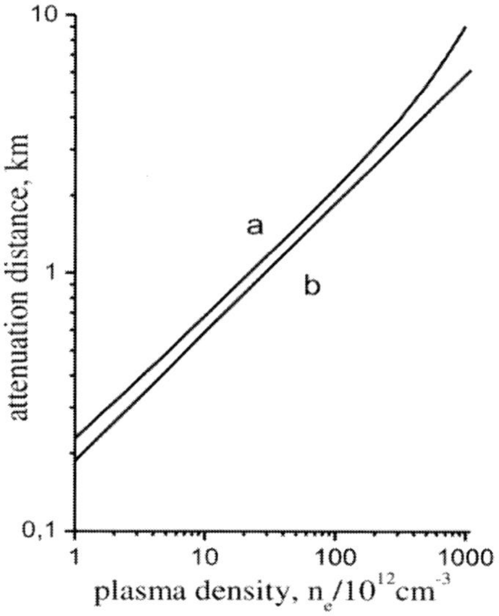

Figure 1b. Characteristic attenuation length of H_{01} mode vs plasma density calculated according analytical formula (3) (curve b) and numerical solution of dispersion equation (1) with Coulomb collisions taking into account (curve a).

At the figure 1 the characteristic propagation distance $(\mathrm{Im}\ h)^{-1}$ for the axially-symmetric sliding modes H_{01} and E_{01} is shown as a function of the plasma wall density. The wavelength of the microwave signal is λ = 8 mm and the inner waveguide radius is R = 30 cm. Calculations are made using the above analytical approximations along with the numerical solution of the dispersion equations (1), (2). The electron elastic transport collision rate is modeled as $\nu=\sqrt{\nu_T^2+\nu_C^2}$, where

$\nu_T \approx 10^{12}$ s^{-1} is the characteristic electron – neutral collision rate in air and ν_C is the Coulomb electron – ion collision rate. Analytical relations (3), (4) describe the roots of the dispersion equations reasonably well, so one can use them to estimate the dispersion parameters of the sliding mode plasma waveguides.

3. Propagation of the phase-modulated microwave pulse in the waveguide

In analogy to conventional nonlinear optics [19], we expect that the initially phase-modulated microwave pulse will exhibit compression or stretching when propagates along the sliding mode plasma waveguide due to its dispersion properties. To demonstrate this pulse compression effect, we will develop a simplified approach assuming the transverse shape of the sliding mode is constant whereas the amplitude and phase of the microwave pulse are the slow-varying functions along the propagation distance. Within this SVAP (slow-varying amplitude and phase) approximation, the Maxwell equations lead in the conventional manner to the following equation which describes evolution of the pulse amplitude $A(t,z)$

$$\left(\frac{\partial}{\partial z}+\frac{u'}{|u|^2}\frac{\partial}{\partial t}-i\frac{k_2}{2}\frac{\partial^2}{\partial t^2}\right)A(t,z)-\left(i\delta\frac{\partial}{\partial t}\right)A(t,z)=0\,, \tag{5}$$

here z is the co-ordinate along the waveguide's axis, $u=\left(\partial\omega/\partial h\right)_{|\omega_0}=u'+iu''$ and $k_2=\left(\partial^2 h/\partial\omega^2\right)_{|\omega_0}=k'_2+ik''_2$ are the complex parameters characterizing the dispersion properties of the waveguide, $\delta=u''/|u|^2$. We assume the initial condition of equation (5) is of the form $A(t,z=0)=C\exp\left[-(\tau_0^{-2}+i\alpha_0)t^2/2\right]$, τ_0 and α_0 are the initial pulse width and chirp parameters, respectively. The above consideration makes it possible to approximate the real waveguide's dispersion by the analytical solutions (3), (4). In the domain of plasma waveguide's parameters of interest, we find that the value of u'' is sufficiently small thus $\delta<<1$, and we can use the perturbation approach to solve the equation (5) assuming

$$A(t,z)=A_0(t,z)+\delta A_1(t,z)+\delta^2A_2(t,z)+\ldots \tag{6}$$

Restricting ourselves by the first order perturbation, we find from equation (6) the following system of equations

$$\left[\frac{\partial}{\partial z}+\frac{u'}{|u|^2}\frac{\partial}{\partial t}-i\frac{k_2}{2}\frac{\partial^2}{\partial t^2}\right]\begin{pmatrix}A_0\\A_1\end{pmatrix}=i\frac{\partial}{\partial t}\begin{pmatrix}0\\A_0\end{pmatrix} \tag{7}$$

Introducing $\eta=t-z/v_g$ the pulse eigen time variable, $v_g=|u|^2/u'$ is the effective group velocity one can find the solution:

$$A_0(\eta,z)=Cf^{-1/2}\exp\left(-(\tau_0^{-2}+i\alpha_0)\eta^2/2f\right), \tag{8.1}$$

$$A_1(\eta,z)=-i(\tau_0^{-2}+i\alpha_0)(z\eta/f)A_0(\eta,z)\,. \tag{8.2}$$

Here $f(z)=1+ik_2(\tau_0^{-2}+i\alpha_0)z$ is the microwave pulse envelope function. Solution (8) is valid until the first-order correction is small with respect to zero-order solution, and one can find the correspondent condition for the propagation distance:

$$\left(\frac{\sqrt{\tau_0^{-4}+\alpha_0^2\delta^2 z^2}}{e|f(z)|}\right)^{1/2} \ll 1 \tag{9}$$

At the figure 2, the dependence of the peak intensity of the phase- modulated microwave pulse on the propagation distance is shown. It is assumed the long –distance sliding mode plasma waveguide is of the radius $R = 30\,\text{cm}$ and the plasma density is $n_e = 1.5\times10^{14}\,\text{cm}^{-3}$. For the microwave pulse of the wavelength $\lambda = 8\,\text{mm}$ we find from equals (3),(4) the real and imaginary parts of the group velocity dispersion coefficient. These parts are $k_2' = \left(\partial^2 h'/\partial\omega^2\right)_{\omega_0} \approx -3.72*10^{-26}\,\text{s}^2/\text{cm}$ and $k_2'' = \left(\partial^2 h''/\partial\omega^2\right)_{\omega_0} \approx 1.4*10^{-27}\,\text{s}^2/\text{cm}$. The calculations are made for different pulse chirp parameters α_0 which are optimum for the particular waveguide's lengths of 1km, 1.5 km and 2 km, respectively. We have to point out that for the optimum chirp parameters the ratio $\sqrt{\alpha_0}/\omega_0 \approx 0.06 \div 0.08$ is small enough so that the SVAP approximation is valid.

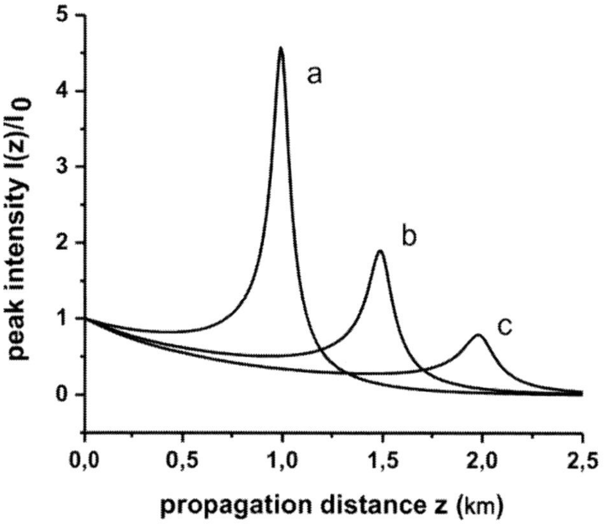

Figure 2. Peak intensity of the microwave pulse vs the propagation distance at different chirp parameters (a) $\alpha_0 = -2.7\times10^{20}\,\text{s}^{-2}$, (b) $\alpha_0 = -1.8\times10^{20}\,\text{s}^{-2}$, (c) $\alpha_0 = -1.35\times10^{20}\,\text{s}^{-2}$.

4.Conclusions

In conclusion, we have considered the propagation of the phase-modulated microwave pulse in the sliding-mode plasma waveguide in air. It is shown that with the proper choice of the chirp parameter the peak intensity of the microwave pulse can be significantly enhanced with respect to the non-phase-modulated pulse thus compensating the attenuation by the pulse compression. Calculations show that the microwave pulse peak intensity at the end of plasma waveguide can be made larger than the initial peak intensity, even for the plasma waveguides of a few kilometres length.

Acknowledgments
This work is supported by Russian Foundation for Basic Research (grants #11-02-01524, #11-02-01414), by the "Dynasty" Foundation (program for support of students) and by the LPI Educational Complex and Special Program of RAS Presidium for young scientist support.

References

[1] Chin S L 2010 *Femtosecond Laser Filamentation* (N.Y.:Springer)
[2] Couairon A, Mysyrowicz, A 2007 *Phys. Rep.* **441** 47
[3] Zhao X M, Wang Y C, Diels J-C, and Elizondo J 1995 *IEEE J. Quantum Electron.* **31** 599
[4] Tzortzakis S, Franco M A, Andre Y-B, Chiron A, Lamouroux B, Prade B S, and Mysyrowicz A 1999 *Phys. Rev. E* **60** R3505
[5] Rodriguez M, Sauerbrey R, Wille H, Wöste L, Fujii T, André Y-B, Mysyrowicz A, Klingbeil L, Rethmeier K, Kalkner W, Kasparian J, Salmon E, Yu J, Wolf J-P 2002 *Opt. Lett.* **27** 772
[6] Ionin A A, Kudryashov S V, Levchenko A O, Seleznev L V, Shutov A V, Sinitsyn D V, Smetanin I V, Ustinovsky N N, and Zvorykin V D 2012 *Appl. Phys. Letts.* **100** 104105
[7] Châteauneuf M, Payeur S, Dubois J, and Kieffer J-C 2008 *Appl. Phys. Lett.* **92** 091104
[8] Zvorykin V D, Levchenko A O, Smetanin I V, and Ustinovskii N N 2010 *JETP Lett.* **91** 226
[9] Penano J, Sprangle P, Hafizi B, Gordon D, Fernsler R, Scully M 2012 *J. Appl. Phys.* **111** 033105
[10] Dormidondov A E, Valuev V V, Dmitriev V L, Shlenov S A, Kandidov V P 2007 Proc. SPIE **6733** 67332S
[11] A. Marian, M. ElMorsli, F. Vidal, Payeur S, Châteauneuf M, Théberge F, Dubois J, Kieffer J-C 2013 *Phys. Plasmas* **20**, 023301
[12] Musin R R, Schneider M N, Zheltikov A M, and Miles R B 2007 *Appl. Opt.* **46** 5593
[13] Ren Y, Alshersby M, Qin J, Hao Z, Lin J 2013 *J. Appl. Phys.* **113** 094904
[14] Alshersby M, Hao Z, Lin J 2013 *Appl. Phys. B* **111** 509
[15] Zvorykin V D, Levchenko A O, Molchanov A G, Smetanin I V, and Ustinovsky N N 2010 *Kr. Soobshch. Fiz. FIAN* **37** No.2 49 [*Bull. Lebedev Physical Inst.* **37** (2) 60]
[16] Zvorykin V D, Levchenko A O, Smetanin I V, and Ustinovskii N N 2011 *Il Nuovo Cimento* **34C** N4 469
[17] Zvorykin V D, Levchenko A O, Shutov A V, Solomina E V, Ustinovskii N N, and Smetanin I V 2012 *Phys. Plasmas* **19** 033509
[18] Shutov A V, Smetanin I V, Ionin A A, Levchenko A O, Seleznev L V, Sinitsyn D V, Ustinovsky N N, and Zvorykin V D 2013 *Appl. Phys. Lett.* **103** 034106
[19] Akhmanov S A, Vysloukh V A, Chirkin A S Chin S L 1992 *Optics of Femtosecond Laser Pulses* (N.Y.:AIP)

Journal of Physics: Conference Series **572** (2014) 012054 doi:10.1088/1742-6596/572/1/012054

X-ray diagnostics for investigation of nanosecond laser plasma

A A Fronya[1,2,3], V A Burakov[1], A A Kologrivov[1], V N Puzyrev[1], A N Starodub[1], O F Yakushev[1]

[1] P.N. Lebedev Physical Institute of RAS, Leninsky pr. 53, 119991, Moscow Russia
[2] National Research Nuclear University MEPhI (Moscow Engineering Physics Institute), Kashirskoe sh. 31, 115409, Moscow Russia

E-mail: nastya708@yandex.ru

Abstract. In this paper, the object of study is a laser plasma produced by nanosecond (~ 2.2 ns) laser low coherence radiation interacted with different solid targets. In the irradiation of target by a laser pulse with intensity of ~ 10^{12} - 10^{14} W/cm^2, the temperature of the formed plasma can be up to a few keV, and the maximum emissivity falls on the X-ray spectrum region. Developed X-ray diagnostics complex allows one to study of luminosity distribution of plasmas with high spatial, spectral and temporal resolution.

1. Introduction

The study of plasma sources with a typical lifetime of a few nanoseconds and a maximum emissivity attributable to X-ray and VUV regions is an interesting and actual issue. Currently, the sources of short-wave radiation (VUV and X-ray spectral range) represent considerable interest as a manipulator of light with a wavelength comparable to atomic dimensions [1], and also in research on controlled thermonuclear fusion , in particular LTF [2, 3]. In these cases laser-plasma light sources are promising, they have a fairly compact dimensions, allow to work in a wide range of spectrum and intensity.

Investigation of the properties of laser-plasma sources requires the development of reliable methods of diagnosis. Comprehensive diagnostics of spatial, angular, temporal and spectral characteristics of the plasma source in X-ray, as well as in the soft X-ray and VUV, allow to obtain information about the efficacy and mechanisms of absorption of laser energy, the distribution of the plasma in space, as well as the role and nature of various nonlinear processes developing in the interaction of radiation with plasma.

The aim of the presented work is creating a diagnostic system for the detection of radiation from a laser plasma in X-ray and VUV spectral ranges, the experimental investigation of laser-target interaction, and the analysis of the experimental data.

2. Experimental setup

The laser-matter interaction experiments have been performed on the "Kanal-2" laser facility [4] LPI of RAS. Experimental facility includes Nd-laser, amplification system for the laser radiation and the vacuum chamber with a set of diagnostic equipment.

---

[3] nastya708@yandex.ru

Diagnostic system for the studying of the plasma X-ray consists of the following channels used in the experiments:

- pinhole camera for obtaining the plasma image in the own X-ray radiation ($\lambda \leq 9$ Å);
- Schwarzschild objective [5] for obtaining the plasma image in the own VUV radiation (180 – 200 Å);
- diagnostic channel for registration of continuous X-ray spectra, based on stepwise attenuator consisting of 13 beryllium filters;
- two channels on the basis of Johann spectrograph with a spherical crystal and grazing incidence spectrograph (GIS) for registration of line X-ray radiation spectra [6];
- channel for registration of the time evolution of plasma X-ray radiation;
- channel based on the mass spectrometer;
- calorimetric system for plasma and laser energy characteristics measuring.

Laser beam parameters such as energy, pulse duration and spectral composition also have been registered. The size of the focal spot of the heating radiation at the surface of the target was ~ 170 μm. Laser energy was 1-6 J.

The targets have been irradiated by one laser beam. Foils of Gd, Al, and alloy of Gd+Al (with containing 10 % Gd) have been used as the targets. Gd is interested as radiation source candidate based on "single line" or "single radiating ion", and has strong arrays radiating close to 6.6 – 6.7 nm.

3. Experimental results

For heating radiation wavelength λ_0=1060 nm, and all types of the studied targets value of the critical density is several mg/cm^3. According to the conditions of the experiment the target with supercritical density have been irradiated. But in the interaction processes near target surface plasma cloud is formed with areas of supercritical, critical and subcritical densities.

In the study of X-ray pulse time-base with the electron energy $\geq$ 1500 eV it was established that there is a formation of hot plasma. The existence time of hot plasma of targets is order of the laser pulse duration, and the hard component X-ray tracks changes in the electron temperature of the plasma [7].

Images of plasma have been obtained in hard X-ray spectral range $\lambda \leq 9$Å, and VUV spectral range $\Delta\lambda$=180-200 Å.

From the obtained images of the plasma in the hard X-ray emission it was found that gadolinium, aluminium targets is characterized by a spatial-limited emission region, that size approximately correspond to the focal spot. And this emission region is hot plasma.

Using the diagnostic channel based on a Schwarzschild objective, we studied images of laser plasma in VUV spectral range $\Delta\lambda$=180-200 Å (Figure 1). A comparison of images of the aluminium, gadolinium plasma, and the alloy of aluminium and gadolinium plasma showed that the emission targets containing the gadolinium has heterogeneity (as a "jets"), while the aluminium target emission more homogeneity over the spatial coordinates. And the radiation intensity of aluminium target is less than the radiation intensity of the gadolinium target and alloy aluminium with gadolinium at similar laser energy.

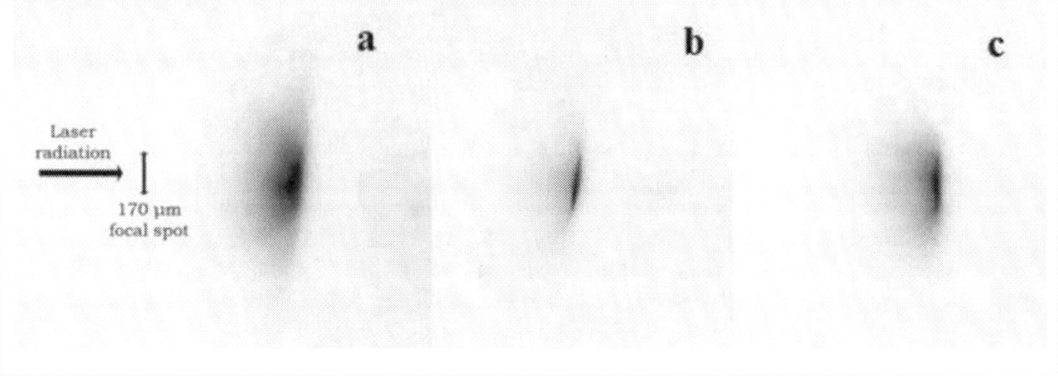

Figure 1. Images of laser plasma VUV spectral range $\Delta\lambda$ = 180-200 Å.
a – Gd target, laser energy 2.5 J; b – Al target, laser energy 2.2 J; c – alloy Gd+Al, laser energy 2 J.

Plasma images in VUV spectral range are also characterized by the presence of area with intense emission ("core") with dimensions close to the focal spot. From the Figure 1 one can see that VUV radiation of plasma is propagated along the surface of the target. And the emission regions in the transverse direction relative to the direction of laser beam propagation have dimensions exceeding the size of the focal spot in several times (3-4).

The results on the ion composition for the gadolinium target are presented on Figure 2. One can see that laser plasma with multiply charged ions, the degree of ionization Z=21 has been formed at small laser energy of 2 J, and the laser intensity $3.5 \cdot 10^{12}$ W/cm^2.

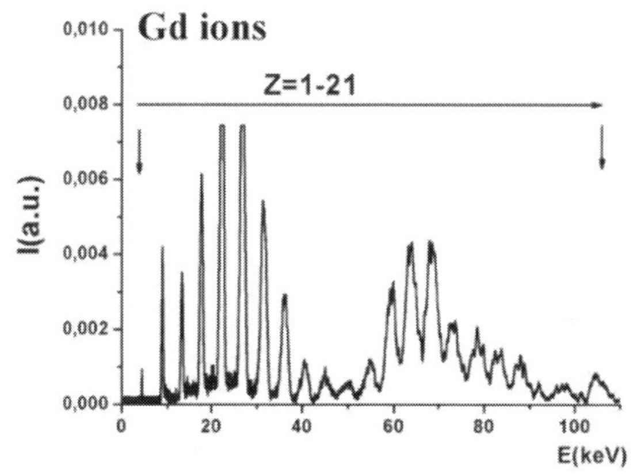

Figure 2. Ions of Gd target laser plasma.

X-ray spectra of multiply charged ions in the 10-200 Å for gadolinium targets have been registered, also the dependence of the intensity of X-ray spectra on the laser pulse energy and its duration have been obtained. One can see that the output intensity of X-ray increases with the intensity of the laser pulse. Furthermore, the output of the plasma X-ray radiation decreases with increasing laser pulse duration. The electron temperature of gadolinium plasma estimates of ~100 eV at an energy of heating laser radiation 6 J.

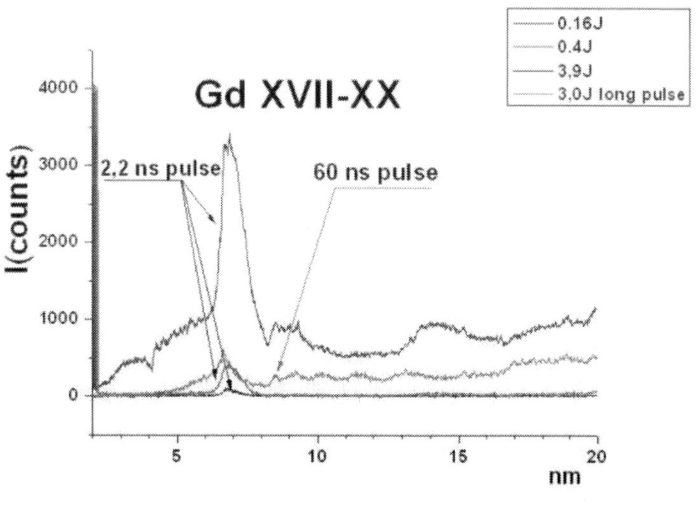

Figure 3. X-ray spectra of Gd target laser plasmas.

4. Conclusion

The realized diagnostic system enables to research of X-ray laser plasma: to study the spectrum of X-ray radiation in a wide range (1-400 Å), the spatial distribution of the laser plasma emitting regions in various spectral ranges (VUV 180-200 Å, the hard X-ray ≤ 9Å), to determine the temporal behaviour of the plasma X-ray pulse with respect to the heating laser pulse, electron temperature of the plasma, and its ion composition.

The experiments on the interaction of nanosecond laser radiation with solid targets of Gd, Al and alloy of Gd+Al have been performed. It was found that the spatial distribution of hard X-rays have a spatial area of intense radiation localized near the focal spot, but the spatial distribution in VUV range have the total size of plasma emitting area exceeds the size of the focal spot. Studying of X-ray spectra and ion composition of created plasma is indicated on formation of multiply-charged plasma with electron temperature is 100eV, it is theoretical temperature needed for excitation of Gd ions radiating at 6.6-6.7 nm.

5. Acknowledgments

The work is partly supported by the Russian Foundation for Basic Researches, grant # 12-02-31441.

[1] Bagchi S, Kiran P, Yang K, Rao A M, Bhuyan M K, Krishnamurthy M, and Kumar G R 2011 *Phys. Plasmas* **18** 014502

[2] Remington B A, Kane J, Drake R P, Glendinning S G, Estabrook K, London R, Castor J, Wallace R J, Arnett D, Liang E, McCray R, Rubenchik A, and Fryxell B 1997 *Phys. Plasmas* **4** 1994

[3] Gavrilov V V, Gol'tsov A Yu, Koval'skiy N G, Koptyaev S N, Magunov A I, Pikuz T A, Skobelev I Yu, and Faenov A Ya 2001 *Quantum electronics* **31** 1071

[4] Fedotov S I, Feoktistov L P, Osipov M V, and Starodub A N 2004 *Journal of Russian Laser Research* **25** 79

[5] Artyukov I A, Borisenko N G, Vasin B L, Vinogradov A V, Gromov A I, Merkuliev Yu A, Osipov M V, Puzyrev V N, Starodub A N, Fronya A A, Chernodub M L, and Yakushev O F 2011 *Proceedings of the Samara Scientific Center of RAS* **13** 617

[6] Burakov V A, Kologrivov A A, Osipov M V, Puzyrev V N, Starodub A N, Fronya A A, and Yakushev O F 2013 *Proceedings of the Samara Scientific Center of RAS* **15** 104

[7] Anan'in O B, Afanas'ev Yu V, Bykovskiy Yu A, and Krokhin O N 2003 *Laser plasmas. Physics and applications* (Moscow: Science)

Journal of Physics: Conference Series **572** (2014) 012055 doi:10.1088/1742-6596/572/1/012055

Numerical modelling of physical processes in a ballistic laboratory setup with a tapered adapter and plastic piston used for obtaining high muzzle velocities

N V Bykov

Bauman Moscow State Technical University, Dorodnicyn Computing Centre of RAS

bykovnv@bk.ru

Abstract. Numerical modelling of a ballistic setup with a tapered adapter and plastic piston is considered. The processes in the firing chamber are described within the framework of quasi-one-dimensional gas dynamics and a geometrical law of propellant burn by means of Lagrangian mass coordinates. The deformable piston is considered to be an ideal liquid with specific equations of state. The numerical solution is obtained by means of a modified explicit von Neumann scheme. The calculation results given show that the ballistic setup with a tapered adapter and plastic piston produces increased shell muzzle velocities by a factor of more than 1.5–2.

1. Introduction

Laboratory installations designed to obtain high velocities for the purpose of studying high-velocity impact processes have broad applications in aeroballistics and space research [1]. Such installations especially have to be economic and easy to use. For this purpose, light-gas guns are now the most widespread, but they do not yet meet the listed requirements. Therefore, only a very small number of laboratories have such devices.

A ballistic setup with a tapered adapter and plastic piston can provide one possible alternative to light-gas guns. The barrel of such a setup consists of two cylindrical sections of diameters d_1 and d_2, joined by a tapered (or profiled) adapter (Figure 1). The accelerated body (assembly) at the initial time consists of a shell (4 in Figure 1) (as a rule, it is a compact element, such as a ball), the plastic piston (3) (for which polyethylene is used) and the inertial pallet (2). In a firing chamber (1) there is a propellant. At the first stage, the assembly accelerates in the first cylindrical section as a unit according to the classical artillery scheme. At this stage, the limit speed for classical ballistic installations can be reached. Then the assembly moves into the tapered channel (5), where the piston experiences plastic deformation. As a result, the forward part experiences additional acceleration (this process is known in light-gas guns as the hydrodynamic effect [1]). After leaving a barrel, the piston and the shell are separated, and the inertial pallet stops part of the propellant gases. Thus, the shell receives an additional increment of speed, which can exceed 50–100% of the shell speed at the entrance to the tapered section.

The design of the laboratory installation allows replaceable tapered sections to be used. Therefore, by varying the length of the piston and its material, as well as the length of the cone and diameter of the output cylindrical section, it is possible to produce various values of muzzle velocities.

In the present paper, the physical principles of the ballistic setup operation on the basis of a numerical solution to the one-dimensional gas dynamics equations are considered.

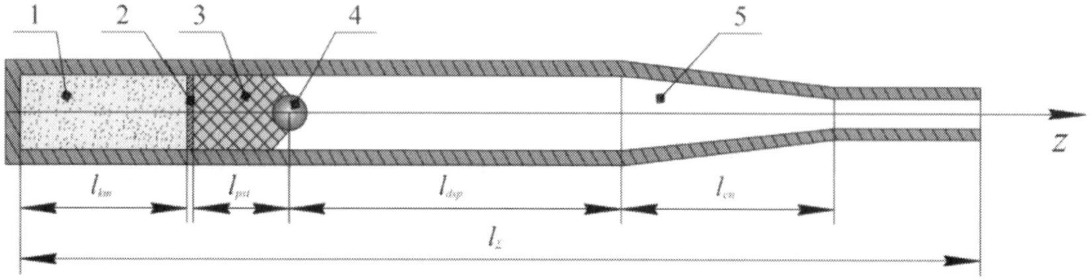

Figure 1. The scheme of ballistic setup; 1 – propellant, 2 – inertial pallet, 3 – plastic piston, 4 – shell, 5 – tapered section.

2. Theory

2.1. Mathematical model

Two-phase numerical modelling of the operation of guns has been performed in many studies [2], but in our case the detailed description of the combustion process in a firing chamber has no crucial importance. Therefore, we will limit our consideration of combustion processes to simple models. The motion of gas in the chamber is described by means of quasi-one-dimensional gas dynamics equations [3], to which the equations considering propellant burn within the framework of a geometrical law are attached [4]

$$\frac{\partial(\mathbf{q}S)}{\partial t}+\frac{\partial(\mathbf{f}S)}{\partial z}=\mathbf{h}, \tag{1}$$

$$\mathbf{q}=\{\rho,\rho u,\rho E,\rho w\},$$

$$\mathbf{f}=\{\rho u,p+\rho u^2,\rho u(E+p/\rho),\rho uw\},$$

$$\mathbf{h}=\{0,p\,\partial S/\partial x,0,(v_1/e_1)\rho Sp\}.$$

Here, t is the time, z is the dimensional coordinate, u, ρ, p and E are the speed, density, pressure and a specific energy of gas, respectively, S is the variable cross sectional area of the barrel bore, f = 1 MJ/kg is the propellant's impetus, v_1 and $2e_1$ are the propellant's burning rate constant and web thickness, respectively and w is the propellant's relative burnt thickness.

System (1) is supplemented with the equation of state

$$p=\frac{(k-1)\varepsilon-(1-\psi)f}{1/\rho-(1-\psi)/\delta-\alpha\psi}. \tag{2}$$

Here $\varepsilon=E-u^2/2$ is an internal energy per unit mass, $\psi=\kappa w(1+\lambda w)$ is the propellant's burnt fraction, k = 1.24 is an adiabatic index, α = 0.001 m^{-3} is the covolume, δ = 1600 kg/m3 is the mass density of the propellant and κ and λ are the propellant's characteristic parameters.

With the high pressures accompanying the process of a shot, it is possible to approximately consider the plastic piston as a compressible ideal liquid. The equation of state of the plastic piston is taken in the following form [5]

$$p=BR(R-1)/(C-R)^2, \tag{3}$$

where $R=\rho/\rho_0$ is the compression rate, ρ_0 is the density of the material at zero pressure and B and C are the empirical constants characterizing a concrete material. For polyethylene, B = 1.19 GPa and C = 1.73.

In one-dimensional gas dynamics problems, it is more convenient to use Lagrange mass coordinates, so we rewrite (1) in the following form

$$\frac{\partial}{\partial t}\left(\frac{1}{\rho S}\right)=\frac{\partial u}{\partial \xi}, \qquad (4)$$

$$\frac{\partial u}{\partial t}+S\frac{\partial p}{\partial \xi}=0, \qquad (5)$$

$$\frac{\partial \varepsilon}{\partial t}+p\frac{\partial}{\partial \xi}(uS)=0, \qquad (6)$$

$$\frac{\partial z}{\partial t}=\frac{p}{J_k}. \qquad (7)$$

Here ξ is a Lagrangian mass coordinate, $d\xi/dz=\rho S$.

2.2. *Initial and boundary conditions*

At the initial time, the following propellant ignition conditions are used: $p=p_0$, $u=0$, $\rho=\Delta$ and $w=0$, where $p_0=5$ MPa is the pressure of propellant ignition and Δ is a charge density.

The nonpenetration condition is used at the bottom of the firing chamber (z = 0). The nonpenetration boundary conditions are also applied at the boundaries with the piston and the shell. The velocity and coordinate of the shell are determined by the equation of motion

$$\varphi m_s(du_s/dt)=p_s S, \qquad (8)$$

where m_s and u_s are the mass and velocity of the shell, respectively, p_s is the pressure on the shell and φ is a coefficient accounting for secondary energy losses. The additional condition of assembly immovability is used until the chamber pressure reaches 60 MPa.

2.3. *A numerical scheme*

For the numerical solution of system (4) - (8), the *explicit Neumann's scheme (known as "cross")* [6] was used. Non-conservatism of the scheme was compensated for by reduction of a spatial grid step. The propellant section was divided into 100 spatial cells, and the plastic piston section was divided into 80 cells. The time step was calculated from the Courant–Friedrichs–Lewy stability condition.

The preliminary calculations for other numerical schemes (Godunov and implicit Neumann scheme) have shown that they do not differ fundamentally from the explicit scheme, but are much more labour intensive.

3. Results and discussion

Calculations of a series of shots from a ballistic setup with a tapered section were carried out. The calculation parameters are specified in Table 1. In all calculations, the pallet mass is taken to be 0.005 kg, the mass of a shell, $m_s=0.01$ kg and the diameter of the first cylindrical section is $d_1=30$ mm. The diameter of the second cylindrical section is defined by means of the coefficient of the barrel narrowing $\gamma=d_2/d_1$. The shell muzzle velocity u_{sm} and the maximum pressure p_{max} are given as output data in Table 1. As the numerical analysis shows, the maximum pressure is usually not reached at the bottom of the chamber (as for a classical artillery system) or at the bottom of the shell, but falls on the internal section of the plastic piston when it passes the tapered section.

Table 1. Numerical modelling parameters

case	l_{km}, *m*	l_{pst}, *m*	l_{dsp}, *m*	l_{cn}, *m*	γ	Δ, *kg/m*^{3}	I_f, *MPa·s*	κ	λ	p_{max}, *MPa*	u_{sm}, *m/s*
1	0.20	0.30	1.50	0.50	2/3	700	0.50	1.0	0.0	143	992
2	0.25	0.40	1.35	0.80	0.6	700	0.30	0.5	0.1	375	1560
3	0.25	0.25	1.55	0.9	0.5	600	0.25	0.6	0.2	396	1830
4	0.35	0.50	1.60	1.0	0.5	800	0.25	0.6	0.2	2802	3972
5	0.45	0.40	1.65	1.0	0.5	600	0.25	0.6	0.2	3420	2953

For a clear demonstration of the discussed effects, we will consider the results of one calculation in detail (case 1 in table 1). This case is chosen because the calculation results are more obvious with these parameters. Figure 2 gives calculated curves of the shell velocity versus the shell coordinate with and without a tapered section. These curves are correlated to the scheme of the ballistic setup. From Figure 2, it follows that the shell receives a considerable acceleration when passing a tapered section (curve I), which doesn't occur if the barrel is purely cylindrical (curve II).

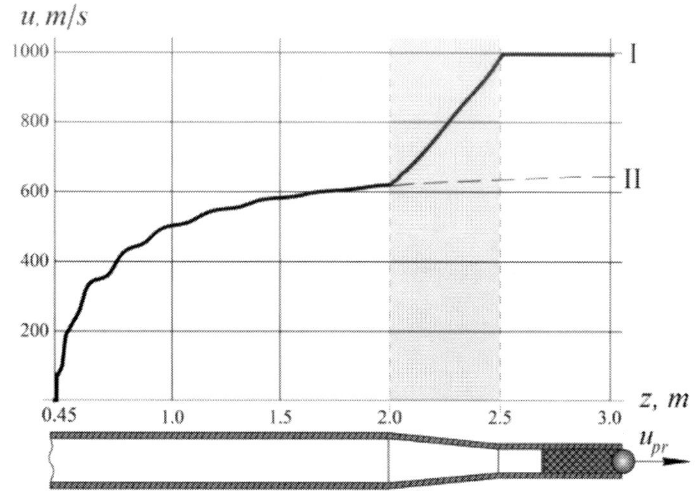

Figure 2. Shell muzzle velocity versus shell coordinate; curve I refers to a calculation with a tapered section and curve II refers to a calculation without a tapered section.

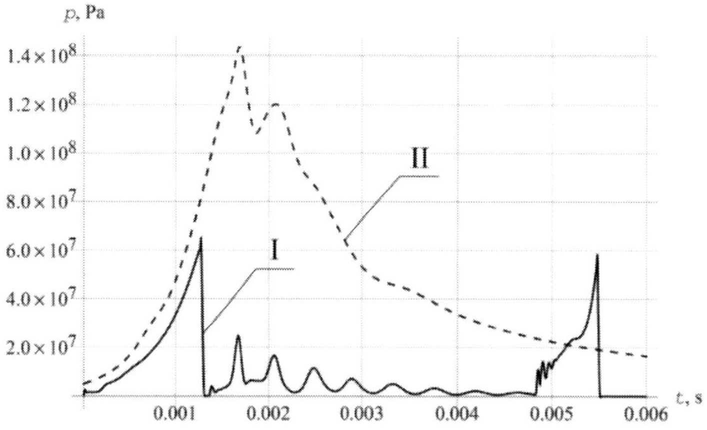

Figure 3. Pressure on the shell (solid curve I) and pressure on the bottom of a firing chamber (dashed curve II) versus time.

Figure 3 shows curves representing the pressure on the shell and on the bottom of the firing chamber versus time. In this case, the maximum pressure is reached at the bottom of the chamber. In other cases, the maximum pressure is attained in the tapered section, because the pressure in the piston is much higher due to more rigid conditions of the shot. Proceeding from this result, one can conclude that the main drawback of the ballistic setup with the tapered section is a high level of pressure in the tapered section. It can be avoided in several ways. First, it is possible to make replaceable tapered sections. Secondly, it is possible to reduce pressure, by making a profiled section instead of a tapered one, or by making several tapered sections that are separated by a certain distance.

The performed calculations show that the ballistic scheme presented produces muzzle velocities in the range from 1500 m/s to 2500 m/s and higher. Ballistic setups with tapered sections can now find applications as rather inexpensive means of obtaining high velocities for high velocity impact research.

This work was supported by a grant of the Russian Foundation for Fundamental Research (Grant No. 12-08-31408 mol_a).

References

[1] Zlatin N A, Krasilschikov A P, Mishin G I, Popov N N 1974 *Ballistic systems and their use in experimental studies* [in Russian] eds. N A Zlatin and G I Mishin (Nauka, Moscow)
[2] Jaramaz S, Mickovic´ D, Elek P 2011 Two-phase flow in a gun barrel: theoretical and experimental studies Int J of multiphase flow **37** 475-87
Kasimov V Z, Ushakova O V, Khomenko Yu P 2003 Numerical modeling of interior ballistics processes in light gas guns *J. of App. Mech. and Tech. Phys.* **44** 612-9
[3] Landau L D and Lifshitz E M 1987 *Fluid Mechanics* (Butterworth-Heinemann)
[4] Carlucci D E and Jacobson S S 2008 *Ballistics: theory and design of guns and ammunition* (CRC Press/Taylor & Francis Group)
[5] Mills E J 1965 Hugoniot Equations of State for Plastics: A Comparison *AIAA*, **3**, 742-3
[6] Kalitkin N N 1975 *Numerical methods* [in Russian] (Nauka, Moscow)

Numerical simulation of turbulent flow in a cyclonic separator

Dmitry Bogdanov and Sergey Poniaev

Division of Plasma Physics, Atomic Physics and Astrophysics, Ioffe Physical Technical Institute, 26 Polytekhnicheskaya, St Petersburg 194021, Russian Federation

E-mail: dimyriy.bogdanov@gmail.com

Abstract. Abstract. Numerical simulation of a turbulent flow of air with dispersed particles through a cyclonic separator is presented. Because of a high streamline curvature in the separator it is difficult to simulate the flow by using the conventional turbulent models. In this work the curvature correction term was included into the $k-\omega-SST$ turbulence model implemented in the OpenFOAM® software. Experimental data and results of numerical simulation by the commercial ANSYS Fluent® solver for a turbulent flow in a U-duct were used to validate the model. The numerical simulation of the flow in the cyclonic separator demonstrates that the implemented turbulence model successfully predicts the cyclonic separator efficiency.

1. Introduction

Cyclonic separators are widely used for dispersed phase separation from the gas [1] . One of their important parameters is the efficiency which is the ratio between the number of filtered particles and the total number of particles injected into a cyclonic separator. However, it is extremely difficult to predict the separator efficiency by using numerical simulation of the turbulent flow because conventional eddy-viscosity models cannot adequately describe the flow [2] because of a high streamline curvature and, more specifically, inadequate calculation of the turbulence kinetic energy production. The kinetic energy production can be corrected using the Shur-Spalart curvature correction function for the Spalart-Allmaras turbulence model reformulated in [3] in terms of the $k-\omega-SST$ turbulence model.

2. Cyclonic separator model

A typical scheme of the air flow in cyclone is presented in Figure 1 [1]. The air with dispersed particles enters the cyclone through an air inlet. Under the centrifugal forces heavy particles move to the boundary layer on a sidewall. The influence of the air flow on the particles which have moved to the boundary layer is relatively small compared with the influence of the gravity force and hence particles fall to the dust chamber. The air leaves the cyclone through an air outlet.

3. SST with curvature correction model formulation

As stated above, a correction of the SST turbulence model is needed because of a high streamline curvature. In this paper the correction function (1) suggested in [2] is used as a multiplier of the production term in the Spalart-Allmaras eddy viscosity transport equation. In [3], it is suggested that this function can be used with respect to the SST model as follows

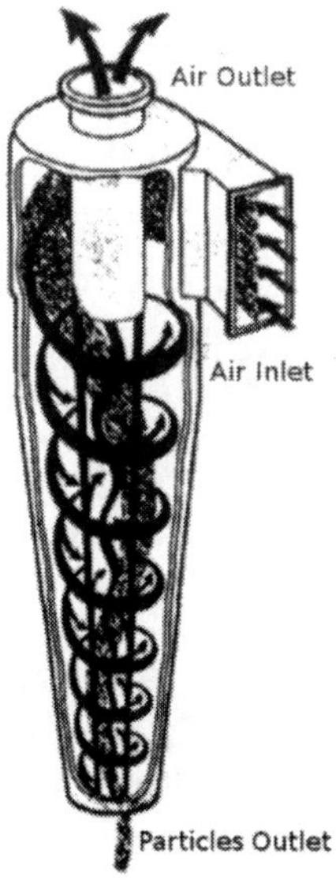

Figure 1: Scheme of an air flow inside cyclonic separator.

$$f_{rotation} = (1 + c_{r_1}) \frac{2r^*}{1 + r^*} [1 - c_{r_3} \tan^{-1}(c_{r_2} \tilde{r})] - c_{r_1} \quad (1)$$

$$\frac{\partial(\rho k)}{\partial t} + \frac{\partial(\rho u_j k)}{\partial x_j} = P_k f_{r_1} - \beta^* \rho k \omega + \frac{\partial}{\partial x_j} \left[\mu_{eff} \frac{\partial k}{\partial x_j} \right] \quad (2)$$

$$\frac{\partial(\rho \omega)}{\partial t} + \frac{\partial(\rho u_j \omega)}{\partial x_j} = \alpha \frac{\rho P_k}{\mu_t} f_{r_1} - \beta \rho \omega^2 + \\ +2(1 - F_1() \frac{\rho \sigma_{\omega_2}}{\omega} \frac{\partial k}{\partial x_j} \frac{\partial \omega}{\partial x_j} + \frac{\partial}{\partial x_j} \left[\mu_{eff} \frac{\partial \omega}{\partial x_j} \right] \quad (3)$$

$$f_{r_1} = \max\left[\min(f_{rotation}, 1.25), 0\right] \quad (4)$$

$$r^* = \frac{S}{\Omega}, \quad \mu_{eff} = \mu + \sigma_{k/\omega} \mu_t \quad (5)$$

$$\tilde{r} = 2\Omega_{ik} S_{jk} \left[\frac{DS_{ij}}{Dt} + (\varepsilon_{imn} S_{jn} + \varepsilon_{jmn} S_{in}) \Omega_m^{rot} \right] \frac{1}{\Omega D^3} \quad (6)$$

$$S_{ij} = \frac{1}{2} \left(\frac{\partial u_i}{\partial x_j} + \frac{\partial u_j}{\partial x_i} \right) \quad (7)$$

$$\Omega_{ij} = \frac{1}{2} \left(\left(\frac{\partial u_i}{\partial x_j} - \frac{\partial u_j}{\partial x_i} \right) + 2\varepsilon_{mji} \Omega_m^{rot} \right) \quad (8)$$

$$S^2 = 2S_{ij} S_{ij}, \quad \Omega^2 = 2\Omega_{ij} \Omega_{ij} \quad (9)$$

$$D^2 = \max(S^2, 0.09\omega^2) \quad (10)$$

$$c_{r_1} = 1.0, \quad c_{r_2} = 2.0, \quad c_{r_3} = 1.0$$

where u_i represents velocity components, ρ – gas density, k – turbulence kinetic energy, ω – specific dissipation rate, S_{ij} – strain rate tensor, Ω_{ij} – vorticity rate, P_k – production term for SST model, f_{r_1} – limited correction function for the SST model, $f_{rotation}$ – correction function for the Spalart-Almaras model, μ_t – turbulence eddy viscosity and DS_{ij}/Dt represents the components of the Lagrangian derivative of the strain rate tensor. α, $beta$, β^*, C – standard SST model constants, c_{r_1}, c_{r_2}, c_{r_3} – SST-CC model constants, F_1 – additional function from the standard SST model [4].

4. Model validation

The turbulence model described in Section 3 was implemented using the OpenFOAM mathematical library. Model validation was performed using the Monson [5] experimental data for a turbulent flow in a U-duct and numerical simulation in ANSYS Fluent. Geometric and flow parameters for the U-duct are listed in Table 1. The flow scheme is presented in Figure 2. The inlet profile used as the inlet boundary condition was obtained from preliminary computations of the turbulent flow in a plane channel, because in the Monson experiment [5] the flow was assumed to be fully developed in the inlet section. Numerical simulation of the flow was performed using both standard SST and SST with the curvature correction term (SST-CC) in both solvers.

Results of numerical simulation for velocity projections U_x and U_y in different cross-sections of the channel are shown in Figures 3 to 6. Figure 3 shows velocity profiles before U-turn. Because the flow curvature in this cross section is negligibly small, both models (SST and SST-CC) in both solvers give the same results which are in very good agreement with experimental data. Figure 4 shows velocity profiles at the center of the U-turn. As one can see from Figure 4, the OpenFOAM results are closer to the experimental values, and the results with the SST-CC model are better than those with the SST model. Figure 5 shows velocity profiles immediately after U-turn. It is evident that SST-CC is much better than SST for both solvers. OpenFOAM results are better on the inner wall of the channel and worse on the outer wall as compared with Fluent. Figure 6 shows velocity profiles in the cross-section at the distance of the one caliber after the U-turn. The flow curvature is small here, so the difference between SST-CC and SST is insignificant and, like in the previous cross-section, the OpenFOAM results are better on the inner wall of the channel and worse on the outer wall, compared with Fluent.

Table 1: Geometric and flow parameters for air flow in U-duct.

Channel height, H	$3.81cm$
Channel length, L	$10H$
Inner radius, R_i	$1.91cm$
Outer radius, R_o	$5.72cm$
Av. velocity at inlet U_{in}	$30.1m/s$
Av. temperature at inlet, T_{in}	$264K$
Pressure at outlet, p_{out}	$1.15atm$
Reynolds number, Re	10^5

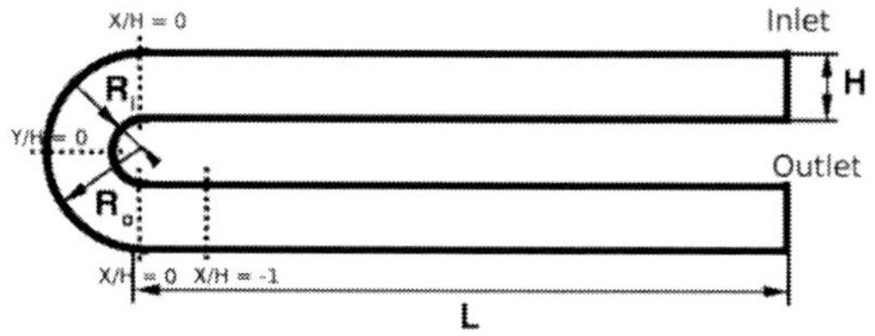

Figure 2: Scheme of an air flow for the Monson U-duct experiment simulation.

5. Results

In this section we consider the flow in Stairmand cyclonic separator [6] shown in Figure 7. Geometric parameters and boundary conditions are presented in Table 2 and Table 3 respectively. The inlet velocity profile was obtained from computation of a fully developed turbulent flow in a square-section channel. Numerical simulation was for three different inlet velocities and three particle diameters. The cyclone efficiency η (the ratio between the particles filtered by the cyclone and the number of injected particles) was calculated for all cases. Results of the numerical simulation are listed in Table 4. As can be seen from Table 4, results of the numerical simulation are in good agreement with the experimental data.

As expected, the cyclone efficiency decreases with decreasing particle diameter. For a particle diameter of $10^{-7}m$ the cyclone is not applicable because only a small number of particles is filtered out in the cyclone. For a particle diameter of $10^{-5}m$ the cyclone efficiency is almost 100% i.e. all particles are filtered out. The cyclone efficiency also falls with decreasing inlet velocity due to a decreasing influence of the centrifugal force. Particle distribution in the cyclone

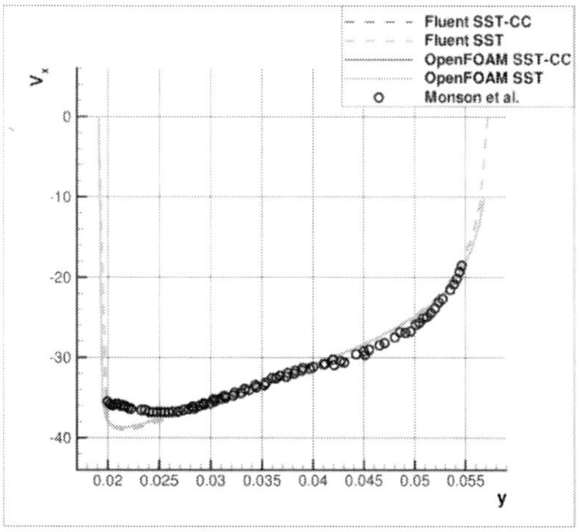

Figure 3: U_x velocity component distribution in $x/H = 0$ cross-section (top channel).

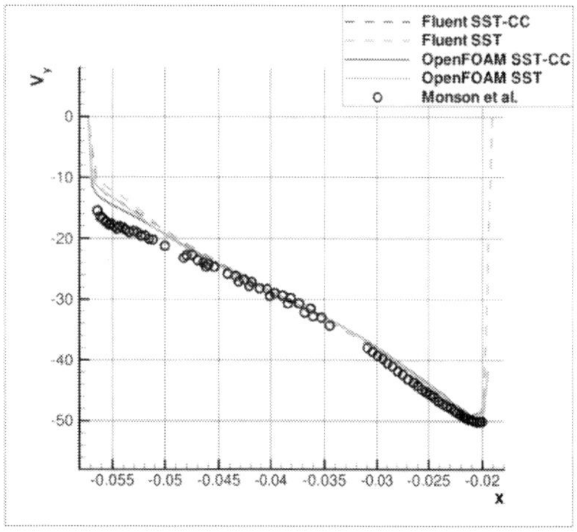

Figure 4: U_y velocity component distribution in $y/H = 0$ cross-section.

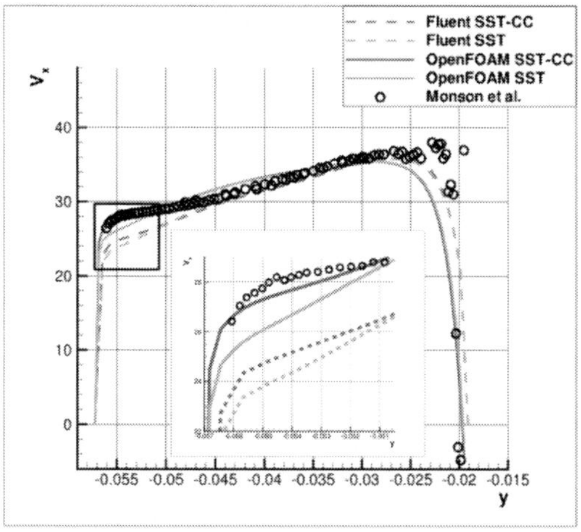

Figure 5: U_x velocity component distribution in $x/H = 0$ cross-section (lower channel).

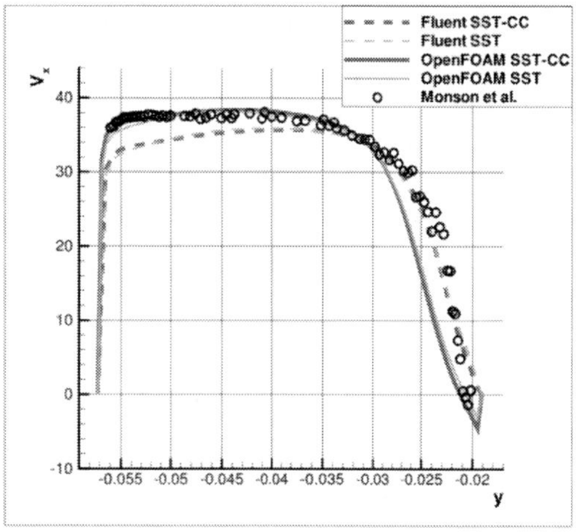

Figure 6: U_x velocity component distribution in $x/H = 1$ cross-section.

is presented in Figure 8 for two particle diameters. It can be seen from Figure 8 that particles with diameter $10^{-7}m$ distributed almost uniformly inside the cyclone which means that the influence of centrifugal forces is not strong enough to filter out a significant fraction of particles. On the opposite, particles with diameters $10^{-5}m$ are distributed mostly inside the boundary layer on the cyclone sidewall and in the lower part of the cyclone and can be filtered out.

Therefore it can be concluded that the curvature correction function suggested in [2], reformulated in [3] and implemented using OpenFOAM in this paper can be successfully used for simulation of turbulent flows with a high streamline curvature in a cyclonic separator.

Table 2: Cyclone geometry parameters.

Cylinder diameter,	$D = 0.205m$
Outlet diameter,	$D_e = 0.5D$
Inlet channel height,	$a = 0.5D$
Inlet channel width,	$b = 0.2D$
Inlet channel length,	$h_e = 0.75D$
Total filter height,	$H = 4.0D$
Cylinder height,	$h = 1.5D$
Lower section diameter,	$B = 0.36D$
Dust height,	$h_d = 0.25D$
Dust diameter,	$D_d = 0.75D$

Table 3: Cyclone flow parameters.

Inlet velocity,	$U_{in} = 5, 10, 15, 20m/s$
Inlet temperature,	$T_{in} = 300K$
Particles temperature,	$T_{p_{in}} = T_{in}$
Particles inlet velocity,	$U_{p_{in}} = U_{in}$
Outlet pressure,	$P_{out} = 1atm$
Wall heat transfer,	$q_w = 0$
Particles diameters,	$d_p \sim 10^{-5}m, 10^{-6}m, 10^{-7}m$

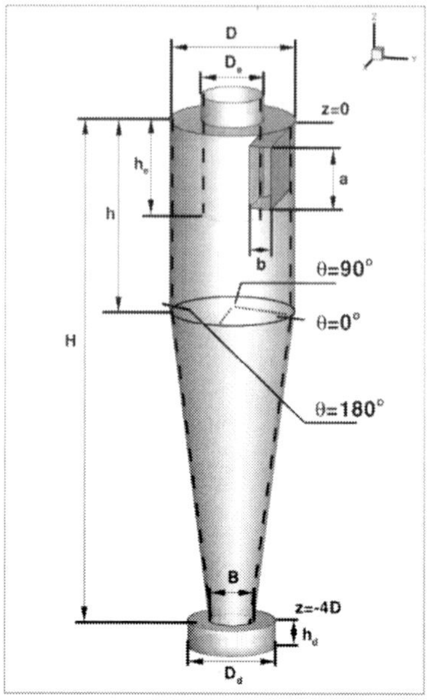

Figure 7: Stairmand cyclone geometry scheme.

Table 4: Cyclonic separator efficiency η comparison.

Flow parameters	η, Numerical simulation	η, Experiment
$U_{in} = 20m/s, d = 5 \cdot 10^{-5}m$	100%	100%
$U_{in} = 20m/s, d = 5 \cdot 10^{-6}m$	93%	90%
$U_{in} = 20m/s, d = 5 \cdot 10^{-7}m$	27%	10%
$U_{in} = 15m/s, d = 10^{-5}m$	80%	90%
$U_{in} = 10m/s, d = 10^{-5}m$	72%	85%
$U_{in} = 5m/s, d = 10^{-5}m$	75%	80%

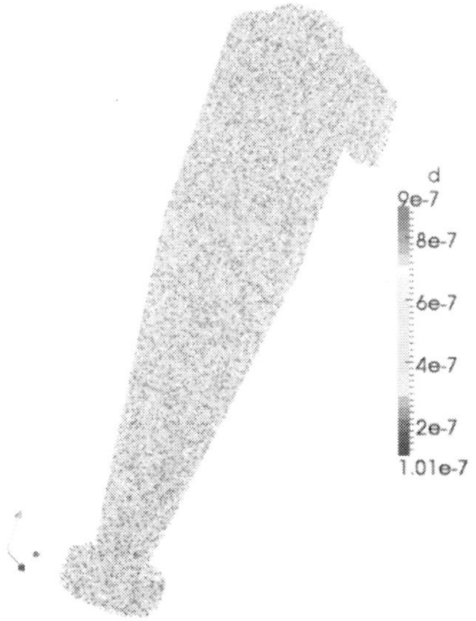

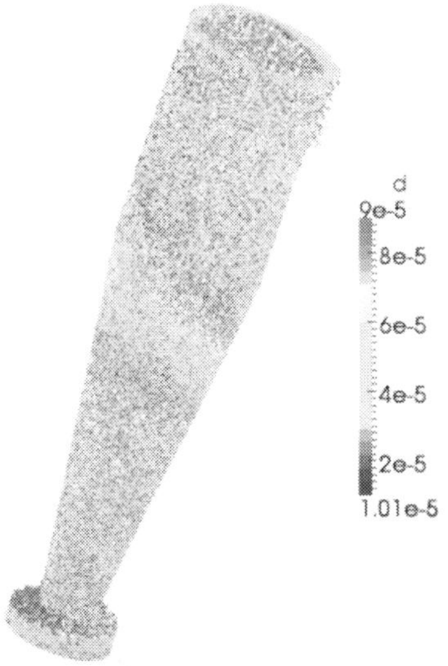

Figure 8: Particles distribution in cyclone for the diameters $d_p \sim 10^{-7}m$ (on the left) and $d_p \sim 10^{-5}m$ (on the right) and the average inlet velocity $U_{in} = 20m/s$.

6. Conclusion

A modified SST model with the curvature correction term was implemented in OpenFOAM. The model validation for the U-duct channel shows a significant improvement of the velocity profiles prediction as compared with the non-modified SST model.

Simulation of the cyclonic separator efficiency shows that the implemented model can be successfully used for this kind of simulations.

References

[1] Uzhov V 1970 *Yaroslavl*
[2] Spalart P and Shur M 1997 *Aerospace Science Technology* **15**
[3] Smirnov P and Menter F 2009 *Journal of Turbomachinery* **131**
[4] Menter F and Esch T Nov 2001 *16th Brazilian Congress of Mechanical engineering (COBEM)*
[5] Monson D, Seegmiller H, McConnaughet P and Chen Y 1990 *AIAA Paper* **21**
[6] Stairmand C 1951 *Trans. Inst. Chem. Eng.*

Computation of wall bounded flows with heat transfer in the framework of SRS approaches

M S Gritskevich[1], **A V Garbaruk**[1], **F R Menter**[2]

[1]St. Petersburg State Polytechnical University, 195251, St. Petersburg, Russia
[2]Software Development Department, ANSYS, 83714, Otterfing, Germany

E-mail: gritckevich@gmail.com

Abstract. A detailed assessment of Scale Adaptive Simulation (SAS) and Improved Delayed Detached Eddy Simulation (IDDES) is performed for prediction of heat transfer for several wall bounded flow. For that purpose a zero pressure gradient boundary layer, a backward facing step, and a thermal mixing in a T-Junction test cases are considered. The results, obtained with the use of ANSYS-FLUENT, show that both approaches are capable to predict both mean and RMS velocity and temperature with sufficient accuracy.

1. Introduction

A prediction of heat transfer for unsteady wall bounded flows is a challenging task for Computational Fluid Dynamics (CFD). CFD methods based on Unsteady Reynolds Averaged Navier-Stokes (URANS) equations, which are typically used in industrial applications, have difficulties in providing accurate mean solution and are almost incapable to predict RMS velocity and temperature. Thus, in many cases, high turbulent viscosity, predicted by URANS models, suppresses development of any transient flow and a steady-state solution is obtained. From the other point, Scale Resolving Simulations (SRS) are able to provide such information with sufficient accuracy but requires large computational resources, which are beyond the capabilities of modern computers. Nevertheless, recent studies (e.g. [1]) have shown promising results for such flows with the use of some advanced SRS approaches. However, a detailed validation of these methods is still required in order to determine their accuracy and range of validity. For that purpose, two modern SRS approaches are considered, namely Scale Adaptive Simulation (SAS) [2] and Improved Delayed Detached Eddy Simulation (IDDES) [3], which are used in conjunction with inflow turbulent content generated with either Vortex Method (VM) [4] or Generator of Synthetic Turbulence (GST) [5,6].

2. Results and Discussions

Three test cases are considered within the paper. A zero pressure gradient boundary layer is a benchmark test case, for which a number of correlations for both velocity and temperature are available for different Reynolds and Prandtl numbers (e.g. [7]). A backward facing step flow [8] includes both separation and reattachment and is often used as a stress test for SRS approaches. Characteristic feature of a thermal mixing in a T-Junction flow [9] is a so-called "thermal striping" effect, which arise due to temperature fluctuation and, therefore, cannot be handled by conventional URANS approaches [1].

All the simulations within the paper have been carried out with the use of ANSYS-FLUENT [10]. Within this code a finite volume method on unstructured grids with a cell-centered data arrangement is applied to governing transient equations. Inviscid fluxes are approximated with the use of second order centered scheme for momentum equations and with the use of second order upwind scheme for temperature and turbulence equations. Time derivatives are discretized with the use of three-level second order backward scheme.

2.1. Zero Pressure Gradient Boundary Layer

Simulations of this flow are carried out at two Reynolds numbers $Re_\delta=\rho\cdot U_0\cdot\delta_0/\mu=130$ and $Re_\delta=1300$ (U_0 is a free stream velocity, δ_0 is a boundary layer thickness at the inlet section, ρ is a constant density, and μ is a constant dynamic viscosity). For each Reynolds number two Prandtl numbers $\mathrm{Pr}=\mu\cdot C_p/\lambda=0.64$ and Pr=6.4 are considered (C_p is a constant specific heat capacity and λ is a constant thermal conductivity). The free-stream flow and the wall are maintained at constant temperature of $T_0=300$ K and $T_0+\Delta T$ ($\Delta T=10$ K) respectively.

A computational domain (Figure 1) is a box with dimensions of $31.25\delta_0\times12.5\delta_0\times3.875\delta_0$ in streamwise, wall-normal and spanwise directions respectively. A computational grid (Figure 1) has $1.1\cdot10^6$ cells for $Re_\delta=130$ and $1.3\cdot10^6$ cells for $Re_\delta=1300$. The grid steps are equal to $\Delta x=0.125\cdot\delta_0$ and $\Delta z=0.0625\cdot\delta_0$ in streamwise and spanwise directions respectively. In wall units this corresponds to $\Delta x^+=60$ and $\Delta z^+=30$ for $Re_\delta=130$ and to $\Delta x^+=450$ and $\Delta z^+=225$ for $Re_\delta=1300$. The grid has a dense clustering towards the wall to satisfy $\Delta y^+<1$ for both Reynolds numbers. A non-dimensional time step of $\Delta t=0.03\cdot\delta_0/U_0$ is applied, which ensures the CFL number to be less than one in the entire domain. Unsteady statistics is obtained by averaging of instantaneous flow fields over 5000 time steps.

Boundary conditions are shown in Figure 1. Inlet profiles are obtained from a precursor SST-RANS simulation of a boundary layer flow up to specified δ_0 and are then used for generation of inflow turbulent content with the use of GST [5,6]. A symmetry condition is specified on the upper boundary. No-slip condition with a constant temperature of $T_0+\Delta T$ is specified on the wall. Periodic condition is applied in spanwise direction. On the outlet boundary a constant pressure is specified, while the rest quantities are extrapolated from the adjacent cells.

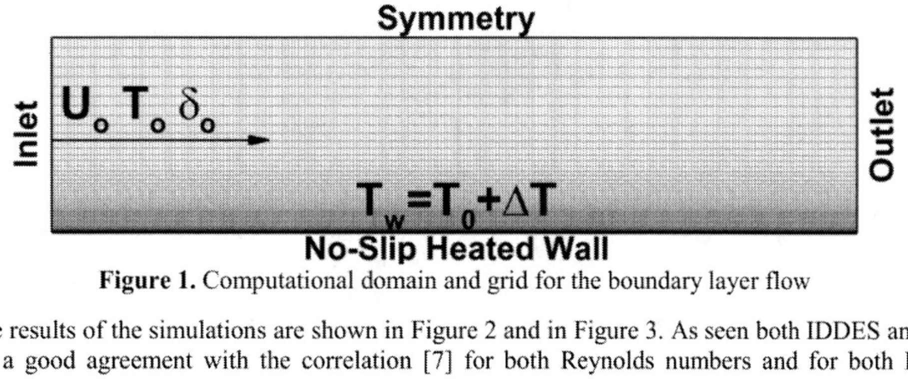

Figure 1. Computational domain and grid for the boundary layer flow

The results of the simulations are shown in Figure 2 and in Figure 3. As seen both IDDES and SAS are in a good agreement with the correlation [7] for both Reynolds numbers and for both Prandtl numbers. Therefore, both approaches are capable to predict this flow with sufficient accuracy.

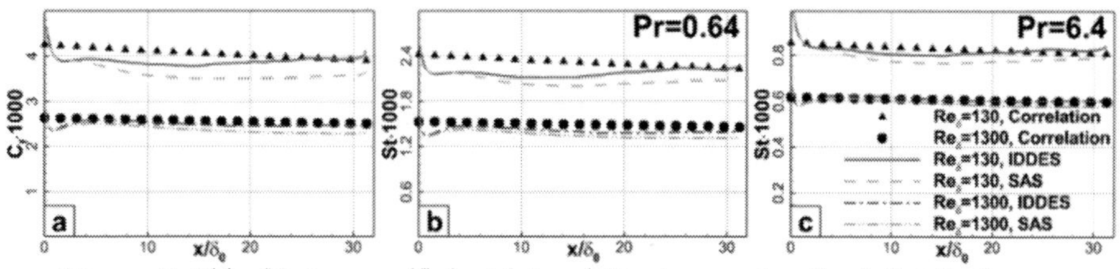

Figure 2. Skin friction coefficient (a) and Stanton number (b, c) distributions at different Reynolds and Prandtl numbers for the boundary layer flow

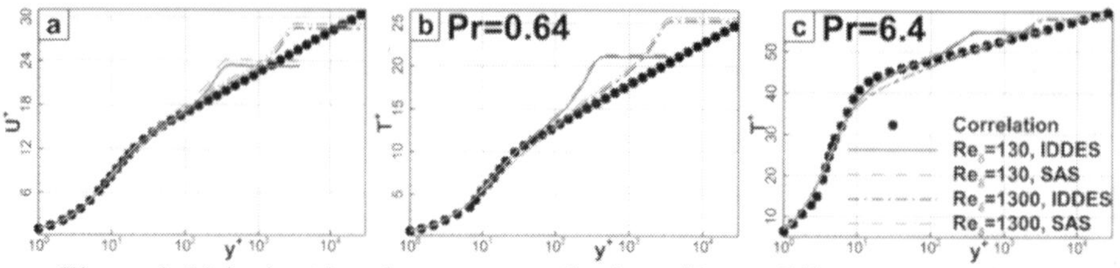

Figure 3. Velocity (a) and temperature (b, c) profiles at different Reynolds and Prandtl numbers for the boundary layer flow

2.2. Backward Facing Step

This flow has been experimentally studied by Vogel and Eaton [8] for $Re=\rho \cdot U_0 \cdot H/\mu=28000$ (U_0 is a bulk velocity at the inlet section, H is a step height, ρ is a constant density, and μ is a constant dynamic viscosity) and for $Pr=\mu \cdot C_p/\lambda=0.7$ (C_p is a constant specific heat capacity and λ is a constant thermal conductivity). An expansion ratio of the upstream plane channel is 1.25, in which a flow is held at constant temperature of T_0=300 K. The lower wall is heated with a constant flux of $Q_w=1.44 \cdot \lambda \cdot T_0/H$, while on the rest walls adiabatic condition are maintained.

A computational domain (Figure 4) for this test case is similar to those used in [3]. In streamwise direction is extends from $-3.8 \cdot H$ to $20 \cdot H$ ($x/H=0$ corresponds to the step location), while in spanwise direction it has a size of $4 \cdot H$. A computational grid (Figure 4) consists from $2.8 \cdot 10^6$ hexahedral cells. The maximum grid steps in streamwise and spanwise directions are equal to $0.1 \cdot H$ and to $0.05 \cdot H$ respectively, which corresponds to $\Delta x^+=200$ and to $\Delta z^+=100$ in wall units. The gird steps in wall normal direction correspond to $\Delta y^+<1$ in the entire domain. A non-dimensional time step of $\Delta t=0.02 \cdot H/U_0$ ensures the CFL number of less than one. Instantaneous flow fields are averaged over 5000 time steps and a sufficient time sample for unsteady statistics is achieved.

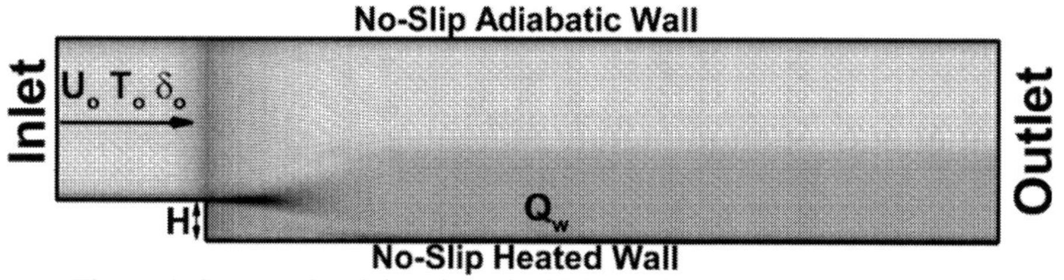

Figure 4. Computational domain and grid for the backward-facing step flow

Boundary conditions are specified as follows (Figure 4). The inflow profiles are obtained from a precursor SST-RANS simulation of a developing plane channel flow up to experimental boundary layer thickness of $\delta_0=H$ and are then used for the GST generator [5,6]. Periodic condition is specified in spanwise direction. No-slip condition is specified on solid walls. On the lower wall a constant heat flux of Q_w is specified, while on the rest walls adiabatic condition is applied. On the outlet boundary a constant pressure is specified, while the rest quantities are extrapolated from the domain.

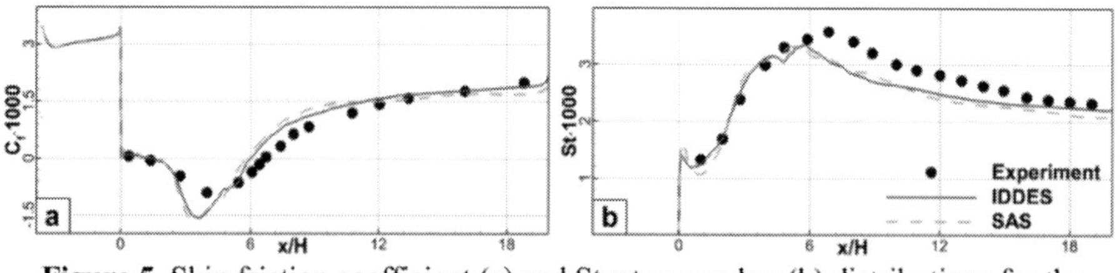

Figure 5. Skin friction coefficient (a) and Stanton number (b) distributions for the backward facing step flow

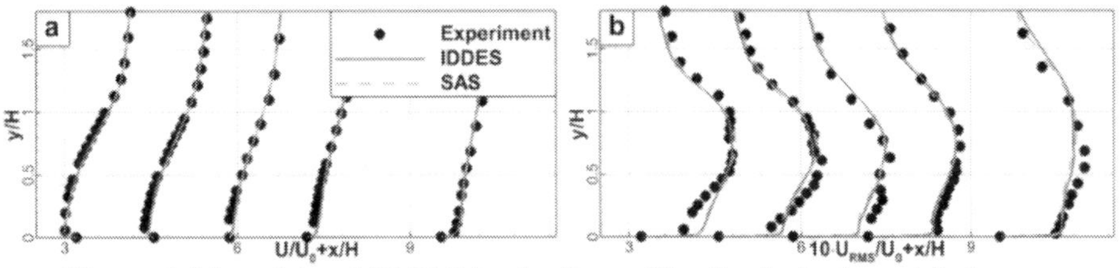

Figure 6. Mean (a) and RMS (b) velocity profiles for the backward facing step flow. The profiles are plotted at x/H=3.2, 4.55, 5.86, 7.2, and 9.53

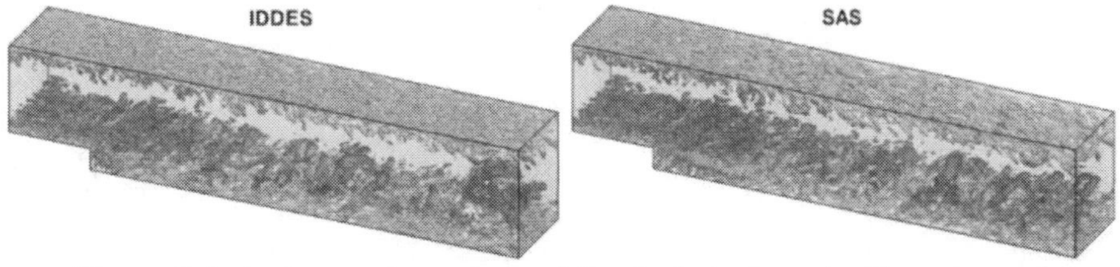

Figure 7. Instantaneous isosurfaces of Q-criterion coloured with contours of velocity magnitude for the backward facing step flow

As seen from distributions of skin friction coefficient and Stanton number (Figure 5) both considered models yield virtually identical results and agree fairly well with the experimental data [8]. Similar trend is also observed for mean and RMS velocity profiles (Figure 6) and for instantaneous isosurfaces of Q-criterion (Figure 7). Thus, both approaches are capable to predict both flow and heat transfer with sufficient accuracy.

2.3. *Thermal mixing in T-Junction*

A thermal mixing of water in a T-Junction has been experimentally studied in [9]. The setup (Figure 8) consists of vertical and horizontal pipes with an inner diameter of D_0 and $1.4 \cdot D_0$ respectively. The length of the straight pipes upstream of the T-Junction is more than $110 \cdot D_0$ for the horizontal pipe and about $20 \cdot D_0$ for the vertical pipe. A constant mass flow and temperature is kept throughout the experiment in both pipes with velocity and temperature equal to U_0 and T_0=309 K in the vertical pipe and to $0.76 \cdot U_0$ and T_0-ΔT (ΔT=17 K) in the horizontal pipe. These parameters correspond to $Re=\rho \cdot U_0 \cdot D_0/\mu=8 \cdot 10^4$ (ρ is a constant density and μ is a dynamic viscosity, calculated based on a temperature field according to [9]) and to $Pr=\mu \cdot C_p/\lambda=5$ (C_p is a constant specific heat capacity and λ is a constant thermal conductivity).

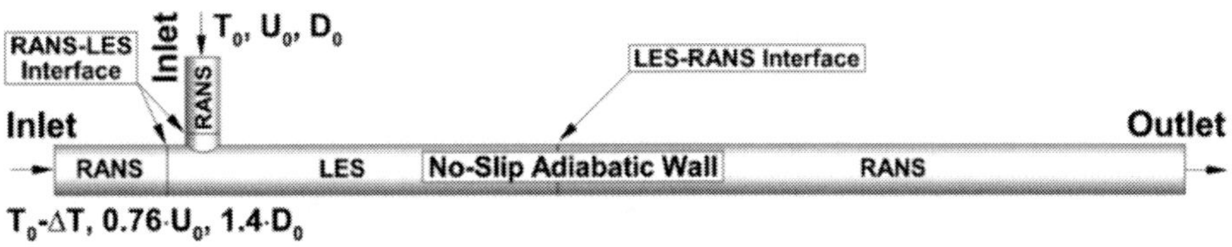

Figure 8. Computational domain for the T-Junction flow

Parameters of computational domain and grid are similar to those used in [1]. The computational domain is shown in Figure 8. The inlet sections are located at z/D_0=3.1 in the vertical pipe and at x/D_0=-4.2 and the horizontal pipe. The outlet section is located at x/D_0=28.0. The computational grid consists of about $4.9 \cdot 10^6$ hexahedral cells. The maximum grid steps in axial (Δ_a) and circumferential (Δ_c) directions are Δ_a/D_0=0.036 and Δ_c/D_0=0.018 respectively, which corresponds to Δ_a^+=7500 and Δ_c^+=4500 in wall units. The size of adjacent to the wall cells corresponds to Δy^+<1 in most of the domain. The used time step is equal to Δt=$0.016 \cdot D_0/U_0$ and corresponds to maximum CFL number of around 4 near the junction. Unsteady statistics is obtained by averaging of instantaneous flow fields over 40000 time steps.

Boundary conditions are shown in Figure 8. The inflow profiles are obtained from a precursor SST-RANS simulation of a pipe flow. For the horizontal pipe a periodic pipe flow is calculated, while for the vertical pipe a calculation of a developing pipe flow up to the experimental value of boundary layer thickness equal to δ_0=0.22·D_0 is performed. No-slip adiabatic condition is applied on pipe walls. A constant pressure is specified on the outlet boundary and the rest quantities are extrapolated from the domain.

For this flow SAS and IDDES are used within an Embedded LES (ELES) approach, according to which the computational domain is decomposed to RANS and LES subdomains as shown in Figure 8. As seen two RANS-LES interfaces at z/D_0=-0.7 and at x/D_0=-1.4 in the vertical and horizontal pipes respectively and one LES-RANS interface at x/D_0=9.8 emerge due to such decomposition. On the RANS-LES interfaces either VM [4] or GST [5,6] are used for generation of turbulent content, while no additional treatment is employed on the LES-RANS interface.

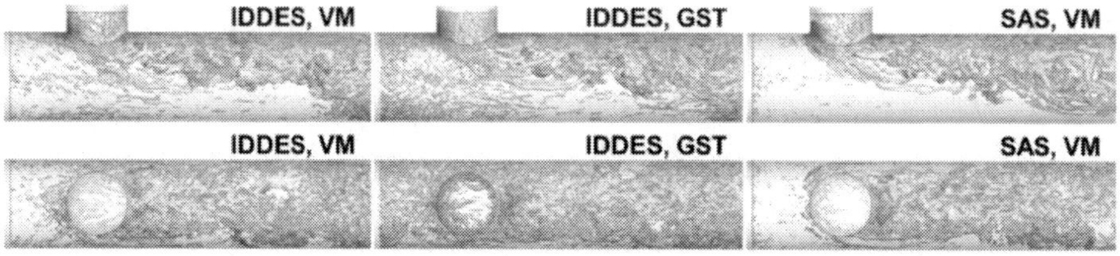

Figure 9. Instantaneous isosurfaces of Q-criterion coloured with contours of velocity magnitude for the T-Junction flow

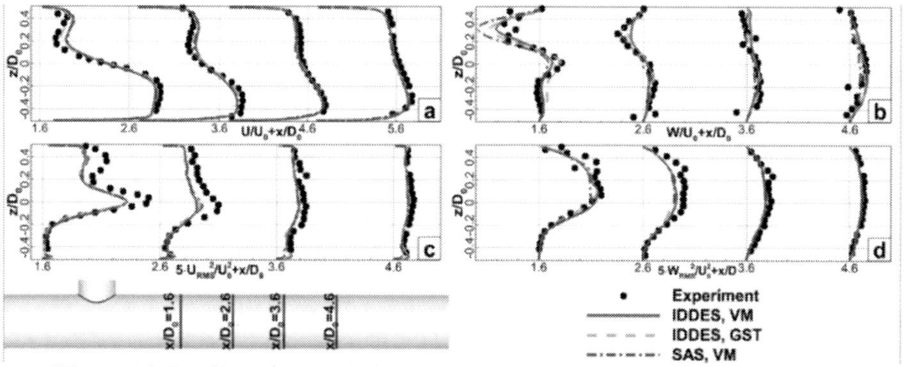

Figure 10. Profiles of mean (a, b) and RMS (c, d) values of velocity for the T-Junction flow. The profiles are show at x/D_0=1.6, 2.6, 3.6, 4.6

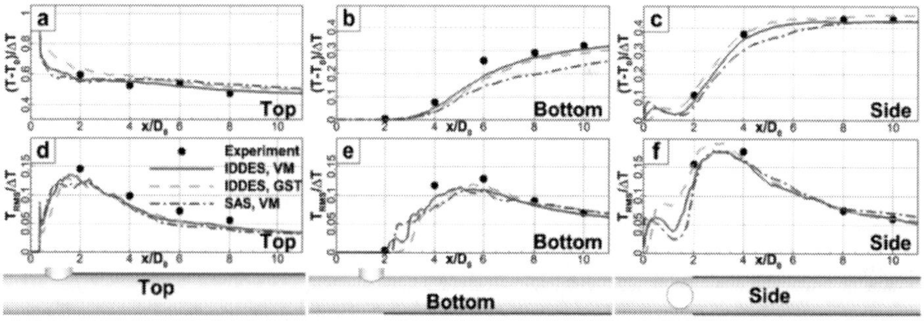

Figure 11. Distributions of mean (a-c) and RMS (d-f) temperature at top, bottom and side sections for the T-Junction flow

As seen from the calculations by SAS and IDDES approaches, instantaneous turbulent structures (Figure 9) are created on the RANS-LES interfaces and are developing downstream. Herewith, the structures obtained with SAS are relatively larger, that those of IDDES. Nevertheless, the presented profiles of mean and RMS velocity (Figure 10) show that both approaches yield virtually identical results that agree fairly well with the experimental data [9]. As regards the distributions of mean and RMS temperature on pipe walls (Figure 11) again good agreement with the experimental data is achieved for all the considered approaches, however, a distribution of mean temperature at bottom section is slightly underestimated for SAS with VM. Nevertheless, all the considered approaches are capable to predict this flow with sufficient accuracy.

3. Summary
A comparative assessment of two modern SRS approaches has been performed. It has been shown, that both approaches are able to predict both mean and RMS values of temperature and velocity when are used in conjunction with inflow turbulent content generators. Among the considered synthetic turbulence generators no advantage has been observed and both VM and GST are able to provide appropriate inflow turbulent content for all the considered flows.

Acknowledgement
The authors from St. Petersburg are partially supported by the Russian Foundation for Basic Research (grants No. 12-08-00256 and No. 14-08-31121)

References
[1] Gritskevich M, Garbaruk A V., Frank T and Menter F R 2012 Investigation of the Thermal Mixing in a T-Junction Flow with Different SRS Approaches *Proceedings of CFD4NRS-4 The Experimental Validation and Application of CFD and CMFD Codes in Nuclear Reactor Technology* pp 1–11
[2] Menter F R and Egorov Y 2010 The Scale-Adaptive Simulation Method for Unsteady Turbulent Flow Predictions. Part 1: Theory and Model Description *Flow Turbulence and Combustion* **85** 113–38
[3] Gritskevich M S, Garbaruk A V., Schütze J and Menter F R 2012 Development of DDES and IDDES Formulations for the k-ω Shear Stress Transport Model *Flow Turbulence and Combustion* **88** 431–49
[4] Mathey F 2008 Aerodynamic noise simulation of the flow past an airfoil trailing-edge using a hybrid zonal RANS-LES *Computers & Fluids* **37** 836–43
[5] Adamian D Y and Travin A K 2011 Improved version of the synthetic eddy method for setting nonstationary inflow boundary conditions in calculating turbulent flows *High Temperature* **49** 704–11
[6] Gritskevich M S and Garbaruk A V 2012 Embedded large eddy simulation with the use of a volume source of turbulent fluctuations *St. Petersburg State Polytechnical University Journal. Physics and Mathematics* **1(141)** 27–36
[7] Kader B A and Yaglom A M 1972 Heat and Mass Transfer Laws for Fully Turbulent Wall Flows *International Journal of Heat and Mass Transfer* **15** 2329–51
[8] Vogel J C and Eaton J K 1985 Combined heat transfer and fluid dynamic measurements downstream of a backward-facing step *ASME Journal of Heat Transfer* **107** 922–9
[9] OECD/NEA 2009 *OECD/NEA: T-Junction Benchmark Specifications, OECD/NEA & Vattenfall*
[10] Mathur S R and Murthy J Y 1997 A pressure-based method for unstructured meshes *Numerical Heat Transfer* **32** 195–215

Scale resolving simulations of the water flow through a rod bundle with split-type spacer grid

A A Matyushenko, A V Garbaruk
Saint-Petersburg State Politechnical University
Russia, Saint-Petersburg, 195251, Politechnicheskaya, 29

Abstract. Unsteady isothermal turbulent flow in periodic sub-channel of 5x5 rod bundle with split-type spacer grid is investigated using modern Scale Resolving Simulation methods: Wall Modelling Large Eddy Simulation (WMLES) and Scale Adapting Simulation (SAS). Solutions obtained with both methods show complicated swirling flow dissipating over the distance from spacer grid and are in a good agreement with the experimental data except an area in close vicinity of the vanes.

1. Introduction

Spacer grids with mixing vanes are widely used in pressurized water nuclear reactors for heat transfer enhancement and coolant flow optimization provided for a nuclear fuel assembly. In recent years Computational Fluid Dynamics (CFD) is increasingly used for design of such grids, analysis of complex flow and heat transfer of relevance to nuclear plant safety. From a thermal hydraulic standpoint, accurate prediction of the velocity and turbulence fields in a rod bundle with spacer grids with mixing vanes is a challenging task. In engineering practice, such tasks are usually solved using steady or unsteady Reynolds Averaged Navier-Stokes equations, (RANS or URANS) which are not always capable to predict such complex strong swirling flow. Scale Resolving Simulation (SRS) methods potentially are more effective; however, the experience of their usage for flow through the rod bundle with spacer grids with mixing vanes is not sufficient.

The aim of present work is an assessment of modern SRS methods for prediction of complex swirling flow through rod bundle with split-type spacer grid with the use of ANSYS FLUENT CFD code. CFD analysis is based on experiments performed in the Korea Atomic Energy Research Institute (KAERI) at Measurements & Analysis of Turbulence in Subchannels – Horizontal (MATiS) test facility [1].

2. MATiS-H test facility

The main test section of the MATiS-H test facility consists of a 170x170 mm^2 square channel which contained a 5x5 rod bundle array installed in a horizontal position (Figure 1). For a fine-scale examination of the lateral flow structure in a sub-channel geometry and for increasing of LDA measurement resolution, the size of the 5x5 rod bundle array is enlarged in 2.67 times and comprises of 25 rods of 25.4 mm outer diameter and 4m length in a regular matrix arrangement with a rod pitch of P = 33.12 mm and a wall pitch of 18.76 mm. Consequently, the hydraulic diameter of the channel cross-section, which considers the flow area and the wetted perimeter in a square duct including the 5x5 rod bundle, is D_H = 24.27 mm. Split-type spacer grid is mounted to the rods with the use of so-called "buttons" (small cylinders). The grid can be moved in the axial direction along the 5x5 rod bundle array, in order to perform velocity measurements at different cross sections downstream of the tips of the mixing vanes of the spacer grid ($0.5D_H$, $1.0D_H$, $4.0D_H$ and $10.0D_H$) based on the test specification, while a measurement cross section remains in a fixed position.

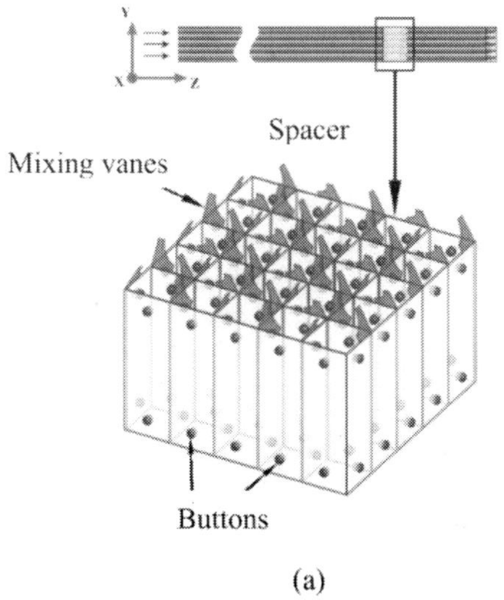

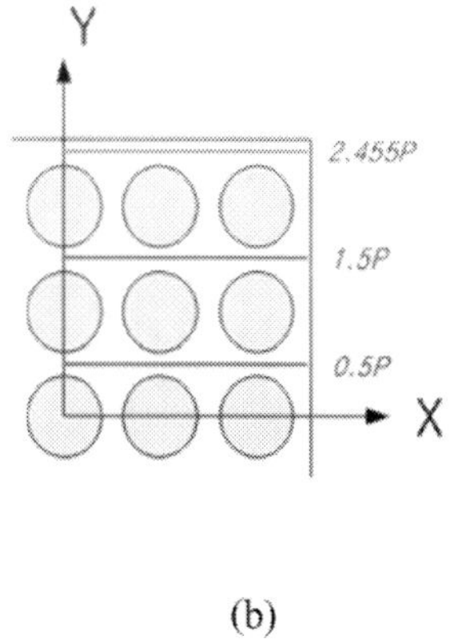

Figure 1. Scheme of the main section of MATiS-H test facility (a) and location of measurement cross sections (b)

Water at normal pressure and temperature conditions is used for all the MATiS-H experiments. The mass flow rate in all experiments is 24.2 kg/s resulting in a bulk velocity of 1.5 m/s corresponding to the Reynolds number based on the hydraulic diameter of Re = 50250. The water flow enters a lower plenum of the horizontal test section with an installed flow breaker. Then the flow passes two flow straighteners, which are installed in order to homogenize the flow and to accelerate the formation of a fully developed flow in the 5x5 rod bundle. The second of those flow straighteners is installed at a distance of $100D_H$ upstream of the spacer grid. Therefore it is assumed, that flow is fully developed at the inlet section of the spacer grid.

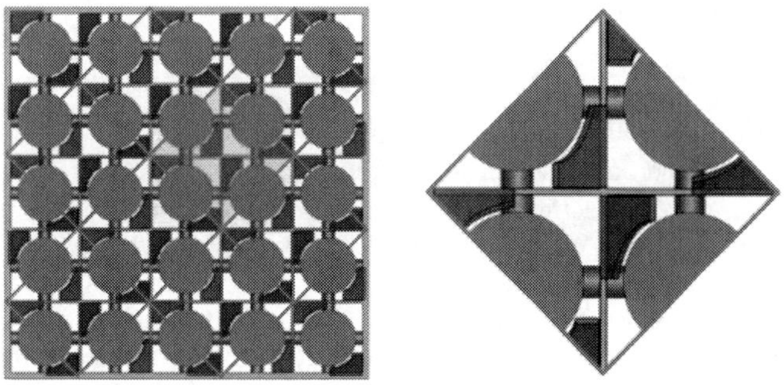

Figure 2. Full configuration of split spacer grid (left) and periodic sub-channel(right)

The most part of the full split-type spacer grid configuration except an area near the duct wall consists of periodic sub-channels (Figure 2). Note that fully periodic flow in the sub-channel should capture all specific features of this flow.

3. Problem definition

A computational domain consists of a section with a length of $4D_H$ located upstream of the spacer grid, the spacer grid itself and a section with length more than $10.0D_H$ downstream of the spacer grid. Topology of the computational domain and applied boundary conditions are shown in Figure 3.

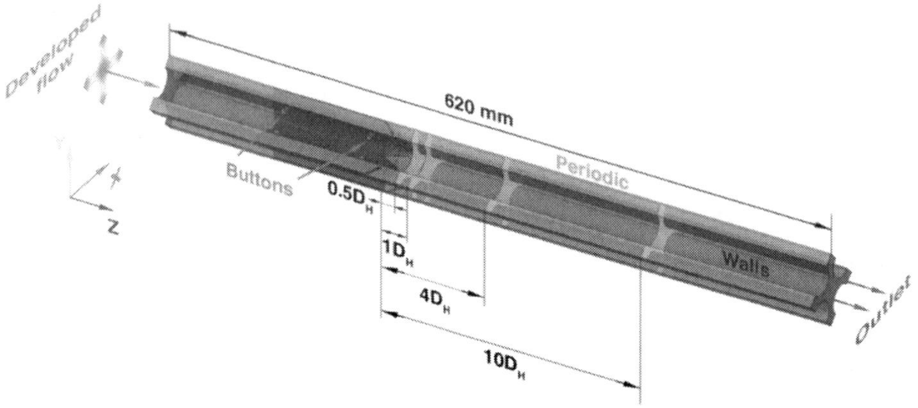

Figure 3. Computational domain

The geometry and a mesh are created with the use of pre-processing applications within the ANSYS Workbench platform (ANSYS Design Modeler and ANSYS Workbench Meshing). The hexahedral mesh is preferable from a CFD accuracy point of view, but it is judged to be less flexible and too expensive to create a high quality mesh. Instead, hybrid meshes are created with the use of ANSYS Workbench Meshing. In the framework of this strategy the flow domain is decomposed into three zones: upstream of the spacers, spacers and downstream of the spacers. At the first stage hexahedral high quality mesh suitable for SRS methods (streamwise and spanwise mesh steps are less than 0.1 of boundary layer thickness) is created in the downstream section. During the second stage tetrahedral mesh with prismatic layers near the walls is built in the spacer grid based on the first stage mesh. Finally upstream mesh is obtained as a sweep of the grid surface at the beginning of the spacer grid. Total mesh parameters are shown in Table 1.

Table 1. Mesh parameters

Number of cells	Max. Y^+	Mean Y^+	Min cell size, mm	Max cell size, mm
40M	5	1.5	0.001	1

Time step of $\Delta t = 0.1$ ms is sufficient to resolve the smallest turbulent structures downstream of the spacer grid (CFL number based on the streamwise grid is of the order of 1).

Water properties at 35°C are specified as a constant property incompressible liquid. SRS methods are performed with the algebraic Wall-Modelling LES (WMLES) [2] model and Scale Adapting Simulation (SAS) [3].

On solid walls automatic near wall treatment is used. Constant static pressure level is specified on the outlet boundary. On the inlet boundary fully developed velocity profiles obtained with the use of SST RANS in combination with Vortex Method (VM) [4] are specified. Other than that, in order to check the influence of the inflow turbulent content on the flow downstream of the spacer grid a SAS computation without VM is also performed.

4. Flow structure

The considered flow through the rod bundle is extremely unsteady. Resolved turbulence upstream of the vanes consists of interaction of the inlet turbulence and a vortex shedding in the wake of the buttons. However, the most significant unsteadiness showed in Figure 4 is provided by the mixing vanes of the spacer grid. Effect of the vanes results in strongly swirling flow in the centre of the sub-channel and in two additional vertices in the rod gap dissipating downstream of the grid (Figure 5).

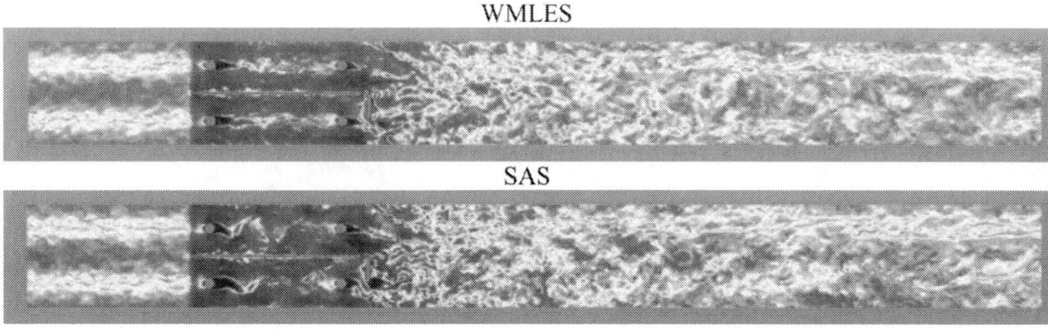

Figure 4. Flow structure visualized by streamwise velocity contours

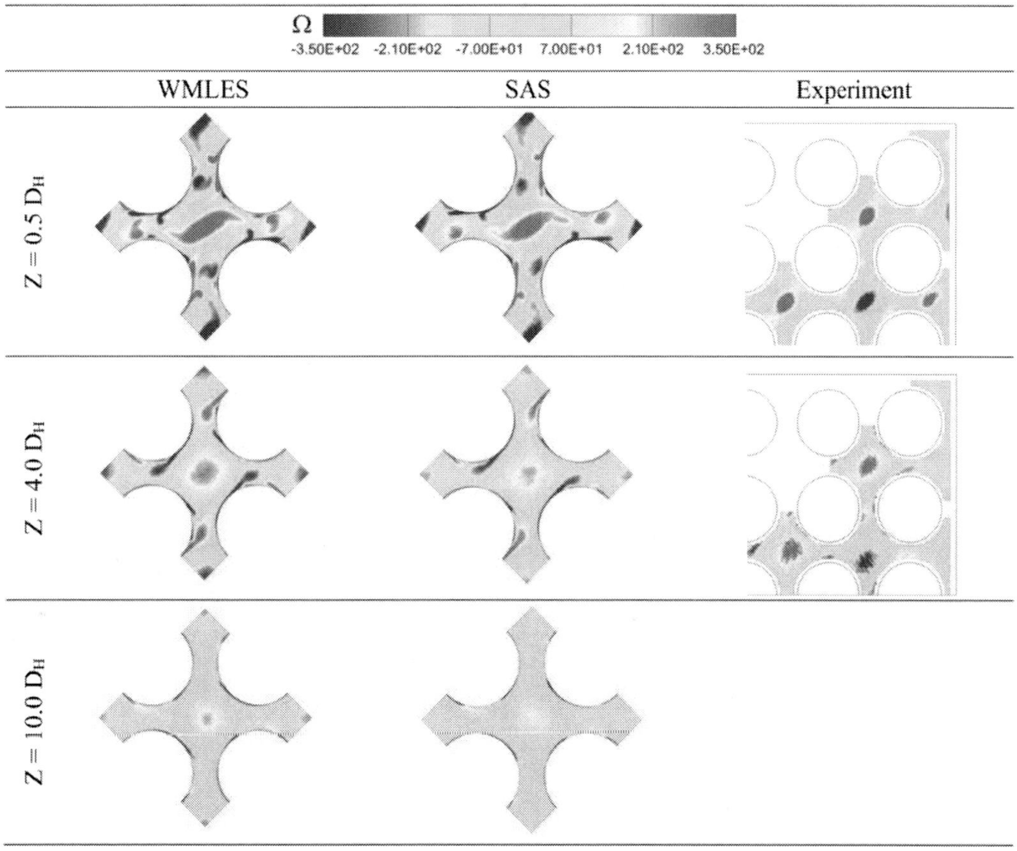

Figure 5. Mean z-vorticity contours for different turbulence models at different sections

5. Comparison of CFD results with experimental data

Time-averaged velocity profiles and turbulence characteristics are compared with the experimental measurements at Y = 0.5P line (Figure 1b) at different cross sections downstream of the spacer grid (Figure 6). Both models predict the main vertices quite well. However, in the nearest section ($Z = 0.5D_H$) there is a visible superiority of SAS over WMLES especially in the level of resolved turbulence. The reason can be in algebraic nature of WMLES. Prandtl model near the walls is used for this method which can be incapable to predict such a complex flow. Further downstream the difference decreases and solutions of both models become similar and correspond to the experimental measurements fairly well.

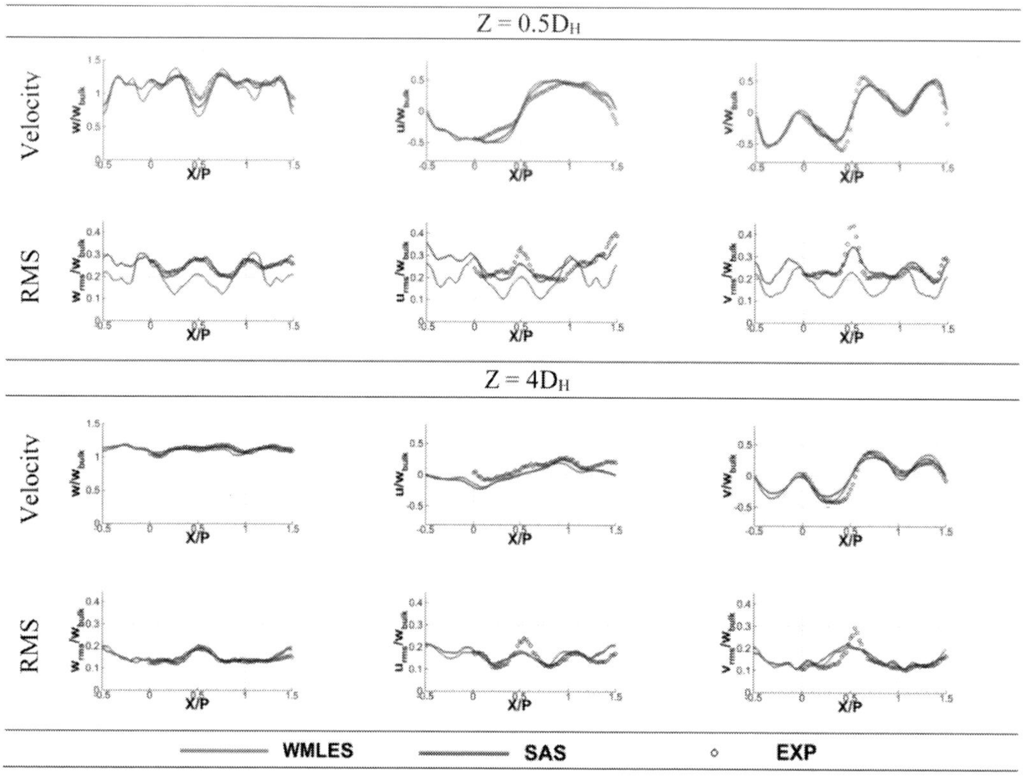

Figure 6. Time-averaged velocity profiles and RMS velocity fluctuations at Y = 0.5P line at different cross sections

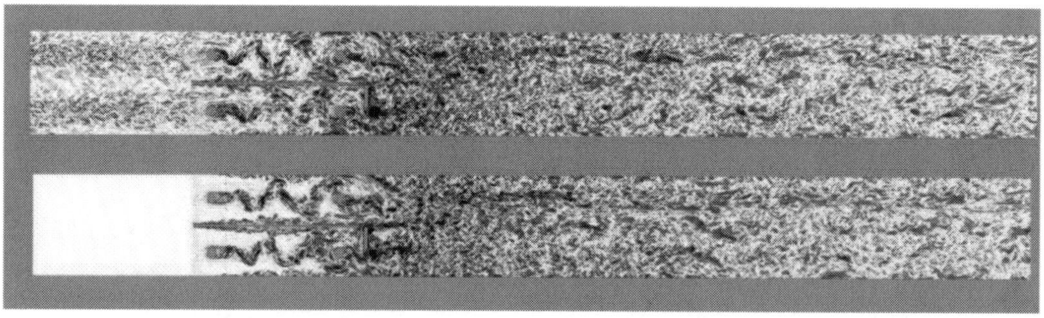

Figure 7. Instantaneous SAS vorticity contours with (top) and without (bottom) inlet turbulent content

6. Influence of inlet turbulent content

An effect of turbulent content specification at the inlet section on vorticity fields is shown in Figure 7. Turbulent structures created by VM have an influence on the flow before and inside the spacer grid. However, downstream of the mixing vanes turbulent structures are very similar for both types of boundary conditions. Comparison of time-averaged characteristics at $Z = 0.5D_H$ section near the mixing vanes (Figure 8) shows similar results for both computations. So, the tubulisation by the mixing vanes does not depend on presence of inlet turbulence.

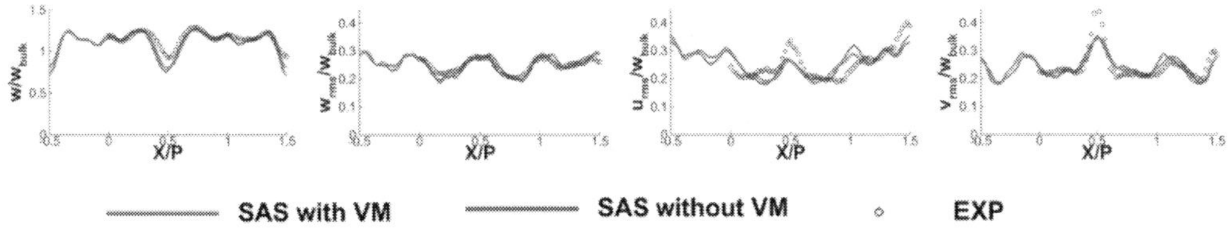

Figure 8. Time-averaged and RMS velocity profiles at $Y = 0.5P$ line at $Z = 0.5D_H$ section

7. Conclusions

The flow through the periodic sub-channel of the rod bundle array with split-type spacer grid has been investigated using ANSYS FLUENT. Flow predictions carried out using WMLES and SAS method show unsteady complicated swirling flow dissipating over the distance from the spacer grid with mixing vanes.

Both methods show fairly good agreement with KAERI MATiS-H benchmark data in prediction of main vortex structures. SAS excels WMLES in prediction of streamwise velocity and RMS values of velocity fluctuations near the mixing vanes. However, in the rest downstream sections both methods yield very close solutions which are in a good agreement with the experiment.

Computation with SAS without VM shows that flow and time-averaged characteristics downstream of the spacer grid is insensitive to turbulent content on the inlet boundary. Note, that for this task WMLES requires generation of turbulent content and should not be used without VM.

Acknowledgements

This work is partially supported by the Russian Foundation for Basic Research (grant No. 12-08-00256)

References

[1] OECD/NEA. 2011. MATiS-H Benchmark. *Final Benchmark Specifications, pp. 44.*

[2] Shur M. L., Spalart P. R., Strelets M. K. & Travin A. K. 2008. A hybrid RANS-LES approach with delayed-DES and wall-modelled LES capabilities. *International Journal of Heat and Fluid Flow, 29(6), 1638-1649. Elsevier Inc.*

[3] Menter F. R. & Egorov Y. 2006. SAS turbulence modelling of technical flows. *Direct and Large-Eddy Simulation VI, 687–694.*

[4] Mathey F. 2008. Aerodynamic noise simulation of the flow past an airfoil trailing-edge using a hybrid zonal RANS-LES. *Computers & Fluids, Volume 37, pp. 836–843.*

Flow of chemically reacting carbon vapor

Yu A Shustrov[1,2], **S A Poniaev**[1], **S V Bobashev**[1], **B G Zhukov**[1]

[1] Ioffe Physico-technical Institute, St. Petersburg, Russia

E-mail: shustrovya@gmail.com

Abstract. Numerical simulation of a chemically reacting flow of an expanding gas flow consisting of carbon C and its clusters (C2 - C5) is presented. It is shown how the chemical composition of the mixture flow changes when it expands and its temperature decreases. The temperature range in which the nonequilibrium chemical model including separate direct and inverse cluster formation reactions should be used is determined. Variations in the effective adiabatic index in the C-C5 mixture and along the nozzle axis are given.

1. Introduction

Flows with chemical reactions, including cluster formation reactions, are of great interest because of use of high-speed plasma jets for material testing and surface modification [1,2]. Typically such jets are produced in plasmatrons. A novel method for obtaining pulsed plasma jets involves the use of an electromagnetic rail generator (railgun) [3]. Of major importance is the fact that the plasma composition in the railgun channel is determined by the material of electrodes (rails) because of erosion that occurs when a plasma piston moves along it. By using a railgun with graphite electrodes, supersonic plasma jets with the velocities as high as 14 km/s [4] can be obtained. Such a high-speed carbon plasma jet can be used to form carbon coatings at different materials.

After an expanding plasma jet leaves the railgun channel, it experiences a considerable temperature drop due to expansion. This leads to changes in its chemical composition. The cluster formation process starts. This is an exothermal process, and the reacting jet is cooled to a lesser degree than a jet of a nonreacting gas. Because of a short outflow time a great role in the jet is played by nonstationary chemical processes. In this paper we report numerical simulation of a nonstationary flow of the mixture consisting of carbon C and its clusters C2 - C5.

2. Numerical model

We consider flows in the absence of external heat exchange and take into account the thermal effect of chemical reactions. In this case the Euler equations in the form of conservation laws in the rectangular coordinate system will acquire the form

[2] Author to whom any correspondence should be addressed.

$$\begin{aligned} &\frac{\partial \rho}{\partial t} + \nabla(\rho \vec{V}) = 0 \\ &\frac{\partial}{\partial t}(\rho \vec{V}) + \nabla(\rho \vec{V}\vec{V}) = -\nabla p \\ &\frac{\partial}{\partial t}(\rho h_s) + \nabla(\rho \vec{V} h_s) = W_T \\ &\frac{\partial}{\partial t}(\rho Y^{(s)}) + \nabla(\rho \vec{V} Y^{(s)}) = W_s \end{aligned} \qquad (1)$$

where ρ is the flow density, $\vec{V}$ is the velocity vector, p is the pressure, h_s is the enthalpy, W_T is the change in the mixture energy due to chemical reactions, $Y^{(s)}$ is the mass fraction of mixture component s, and W_s is the change of the s component of the mixture due to chemical reactions.

The heat capacity of each component of the mixture and also enthalpy and entropy are calculated by using relevant polynominals as functions of T

$$\begin{aligned} &\frac{C^0_{pk}}{R} = a_{1k} + a_{2k} T_k + a_{3k} T_k^2 + a_{4k} T_k^3 + a_{5k} T_k^4 \\ &\frac{H^0_k}{RT_k} = a_{1k} + \frac{a_{2k}}{2} T_k + \frac{a_{3k}}{3} T_k^2 + \frac{a_{4k}}{4} T_k^3 + \frac{a_{5k}}{5} T_k^4 + \frac{a_{6k}}{T_k} \\ &\frac{S^0_k}{R} = a_{1k} \ln T_k + a_{2k} T_k + \frac{a_{3k}}{2} T_k^2 + \frac{a_{4k}}{3} T_k^3 + \frac{a_{5k}}{4} T_k^4 + a_{7k} \end{aligned} \qquad (2)$$

Coefficients $a_{1k}...a_{7k}$ for chemical components are taken from the NIST database [5] and tables of thermodynamic properties of different species [6].

To calculate the dependence of the reaction rate constant on temperature, a standard Arrhenius equation is used

$$k_i = A_i \exp\left(-T_a / T\right) \ . \qquad (3)$$

The reactions in carbon vapor used in this simulation are listed in Table 1. Formation and dissociation of carbon cluster C2-C5 are included in this set. The coefficients for all of these reactions are also given in Table 1.

Table 1. Reaction coefficients for carbon cluster formation (f) and (r) reactions [5].

N	*Reaction*	*A, (m³/kmol-s), (1/s)*	*T_a, K*
1	C + C → C2	$2 \cdot 10^{11}$	0
1r	C2 → C + C	$2.9 \cdot 10^{15}$	74000
2	C2 + C → C3	$2 \cdot 10^{11}$	0
2r	C3 → C2 + C	$1.9 \cdot 10^{16}$	84814
3	C2 + C2 → C3 + C	$2 \cdot 10^{12}$	8616
3r	C3 + C → C2 + C2	$1.3 \cdot 10^{13}$	22142
4	C2 + C2 → C4	$2 \cdot 10^{11}$	0
4r	C4 → C2 + C2	$3.6 \cdot 10^{16}$	66474

5	C3 + C → C4	$2 \cdot 10^{11}$	0
5r	C4 → C3 + C	$5.6 \cdot 10^{15}$	52900
6	C + C4 → C5	$2 \cdot 10^{11}$	0
6r	C5 → C4 + C	$1.1 \cdot 10^{16}$	80482
7	C2 + C3 → C5	$2 \cdot 10^{11}$	0
7r	C5 → C2 + C3	$2.2 \cdot 10^{16}$	62142

3. Results

First of all let us consider equilibrium molar concentrations of C atom and C2-C5 carbon clusters at different temperatures. Previously such an estimation were made by using not a forward and reversa reaction but by using equilibrium constant [8,9]. Fig. 1a and 1b show such dependence for two pressures. It is clearly seen that for a lower pressure ($p = 10^4$ Pa) there is only C5 clusters up to 1500 K. A further temperature increase leads to changes in mixture composition and when temperature becomes higher than 5500 K only C atoms exist in the mixture. For a higher pressure ($p = 10^6$ Pa) the dependence is similar, the only difference is that the change of the mixture composition from C5 to C occurs in a broader temperature range (1500 - 8000 K). It is also interesting to note that there are few C4 clusters at any temperature in the equilibrium case. C3 and C2 cluster concentrations have maxima at different temperatures and the maximum concentration of C2 clusters are lower than that of C3 clusters for both pressures. The process of carbon mixture composition variation occurs in the following way: if we start with a low (less than 1500 K) temperature, only C5 clusters exist, after 1500 K the processes of C5 cluster dissociation and C3 cluster formation start. Near the maximum value of C3 cluster molar concentration (3000 K for 10^4 Pa and 4000 K for 10^6 Pa) the process of C2 cluster formation begins and C atoms appear in the mixture. When the temperature becomes higher than 3000 K for 10^4 Pa and 4000 K for 10^6 Pa, the number of C atoms increases and the amount of C2 clusters at first increases and beginning from 4000 K for 10^4 Pa and 5500 K for 10^6 Pa decreases. Only the amount of C atoms increases after these temperatures. Finally only C atoms exist in the mixture. No doubt, ionization of C atoms starts at higher temperatures but the consideration of this effect is out of the scope of this paper. In any case the ionization becomes important at the temperatures higher than 10000 K.

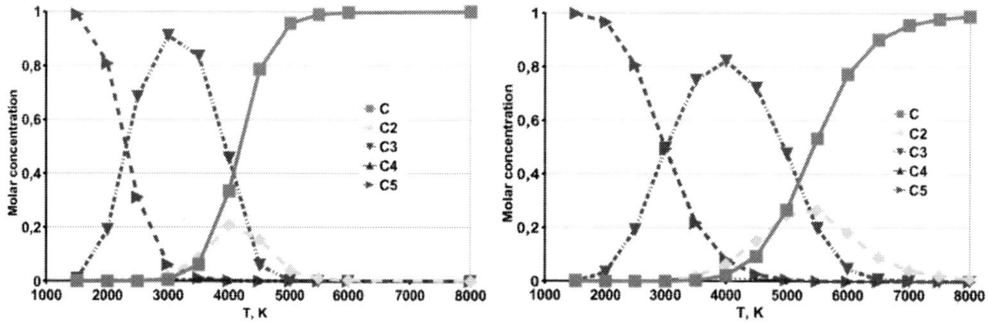

Figure 1. Molar concentrations of C atoms and C2-C5 clusters at different temperatures under equilibrium conditions. (a) pressure $p = 10^4$ Pa, (b) pressure $p = 10^6$ Pa

The expanding flow parameters can be solved analytically in the case of a constant adiabatic index. In order to see whether this is applicable to our case, i.e., carbon vapor flows, we need to calculate the range in which the adiabatic index of the C-C5 mixture changes at different temperatures. Fig.2 shows variations in the adiabatic index under equilibrium conditions for two pressures. It can be seen that there is a strong change in the adiabatic index. This is caused by changes in the mixture composition (see Fig.1). At low temperatures, when the mixture consists of C5 clusters alone, the adiabatic index is about 1.1 and it increases up to 1.65 with increasing temperature. It can also be seen from Fig.2 that the adiabatic index of the mixture exhibits different behaviors for different pressures.

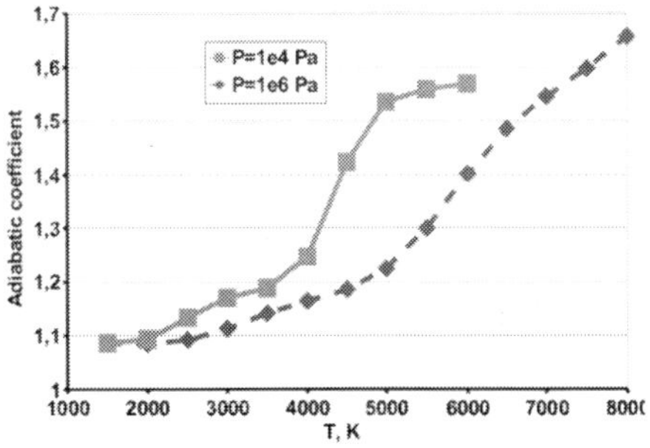

Figure 2. Adiabatic index for C-C5 mixture at different temperatures under equilibrium conditions.

To show how a chemical transformation of the C-C5 mixture affects the expanding flow parameters, we simulated a flow in the axis-symmetric nozzle. The nozzle critical cross-section had a radius of 1.45 mm and nozzle wall angle was 12°. Fig.3 shows the geometry of the calculation zone and a computational mesh. Two simulations were made to show how inlet values of temperature affect the flow parameter distribution along the nozzle axis. For the first simulation, the temperature at the critical cross section of the nozzle was T*=4200K and for second simulation T*=6000K, the pressure was 10^4 Pa for both cases. It can be seen from Fig.1 that for the case of T*=6000 K only C atoms exist in the critical cross-section, and for T*=4200 K the mixture includes C atoms, C2 and C3 clusters. The code based on the OpenFOAM [10] platform was used for these simulations.

Fig.4 shows the adiabatic index variation along the nozzle axis for the case T* = 6000 K. It can be seen that there is a decrease in the adiabatic index along the axis. If we compare this figure with Fig.2, we can conclude that we see a typical behaviour of the adiabatic index, i.e. its fall with decreasing temperature.

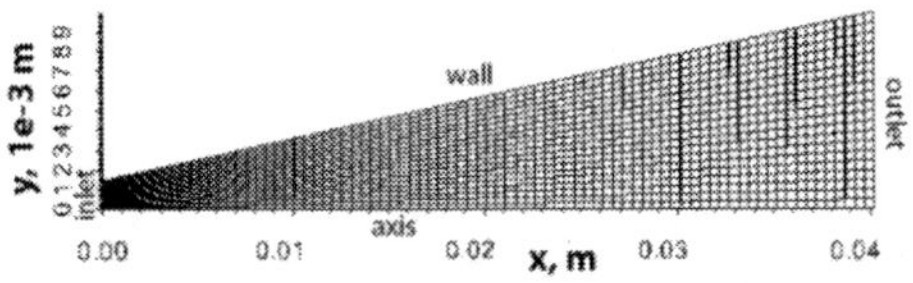

Figure 3. Geometry, mesh and boundary types of computational domain.

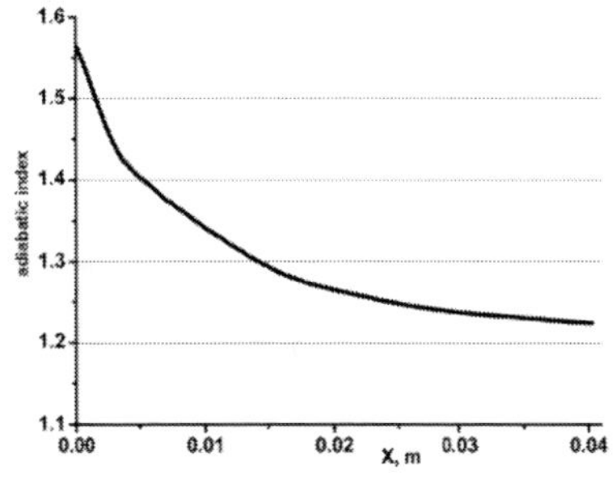

Figure 4. Adiabatic index for C-C5 mixture along the nozzle axis for the simulation with temperature in critical nozzle cross-section T*=6000 K.

Fig.5 shows the change in the normalized temperature T/T* along the nozzle axis. The x-axis is the normalized nozzle cross-section A/A*. To show how the temperature decreases at a constant adiabatic index γ, the lines for γ = 1.1, 1.2, 1.3, 1.4, 1.5 1.6 and 1.7 are plotted in the figure. There are also two lines for our simulation at T*=4200K and T*=6000K, i.e., for a varying adiabatic index. It is seen that there is a difference between two simulations. The normalized temperature for the case T*=4200K decreases faster than for the case T*=6000K.

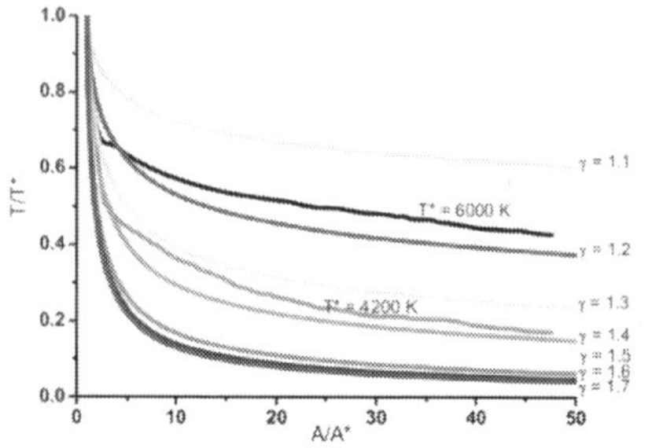

Figure 5. Normalized temperature T/T* as a function of normalized nozzle cross-section A/A*.

If we compare the lines for the constant adiabatic index, it can be seen that at the initial stage of expansion the temperature drop in both simulations corresponds to a relatively high adiabatic index (1.3 for T*=6000 K and 1.5 for T*=4200K). After the initial stage of expansion the adiabatic index decreases and becomes about 1.2 for T*=6000 K and 1.4 for T*=4200 K.

4. Conclusion

Numerical simulation of a chemically reacting flow of an expanding carbon vapour consisting of carbon C and its clusters C2-C5 has been carried out. It has been shown that it is necessary to take into account the nonequilibrium chemical model including separate direct and inverse cluster formation reactions when temperature of the mixture is higher than 1500 K and lower than 5000 K for the mixture pressure lower than 10^4 Pa. It has also been shown that the mixture parameters (e.g. temperature) during expansion of the flow in the nozzle are strongly affected by the parameters at the nozzle inlet. For the case of temperature in the critical cross-section of the nozzle T*=6000 K the adiabatic index along the nozzle axis changes in the range 1.2-1.6.

Acknowledgements

The work was supported by RFBR grant 12-08-01050-a and Program of RAS Presidium P02 and P25.

References

[1] Wana Y P, Sampatha S, Prasada V, et al. 2003 Journal of Materials Processing Technology vol 137 p 110
[2] Rao N P, Tymiak N, Blum J, Neuman A, Lee H J, Girshick S L, McMurry P H, and Heberlein J, 1998, *Journal of Aerosol Science* vol. 29 no 5–6 p 707
[3] Zhukov B G, Kurakin R O, Sakharov V A, Bobashev S V, Ponyaev S A, 2013 Technical Physics Letters vol 39(6) p 569
[4] Bobashev S V, Zhukov B G, Kurakin R A, Ponyaev S A, Reznikov B I, Rozov,SI, 2010, Tech. Phys. Lett. v.36 1 p 72
[5] NIST Standard Reference Database http://webbook.nist.gov/chemistry/
[6] Gurvich L V, Veyts I V, and Alcock C B, *Thermodynamic Properties of Individual Substances*, 4th Ed. (CRC, Boca Raton, 1994; Nauka, Moscow, 1982), Vol. 4.
[7] Krestinin A V, Moravskii A P, and Tesner P A, 1998 Khimical Fizica vol 17 **p** 71
[8] Reznikov B I, Alekseev N I, Bobashev S V, Zhukov B G, Ponyaev S A, Kurakin R O, Rozov S I, 2011 Tech. Phys. vol 56 8 p 1106-1111
[9] Alekseev N I, Reznikov B I, Zhukov B G, Bobashev S V, Ponyaev S A, Kurakin R O, 2012 Journal of Engineering Physics and Thermophysics vol 85 (3) p 593
[10] OpenFOAM CFD software package www.openfoam.org

16th Russian Youth Conference on Physics and Astronomy (PhysicA.SPb/2013) IOP Publishing
Journal of Physics: Conference Series **572** (2014) 012060 doi:10.1088/1742-6596/572/1/012060

Two-dimensional hybrid model for a glow discharge: comparison with fluid and kinetic (particle) models, reliability and accuracy

E Eylenceoğlu and I Rafatov
Middle East Technical University, Ankara, Turkey
E-mail: eender@metu.edu.tr

Abstract. We developed and tested a two-dimensional Monte Carlo – fluid hybrid numerical code for the DC glow discharge simulations. The model is based on the separation of electrons into two parts, namely, the low energetic (slow) and high energetic (fast) groups. Ions and slow electrons are described within the fluid model using the drift-diffusion approximation for particle fluxes. Electrostatic field is obtained from the solution of Poisson equation. Fast electrons, represented by the appropriate number of super particles emitted from the cathode, are responsible for ionization processes in the discharge volume. Test calculations were carried out for the argon plasma. The vortex current formation in a DC discharge is observed in the case of rectangular geometry.

1. Introduction

The physics of gas discharge phenomena, natural or man-made, is based on the dynamics of electrons and ions, created by the ionization of neutral gas particles. Gas discharges demonstrate certain nonlinear properties, among which the formation of sheath layers near the electrodes, existence of specific regions inside the discharge volume glowing with different intensities, secondary electron emission from electrodes by the impact of energetic charged particles. Furthermore, nonlocal transport characteristics of electrons in high electric field regions make this phenomena complicated to investigate. Numerical modeling techniques used to study the gas discharges include fluid, hybrid, and kinetic (particle) methods [1, 2]. Within the fluid method, the particle species (electrons, ions, and neutrals) are considered as fluids described by the continuity equations with drift-diffusion approximation for particle fluxes. Fluid models usually employ a local field approximation (LFA), according to which transport coefficients as well as the ionization rate are determined as functions of the reduced electric field [3, 4, 5]. Fluid methods in general are capable to describe basic qualitative properties of plasma. Within the kinetic (particle) methods, which provide quantitatively accurate results, particle species are simulated by the suitable number of super particles [6, 7, 2]. Hybrid methods are partially fluid and partially kinetic [8, 9, 10, 11].

In this work we present a two-dimensional Monte Carlo – fluid hybrid model for a rectangular discharge geometry filled with argon gas. The details of the model are described in section 2. Numerical results and discussion are presented in section 3. Finally, in section 4, a brief summary of the study is given.

2. Model

The model is based on separation of the electrons into two groups, namely, the low energetic (slow) and high energetic (fast) electrons [12]. Ions and slow electrons are described within the fluid model using drift-diffusion approximations for particle fluxes [4]. Electric field is obtained from solution of the Poisson equation. Dynamics of fast electrons and the effect of collisions are simulated by the Monte Carlo Collision (MCC) method that allows to take the nonlocal transport of electrons in the cathode region properly into account [12, 9]. Fast electrons are responsible for ionization processes in the discharge volume. These are simulated by a suitable number of super particles (e.g. 500) emitted from cathode into the discharge volume. Electron elastic, excitation and ionization collisions with the background neutral gas particles are included in the model, while the electron–electron and electron–ion collisions are ignored. The occurrence, type and scattering angles of the collisions are determined by random numbers uniformly distributed between 0 and 1.

2.1. Fluid model

Fluid model includes continuity equations for electrons and ions [13],

$$\frac{\partial n_e}{\partial t} + \nabla \cdot \mathbf{\Gamma}_e = S_e, \tag{1}$$

$$\frac{\partial n_i}{\partial t} + \nabla \cdot \mathbf{\Gamma}_i = S_i, \tag{2}$$

coupled with the Poisson equation for the electric field,

$$\nabla^2 V = -\frac{e}{\varepsilon_0}(n_i - n_e). \tag{3}$$

Here n denotes the particle density, $\mathbf{\Gamma}$ the flux density, V the electric potential. Subscripts e and i refer to electrons and ions. S the ionisation source term obtained from the MCC simulation of fast electrons. Flux densities are expressed in the drift-diffusion form [4],

$$\mathbf{\Gamma}_{i,e} = -D_{i,e}\nabla n_{i,e} \pm \mu_{i,e}\mathbf{E}n_{i,e}, \tag{4}$$

where D is the diffusion coefficient, μ the mobility coefficients, and $\mathbf{E}$ the electric field vector.

2.2. MCC Simulation of Fast Electrons

The fast electrons are traced in the discharge volume as they move in the electric field and make random collisions (elastic or inelastic) with the background neutral gas atoms. The velocities $\mathbf{v}$ and positions $\mathbf{r}$ of these particles between collisions are obtained by using dynamical equations,

$$\mathbf{v}_i^{k+1} = \mathbf{v}_i^k + \mathbf{a}_{eff,i}^k \Delta t, \tag{5}$$

$$\mathbf{r}_i^{k+1} = \mathbf{r}_i^k + \mathbf{v}_i^k \Delta t + \frac{1}{2}\mathbf{a}_{eff,i}^k \Delta t^2. \tag{6}$$

Here indices i and k refer to the i'th electron and the time level. Δt is the time step size. In the case of one-dimensional motion effective acceleration $\mathbf{a}_{eff}$ is determined from

$$\mathbf{a}_{eff} = -\frac{e}{m_e}(\mathbf{E} + \frac{1}{2}v_x \Delta t \frac{d\mathbf{E}}{dx}). \tag{7}$$

Here e is the electron charge, m_e the electron mass, E the electric field magnitude, and v_x the x component of the velocity.

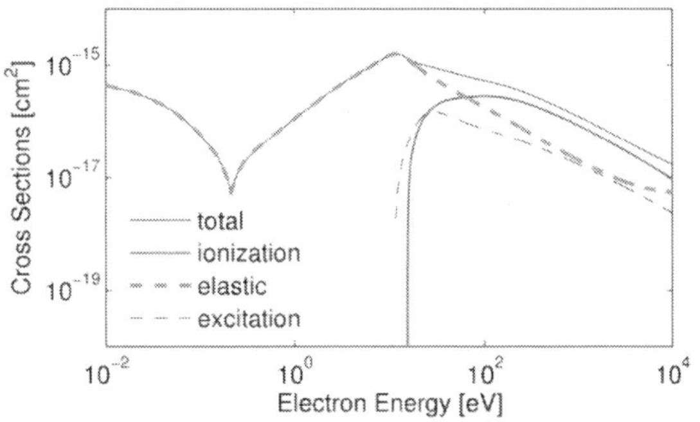

Figure 1. Elastic, excitation, ionization and total cross-sections for electrons in argon gas [14].

For a small time step Δt the collision probability of electrons is expressed in the form

$$P_c = 1 - e^{-N_g \sigma_{tot}(v_r) v_r \Delta t}, \tag{8}$$

$$\sigma_{tot} = \sigma_{el} + \sigma_{ex} + \sigma_{iz}, \tag{9}$$

where N_g is background neutral gas density, σ_{tot} is velocity (or energy) dependent total collision cross section, v_r is a relative velocity. The cross-section set is presented in figure 1. For the collisions between fast electrons and neutral gas particles, these velocities coincide with the electron velocities. For a group of electrons, the probability of collisions can be obtained by averaging over distribution and expressed in the form

$$P_c = 1 - e^{-N_g < \sigma_{tot}(v) v > dt}. \tag{10}$$

When multiplied with the total number of electrons in the simulation, this probability gives the total number of electrons that collide with the background gas atoms. The criterions of making a collision and a specific type of collision (elastic, ionization or excitation) are determined by the random numbers uniformly distributed between 0 and 1 [10, 2]. After all fast electrons within MCC cycle have escaped from the volume or converted to the slow electrons, the ionization events data is interpolated into the grid points in order to determine the ionization rate profile. The ionization source profile is weighted to the actual value using the equation [11, 17]

$$S_{i,e} = \frac{j}{(1 + 1/\gamma)\Delta x} \frac{N_{i,e}}{N_0}, \tag{11}$$

where j is the electric current density at the cathode, γ the secondary electron emission coefficient, and Δx the grid cell size. $N_{i,e}$ are the ionization profiles obtained from MCC cycle. N_0 is the initial number of super electrons emitted from cathode in the MCC cycle.

2.3. Geometry and Boundary Conditions

The calculations are carried out in the a rectangular region with a uniform grid in each direction. Cathode and anode locate at the left and right border respectively.

Boundary condition for the electric potential are the grounded potential on the cathode and applied voltage V_a on the anode. On the dielectric side walls, which are perfect absorbers, electric field is determined by the Gauss law, $\hat{\mathbf{n}} \cdot \mathbf{E} = -\sigma/\varepsilon_0$. The charge accumulation process is expressed by the equation

$$\frac{d\sigma}{dt} = e(\mathbf{\Gamma}_i - \mathbf{\Gamma}_e) \cdot \hat{\mathbf{n}}, \tag{12}$$

where σ is the surface charge density, $\hat{\mathbf{n}}$ is outward normal to the wall surface.

Boundary conditons for particle species on the electrodes and side walls are given by the directed fluxes. For electrons [15],

$$\hat{\mathbf{n}} \cdot \mathbf{\Gamma}_e = \frac{1}{4} n_e v_e - \alpha_e \mu_e n_e (\hat{\mathbf{n}} \cdot \mathbf{E}) - \beta_e \gamma \mid \hat{\mathbf{n}} \cdot \mathbf{\Gamma}_i \mid, \tag{13}$$

and for ions,

$$\hat{\mathbf{n}} \cdot \mathbf{\Gamma}_i = \frac{1}{4} n_i v_i + \alpha_i \mu_i n_i (\hat{\mathbf{n}} \cdot \mathbf{E}). \tag{14}$$

Here, $v_{i,e} = \sqrt{8\pi T_{i,e}/em_{i,e}}$ are the electron and ion thermal speeds. The last term in equation (13) describes the secondary electron emission from the walls bombarded by energetic ions, $\beta_e = 1$ on cathode and 0 otherwise. $\alpha_{i,e}$ controls the directed convective fluxes to the walls: it equals 1 if the flux is directed into the wall and 0 otherwise.

2.4. Plasma Parameters

Model is simulated in argon gas at pressure $p = 1$ Torr with a constant temperature $T_g = 0.025$ eV. $L_x = 1$ cm and $L_y = 1$ cm. We used constant transport coefficient for slow electrons, $\mu_e = 3 \times 10^5$ cm^{2}s^{-1}V and $D_e = \mu_e T_e$, where $T_e = 1$ eV [17]. Ions are assumed to be in thermal equilibrium with background gas, $T_i = T_g = 0.025\ eV$. Mobility and diffusion coefficients of ions are determined as functions of reduced electric field as in Ref. [16] and [18].

3. Numerical Results and Discussion

3.1. 1D Results: Comparison and Validation

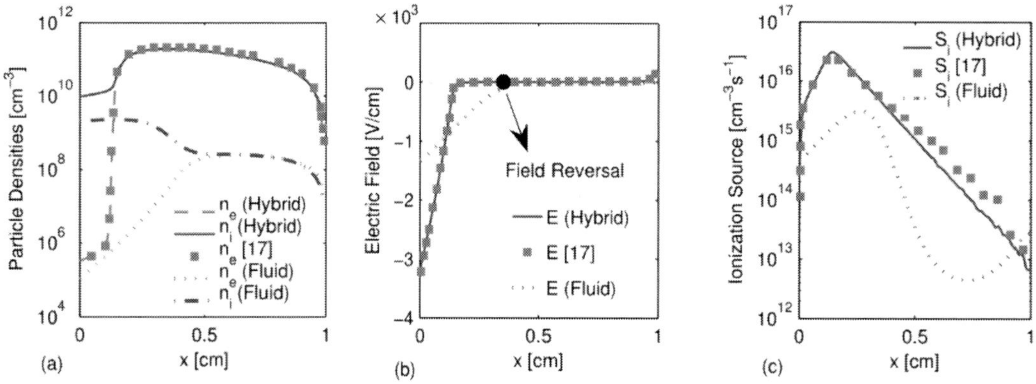

Figure 2. Comparison of 1D results with [17]. (a) electron and ion densities, (b) electric field, (c) ionization source profiles. $p = 1$ Torr, $T_e = 1$ eV, $\gamma = 0.06$, $L = 1$ cm, $V_a = 250$ V.

We compared our 1D simulation results (electron density, electric field and ionization source profile) in figure 2 with results of Ref. [17]. For electrons and ions zero density boundary conditions are used on the anode, zero gradient for ions and secondary electron emission condition for electrons on the cathode. Results are in good agreement. Figure 2 contains also fluid model results obtained under the same parameter regime. Results show that fluid method cannot predict correctly the charged particle densities, electric field, and ionization source profile. Moreover, the field reversal, which occurs at $x = 0.36$ cm within the hybrid model, is not obtained by the fluid model.

3.2. 2D Results

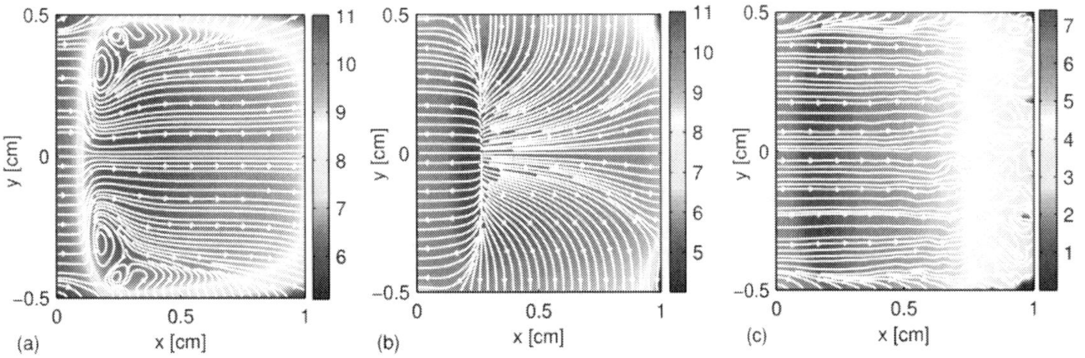

Figure 3. 2D hybrid model. (a) electron density (colored) and current flow (white streamlines), (b) ion density (logarithmic scale) and current flow, (c) fast electron density and current flow. $p = 1$ Torr, $T_e = 1$ eV, $\gamma = 0.06$, $L_x = 1$ cm, $L_y = 1$ cm, $V_a = 250$ V.

2D hybrid model results are presented in the figure 3. As can be seen from the electron and ion density profiles in panels 3a and 3b, strong domination of electron flux directed to the walls over the ions flux leads to formation of sheath reagins near the walls. Maximum density values locate on the symmetry axis as expected. The potential and electric field profiles represent a usual pattern on the cathode region with a high voltage drop and electric field in this region.

3.3. Electric Current Flows

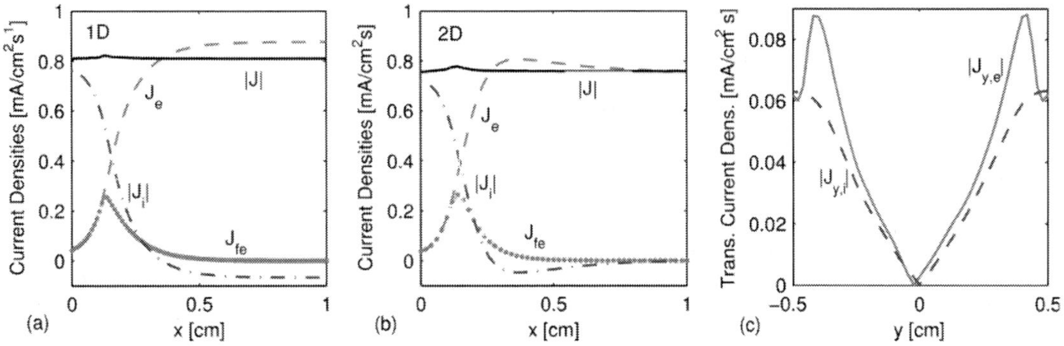

Figure 4. Electric current densities of electron and ions. (a) 1D, (b) axial profiles in 2D, (c) transversal profile through point $x = 0.5$ cm in 2D.

An interesting result is presented in figure 3: vortex current formations in the DC discharge. Figure 3a illustrates the total electron current flow, 3b the ion current flow, and 3c the fast electron current flow. The vortex current phenomena was predicted theoretically and identified by using numerical calculations in [19] in the case of an ICP (Inductively Coupled Plasma) discharge. As it follows from figure 3a and 3c, these formations is mostly due to the electron current flow.

Figure 4 represents the electron, ion, and total electric current profiles. Flat line for the total current density in figure 4a supports the correctness of the simulations. The same behaviour occurs in 2D case (figure 4b), disturbed by the fast electron flux in the cathode sheath. Figure 4c demonstrates the nonambipolar behavior of electron and ions currents in the transversal direction.

4. Conclusion

We developed and tested 2D Monte Carlo – fluid hybrid numerical model for a DC glow discharge. 1D calculations were also implemented in order to carry out the validation of the numerical code. Basic properties of glow discharge plasma sustained in argon gas (such as spatial profiles of charged particle densities, electric field, and ionization source) were studied. Simulations showed that the hybrid numerical model with fast electron group responsible for the ionization processes in the volume can be used to describe the nonlocal characteristics of plasmas of DC discharge. Vortex current formations in the total electric current flow were observed.

Acknowledgments

The work was supported by research grants 210T072 and 212T164 from the Scientific and Technical Research Council of Turkey (TUBITAK).

References

[1] Lister G G 1992 *J. Phys. D: Appl. Phys.* **25** p 1649-1680
[2] Donko Z 2011 *Plasma Sources Sci. Technol.* **20** p 024001
[3] Raizer Y P 1991 *Gas Discharge Physics* (Berlin: Springer-Verlag) p 1-67
[4] Surzhikov S T and Shang J S 2004 *Journal of Computational Physics* **199** p 437-464
[5] Rafatov I R, Bogdanov E A and Kudryavtsev A A 2012 *Physics of Plasmas* **19** p 033502
[6] Birdsall C K 1991 *IEEE Trans. Plasma Sci.* **19** p 65-85
[7] Vorbencoeur J P 2005 *Plasma Phys. Control. Fusion* **47** p 231-260
[8] Fiala A, Pitchford L C and Boeuf J P 1994 *Phys. Rev. E* **49** p 5607-5623
[9] Surendra M, Graves D B and Jellum G M 1990 *Phys. Rev. A* **41** p 1112
[10] Bogaerts A and Gijbels R 1995 *Phys. Rev. A* **52** (5) p 3743
[11] Donko Z, Hartmann P and Kutasi K 2006 *Plasma Sources Sci. Technol.* **15** p 178-186
[12] Boeuf J P and Pitchford L C 1991 *IEEE Trans. Plasma Sci.* **19** p 286
[13] Lieberman M A and Lichtenberg A J 2005 *Principles of Plasma Discharges and Materials Processing* (New Jersey: John Wiley Sons) p 14-80
[14] PHELPS database, http://www.lxcat.laplace.univ-tlse.fr, retrieved August 27, 2013
[15] Hagelaar G J M and Pitchford L C 2005 *Plasma Sources Sci. Technol.* **14** p 722
[16] McDaniel E W 1964 *Collision Phenomena in Ionized Gases* (Wiley, New York) chap. 9
[17] Derzsi A, Hartmann A, Korolov I, Karacsony J, Bano G and Donko Z 2009 *J. Phys. D: Appl. Phys.* **42** p 225204
[18] Frost L S 1957 *Phys. Rev.* **105** p 354
[19] Bogdanov E A, Chirtsov A S and Kudryavtsev A A 2011 *PRL* **106** p 195001

Blackout mitigation in a plasma layer near a high-speed body in ExB fields

S A Poniaev[1,3], **Y A Kurakin**[1], **A A Schmidt**[1], **S V Bobashev**[1], **L Steffens**[2], **B Esser**[2], **A Gulhan**[2]

[1] Ioffe Physico-technical Institute, St. Petersburg, Russia
[2] Institute of Aerodynamics and Flow Technology, DLR, Cologne, Germany

E-mail: serguei.poniaev@mail.ioffe.ru

Abstract. The space vehicle deceleration at reentry into the atmosphere leads to aerodynamic heating and, hence, ionization of near-wall plasma layers. The plasma layer causes radio communication blackout. This report discusses the possibility to provide a blackout mitigation by changing the plasma parameters near the body with the help of the magnetohydrodynamic (MHD) effects. The numerical simulations were carried out for the conditions corresponding to those at the DLR test facility.

1. Introduction

During interplanetary missions or re-entry flights to the earth the communication between the spacecraft and control station on the earth may interrupt. In the frame of the project ' „Communication Blackout Migration for Spacecraft – COMBIT“ the scientists of two Helmholtz Centers – German Aerospace Center (DLR) in Cologne and Karlsruhe Institute of Technology (KIT) – develop in co-operation with the scientists of the IOFFE Physical-Technical Institute of the Russian Academy of Sciences in Saint-Petersburg a new method to solve this problem.

When a vehicle enters the radio blackout phase, it loses all communication including GPS signals, data telemetry, and voice communication. Blackout occurs when the radio waves used for communication between ground stations and satellites are attenuated and/or reflected by the plasma layer that is created during hypersonic/re-entry flight. This phenomenon was well known at the beginning of space exploration when space capsule would experience several minutes of radio blackout during re-entry. Nearly 50 years later, hypersonic capsules still experience radio blackout during re-entry.

Maximum heating at the reentry trajectories of reusable space vehicles occurs at the heights ranging from 80 to 60 km, the temperature in the near-wall layers can reach in this case 10^4 K. In this situation the electron density can reach $n_e \approx 10^{17} - 10^{20}$ m^{-3}, and the characteristic plasma frequency in the near-wall layer of the ionized gas which is equal to $f_p = 0.9\times10^4\sqrt{n_e}$ is in the range 2.85 – 90 GHz [1-3]. This causes reflection and attenuation of radio signals the frequencies of which are lower than the plasma frequency, and the plasma layer near the vehicle plays the role of a screen that blocks transmission of electromagnetic waves with the frequencies used for communication with space

[3] Author to whom any correspondence should be addressed.

vehicles (the so-called communication blackout occurs). The time of blackout at the stage of vehicle reentry into the atmosphere can be as long as 20 min.

The COMBIT team tries to solve the radio blackout problem by reducing the electron number density of the plasma layer. This should be achieved by means of magneto-hydrodynamic (MHD) effects [4,5]. A combined magnetic and electrical field created close to the antenna should deflect charged particles in such a way that a local gas layer with low plasma frequency forms and the communication between the control station and spacecraft is guaranteed.

2. Planned experiments

DLR is responsible for the design and conducting experiments in the arc heated facility L2K, which allows producing ionized hypersonic flow with a total temperature of several thousand degrees. At such temperatures the gas is composed of molecules, atoms, ions and electrons.

In L2K [6,7] the ionized high temperature gas is test model is accelerated to hypersonic speeds using a supersonic nozzle. Fig.1 shows a schematic of L2K facility. Inside the extension part of nozzle and in the free stream the velocity but the temperature, pressure and density decrease. But in front of the model a bow shock forms and decelerates the flow. The shock heats the gas to temperatures, at which the flow is ionized. The magnet is mounted in the middle part of the model. Two electrodes for producing electrical field are integrated on the model surface upstream and downstream of the magnet. The sending antenna inside the model is placed close to the downstream electrode. Its signal has to pass the ionized flow before reaching the receiving antenna, which is placed outside the flow.

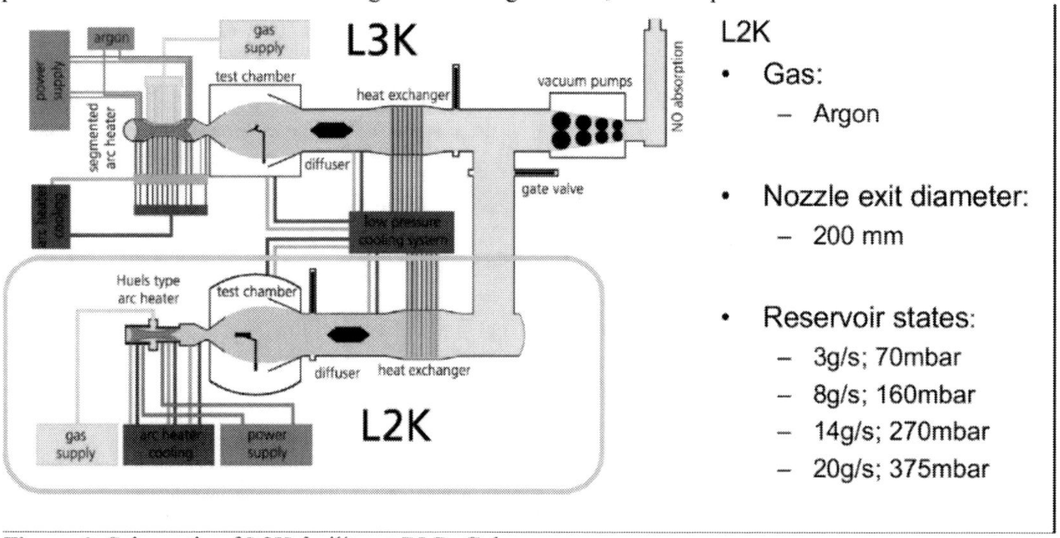

Figure 1. Schematic of L2K facility at DLR, Cologne

In L2K the ionization of the flow is weaker compared to the flight conditions with black-out effects. Therefore the experiments have to be carried out at higher intensity of the magnetic field, which cannot be realized with permanent magnets. Only cooled superconductive magnets are able to achieve magnetic field intensities beyond 2 Tesla on the model surface. The compact size superconducting magnet of KIT is cooled down to 4 Kelvin, i.e. -269°Celcius using Helium as coolant. Another very challenging requirement is the sufficient and reliable thermal isolation of the magnet from the high temperature environment of the flow.

During preliminary tests it could be demonstrated that even in the weakly ionized flow of L2K the attenuation of the communication signal can be measured clearly. These tests showed also some positive effect of the permanent magnets with a magnetic field strength of 1 Tesla are measurable. The main tests with superconducting magnets are planned in end of 2014.

3. Numerical simulation

The goal of our study was to estimate whether it is possible to affect the flow structure by the MHD methods with the aim of achieving a local decrease in the electron concentration near the antenna. To this end, by using external magnetic and electric fields produced by a coil placed in the body and a system of electrodes situated at its surface (see Fig. 2), a region of electromagnetic interaction between the plasma flow and external $\vec{E}\times\vec{B}$ fields is generated. Two fundamentally different electrode orientations are considered: 1) electrodes are orientated perpendicularly to the incoming flow (Fig. 2 (a)), 2) electrodes are parallel to the incoming flow (Fig. 2 (b)).

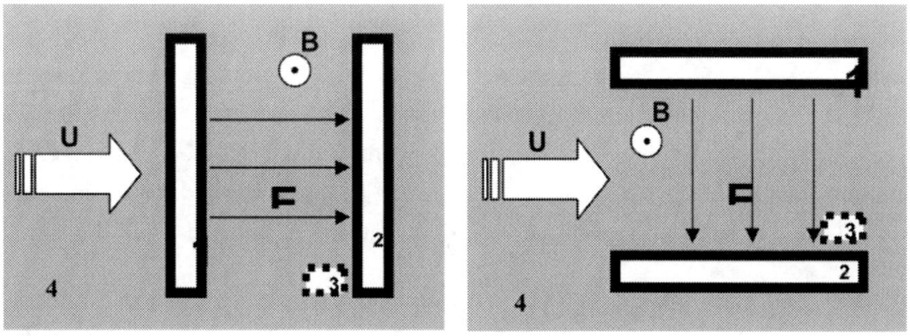

Figure 2. Scheme of electrodes on the model surface and magneticoro field for creation of «magnetic window»: (a) – electrodes (1 is anode and 2 is cathode) are perpendicular to the flow; (b) – electrodes (1 and 2) parallel to the incoming flow; 3 is possible antenna position, 4 – surface of model.

To carry out the estimates, the model of a three-component plasma consisting of atoms, ions, and electrons was used. A stationary flow along the plane simulating the space vehicle surface is considered. The model is based on the following assumptions: 1) the neutral component moves with a constant uniform velocity; 2) the gradient of electron pressure in the plasma layer of interest can be neglected; 3) the plasma layer has a small thickness, all MHD processes occur inside this layer, the electric field in the immediate vicinity of the electrodes is uniformly distributed throughout the layer thickness, there is no electric field across the layer, and all plasma parameters are constant across the layer; 4) the magnetic induction vector is directed perpendicularly to the space vehicle surface across the plasma layer.

Thus the problem of the motion of charged particles is reduced to the solution of Ohm's law which allows one to obtain distributions of the potential and total current. The distribution of the ion concentration is obtained by using the continuity equation for ions.

If we use these assumptions and take into account the ion slip and Hall effect, the equation for the generalized Ohm's law is

$$\vec{j}=\sigma\frac{(1+s)\left(\vec{E}+\vec{U}\times\vec{B}+\dfrac{\nabla p_e}{en_e}\right)-\beta_e\left(\vec{E}+\vec{U}\times\vec{B}+\dfrac{\nabla p_e}{en_e}\right)\times\vec{b}}{(1+s)^2+\beta_e^2},$$

where $u\approx(n_iu_i+n_au_a)/(n_i+n_a)$, $U_k=u_k-u$, $\vec{b}=\vec{B}/B$, indexes e, i, a refer to electrons, ions and atoms, respectively, σ, s, β_e are the coefficient of electric conductivity of electrons, coefficient of ion slip and Hall parameter for electrons, respectively. For the flow under consideration these parameters can be estimated as $\beta_e \approx 600 \cdot B >> 1$, $s \approx 200 \cdot B >> 1$.

If we take into account that in our case $t_0 >> v_{ia}^{-1}$ and $m_e v_{ie} << m_a v_{ia}$ and also that $u \approx u_a$ at a low ionization degree, and $U_a \approx 0$, and also that $n_e kT_a << n_e m_a v_{ia} u_0$, we obtain the expression for the ion velocity

$$U_i = \frac{\sigma_{i\perp} E' + \sigma_{iH} \left(b \times E' \right)}{e n_e},$$

where $\vec{E}' = \vec{E} + \vec{U} \times \vec{B}$, σ_i is the coefficient of electric conductivity of ions, υ_{kl}, is the collision frequency of particles of k and l classes, and m_l is the mass of l class particle. Therefore, we can find the ion velocity $u_i = u + U_i$. The distribution of ion concentrations can be determined by solving the continuity equation

$$\nabla (n_i u_i) = 0.$$

The ion concentration at the inlet boundary of the computational domain was taken to be $n_i = 2.81 \cdot 10^{17}$ m^{-3}, the velocity of the incoming flow was 2250 m/s. The computational domain size considerably exceeded the characteristic size of the interelectrode gap in order to fulfill the condition $U_i \approx 0$ at the boundaries.

Fig 3 shows distribution of ion concentration for three cases: (i) the direction of the plasma flow coincides with direction of the applied electric field (potential difference is 100 V) (Fig.3(a)); (ii) the direction of the applied electric field (potential difference is 100 V) is opposite to the direction of the plasma flow (Fig.3(b)) and (iii) direction of the applied electric field is perpendicular to the direction of the plasma flow (potential difference is 900 V) (Fig.3(c)). Because of the absence of charge separation in the flow the distributions of ion concentration is equal to the distribution of electron concentration, which is responsible for the effects of communication blackout.

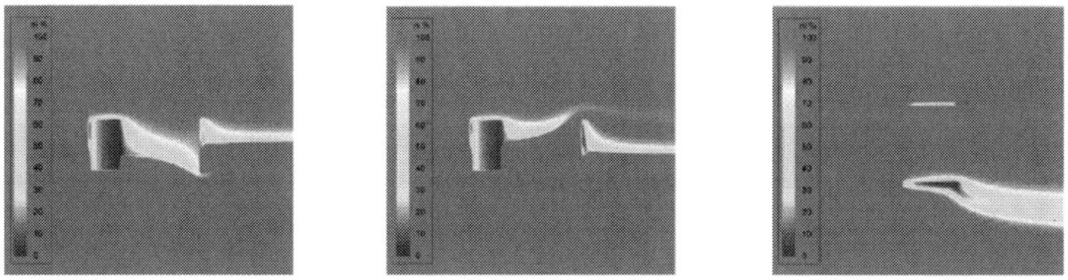

Figure 3. Distribution of normalized ion concentration near the model surface: (a) – are perpendicular to the flow; (b) – electrodes are perpendicular to the flow, electric field have the other direction than is case (a); (c) – electrodes are parallel to the flow.

It can be seen that in the first two cases a noticeable decrease in the ion concentration is observed immediately after the first electrode irrespective of whether it is anode or cathode. The applied field direction plays a minor role because the intensity of the electric field that arises in the plasma moving in the magnetic field is much higher than that of the applied field. In the configuration where electrodes are normal to the flow the plasma flow is decelerated, and the applied electric field direction affects only the size and direction of the zone of a reduced ion concentration. In the case when electrodes are parallel to the incoming plasma flow, the electromagnetic interaction causes acceleration of the plasma flow. The zone of reduced ion concentration is near the cathode, it extends down the flow near the outer side of the cathode.

These results allow us to formulate preliminary recommendations on formation of the external field configuration for affecting plasma flow parameters.

In particular, when electrodes are parallel to the plasma flow and the MHD interaction provides its acceleration, the most suitable place for the antenna can be out of the interelectrode gap.

Acknowledgements
The work was supported by RFBR grant 12-01-91320-SIG_a.

References

[1] Boyd I D, 2007, AIAA 2007-206
[2] Hartunian R A, Stewart G E, Fergason S D, Curtiss T J, and Seibold R W, 2007 Aerospace report ATR 2007(5309)-1
[3] Lehnert R, Rosenbazlm B, 1965 NASA TN D-2732
[4] Kim M, Keidar M, Boyd I B 2009 AIAA 2009-1232
[5] Kim M, Boyd I D, Keidar M, Jones C H, Beilis I, Pekker L, Raitses Y, Levchenko I, 2010, Journal of Spacecraft and Rockets vol 47 p 29
[6] Gülhan A, Esser B, and Koch U, 2001 Journal of Spacecraft and Rockets vol. 38 no 2 p 199
[7] Gülhan A, and Esser B, 2002 Advanced Hypersonic Test Facilities, Progress in Astronautics and Aeronautics vol 198 pp 375-403

Investigation into the processes of atom redistribution encountered during of the formation of metal layers on the surface of aluminum gallium nitride

I A Lamkin[1], S A Tarasov[1], A A Petrov[1], E A Menkovich[1], A V Solomonov[1], S Yu Kurin[2]

[1]Saint-Petersburg Electrotechnical University "LETI", Prof. Popova 5, St. Petersburg 197376, Russia
[2]GaN-Crystals Ltd., 27 Engels Ave, St. Petersburg 194156, Russia,

E-mail: ialamkin@mail.ru

Abstract.. The results of an investigation into the properties of double-layer metal contacts to the epitaxial layers of $Al_xGa_{1-x}N$ solid solutions for ultraviolet photodetectors and UHF electronic devices are presented. The processes of atom redistribution during the formation of metal layers and their subsequent annealing to form low resistance contacts were studied by Auger electron spectroscopy. The results obtained allowed us to develop a technique for creating ohmic contacts having low resistance value.

1. Introduction

The metal contacts play an important role in the operation of modern optoelectronic and UHF devices based on nitrides and other wide-bandgap semiconductors [1-3]. Atom redistribution during the formation of these layers, as well as further thermal effects, have a strong influence on the devices characteristics. The parameters of metal layers are of special importance for the creation of photodetectors based on Schottky barrier [4], allowing for the control of not only their electrical characteristics, but also the degree of selectivity of the spectral sensitivity. The current-voltage (I-V) characteristics and the contact resistance of metal-semiconductor (MS) structure strongly depend not only on the metal contact, but also on the processes occurring at the MS interface during the preparation of the epitaxial layer surface, metal deposition and subsequent high-temperature treatment [5].

Currently during the formation of ohmic contacts to gallium nitride (GaN) and aluminum gallium nitride solid solutions ($Al_xGa_{1-x}N$) the technologies of vacuum thermal evaporation or electron beam-induced deposition of thin metal layers or multilayer metal compositions are used. Typically used metals are Ti [6], Al [7 - 9], Ti/Al [10, 11], Ti/Ag [12], Ti/Al/Ni/Au [13] with subsequent high-temperature annealing in nitrogen (N_2) ambient [14, 15]. It was shown [13] that in four-layer contacts, e.g. Ti/Al/Ni/Au, Ni acts as a blocking layer that prevents diffusion of Au into Ti and Al layers. Such metal can be Ti, Ni, Mo or Pt [16]. Au top layer facilitates unwelding of gold contacts when attaching photosensitive chip to package contacts and also performs a protective function [14]. The best results in the formation of ohmic contacts to $Al_{0,3}Ga_{0,7}N$ were achieved using double-layer Ti/Al contacts with subsequent "fast" (90 s) annealing in N_2 ambient at high temperature of 800 - 900 °C [8], the resistance value was equal to 10^{-4} ohm · cm^2.

2. Sample and experimental technique

We investigated the samples grown by two approaches: molecular beam epitaxy with N_2 plasma activation under metal-rich conditions and hydride vapor phase epitaxy [17] on c-Al_2O_3 substrates. The contacts were formed by vacuum thermal evaporation at pressure of residual gases of 10^{-5} Torr. As structures Ti/Al double-layer compositions were created and studied. The thickness of metal layers was in the range of 15-75 nm. The best results were achieved for the Ti layer thickness of 15 nm, and the top Al layer thickness of 35 nm. Before deposition of the metal contacts to the surface, the structures were cleaned using various chemicals, in particular, H_2O_2, CCl_4, HCl, followed by structures washing in distilled water. For better metal adhesion the structures were heated to a temperature of 300°C during the deposition. To obtain ohmic I-V characteristics the contacts were annealed in vacuum at different residual gas pressures for 1-30 minutes.

3. Experimental results and discussions

Figure 1 shows the I-V characteristics of the Ti/Al contacts to n-type $Al_xGa_{1-x}N$ with aluminum content x = 0.08. One can see that the I-V characteristics of the contacts are non-linear both for untreated structures and structures annealed at relatively low temperatures. Nonlinearity decreases with the increase of temperature, and at 750°C contacts become ohmic. Further increase of the annealing temperature leads to a slight increase slope, but this increase is insignificant.

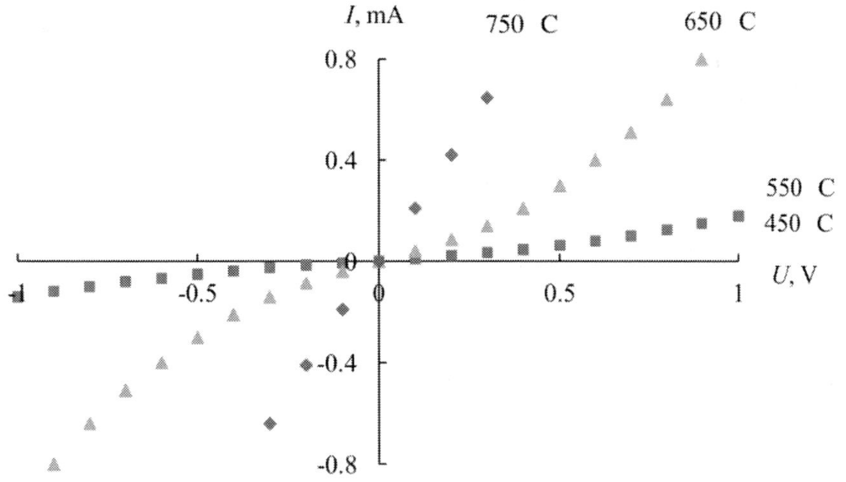

Figure 1. Current-voltage characteristics of the ohmic Ti / Al contacts to the $Al_{0.08}Ga_{0.92}N$ layer annealed at the different temperatures for 10 minutes.

Several samples were investigated by Auger electron spectroscopy (AES) to study the distribution of metal atoms in the surface region of the ohmic contacts and the influence of high temperature annealing on this distribution. Figure 2 shows the Auger spectrum of the Al/Ti/$Al_{0.5}Ga_{0.5}N$ structure after deposition of metals. One can see that although there is a small concentration of oxygen on the surface of the structure, its concentration is not high, and the penetration depth is very low. This may indicate high quality of the created structure. As well, the Al layer is also clearly visible on the graph; its thickness was originally equal to 35 nm. The sublayer for Al is a 15 nm thick titanium layer . However, the results of AES show that there is a partial mixing of titanium layer with aluminum top layer. Titanium also penetrates into the $Al_{0.5}Ga_{0.5}N$ epitaxial layer.

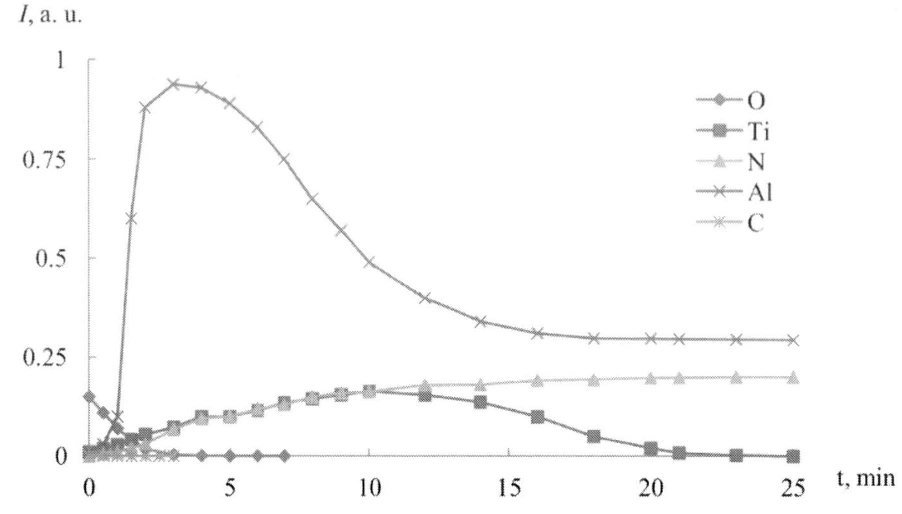

Figure 2. AES depth profile of the structure $Al/Ti/Al_{0.5}Ga_{0.5}N$ as deposited.

Figure 3 shows the AES-profile of the $Al/Ti/Al_{0.5}Ga_{0.5}N$ structure after annealing for 10 min at 750 °C in vacuum at a residual gas pressure of 10^{-5} Torr. One can see that both concentration and penetration depth of oxygen are increased, but still remain insufficient to affect the quality of the structure. The same is true for a small concentration of carbon.

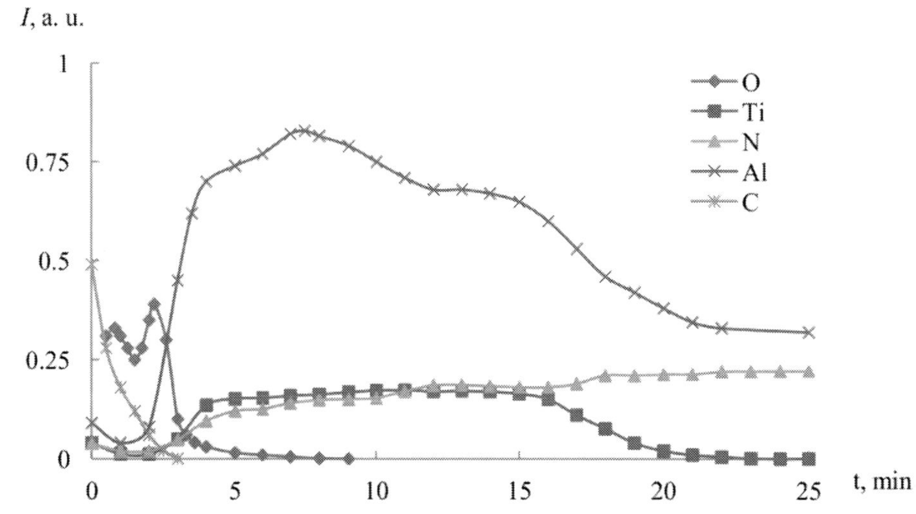

Figure 3. AES depth profile of the structure $Al/Ti/Al_{0.5}Ga_{0.5}N$ after annealing at 750 °C.

The important effect observed from the results of ASE in the annealed structures is a penetration of aluminum into the depth of the structure followed by nitrogen release from the surface region. As a result, profiles of N, Ti and Al in the top of the structure have almost the same depth. This may indicate the formation of titanium aluminum nitride composition that leads to improvement of ohmic I-V characteristics.

The analysis of investigated characteristics allowed us to formulate some conclusions on redistribution processes of atoms in the studied structures. In our opinion, during annealing there are two basic phenomena that determine the formation of ohmic contact. The first of them is a release of nitrogen by titanium from epitaxial layer [18]. Nitrogen diffuses actively into the metal layers resulting in relatively uniform distribution within. Titanium is also redistributed in the metal layer and the shape of its profile is similar to the profile of nitrogen (Fig. 4). This could indicate phase formation on the surface of semiconductor that is similar in composition to titanium nitride. Titanium does not penetrate into the depth of semiconductor interacting only with nitrogen. It is possible that a thin layer with a higher concentration of titanium and nitrogen is formed on the surface of semiconductor (some increase in the profile of both atoms for 15 s is observed). It is important that in this case nitrogen is removed from the near-surface layer of semiconductor that leads to the formation of nitrogen vacancies therein. Such vacancies behave as donors [19, 20]. Therefore, this process may be considered as subsurface doping of semiconductor that narrows the barrier at the MS interface to the tunnel-transparent thickness. It is necessary to note that in our approach there is no other source of nitrogen except the semiconductor because annealing is performed in vacuum. This leads to more intensive nitrogen release out of the semiconductor and presumably higher concentration of vacancies than in the normal case and thus lower resistance.

The second process is a redistribution of aluminum in metal layer until the profile approaching the U-shaped form is obtained. The boundaries of new aluminum layer similar to the boundaries of titanium (titanium nitride) layer. This indicates almost uniform distribution of the aluminum particles in the contact layer. Most likely, the aluminum particles assist charge carriers in overcoming the contact layer, possibly by a process similar to the hopping conduction. Therefore, the use of titanium single layer in the structure does not allow for obtaining low contact resistance even at prolonged annealing time. In this case nitrogen is released, nitride with a transparent barrier at the boundary is formed, but its resistance is high. Aluminum reduces the resistance. It is possible that the presence of aluminum additionally reduces the height of the barrier at the boundary, may be due to formation of $TiAl_3$ composition. The ohmic contact with low resistance is formed when titanium is not actively interacting with the nitrogen.

Both considerable time and high temperature are required for above-mentioned redistribution of atoms and formation of nitride-based and intermetallic compositions. This determines the optimal annealing time of 10 min and temperature of 750°C. Since this achieves the required uniformity of the distribution of atoms in the contact, a further increase of these parameters does not lead to the resistance improvement. Moreover, there may be some degradation of the contacts because of their partial evaporation and contamination. The optimal thickness of the layers is determined by the concentration of atoms required to form these compositions and for correct distribution of aluminum in the contact layer.

4. Conclusion

We have studied the processes of atom redistribution that occur during the formation of metal compositions on the surface of epitaxial layers of $Al_xGa_{1-x}N$ solid solutions of n-type conductivity taking into account the influence of high-temperature treatment. The ohmic behavior of I-V characteristics can be achieved by annealing in vacuum for 10 min and at a temperature no less than 750°C for double-layer Ti/Al compositions. Investigations by AES revealed that annealing leads to the release of nitrogen from the epitaxial layer and the formation of titanium nitride composition in the surface layer accompanied by the appearance of nitrogen vacancies. Lack of nitrogen in the annealing chamber may contribute to this process. In addition, the redistribution of aluminum in the surface layer with a possible formation of intermetallic composition is observed. These processes contribute to a significant reduction of the contact resistance.

Based on conducted research the technology for creation of Ti/Al/n-$Al_xGa_{1-x}N$ ohmic contacts by vacuum thermal evaporation approach was developed. The optimal thickness of Ti and Al layers was determined (15 nm and 35 nm, respectively) as well as time parameters and temperature parameters of annealing. The main feature of the technology is the creation of ohmic contacts with subsequent annealing in vacuum instead of nitrogen or argon. Achieved contact resistance is equal to $8 \cdot 10^{-5}$ ohm $\cdot$ cm^2.

References

[1] Menkovich E A, Tarasov S A, Lamkin I A Luminescence of nanostructures based on semiconductor nitrides *Functional Materials* vol. 19, № 2, p. 233, 2012.
[2] Kurin SYu, Antipov A A, Roenkov A D, Barash I S, Helava H I , Menkovich E A, Tarasov S A, Lamkin I A, Shmidt N M, Makarov Yu N UV LEDs for high-current operation *Journal of Physics: Conference Series* vol. 461. p. 012028, 2013.
[3] Lamkin I, Tarasov S Ultraviolet photodiodes based on AlGaN solid solutions *Journal of Physics: Conference Series* vol. 461. p. 012025, 2013.
[4] Pikhtin A N, Tarasov S A, Kloth B Ag-GaP Schottky photodiodes for UV sensors *IEEE Transactions on Electron Devices* vol. 50, № 1, p. 215, 2003.
[5] Pikhtin A N, Tarasov S A, Kloth B New values of the Ag-n-GaP potential barrier *Technical Physics Letters* vol. 28, № 10, p. 872-873, 2002.
[6] Forest J S, Moustakas T D Electrode material and electrode for III-V group compound semiconductor *Applied Physics Letters*. 1993. №. 62. P. 2859.
[7] Van Daele B., Van Tendeloo G., Ruythooren W., Derluyn J., Leys M.R., Germain M. The role of Al on Ohmic contact formation on n-type GaN and AlGaN/GaN Letters. 2006. № 6(87). P. 061905.
[8] Luther B. P., Mohney S. E., Jackson T. N., Asif Khan M., Chen Q., Yang J. W. Investigation of the mechanism for Ohmic contact formation in Al and Ti/Al contacts to n-type GaN *Applied Physics Letters*. 1997. № 1(70). P. 57-59.
[9] Smith L. L., Davis R. F., Kim M. J., Carpenter R. W., Huang Y. Microstructure, electrical properties, and thermal stability of Al Ohmic contacts to n-GaN *Journal of Materials Research*. 1996. № 9(11). P. 2257-2262.
[10] Morkoc H., Strike S., Gao G.B., Lin M.E., Sverdlov B., Burns M.. *Journal Applied Physics*. 1994. № 76(3). P. 1363.
[11] Lin M.E., Ma Z., Huang F.Y., Fan Z.F., Allen L.H. and Morkoc H. *Applied Physics Letters*. 1994. № 64(8). P. 1003.
[12] Guo J.D., Lin C.I., Feng M.S., Pan F.M., Chi G.C., Lee C.T. *Applied Physics Letters*. 1996. № 68(2). P. 235.
[13] Zhifang F., Noor Mohammad S., Kim W., Ozgur Aktas, Botchkarev E., Morkoc H. *Applied Physics Letters*. 1996. № 68(12). P. 1672.
[14] Кузнецов Г.Д., Сушков В.П., Кушхов А.Р., Ермошин И.Г., Билалов Б.А. Омические контакты к GaN *Материалы электронной техники*. 2009. № 3. С. 4-13.
[15] Qin Z.X., Chen Z.Z., Tong Y.Z., Ding X.M., Hu X.D., Yu T.J., Zhang G.Y. Study of Ti/Au, Ti/Al/Au, and Ti/Al/Ni/Au ohmic contacts to n-GaN *Appl. Phys. A*. 2004. V. 78. P. 729-731.
[16] Jung Young-Ro, Lee Jae-Hoon, Kim Jung-Kyu, Lee Young-Hyun, Lee Myoung-Bok, Lee Jung-Hee, Hahm Sung-Ho. Pt/AlGaN Metal Semiconductor Ultra-Violet Photodiodes on Crack-Free AlGaN Layers *Journal Applied Physics*. 2003. № 42. P. 2349-2351.
[17] Kurin S, Antipov A, Barash I, Roenkov A, Helava H, Tarasov S, Menkovich E, Lamkin I, Makarov Yu. CHVPE growth of AlGaN-based UV LEDS *Physica Status Solidi (C) Current Topics in Solid State Physics* vol. 10, № 3, p. 289-293, 2013.
[18] J. K. Kim, H. W. Jang, J. L. Lee Mechanism for Ohmic contact formation of Ti on n-type GaN investigated using synchrotron radiation photoemission spectroscopy *Journal of Applied Physics*. V. 91, 9214, 2002.
[19] J. Neugebauer and C. G. Van de Walle, Atomic geometry and electronic structure of native defects in GaN *Physical Review B*. Vl. 50, 8067, 1994.
[20] D. C. Look, G. C. Farlow, P. J. Drevinsky, D. F. Bliss, J. R. Sizelove On the nitrogen vacancy

in GaN *Applied Physics Letters.* V. 83, 3525, 2003.

Journal of Physics: Conference Series **572** (2014) 012063 doi:10.1088/1742-6596/572/1/012063

Research of the solar-blind and visible-blind photodetectors, based on the AlGaN solid solutions

I A Lamkin[1], S A Tarasov[1], A A Petrov[1], E A Menkovich[1], A V Solomonov[1], S Yu Kurin[2]

[1]Saint-Petersburg Electrotechnical University "LETI", Prof. Popova 5, St. Petersburg 197376, Russia
[2]GaN-Crystals Ltd., 27 Engels Ave, St. Petersburg 194156, Russia,

E-mail: ialamkin@mail.ru

Abstract. The paper is devoted to the development and study of solar-blind and visible-blind photodetectors. We report on the spectral characteristics of the ultraviolet photodetectors based on Shottky barrier to the epitaxial layers of the n-$Al_xGa_{1-x}N$ solid solutions. The use of Schottky barrier photodiodes is advantageous since it does not require the growth of additional epitaxial layer of *p*-type conductivity.

1. Introduction

Currently ultraviolet (UV) photodiodes are of great interest. There are several types of such photodiodes. Let us consider several variants and their main drawbacks. SiC-based photodiodes have a relatively narrow range of sensitivity. GaP photodiodes have a good signal to noise ratio but it is impossible to create solar-blind and visible-blind photodiodes based on them without the use of filters. Si-based photodiodes also requires expensive optical filters for registration shortwave signals against the backdrop of strong illumination by visible light. From this point of view, the most promising material for short-wavelength photodiodes is a solid solution $Al_xGa_{1-x}N$ [1]. In particular, the material system AlN-GaN forms a series of direct-gap solid solutions that allow creating photodiodes with sharp wavelength edge of photosensitivity in a given region, including solar- and visible-blind ones. In addition, the nitrides have high chemical resistance and good thermal stability which allows devices based on them to be used in extreme conditions [2].

Solar-blind and visible-blind photodiodes are used in the wavelength ranges less than 290 nm and 360 nm, respectively, and have no sensitivity outside these spectral ranges. For example, they can be applied in the military industry as solar-blind sensors registering flame radiation from a missile against the backdrop of exceeding background illumination by the visible light. Visible-blind photodiodes can be used for studying ozone layer depletion where the amount of ultraviolet radiation is greatly increased [3, 4]. The use of Si photodiodes for such applications is unreasonable because they are much more responsive to both visible and infrared spectral ranges and require expensive optical filters. Furthermore, it is often impossible to choose an optical filter with transmission in the required spectral range.

The use of Schottky barrier photodiodes is more advantageous, since it does not require an additional epitaxial p-type layer [5, 6]. The paper demonstrates the technology for creating visible-blind and solar-blind nitride-based photodiodes on the basis of a relatively simple and low-cost hydride vapor phase epitaxy (HVPE).

Photodiodes based on Schottky barrier are easier to build than photodiodes based on p-n-junction. [7]. However, such structures do not require light to penetrate into the bulk of the semiconductor as the absorption occurs at the metal-semiconductor interface. However, despite all the advantages of photodiodes based on epitaxial layers of solid solutions $Al_xGa_{1-x}N$ there are still some not fully solved problems. In particular, there are difficulties in growing epitaxial layers of $Al_xGa_{1-x}N$ with high Al content. High dislocation density leads to high values of leakage currents and may affect the short-wavelength edge of the photosensitivity, as well. In this paper, we report on the structures grown by two epitaxial methods: 1. Structures based on the top epitaxial layer $Al_{0.1}Ga_{0.9}N$ were grown by HVPE. 2. $Al_xGa_{1-x}N$ structures with Al content in the range of 0.1 $<x<$0.6 were grown by molecular - beam epitaxy (MBE).

2. Sample and experimental technique

The photodetectors were formed by the epitaxial layers grown by two methods: MBE with plasma activation of nitrogen in the metal-rich conditions, and HVPE [8] on the Al_2O_3 (0001). The use of the Al_2O_3 substrate having a good crystalline affinity for materials III-N as well as relatively thick AlN buffer layers provides a higher quality of the surface structure. The introduction of three-stage AlN buffer layer and AlN/AlGaN short-period lattice at the early growth stages provided satisfactory structural quality of the layers having dislocation density of$10^9 - 10^{10}$ cm^{-2}. The studies of the kinetics of AlGaN layers growth under metal-rich conditions when the total flow ratio of III group elements to the flow of activated nitrogen $F_{III} / F_N \approx 1.5 - 2$ showed that the surface became atomically smooth, even at relatively low growth temperatures of 700 °C. An additional advantage of the metal-rich conditions is the possibility to control Al content in the layers as equation $x = F_{Al} / v_{AlGaN}$ is valid in this mode, wherein F_{Al} - calibrated Al flow, v_{AlGaN} - measured growth rate of the layer which is determined by the active nitrogen flow, i.e. $v_{AlGaN} = F_N$. In the same mode the $Al_xGa_{1-x}N$ ($x = 0 - 0.65$) layers were grown (Fig. 1).

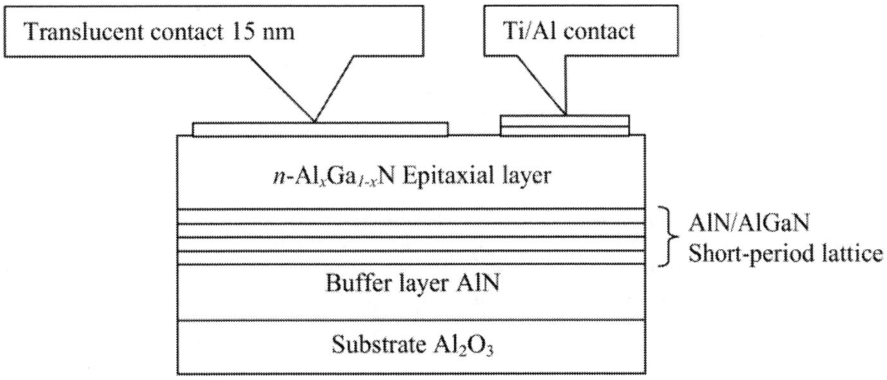

Figure 1. The structure of $Al_xGa_{1-x}N$ samples grown by MBE.

The high dislocation density as well as the high manufacturing costs of layers growing by epitaxial methods (molecular-beam epitaxy, vapor phase epitaxy from metal-organic compounds) create not high-quality and cheap photodetectors. The samples grown by HVPE were studied in order to address above-mentioned issues. This method has a number of technological advantages over MBE and allows significant reduction of the production cost of $Al_xGa_{1-x}N$ solid solutions epitaxial layers. In particular, the HVPE produces epitaxial layers of high crystalline quality at the same time providing the ability to control the rate of growth over a wide range: from a fraction of a micron to tens and hundreds of

Journal of Physics: Conference Series **572** (2014) 012063 doi:10.1088/1742-6596/572/1/012063

micrometers per hour. The dislocation density in the used samples ranged from $9 \cdot 10^7$ cm^{-2} to $8 \cdot 10^8$ cm^{-2}.

The photodetectors based on metal-semiconductor contact were created according to the method for vacuum thermal resistive deposition of metal on the $Al_xGa_{1-x}N$ solid solution epitaxial layers (Fig. 2).

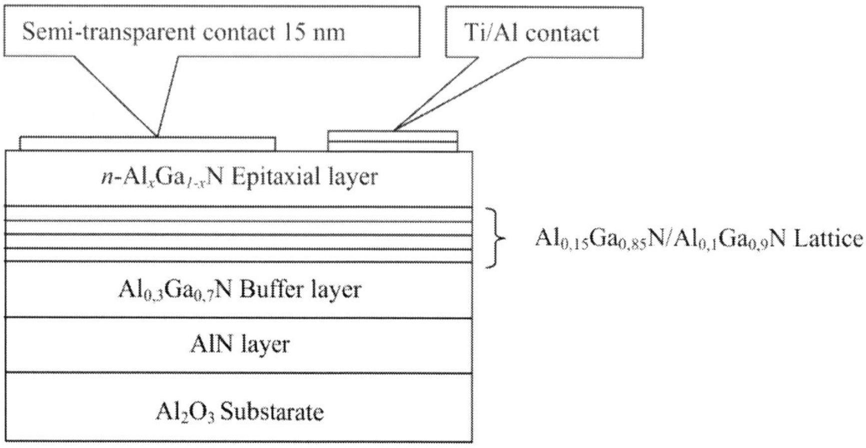

Figure 2. The structure of $Al_xGa_{1-x}N$ samples grown by CHVPE.

The contacts were created by vacuum thermal deposition. The Ti/Al structure, where Ti is a sublayer for Al, was used as an ohmic contact. Such metals as Au, Ni, Ag, In, Sn were deposited up to thickness of 15 nm for the Schottky barrier. During deposition the substrate was heated up to 300°C.

The study of the spectral characteristics was performed with the use of a diffraction grating monochromator having a light source on the basis of a xenon lamp with a bulb absorbing no UV radiation component.

3. Experimental results and discussions

Figures 3 and 4 show the spectral characteristics of UV photodiodes test structures, their long-wavelength sensitivity limit can be determined by the bandgap of the forbidden region of the $Al_xGa_{1-x}N$ solid solution. Fig.3 illustrates the use of the $Al_xGa_{1-x}N$ solid solution with the aluminum content of $x = 0.08$ that allows creating a visible-blind UV detector. Such a detector is sensitive within the 200 - 360 nm wavelength range. Furthermore, it is almost insensitive to radiation wavelength over 370 nm that allows detecting UV radiation under a strong ambient illumination by visible light. Fig. 3 shows that the Au-based Schottky barrier photodiode has a sensitivity from 210 to 360 nm, FWHM of 50 nm and the maximum photoresponse at 320 nm. The use of Ni as a translucent contact has extended the sensitivity range that was within 200 - 360 nm with a maximum photoresponse at 300 nm. The FWHM of the sensitivity spectrum of Ni-$Al_{0.08}Ga_{0.92}N$-based structure was equal to 95 nm. The spectral operational range of the Ni-based photodiode is the same as for Ag photodiode but its FWHM is 10 nm less and is equal to 85 nm. The wavelength of maximum sensitivity is 285 nm (Fig. 3).

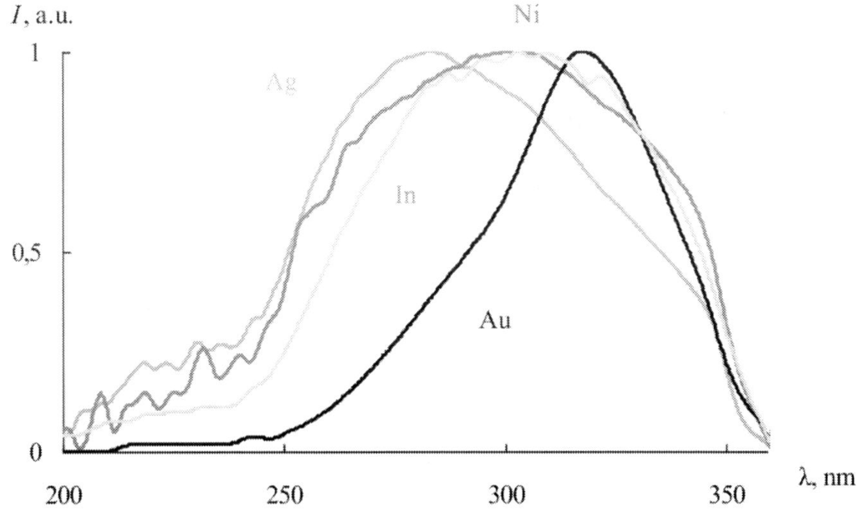

Figure 3. The spectral characteristics of photodetectors based on the Schottky barrier to the $Al_{0,08}Ga_{0,92}N$ with different metal of Schottky barrier.

The increase in the aluminum content enabled the shift of the sensitivity edge to shortwave region, including a "solar-blind" photodetector at a value of $x = 0.38$ (see Fig. 2.4). The red edge of solar-blind UV photodiode is 290 nm. At x values higher than 0.8 the sensitivity range of $Al_xGa_{1-x}N$ structure would be in the vacuum UV r with the wavelength less than 200 nm.

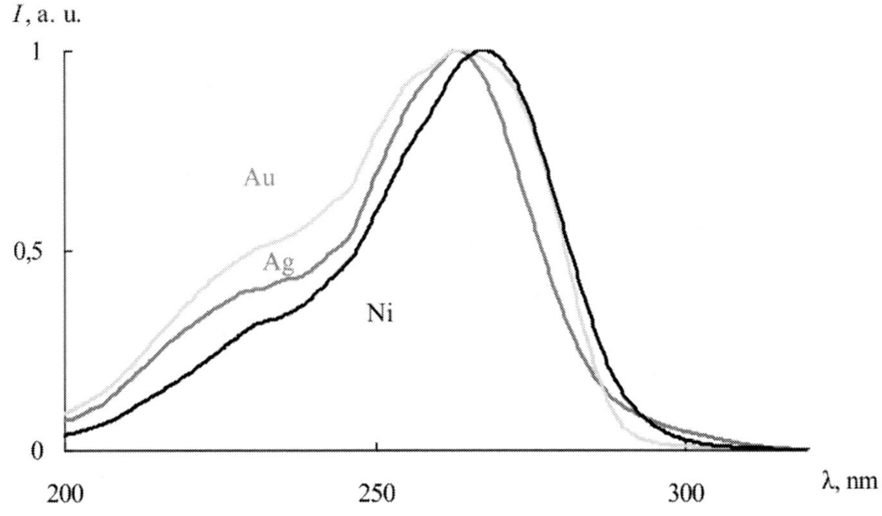

Figure 4. The spectral characteristics of photodetectors based on the Schottky barrier to the $Al_{0,38}Ga_{0,92}N$ with different metal of Schottky barrier.

Figure 4 shows that the FWHM of $Au\text{-}Al_{0.38}Ga_{0.92}N$ structure is50 nm at maximum photosensitivity of 265 nm. The Ni- and Ag-based photodetector has approximately the same value of FWHM (about 35 nm), however, the maximum photosensitivity for Ni-based Schottky barrier is 270 nm and it is shifted to 4 nm relative to the maximum for Ag.

The studies have demonstrated that the greatest sensitivity is revealed by the Au-contact structure (Fig. 5), in particular, the value of the Au-$Al_{0.08}Ga_{0.92}N$ Schottky barrier photoresponse is 9 times higher than that one of the Sn-based diode, and 1.6 times higher than that of Ni-$Al_{0,08}Ga_{0,92}N$.

The quality of the metal-semiconductor interface and the associated surface recombination sufficiently affects the sensitivity of the shortwave edge. In Au-based structures border was the most biased toward higher energies region that indicates the low surface recombination velocity. This fact considerably determines the sensitivity of gold based structures.

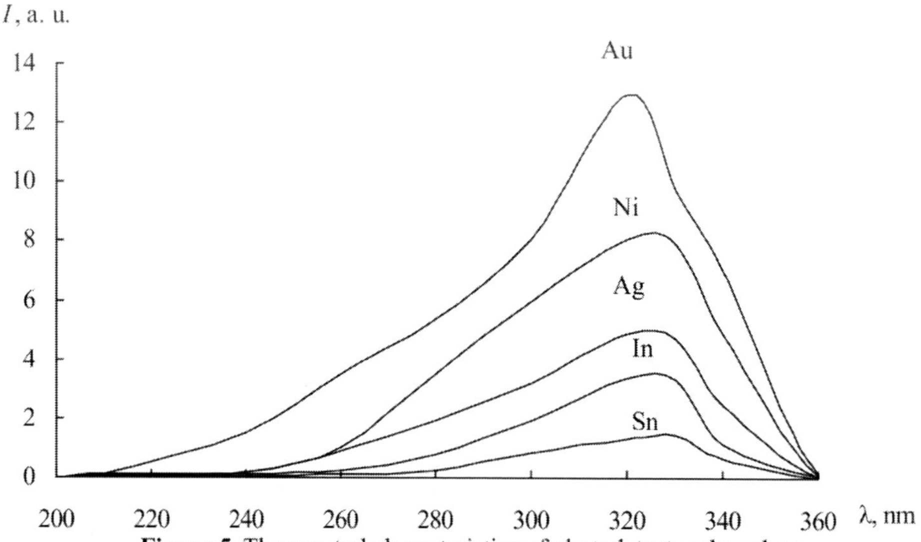

Figure 5. The spectral characteristics of photodetectors based on Schottky barrier to the $Al_xGa_{1-x}N$ with different metal of Schottky barrier.

Pretreatment of the semiconductor surface significantly affects the sensitivity of the photodiodes based on Schottky barrier. To study this effect we washed the structures with acids and alkalies (see Table 1) for 1 minute before metal deposition. Table 1 also shows the influence of the amount of chemical reagent on the value of maximum photosensitivity (I_{max}). Changing the position of the photoresponse maximum (λ_{Imax}) is associated with the composition fluctuations of the solid solution on the substrate.

Table 1. Chemical reagents used for the surface treatment.

Surface treatment	λ_{Imax}, nm	I_{max}, a. u.
$KOH+H_2O$	337	0,224
HF (20%)	338	0,214
HNO_3:HCl	335	0,197
$HCl+H_2O$	339	0,144
CCl_4	336	0,142
H_3PO_4	333	0,088
H_2SO_4:H_3PO_4:H_2O	337	0,072
as grown	334	0,041

It was established that the surface pretreatment in KOH allows the increase in photosensitivity by 5.5 times as compared with untreated surfaces. Close parameters demonstrated structures processed in HF (see Table 1).

4. Conclusion

The experimental samples were created of UV photodetectors based on metal-solid n-$Al_xGa_{1-x}N$ solutions with various x aluminum content. We established that the heat treatment of the gold contact structure provides 4 time increasing in photosensitivity. We observed that the pre-clean parameters of the epitaxial layer surface have the most important influence on the contact characteristics. Heating is essential for a better adhesion of the metal to the surface of the epitaxial layer. We demonstrated that the heating temperature during the deposition of the sample is tol be 300 °C. The contact thickness should be 15 nm in order to meet the translucent effect. It was established that the surface pretreatment in KOH allows the increase in photosensitivity by 5.5 times as compared with untreated surfaces.

The gold translucent contact provided a high photosensitivity along with the other advantages such as oxidation resistance, high conductivity, and performance ability. This leads to the most perspective use thereof in creating the studied photodiodes.

On the basis of n-$Al_xGa_{1-x}N$ solid solution with aluminum content of $x = 0.08$ we succeed in creating a visible-blind UV detector. Such a detector is sensitive in 200 - 360 nm wavelength range. The increase in Al content up to 0.38 allowed us to implement a pilot solar-blind UV photodiode having the spectral sensitivity region that lies within the range of 200 - 290 nm. The photodiode has no sensitivity to the radiation wavelength exceeding 300 nm.

References

[1] I Lamkin, S Tarasov Ultraviolet photodiodes based on AlGaN solid solutions *Journal of Physics: Conference Series.* V. 461. p. 012025, 2013.

[2] E A Menkovich, S A Tarasov, I A Lamkin Luminescence of nanostructures based on semiconductor nitrides *Functional Materials* vol. 19, № 2, p. 233, 2012.

[3] E Monroy, F Calle, J Pau, E Munoz, F Omnes, B Beaumont, P Gibart *Journal of Crystal Growth.* 2001. vol. 230. p. 537. 2001.

[4] S Averin, P Kuznetzov, V Zhitov, N Alkeev *Opt Quant Electron.* 39. P. 181-192. 2007

[5] Pikhtin A N, Tarasov S A, Kloth B Ag-GaP Schottky photodiodes for UV sensors *IEEE Transactions on Electron Devices* vol. 50, № 1, p. 215, 2003.

[6] Pikhtin A N, Tarasov S A, Kloth B New values of the Ag-n-GaP potential barrier *Technical Physics Letters* vol. 28, № 10, p. 872-873, 2002.

[7] S L Rumyantsev, N Pala, M S Shur, R Gaska, M. E. Levinshtein Low-frequency noise in $Al_{0.4}Ga_{0.6}N$-based Schottky barrier photodetectors *Applied Physics Letters.* V. 79. (6). P. 866-858. 2001.

[8] S Kurin, A Antipov, I Barash, A Roenkov, H Helava, S Tarasov, E Menkovich, I Lamkin, Yu Makarov CHVPE growth of AlGaN-based UV LEDS *Physica Status Solidi (C) Current* V. 10, № 3, p. 289-293, 2013.

Journal of Physics: Conference Series **572** (2014) 012064 doi:10.1088/1742-6596/572/1/012064

Electrodynamical properties and modes of finite-width planar ferrite waveguide

A.V. Sadovnikov[1], **K.V. Bublikov**[1]

[1]Saratov State University, Saratov, 83 Astrakhanskaya str., 410012, Russia

E-mail: SadovnikovAV@gmail.com

Abstract. The present work focuses on the detailed study of the processes of electromagnetic wave propagation in finite-width planar ferrite film. The method of computation of electrodynamical characteristics of finite-width ferrite waveguide is proposed. The efficiency of method is estimated. Different boundary conditions are used and compared.

1. Introduction

Planar ferrite waveguide structures fabricated from thin film of yttrium iron garnet (YIG) can be used as basic elements of various functional devices. In spintronics and magnetoelectronics this structures applied for waveguides, couplers, delay lines and filters [1]. The electrodynamical characteristics of YIG-based devices can be changed in a wide range due to the possibility of operation frequency tuning of these devices by an external magnetic field, the existence of variety of the dispersion types of propagating waves (surface and volume, forward and backward). Considering the wave propagation process in the finite-width ferrite thin film waveguides one has to determine the spectrum and the dispersion characteristics of the waveguide modes, group delay, attenuation coefficients and excitation conditions. The excitation spectrum of magnetostatic surface spin waves (MSSW) in regular finite-width ferrite waveguide was widely investigated [2], but the multimode MSSW propagation in planar ferrite waveguides requires more detailed consideration.

2. Numerical modulation of ferrite- based waveguide.

Many theoretical and numerical methods of studying of wave propagation process in thin ferrite waveguide have been developed, including micromagnetic simulations and the plane wave method. In the present work finite element method (FEM) was used to simulate wave equation, obtained directly from Maxwell equations [3,4].

Geometries of computational cell are depicted in figure 1, a, c. The Damon-Eshbach (DE) spin waves geometry was considered in numerical simulation, the external field is oriented along the x-axis. The details of numerical method is presented in work [4]. Perfect electric conductor (PEC) or perfect magnetic conductor (PMC) [3] was placed on the left and right borders of computational cell and PEC was placed on the bottom and top borders. Two main configuration of computational domain was considered: right and left boundaries of YIG film

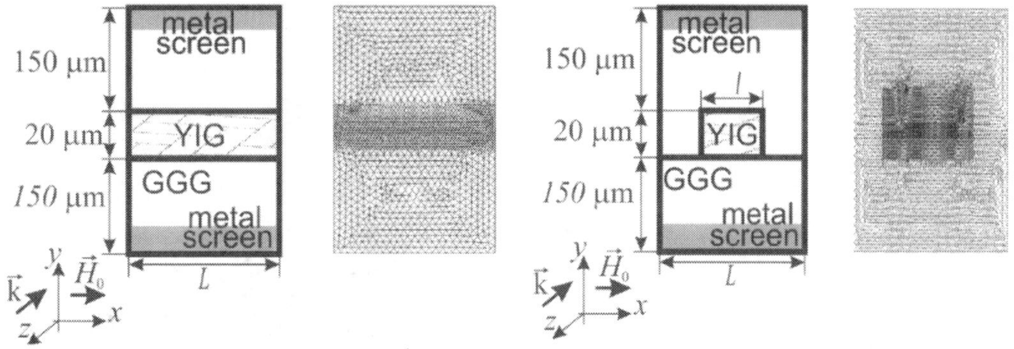

Figure 1. Geometry of computational cell (the waveguide *xy*-section) for different methods of numerical simulation: (a) – (c) and the corresponding meshes for FEM solver (b,d).

The schematic pictures of considered structures are shown in Figure 1 a, c. The ferrite film bounded on the left and right side with PMC or PEC (figure1, a). The YIG film which is not bounded by the PMC or PEC is shown in figure 1, c. This structure will be referred to as open YIG wave guide because the magnetic and electric fields mainly concentrated in the area of ferrite film. The corresponding meshes used in FEM simulation are shown in figure 1, b, c. The width of the computational cell is L, the width of the open waveguiding YIG structure is l. The parameters L and l were varied in the numerical modeling. The upper and lower boundaries of the computational areas were chosen to be a PEC.

The value of the external bias magnetic field was H_0=300 Oe, saturation magnetization of YIG film was $4\pi M_0$=1750 G. The results of dispersion computation of structure depicted in figure 1, a with PMC at the left and right borders of computational area (coincided with left and right borders of YIG film) are presented in figure 2.

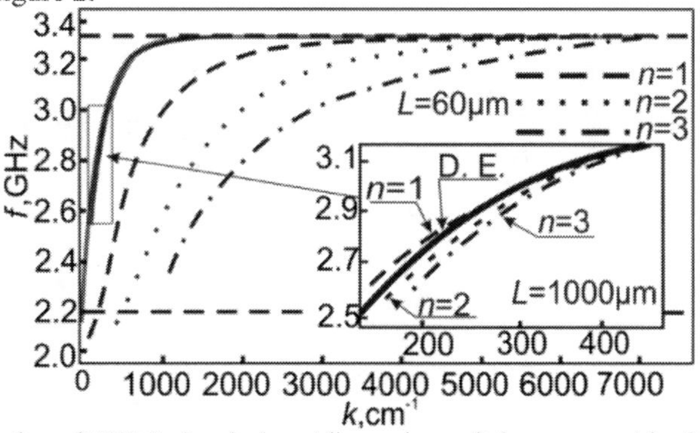

Figure 2. The results of FEM simulation (dispersion of the waveguide from figure 1, a) with PMC. D.E. means the Damon-Eshbach magnetostatic wave.

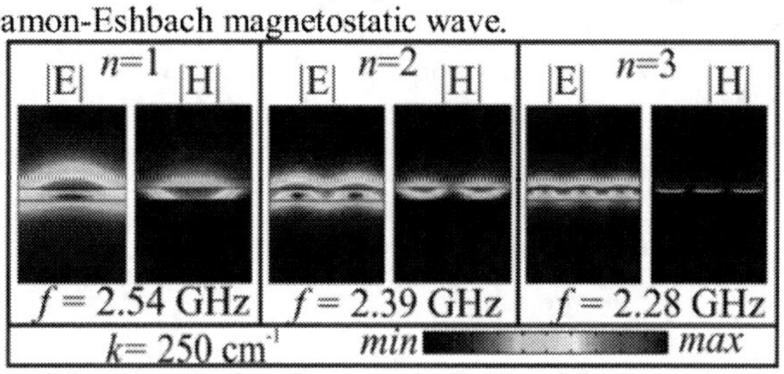

Figure 3. Distribution of the electrical and magnetic fields of the waveguide from figure 1, a with PMC as a boundary condition on the left and right sides of the structure.

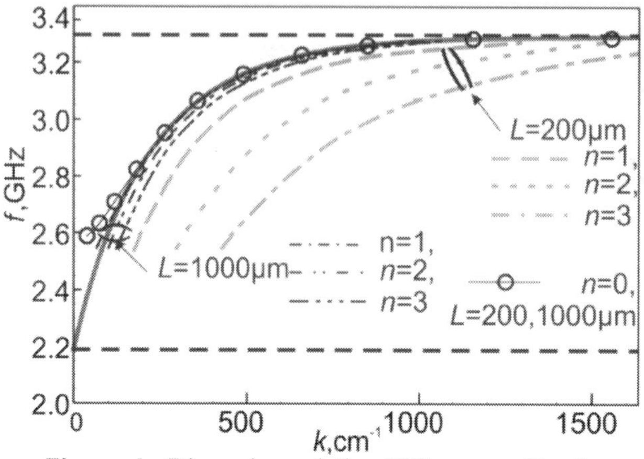

Figure 4. Dispersion of the YIG-waveguide from figure 1 a with PEC boundary conditions.

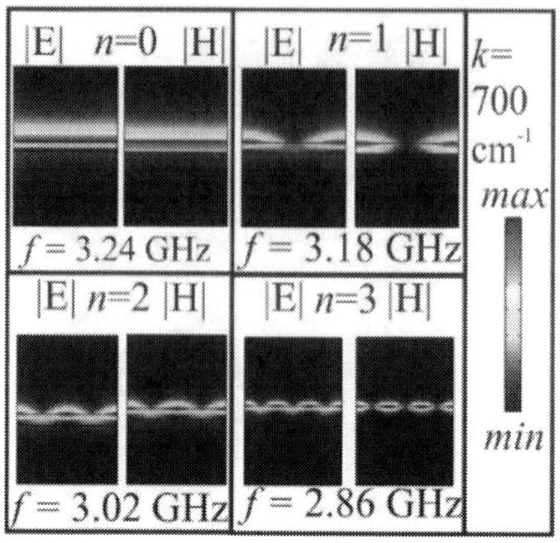

Figure 5. The electrical and magnetic fields distribution in the computation cell with YIG-waveguide from figure 1 a with PEC as a boundary condition on the left and right sides of the structure.

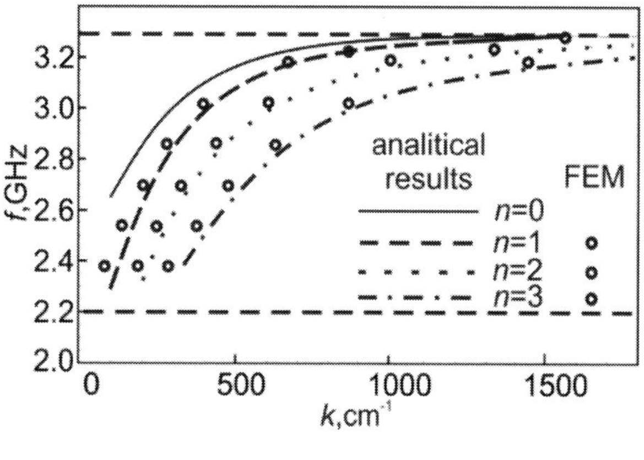

Figure 6. The results of modulation waveguide from figure 1 c with magnetic walls. Continuous line is analytical results from [5].

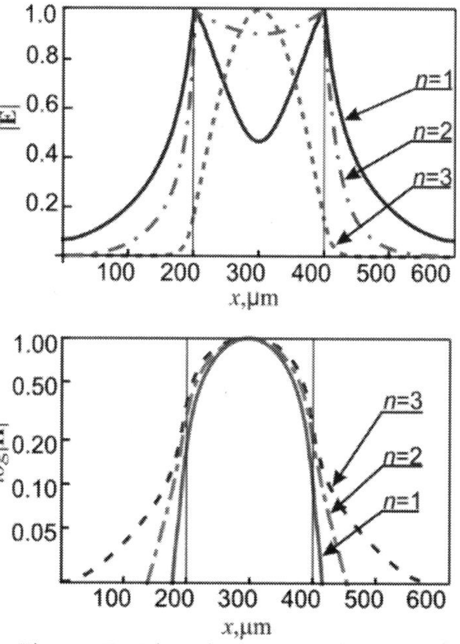

Figure 7. The electrical and magnetic fields distribution along x-axis of YIG-waveguide from figure 1 c. Vertical lines shows edges of YIG film.

Two different values of the width of structure was considered (L=60 μm and L=1000 μm). The multimode nature of distribution of magnetic and electric fields and a good correspondence between the computed dispersion for the structures with L=1000 μm and the Damon-Eshbah [3] magnetostatic wave (which described the unbounded YIG film in x-direction). The considerable transformation of dispersion with a decrease of L are observed in FEM simulation.

The results of numerical simulation of the same structure (figure 1 a) with PEC in the left and right borders of computation cell is depicted in figure 4. In this configuration of computation cell the zero mode of the metallic rectangular waveguide can be obtained in FEM simulation. It has been shown that the dispersion of this mode doesn't change with the changing the width of waveguide and computation cell L therefore.

The distributions of |E| and |H| field in cross-section of YIG-waveguide both for the case of metallic and magnetic walls in the left and right borders of computation cell were shown in figure 3 and figure 5. We can see influence of boundary conditions on distribution of fields. The uniform distribution of of |E| and |H| fields along x-axis for zero mode is shown in figure 5 for wavenumber k=700 cm^{-1}.

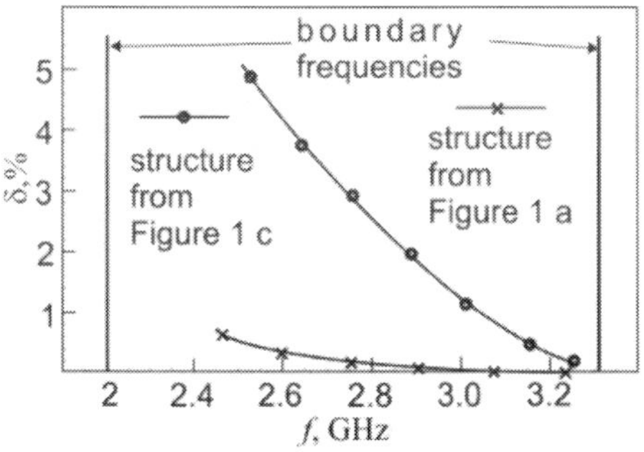

Figure 8. The difference between analytical results [2] and FEM simulation of structure in figure 1 a with PEC and figure 1 c with PMC.

The results of numerical simulation of an open waveguide (figure 1 c) are presented in figure 6 and figure 7. The rapid decrease of the |E| and |H| components near the borders (see figure 6) demonstrates the possibility to consider this structure to be a similar to ferrite film without metallic or magnetic walls on the boundaries. The estimation of difference between analytical results (from work [2]) and FEM simulation of structure in figure 1 a (with PEC) and figure 1 c (with PMC) is depicted in figure 8. For this purpose we will enter variable δ, which is defined as:

$$\delta(f) = \frac{|k_{FEM} - k_{OK}|}{k_{FEM}} \cdot 100\%,$$

where k_{FEM} is the FEM simulation result and k_{OK} is analytical calculation of finite-width dispersion in magnetostatic approximation [2]. The difference for structure, depicted in figure 1 a, is less than 5%, and for structure, depicted in figure 1 c, is less than 1%. In both cases the maximum differences is observed in the region of small wavenumbers. It can be explained with influence of borders on the localized fields at this wavenumbers. This effect is impossible be considered in magnetostatic approach.

3. Summary

The detailed study of the processes of electromagnetic wave propagation in finite-width planar ferrite film was performed with FEM computation. The efficiency of method of electrodynamical characteristics computation was estimated. Different boundary conditions were used and compared in the FEM simulation of Maxwell's system of equations. The surface magnetostatic waves were considered. Decrease the width of a YIG-film leads to the increase of the wavenumber of surface wave at fixed frequency. Electric and magnetic field distribution at different boundary conditions can explain the differences with magnetostatic approach, which are clearly seen at the small wavenumber region (k~100 cm^{-1}).

Acknowledgments
This work is supported by the Russian Foundation for Basic Research (grant No. 13-07-12409, 14-02-00976, 14-07-00273, 14-07-92603).

References

[1] Lenk B, Ulrichs H, Garbs F and Münzenberg M 2011 Physics Reports **507** 107
[2] O"Keffe T W and Patterson R W 1973 J.Appl. Phys. **49** 4886
[3] Silvester P P and Ferrari R L 1996 Finite Elements for Electrical Engineers (3rd edition). Cambridge University Press
[4] Sadovnikov A V and Rozhnev A G 2012 Appl. Nonlinear Dyn. **20** 143
[4]. Damon R W, Eshbach 1961 J R J. Phys. Chem. Solids. **19** 308
[5] Camley R E 1987 Surface Science Reports. **7** 103.

AUTHOR INDEX

AUTHOR INDEX

AUTHOR INDEX